T0319574

E-Paper Displays

Wiley– SID Series in Display Technology

Series Editor: Dr. Ian Sage

E-Paper Displays
Bo-Ru Yang (Ed.)

Liquid Crystal Displays - Addressing Schemes and Electro-Optical Effects, Third Edition
Ernst Lueder, Peter Knoll, and Seung Hee Lee

Flexible Flat Panel Displays, Second Edition
Darran R. Cairns, Dirk J. Broer, and Gregory P. Crawford

Amorphous Oxide Semiconductors: IGZO and Related Materials for Display and Memory
Hideo Hosono, Hideya Kumomi

Introduction to Flat Panel Displays, Second Edition
Jiun-Haw Lee, I-Chun Cheng, Hong Hua, and Shin-Tson Wu

Flat Panel Display Manufacturing
Jun Souk, Shinji Morozumi, Fang-Chen Luo, and Ion Bita

Physics and Technology of Crystalline Oxide Semiconductor CAAC-IGZO: Application to Displays
Shunpei Yamazaki, Tetsuo Tsutsui

OLED Displays: Fundamentals and Applications, Second Edition
Takatoshi Tsujimura

Physics and Technology of Crystalline Oxide Semiconductor CAAC-IGZO: Fundamentals
Noboru Kimizuka, Shunpei Yamazaki

Physics and Technology of Crystalline Oxide Semiconductor CAAC-IGZO: Application to LSI
Shunpei Yamazaki, Masahiro Fujita

Interactive Displays: Natural Human-Interface Techniques
Achintya K. Bhowmik

Addressing Techniques of Liquid Crystal Displays
Temkar N. Ruckmongathan

Modeling and Optimization of LCD Optical Performance
Dmitry A. Yakovlev, Vladimir G. Chigrinov, and Hoi-Sing Kwok

Fundamentals of Liquid Crystal Devices, Second Edition
Deng-Ke Yang and Shin-Tson Wu

3D Displays
Ernst Lueder

Illumination, Color and Imaging: Evaluation and Optimization of Visual Displays
P. Bodrogi, T. Q. Khan

Liquid Crystal Displays: Fundamental Physics and Technology
Robert H. Chen

Transflective Liquid Crystal Displays
Zhibing Ge and Shin-Tson Wu

LCD Backlights
Shunsuke Kobayashi, Shigeo Mikoshiba, and Sungkyoo Lim (Eds.)

Mobile Displays: Technology and Applications
Achintya K. Bhowmik, Zili Li, and Philip Bos (Eds.)

Photoalignment of Liquid Crystalline Materials: Physics and Applications
Vladimir G. Chigrinov, Vladimir M. Kozenkov, and Hoi-Sing Kwok

Projection Displays, Second Edition
Mathew S. Brennesholtz and Edward H. Stupp

Introduction to Microdisplays
David Armitage, Ian Underwood, and Shin-Tson Wu

Polarization Engineering for LCD Projection
Michael G. Robinson, Jianmin Chen, and Gary D. Sharp

Digital Image Display: Algorithms and Implementation
Gheorghe Berbecel

Color Engineering: Achieving Device Independent Color
Phil Green and Lindsay MacDonald (Eds.)

Display Interfaces: Fundamentals and Standards
Robert L. Myers

Reflective Liquid Crystal Displays
Shin-Tson Wu and Deng-Ke Yang

Display Systems: Design and Applications
Lindsay W. MacDonald and Anthony C. Lowe (Eds.)

E-Paper Displays

Edited by

Bo-Ru Yang
State Key Lab of Opto-Electronic Materials & Technologies,
Guangdong Province Key Lab of Display Materials and Technologies,
School of Electronics and Information Technology,
Sun Yat-Sen University, Guangzhou, China

This edition first published 2022

Registered Offices
John Wiley & Sons, Inc., 111 River Street, Hoboken, NJ 07030, USA
John Wiley & Sons Ltd, The Atrium, Southern Gate, Chichester, West Sussex, PO19 8SQ, UK

Editorial Office
The Atrium, Southern Gate, Chichester, West Sussex, PO19 8SQ, UK

For details of our global editorial offices, customer services, and more information about Wiley products visit us at www.wiley.com.

Wiley also publishes its books in a variety of electronic formats and by print-on-demand. Some content that appears in standard print versions of this book may not be available in other formats.

Library of Congress Cataloging-in-Publication Data
Names: Yang, Bo-Ru, 1979– editor.
Title: E-paper displays / edited by Bo-Ru Yang, State Key Lab of Opto-Electronic Materials & Technologies, Guangdong Province Key Lab of Display Materials and Technologies, School of Electronics and Information Technology, Sun Yat-Sen University, Guangzhou, China.
Description: First edition. | Hoboken, NJ : John Wiley & Sons, 2022. | Series: Wiley SID series in display technology | Includes bibliographical references and index.
Identifiers: LCCN 2021061537 (print) | LCCN 2021061538 (ebook) | ISBN 9781119745587 (hardback) | ISBN 9781119745594 (adobe pdf) | ISBN 9781119745600 (epub)
Subjects: LCSH: Electronic paper.
Classification: LCC TK7882.E44 E24 2022 (print) | LCC TK7882.E44 (ebook) | DDC 004.5/6–dc23/eng/20220207
LC record available at https://lccn.loc.gov/2021061537
LC ebook record available at https://lccn.loc.gov/2021061538

Cover Design: Wiley
Cover Image: © Marko Aliaksandr/Shutterstock.com

Set in 9.5/12.5pt STIXTwoText by Straive, Pondicherry, India
Printed and bound by CPI Group (UK) Ltd, Croydon, CR0 4YY

C9781119745587_020822

Contents

List of Contributors

Karlheinz Blankenbach
Pforzheim University
Display Lab
Pforzheim, Germany

Clinton Braganza
Portage Blvd
Kent, USA

Ben Broughton
Oxford, UK

Guy P. Bryan-Brown
Director of Technology
New Vision Display, Inc.
Malvern, UK

Anne Chiang
Chiang Consulting,
Cupertino, USA

Vladimir Chigrinov
State Key Laboratory of Advanced Displays and Optoelectronics Technologies
Hong Kong University of Science and Technology
Hong Kong S. A. R., China

Paul S. Drzaic
Apple, Inc.
CA, USA

Mauricio Echeverri
Kent Displays Inc.
Portage Blvd
Kent, USA

Robert J. Fleming
CTO CLEARink Displays
San Jose, CA, USA

Yoko Fukunaga
JDI Higashiura Plant
Aichi-ken, Japan

Peiman Hosseini
Bodle Technologies
Oxford, UK

J. Cliff Jones
School of Physics and Astronomy
University of Leeds
Cheshire, UK

Norihisa Kobayashi
Graduate School of Engineering
Chiba University
Japan

Hoi-Sing Kwok
State Key Laboratory of Advanced Displays and Optoelectronics Technologies
Hong Kong University of Science and Technology
Hong Kong S. A. R., China

Kristiaan Neyts
Professor and Head of Liquid Crystals and Photonics Group, ELIS Department, Ghent University,
Belgium

Doeke J. Oostra
Etulipa, a Miortech company
The Netherlands

Zong Qin
School of Electronics and Information Technology
Sun Yat-Sen University
Guangzhou, China

Abhishek Srivastava
State Key Laboratory of Advanced Displays and Optoelectronics Technologies
Hong Kong University of Science and Technology
Hong Kong S. A. R., China

Wanlong Zhang
State Key Laboratory of Advanced Displays and Optoelectronics Technologies
Hong Kong University of Science and Technology
Hong Kong S. A. R., China

Series Editor's Foreword

Paper is the medium which has dominated the presentation of both the written word and graphics for two thousand years. The happy partnership of paper with printing was first developed in China before being adopted in the Western world in the Middle Ages and it initiated the first information explosion. Perhaps no other invention has had such a fundamental influence on the development of human society and way of life. Printed books and periodicals are attractive, affordable, and so familiar that we can truly pay attention to the contents alone, and overlook the qualities of the physical page. With this cultural and technical background, it is not surprising that printed paper is widely regarded as a reference point for electronic display technologies and since the introduction of the first flat panel displays, achieving paper-like performance has been a common dream and aspiration of engineers. At the same time, users have bemoaned the shortcomings of electronic displays and the poor reading experience offered by early generations of displays.

In the present volume, Professor Bo-Ru Yang has brought together an outstanding selection of authors to examine the technologies which aspire to mimic the properties of printed paper. It is fair to say that no electronic display can yet rival all the desirable characteristics of paper, and a basic question which arose at an early point in the planning of the volume, was which technologies or aspects of performance should be included. Professor Yang and his team have taken an inclusive approach. In this book the reader will find accounts of displays which offer different combinations of desirable properties including paper-like appearance under ambient light, long-term image storage with zero or ultra-low power consumption, light weight and flexibility, and the ability to accept user input with the ease of pencil on paper. It follows that a wide range of display technologies are included, with an emphasis on device modes which offer ambient light viewability. Several liquid crystal modes, both with and without polarizer, can provide image storage either intrinsic to the display or in ultra-low power drive electronics. Meanwhile, displays based on electrophoretic or electrowetting effects can offer outstanding optical appearance under a wide range of ambient lighting. Innovative and emerging displays are also considered, with a chapter on phase-change displays offering an early view of the potential of this new development. Of course, physics of operation, the fabrication, engineering and especially the addressing characteristics of each display mode can be very different, and these issues are comprehensively covered, with special attention to those aspects of the devices which are less well covered in earlier sources. Other, user-related properties of the devices—the difficulty of providing high quality reflective colour, and the relation of the display performance to human perception of image quality and metrology are carefully considered.

In the 21st century, our priorities regarding electronic displays have changed. Excellent optical performance is today taken for granted while convenience, light weight and long battery life have increasingly driven user approval. Especially, the environmental impact of every activity we undertake, now demands scrutiny. Printed media are major consumers of environmental resources including energy, timber, carbon emissions and chemical residues while electronic devices have the potential to reduce or eliminate these problems—but only if they are responsibly manufactured and accepted by users sufficiently to displace print over a long period of use and wide range of use cases. Paper-like display quality can really make a difference to the world. The present book provides a comprehensive and authoritative overview of the field, authored by leading experts on each aspect of the subject, and I believe it will offer a most valuable source of reference to display professionals and advanced students, for many years to come.

Ian Sage
Malvern

Editor's Preface

With the advent of the Internet of Things (IoT), devices and objects around us are being equipped with built-in sensors and electronics which allow them to analyse and share information in real time, in the manner we associate with intelligent organisms. It follows that vast numbers of devices will be fitted with displays to present the information. The needs of such autonomous miniature devices mean that displays with low power consumption, excellent sunlight readability, and geometrical conformability which are compatible with low-cost fabrication methods, are becoming critical components in future IoT environments. E-paper displays have the inherent advantages of reflective operation, zero-power bistability, and in many cases they can be fabricated by printing-based processes. Therefore, they are regarded as among the most promising display technologies for these applications. Furthermore, many recent applications, such as fixed and mobile signage for transport, advertising billboards, architectural coatings, wrapped vehicles, e-readers, retail labelling, dynamic artworks, and many others, have started to exploit the unique advantages of E-paper.

Unlike LCD and OLED technologies, E-paper display technologies have not up until now been collectively and comprehensively reviewed, and there has been no published source which can provide scientists, engineers, and users with enough broad, insightful, and up-to-date knowledge to support research and development or product integration in this field. To fill this need, the present book represents the achievement of a three-year collaboration with prestigious scholars, experts, and entrepreneurs in E-paper display fields.

In this book, we have tried to cover as extensive a range of E-paper technologies as possible. Started with the development history of electrophoretic displays, followed by the fundamental mechanisms, physical models, driving waveforms, image processing, and advanced structures of electrophoretic displays, we describe the technological details and review the development progress to show how electrophoretic E-papers become commercially successful. In addition, bistable and reflective LCDs for E-paper applications, including Cholesteric LC, Zenithally Bistable Display (ZBD), Memory in Pixel (MIP), and Optically Rewritable (ORW) LCDs are also introduced. After that, the emergent technologies such as electro-wetting, electro-chromic, and phase change materials for E-paper display applications are reported and summarized. In the last part, the special needs for metrology of E-paper displays are explained. This book covers the broad spectrum of state-of-the-art reflective and bistable E-paper technologies, and we believe it will provide an invaluable handbook and reference for researchers, advanced students, and professionals in this field.

1

The Rise, and Fall, and Rise of Electronic Paper

Paul S. Drzaic[1], Bo-Ru Yang[2], and Anne Chiang[3]

[1] *Apple, Inc*
[2] *Professor, School of Electronics and Information Technology, State Key Laboratory of Optoelectronic Materials and Technologies, Guangdong Province Key Laboratory of Display Material and Technology, Sun Yat-Sen University, Guangzhou, China*
[3] *Principal, Chiang Consulting, Cupertino, CA*

1.1 Introduction

For over a thousand years, before the world of electronics, paper was the dominant medium for people to share written and later printed information. People become familiar with paper at an early age, and there is an enormous worldwide infrastructure for the production and distribution of printed material. Despite this huge built-in advantage, paper and print now fall short in providing for many demands of modern life. The past few decades have seen the emergence of electronic networks that transmit vast amounts of information, on-demand, for use in various ways. Electronic displays are a necessary part of this infrastructure, converting bits to photons and serving as the final stage of transmitting information to people.

Over the years, several electronic display technologies have waxed and waned; cathode ray tubes, plasma displays, and super twisted nematic (STN) displays come to mind. A few technologies are dominant; backlit active-matrix liquid crystal displays, active matrix organic light-emitting diode (OLED) displays, and inexpensive, passively addressed liquid crystal displays. Along the way, tens of different types of display technologies have been invented and explored, but ultimately have failed to catch on. A few displays have found a home in niche products and promise greater future application. Reflective displays, particularly electronic paper, are examples that have managed to find a place in the display ecosystem, with unique applications best served by these technologies.

This book aims to update on some of the most exciting new areas in electronic paper technology. This introductory chapter focuses primarily on electrophoretic displays (EPDs) and how they became synonymous with electronic paper. The story starts in the early 1970's, with the proposal and first demonstration of the electrophoretic movement of charged particles to make an optical effect. After intense effort, the technology was mostly abandoned, only to be resurrected by a start-up company, E Ink. Several key, and rather improbable inventions had to be made to develop a technology competitive to the dominant liquid crystal technology. Finally, the right application and ecosystem, the Amazon Kindle electronic book, was necessary to cement commercial success.

The field of reflective displays is very rich. The many other chapters in this book and recent reviews [1, 2] provide a wealth of resources for understanding the many technologies that have been developed in the quest to achieve a paper-like display. In this chapter, we will examine the following:

- A description of print-on-paper and how the optics of real paper compare with potential electronic paper competitors.

E-Paper Displays, First Edition. Edited by Bo-Ru Yang.

- A hierarchical summary of the different technical approaches for reflective displays.
- A detailed look at the historical development of EPDs, starting with the invention of the technology and ending with the introduction of the Amazon Kindle. Looking at the various developments in the context of its times, the EPD story offers some lessons in what it takes for a technology to transition from the laboratory to commercial success.

1.2 Why Electronic Paper?

Electronic paper has undoubtedly caught the imagination of the world. A Google search for electronic paper in September 2021 returns over 12 *billion* hits. This interest reflects people's love affair with paper as a medium for transmitting information. Yet, it is easy to recognise that printing ink onto dead trees is not easily compatible with today's networked world. What are the attributes of print-on-paper that make it so important?

- Paper is a reflective medium that automatically adapts to changing lighting
- Unlike most emissive displays, paper can be easily read in bright sunlight.
- The appearance of paper is relatively constant over different viewing angles, without significant shifts in luminance or color.
- Paper can be lightweight and flexible. The user can easily annotate it. with a pen or pencil.
- Paper is inexpensive
- Paper can be archived.

Nevertheless, physical paper cannot be instantly updated with information from electronic networks or easily serve as an interface with electronic devices. Today's backlit LCDs and OLED displays are ubiquitous as a means of transmitting information, but with the limitations that emissive displays possess, including eyestrain and low visibility in sunlight. Electronic paper can combine the power of electronic devices and networks with all the attributes of paper.

So what strategies can be taken to enable electronic paper? It is instructive to understand the composition and design of print-on-paper and see how many of these properties can be converted to something under electronic control to compete with printed media.

1.3 Brightness, Color, and Resolution

Conventional, non-electronic paper consists of a mat of tightly pressed fibers, most commonly derived from parchment or wood pulp. The combination of fibers and embedded air pockets scatter light and provide the reflective characteristics of paper. Historically, additives to the paper pulp during fabrication have also provided glossiness, color, aid in manufacture, or other desirable characteristics (Figure 1.1).

Compared to a white optical standard, the perceived reflectivity of paper often ranges from 50–80%, but can be even higher. The whiteness or brightness of paper depends on several factors, including the density of fibers, the paper thickness, the presence of additives such as titanium dioxide, clays, or fluorescent agents, and whether the viewing surface is made glossy through calendaring and coatings. The color of light reflected from white paper may differ somewhat from a perfect reflector due to the fluorescing whiteners' presence, or some underlying color absorbance from the paper. The human eye readily accommodates for these changes, though, so the perception of consistent color and lightness of a page relative to its surroundings is easily achieved (Figure 1.2).

Print-on-paper consists of drops of colored ink impregnated into the paper fiber. It is straightforward to devise dyes and pigments that absorb red, green, or blue. To generate the color characteristics of print, the CMYK subtractive color system can be used (Figure 1.3). The colors in print are usually comprised of cyan (absorb red), magenta (absorb green), and yellow (absorb blue) (Figure 1.4). Black pigment (the K in CMYK) is also commonly used, as it is challenging to achieve a neutral black color by mixing cyan, magenta, and yellow.

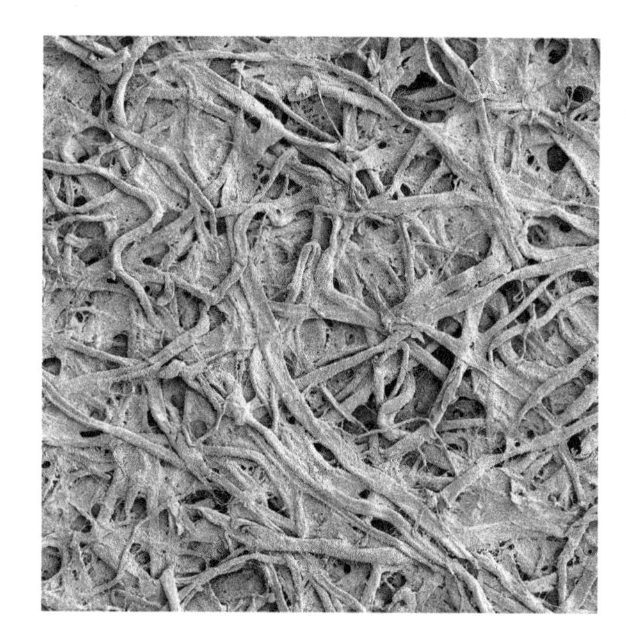

Figure 1.1 Arches 100% cotton rag paper. Scanning electron microscope image @100×. *Source:* http://paperproject.org [3] Used with permission of CJ Kazilek.

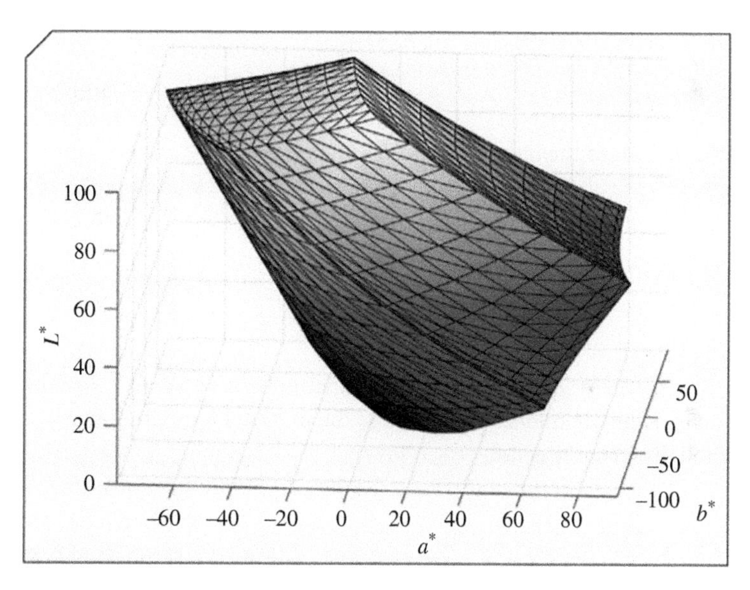

Figure 1.2 CIELAB color system [4] John Wiley & Sons.

Figure 1.3 Color separation of an image into its CMYK components, and the final printed image. The absence of a color is white.

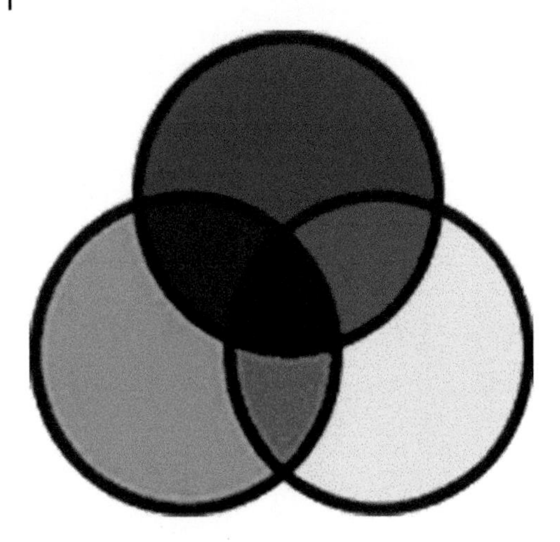

Figure 1.4 Subtractive color mixing. Cyan and magenta overlap to make blue, cyan, and yellow overlap to make green, yellow, and magenta overlap to make red, and all three mixed together provide black.

To achieve a wider color gamut, inkjet printing may use six or more colors to print. Additionally, spot color printing can deposit specialty inks (such as fluorescent pigments) that currently have no analog in emissive displays.

Depending on the industry, a variety of different color spaces and metrics have been developed for printed and reflected color. For example, the CIE 1976 (L*a*b*) color space is widely used to measure reflective colors and print. The human perception of lightness is measured by L*, which roughly scales as the cube root as the reflected luminance level.

For color reproduction in the print industries, the SNAP standard (Specifications for Newspaper Advertising Production) and SWOP standard (Specifications for Web Offset Publications) are widely used [1]. These print standards are rarely applied to electronic displays, though the advent of colored reflective displays approaching print-like appearance could change this situation.

Grayscale in printing is achieved using halftones (Figure 1.5). Each dot on a printed page defines an area where smaller halftone dots are printed. The more halftone dots, the deeper the color or darker the black while white is the absence of halftone dots. The smaller the dots, the higher the resolution. Print is often defined in lines per inch." Some examples of everyday printed objects include:

- Newspaper (monochrome) – 65–100 LPI
- Books and magazines (color) – 120–150 LPI
- Art books (color) – 175–250 LPI
- Photorealistic inkjet printer (color) – 250–300 LPI

Likewise, the resolution-defined dot for color printing consists of multiple prints of smaller dots. The combination of different colors and black and the underlying brightness and color of the paper provide the specific color and lightness of that dot.

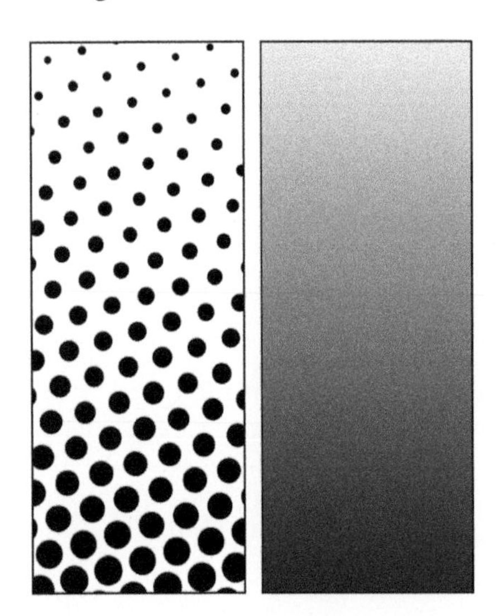

Figure 1.5 Examples of halftone dots of variable size, enabling grayscale [5] / Slippens / Public Domain.

With inkjet printing, commercial printers can also control the size of the drop, such as achieving four different sizes (Figure 1.6). The number of achievable gray levels is a combination of the number and size of the printed dots within the equivalent printed pixel [6, 7].

1.4 Reflectivity and Viewing Angle

An important aspect of paper is that the image printed on the page appears to have constant lightness and color irrespective of the viewing angle and lighting conditions. If the paper is glossy, there may be some glare from the surface, but that reflectance rarely interferes with the user interacting with the page. The near-constant appearance of the printed page is representative of Lambertian reflectance. The underlying paper scatters light, impinging onto the surface from many angles and then reflected uniformly into all spatial angles. Whether the page is illuminated by a collimated source such as a light bulb in a dark room or a more uniform source such as a cloudy sky, the distribution of reflected light from the paper does

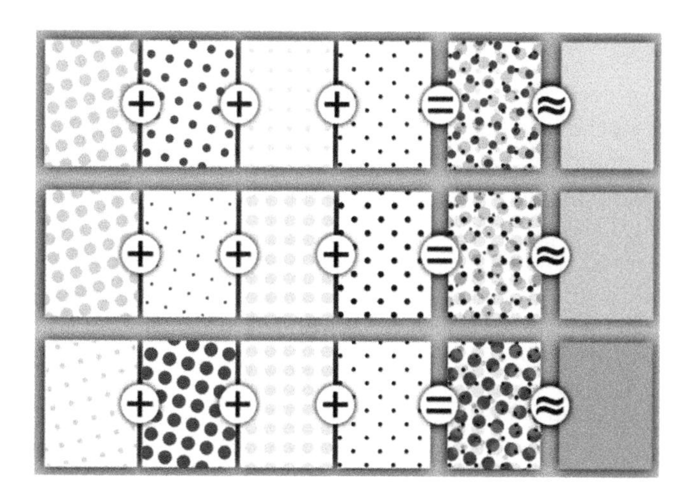

Figure 1.6 Printed color generation through printed CMYK halftones [5] / Slippens / Public Domain.

not change that much. Likewise, the amount of light scattered into the viewer's eyes for a given spot on the paper is also constant with viewing angle [8].

1.5 Translating Print-on-Paper into Electronic Paper

With this background, we can examine why designing an electronic display with the appearance of high-quality print-on-paper is challenging.

1.5.1 Brightness

Electronic displays require a transparent front sheet as the top surface of the display. This top surface (usually plastic or glass) provides a flat dielectric interface with air and degrades the optics of electronic paper in several ways.

- A significant fraction (4–10%) of light is directly reflected off the display without interacting with the display medium. This reflected light is seen as glare and degrades the contrast of the underlying display to the viewer. This reflected light also reduces the available light scattered back to the viewer. Anti-reflective coatings can reduce this reflected light, but with added cost and sometimes compromises in durability.
- Even if there is perfect diffuse reflectance, a significant fraction of that light will be trapped by internal reflectance at the display-air interface [9]. For an EPD where the dielectric phase is somewhere in the range of 1.38–1.45, the maximum amount of light that can be outcoupled from a white scattering layer, with no other absorptive losses, is predicted to be no more than 65%. These incoming and outcoupling losses limit the potential brightness of an embedded Lambertian scatterer within an electronic paper display.

There are strategies to improve the luminance of scattering-based displays by incorporating focusing optical elements with the display itself. Fleming et al. have demonstrated a display that uses a retroreflective prismatic element within the display and a black electrophoretic colloid to control the reflection from this element [10, 11]. While the reflectivity of the micro prism layer will depend on the geometric details of the source illumination, in many lighting conditions, the micro prism itself provides optical gain, and the display can appear "whiter than white."[12]. One display by Fleming et al., when illuminated by moderately collimated light, is reported to have an apparent brightness over 80%, with a 20: 1 contrast ratio. Fleming will describe more information on these types of displays in Chapter 4 of this volume. Further information about the metrology of measuring reflective display and e-paper will be described in Chapter 12.

1.5.2 Grayscale – Analog vs. Digital

High-quality images require grayscale, with 256 levels for each color channel being standard in emissive displays. Today, it is impractical to rely only on halftoning approaches to generate high-quality grayscale, which would require further subdividing each CMYK subpixel into an additional 4, 8, or 16 subpixels to maintain the native resolution. Instead, grayscale is generated using analog techniques, in which any pixel can be programmed to show intermediate reflective states between black and white. While many reflective display technologies, including EPDs, can show continuous grayscale within any pixel, maintaining accurate and uniform grayscale across all pixels is challenging. Analog gray levels are typically restricted to a number that enables uniform luminance without mura artifacts. For example, today's commercial e-paper displays from E ink are limited to 16 analog levels of gray to achieve good uniformity. However the underlying technology is capable of more gray levels [13].

Sophisticated dithering algorithms have been developed to minimize perceptual errors, some of which have been examined for use in electronic displays [14]. Dithering is analogous in many ways to halftones in print. In an electronic display, dithering sacrifices resolution to give the appearance of intermediate gray levels or to suppress non-uniformities between pixels.

1.5.3 An Overview of Approaches to Color Electronic Paper

Many inventive approaches have been proposed to achieve electronic color paper. Some designs take advantage of an electro-optical medium that can change color at the subpixel level intrinsically. Other approaches use color filters with an otherwise colorless black and white effect. Accurate and uniform grayscale is essential for high-quality images and is often very difficult to achieve.

Figure 1.7 illustrates an EPD in combination with color filters [15]. The function of a color filter is to absorb portions of the visible spectrum so that reflected light is colored. RGB or CMY primary colors can blend the primaries to provide an extended color range. Grayscale is adjusted by controlling the state of the reflective medium, similar to gray in black and white displays. These displays inevitably must make trade-offs between color saturation, color space coverage, and brightness. Reasonable brightness is usually achieved at the cost of color purity and a restriction of available colors.

Another approach is the adoption of multiple color particles as electrophoretic elements [16, 17]. Different colors can be generated in each sub-pixel by mixing C, M, Y translucent particles with scattering white particles, as shown in Figure 1.8. This approach was firstly demonstrated by Fujifilm in 2012 [16], and later on by E ink [17], which named it Advanced Color ePaper (ACeP). The colored pigments have different electrophoretic mobilities and responsivities, which facilitate to shuffle the color particles with driving waveforms of different voltage and pulse widths. As shown in Figure 1.8, different primary colors can be presented by positioning the white particle layer above or underneath the translucent C, M, Y particles. Integrating the CMYW particles into a display unit, ACeP can present different colors and gray levels in each subpixel by controlling the driving waveforms.

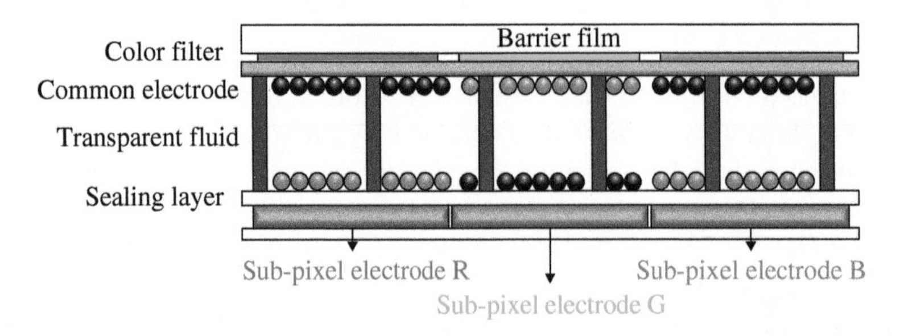

Figure 1.7 Color electrophoretic display using B/W particles and subpixel color filters [15] / John Wiely & Sons.

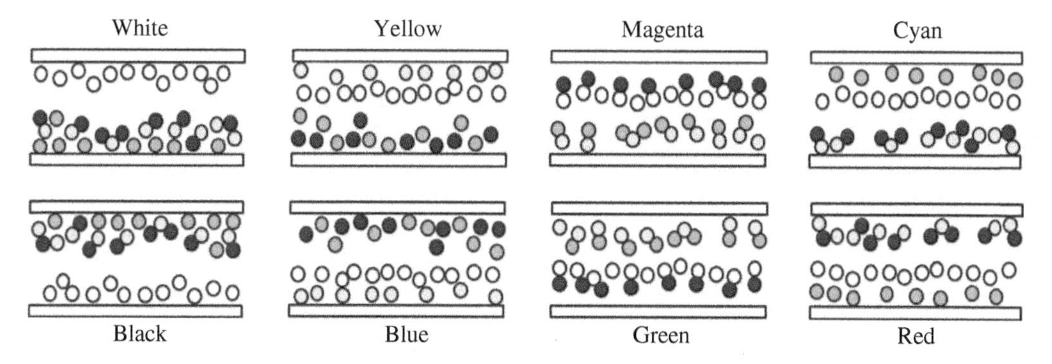

Figure 1.8 Eight primary colors of ACeP by different pigment arrangements [17] / John Wiely & Sons.

These driving waveforms are complex, and frame rates are currently slower than other EPD approaches. Nevertheless, the multiple color particle design dramatically increases the color saturation and reflective luminance achievable compared to the color filter approach. Figure 1.9 compares the color performance of these two types of color EPD prototypes [16, 17].

Many other inventive approaches have tackled the problem of making the electro-optical medium capable of switching color. Controlled lateral migration of colored fluid and particles enables a bi-primary color system. This design allows a single pixel to be changed through mixtures of color states [19].

A variety of other physical phenomena can be used to modulate reflected light electrically. The following table describes several varieties of reflective displays that have been developed over the past several decades. The fact that so many technologies have been pursued shows both the interest in the electronic paper and the difficulty in achieving a paper-like display. We compare the various reflective display technologies in Table 1.1, showing the strengths and weaknesses of the different approaches. More information on these different technologies can be found in reviews [1, 2, 20] and other chapters in this book.

Figure 1.9 Prototypes of colored electrophoretic displays: the left image uses a color filter and front light, while the middle and right images use the multiple color particles design [16–18] A. [18] Ian French, et al. 2020; B.[16] Hiji N, et.al, 2012 / John Wiley & Sons, Inc.; C.[17] Telfer S J, MD Mccreary, 2016 / John Wiley & Sons, Inc.

Table 1.1 Summary of performance and other key factors for monochrome e-Paper technologies.

	Electrophoretic (commercial)	Electrophoretic (eTIR)	In-plane EPD	Electro kinetic	Liquid powder	Electrochromic	Electrowelting	Electro fluidic	MEMs (IMOD)	Cholesteric LC	PDLC	Reflect LCD w\ polarizer	Flip dot	Zenithal bistable device
White, color %R	44% W [21]	60% [10]	75% [23]	~62% [15]	25 ~ 30% [28]	70% [30]	70% W [34]	72% W [37]	61% (Indoor) 80% (Outdoor) [40]	40% [35]	50% W	42% W [43]	80% [45]	39% [48]
Black %R	3% [21]	4% [10]	3% [23]	2%	3 ~ 4%[28]	–	3–5%	<3% [37]	>3% [35]	5% [35]	5%	2%	6%	2% [48]
Contrast ratio	~15 : 1 [21]	15 : 1 [10]	23 : 1 [23]	30 : 1	8 : 1	38.1 : 1 [31]	>35 : 1 [35]	>30 : 1 [37]	15 : 1	8 : 1 [35]	10 : 1	>20 : 1 [43]	13.3 : 1	20 : 1 [48]
Lambertian	Yes	Partial [10]	Yes	Yes	Yes	Yes	Partial	Yes	No	Partial	No	Partial	Yes	Partial[48]
mono SNAP/ SWOP	–	Maybe SWOP [10]	Maybe SWOP [23]	SNAP	–	Maybe SWOP [30]	SNAP [34]	SNAP	SNAP	–	Maybe SNAP	Maybe SNAP [43]	–	–
Driving voltage	15	2–4 [11]	<10	~5 [26]	40–70	1.4 [31]	15 ~ 20	10–20	5–10	(lab) < 4 (product) 25–40 [41]	5	~3 [43]	4 ~ 125[46]	20
Bistable	Yes	Yes [10]	No	No	Yes	Yes [32]	Yes [34]	Yes	Yes	Yes	No (Yes with matrix)	Yes	Yes	yes
Switching speed (msec)	100's	(33 fps) 33 [10]	15 (100 V) 30 (40 V) [24]	<300 [27]	<0.2 [29]	<10 [33]	10	~4 [38]	0.01's	300 (vertical) 5 (in-plane) [41]	51 [42]	48 [43]	~66 [47]	20 ms
Matrix drive	AM	AM [11]	AM, PM	AM	PM	PM	AM	AM, PM	PM	PM	AM	PM	–	AM [49]
Greyscale approach	Pulse	Pulse [10]	Analog/ Pulse	Pulse	Multi-write	–	Analog/Partial filling/Spatial dithering [36]	Pulse	Spatial/ Temporal	Pulse	Analog	Halftone	Pulse	Analog [49]
Grayscale bit level	4	4 [10]	5 (~30) [25]	>4 [27]	2	6 [30]	4	4 [39]	6	4	8 (60 Hz) 1 (6 Hz) [42]	2 [44]	1	4 [49]

Neutral white point	Yes	–	Yes	Yes	Yes	Possible	Yes	Yes	Yes	No	Yes	Possible	–	Yes [48]
Lifespan	Good	–	Unproven	Unproven	Good/ltd.	Ltd./unproven	Unproven	Unproven	Good	Good	Good	Great	Commercial	Good
Power (Static)	None	None [10]	Very low	Very low	None	Very low	None [34]	None	Very low	None	Low	Moderate	None	Very low
Power (Video)	Low	Low [11]	n/a	n/a	Low	Very high	High	High	High	Moderate	Moderate	Low	High	Very low
Years of research	Since 1969 [22]	Reported 2018 [10]	Since 2000 [1]	Since 2009 [1]	Since 2003 [1]	Since 1989 [1]	Since 2003 [1]	Since 2009 [1]	Since 1997 [1]	Since 1994 [1]	In mid 1980s [1]	Since 2001 [1]	–	Since 1995 [48]
Maturity	Many products	Prototype [10]	AM & PM demo	AM demo	ESL product/ large PM	Smart card products	AM demo	Segment product	PM products	PM products	AM prototypes	Products	Products	Products

There are several interesting aspects of Table 1.1 worth noting:

- Several e-paper technologies reach the reflective luminance of monochromatic SNAP (R ~ 60%), and a few of them move toward SWOP (R ~ 76%).
- The required driving voltages for many e-paper technologies are below 15 V, the maximum voltage provided by mainstream active matrix backplanes. As we will describe in a later section of this chapter, active-matrix backplane compatibility is a significant practical advantage.
- Many e-paper technologies have demonstrated a video speed response. However, it is important to note that high-quality video reproduction requires accurate grayscale performance, and accurate color if present. Not all technologies report this metric.

Later chapters in this book provide an up-to-date review of several essential e-paper technologies.

- In Chapters 2 and 3, the fundamental mechanisms, physical models, driving waveforms, and image processing of EPDs will be reviewed.
- Chapter 4 describes the Clear Ink EPD with a total internal reflective structure, enhancing the display's brightness.
- Chapters 5 through 8 describe several different categories of reflective liquid crystal displays. Chapter 5 introduces reflective displays based on cholesteric liquid crystals that rely on Bragg reflection from helical liquid crystal structures. Chapter 6 introduces the zenithal bistable display (ZBD), which utilizes a grating-type surface alignment structure to achieve bistable reflective states. Chapter 7 introduces the memory-in-pixel (MIP) liquid crystal display technology, which realizes a bistable display by building a static memory circuit behind the reflective electrode for each pixel. Chapter 8 introduces optically rewritable (ORW) technology, which uses a photoalignment layer to control the orientation of the liquid crystals and spatially generate bright and dark reflective states.
- Chapter 9 reviews electrowetting display technology, in which an applied voltage causes the physical translation of a dyed fluid within the pixel, with a resulting optical appearance change.
- Chapter 10 discusses electrochromic technology, generating reversible color change through an electrochemical redox reaction.
- Chapter 11 introduces the phase transition material technology. Under optical or electrical energy stimulation, a phase change material cycles between amorphous and crystalline states with a resulting change in reflectivity.
- Chapter 12 reviews the metrology of reflective displays, providing the means to characterize the performance of different technologies.

It is fair to say that the electronic paper technology that has seen the greatest success in transitioning to high volume products is the microencapsulated EPD technology from E Ink Corporation. The next section of this chapter will review the historical development of EPDs, including the many inventions and milestones needed to achieve a competitive device. It took over 35 years to go from the initial concept, proposed by Xerox and Matsushita, to a successful high-volume product, the Amazon Kindle. During this development period, many engineers and scientists' insights and hard work developed a combination of technical achievements and manufacturing scale that could enable competitive display products. Even with the baseline technology resolved, it still took the right product concept ready at the right time to lead to the commercial success of the Amazon Kindle. There are lessons to learn from observing how a laboratory curiosity was transformed into a technology and product that opened up a new industry.

1.6 The Allure of Electronic Paper vs. the Practicality of LCDs

The idea of a personal electronic device allowing user input and a display output screen is widely attributed to work done in the late 1960's by Alan Kay of Xerox PARC. Kay coined the term DynaBook for the apparatus [50]. He also laid out many of the characteristics of the device, including that the visual output should be, at the very least, of higher quality than what can be obtained from newsprint. While electronic paper was not explicitly called out as a display medium, that analogy to newsprint is a natural precursor for an electronic book. Nevertheless, until the late 2000's, liquid crystal displays dominated all implementations of personal electronic devices.

Why did it take so long to develop attractive electronic paper products? One overwhelming factor is the success of the liquid crystal display. LCD technologies started development about the same time as EPDs, and one successful variant (the super twisted nematic, STN) LCD did not appear until over ten years later. Reflective LCDs' suffered from poor brightness, poor contrast, and poor viewing angle characteristics. Nevertheless, LCDs can be matrix addressed to enable high pixel count, have reasonable response time, and be used in transmissive mode. LCDs were seen as "good enough" for many early applications. With volume production came incremental improvements in performance, reliability, and cost year-by-year. For a new electronic paper technology to take hold, it must outperform LCDs in multiple aspects and have an application important enough to attract investment in manufacturing at scale. For decades, no reflective display technology seriously challenged LCDs.

1.7 The Evolution of Electrophoretic Display-Based Electronic Paper

1.7.1 Early History

Electrophoresis is the movement of charged colloidal particles under the influence of an external electric field. Electrophoretic deposition is a well-known technology, with an early patent dating to 1919 [51]. In electrophoretic printing [52], charged colloidal particles suspended in a dielectric fluid were directed to deposit in specific electrode areas by applied electric fields. Various methods were used to fuse the particles to form a permanent coating. Such liquid toner technology was later applied to xerography, where it competed with the more successful dry toner technology developed by companies such as Xerox.

An EPD could harness a similar effect, using a reversible translation of the colloid to form an image visible through a transparent electrode. Perhaps inspired by liquid toner technology, Evans et al. applied for a US patent in 1969 (granted in 1971) on a "Color Display Device," which described an EPD [53]. Many essential elements in an EPD were disclosed, including:

- Charged particles in a dielectric fluid
- The dielectric fluid may be clear or opaque (colored)
- Different colored pigments may be used at the same time
- Charge control agents may be added to the particle suspension to aid in particle charging
- Bistability in the display was claimed, with image retention after the field is removed
- The potential for matrix (X-Y) addressing

Despite the wide-ranging descriptions, no actual working embodiments were described. There does not appear to be any public record of immediate follow-up work by Evans or his coworkers. EPDs did become an important area of development by others at Xerox in the late 1970's, under a different group of researchers and engineers.

Nearly contemporaneously, Ota of the Matsushita Corporation applied for a patent in 1970 (granted in 1972) for an EPD Device [54]. Many of the same concepts described by Evans were disclosed in that patent. Ota also described an embodiment where the electrophoretic fluid was separated into distinct compartments within the display. Compartmentalizing the electrophoretic fluid foreshadowed a critical development area 25 years into the future when the electrophoretic fluid across the display would be separated into tiny, active microcapsules for improved stability.

In 1973, Ota et al. published the first peer-reviewed article on EPDs [55]. This paper disclosed specific examples of the electrophoretic medium and display construction and showed photographs of digital clocks with seven-segment display digits. These were the first public examples of working EPDs.

In the early 1970's, commercial electronic displays typically meant emissive displays such as cathode ray tubes, vacuum fluorescent displays, or Nixie tubes. There was a growing need for low-power reflective displays in consumer devices, but these technologies were still under development and not yet widely used. A 1975 paper by Goodman of RCA Laboratories [56] described the landscape for low power reflective displays: liquid crystals (dynamic scattering and twisted nematic type), electrophoretic, and electrochromic. A wide range of properties was compared within the paper. However, the paper acknowledged that all these technologies were still at an early stage of development with unknown future potential.

An advantage of LCDs is that they could be matrix addressed to enable 48-line displays, whereas the other technologies could not. If the application required more than a limited number of pixels, then LCDs were the only choice at that time. The paper concluded that *"Although (none of the three display types) are yet not suitable for matrix address video displays, LCDs, ECDs, and EPIDs are potentially useful in those applications where low power consumption is important, and matrix address of a large number of elements is not important."* LCDs were being used in *"watches, calculators, and panel displays."* Still, *"Primarily because of the somewhat limited viewability of LCDS, ECDs, and EPIDS are being investigated for utilization. . . ."* EPDs suffered from poor reliability and the need for high operating voltages, but the race was on to overcome these difficulties and supplant low visual-quality LCDs.

1.8 Initial Wave of Electrophoretic Display Development

The 1970's and early 1980's were a period of intense development in EPDs. Multiple companies developed the technology, including Matsushita, Xerox, Philips, Plessey, EPID Inc., Thomson-CSF, Seiko Epson, Exxon, and Copytele. It is instructive to describe what was and what was not accomplished by these pioneering companies.

During this period, the development focus was primarily on single color electrophoretic pigments suspended in a contrasting fluid. Many papers described progress on the electrophoretic suspension and in controlling the switching and memory characteristics of the panel. Some highlights in the technology development are described as follows.

1.8.1 The Electrophoretic Fluid, Thresholds, and Memory

In theory, the notion of charged pigment particles suspended in a dielectric fluid, being attracted or repelled from a transparent conductor to change a display's appearance, seems quite simple. In practice, the electrophoretic fluid was a witch's brew of particles with particular surface chemistries, low viscosity, high resistivity oil, one, or more dyes, surfactants, and charge control agents.

In 1977 Lewis et al. published a study describing several key behaviors of an EPD and the potential chemistry and physics behind them [57]. Topics such as stability against permanent flocculation, the challenges of creating a switching threshold, and the usefulness of various pigment, fluid, and polymer chemistries were described. The following list provides a snapshot of the characteristics of an EPD that Lewis et al. were able to achieve:

- 38 μm cell gap, comprising polymer-coated titania in a diethyl phthalate solution of Sudan Black
- Drive voltage of 15 V
- one second switching time
- Erase/conditioning voltage of 30 V for one second required before addressing
- 1 $\mu A/cm^2$ drive current
- Memory of weeks

Lewis et al. also provided a physics-based analysis for reversible display memory based on the competition of van der Waals attraction and electrostatic repulsion between two particles within the electrical double layer. While this model provided some qualitative guidance, the authors recognized that the complex multi-component composition found in actual displays was beyond the scope of a simple analysis. The authors postulate that steric stabilization by modification of the particle surface could be a helpful approach, which was an important area of future development by many companies.

A 1978 paper by Murau and Singer provided additional insight into the physics behind electrophoretic device stability and memory functions [58]. This paper described a further theoretical framework for particle attractions and repulsions. It also provided dynamic electrical measurements of working displays that demonstrated the complex behavior of multiple charge carriers with different mobilities. They claimed that properly stabilized suspensions could achieve over 250 000 000 switches without severe degradation.

Croucher and Hair later provided an additional detailed description of desirable properties for the electrophoretic fluid, along with examples for many categories [59]. They examined several aspects of dispersion stability, including the tendency of different suspensions to settle (gravity-induced sedimentation), the tendency toward flocculation, and the ease of redispersion of the particles with time. They concluded that steric stabilization of the pigment particles is essential, and described the characteristics of polymers for stable physisorption onto the particles. They also speculated on potential advances, such as chemisorption (chemical attachment) of the stabilizing polymers onto the pigment, and demonstrated the potential of charge control characteristics into the steric stabilizer.

Novotny and Hopper published an analysis of the electro-optical characteristics of an EPD device [60]. The paper described a model where the device's memory characteristics relied on particle-electrode and particle-particle attraction, and observed devices with memories on the order of many hours. They concluded that the wide distribution of such forces would work against the appearance of a voltage threshold for switching.

The lack of a voltage threshold was a major drawback for EPDs, as it reduced the ability of the device to use simple row and column electrodes (passive addressing) to enable displays with a large number of pixels. Other approaches toward enabling x-y addressing are described later in this chapter, but the problem wasn't truly solved until the active matrix addressing became practical in the early 2000's.

1.8.2 Memory Function, Device Stability, and Control Grids

From the earliest times, the memory function for EPDs was noted as an essential feature of these displays. EPD memory was attributed to the EPD particles remaining in place after removing the switching voltage. Long-term memory could lead to a low power device, which would consume power only during switching. Of course, if the particles stick too firmly to the panel electrodes, they could become stuck and degrade the contrast of the device due to lack of switching. Any particle adhesion to the panel electrodes or other particles would need to be finely balanced to offset gravitational settling (which could erase the image) and permanent particle stiction and flocculation (which could cause permanent ghosting).

The location of the pigment particles within the display will vary with the images rendered by the display. The local distribution of pigments would tend to vary from region to region within the display over time. This history dependence could require reset pulses to be applied to the EPD to erase the history. Moreover, the particles could migrate over time within the panel to give a non-uniform distribution of pigment, which could cause a temporary or permanent reduction in contrast. Neither characteristic is attractive in a display and provided serious challenges to developing EPDs.

Despite the strong interest in EPD memory function, there is no complete description of the chemistry and physics of the EPD memory effect in this time. In his original paper, Ota noted that the EPD has a memory function, because the pigment particles deposited on the surface of the electrode remain on the surface even after removal of an applied voltage mainly due to van der Waals attractive force between the pigment particle and the electrode. No specific details beyond this simple description were given, though.

Further mentions of memory function were made by Dalisa and coworkers in two 1977 papers, though again, no detailed description of explanation was provided [61, 62].

To enable matrix addressing, Dalisa and coworkers proposed adding a third electrode layer, a "control grid," between the other two cell electrodes to enable matrix addressing. The electrode rows in the control grid would be perpendicular to the columns in the bottom electrode. While the top electrode would uniformly attract (or repel) the electrophoretic particles, the control grid could provide an additional potential to either enhance or diminish the effective voltage between the top and bottom electrodes. Dalisa et al. claimed that the device demonstrated 5 ms switching per line at ±15 V and a resolution of 5 lines/mm. While the paper contended that further improvements could be made, the dimensions of the control grid and added complexity would limit this approach to low or medium-resolution displays.

Hopper and Novotny published two papers in 1979 that provided detailed descriptions of some EPD systems and treated important topics such as response times, memory, and device addressing [63, 64].

- These papers described device dynamic and optical trade-offs with cell thickness, electric field, particle concentration, and other formulation variables. They noted a complex electrical transient behavior due to the contributions of the many charged components present in the electrophoretic fluid.
- Similar to Ota, the papers described a hypothesis where the device's memory characteristics relied on particle-electrode and particle-particle attraction and observed devices with memories on the order of many hours. They concluded that the wide distribution of such forces would work against the appearance of a voltage threshold for switching. The device switching was dependent on the voltage history of the device. No other physical mechanisms regarding the memory effect were provided, though.
- Hopper and Novotny recognized that passive addressing of EPDs to achieve high pixel count displays was not likely. They noted that thin-film transistor (TFT) active-matrix addressing was possible, though the capability to fabricate useful TFT backplanes was not very developed.
- Hopper and Novotny noted that particle migration and settling where serious problems in EPD displays. Along with Harbor, they used a public invention disclosure to propose that subdividing the EPD panel into subdivisions could potentially resolve this issue [65]. As will be noted later, subdivided EPD panels were reported in 1986, and the microencapsulated EPDs described later in this chapter provided the ultimate segregation for particles.

Murau and Singer published a detailed study examining the mechanisms behind various factors leading to instability in electrophoretic panels [58]. They attributed the steric repulsion due to adsorbed polymers and modeled the balance between these repulsive forces between particles with the electric fields that induce tight packing of particles at the electrode surface. They also described two interesting phenomena that degraded display reliability:

- Particles undergoing vertical switching can migrate laterally outside of the electrode area and then become effectively stuck in field-free regions.
- Electrohydrodynamic fluid instabilities caused by slow-moving background changes in the suspension, leading to undesirable particle clustering

By carefully optimizing several aspects of the electrophoretic fluid, Murau and Singer were able to demonstrate a device that was switched over 2.5×10^8 times without serious degradation.

1.8.3 Optics

While the white TiO_2 pigment is similar to the pigments used in white paint, there are multiple aspects of the EPD that significantly reduce the reflectivity of a display compared to white paint. The reflectivity of a white pigment below a clear, dielectric surface will be reduced due to refraction and total internal reflection within the device and the reduction of illuminating light reflected from the display surface. An EPD device containing a dyed fluid, there will be additional losses due to residual colored dye in the fluid above the electrophoretic particles, even in the bright state. Reducing the dye concentration improved brightness, but at the cost of contrast due to reduced hiding of the white pigments in the dark state (Figures 1.10 and 1.11).

One of the first analyses of this interplay between brightness and contrast was provided by Vance [66]. Vance modeled the reflectivity of TiO_2 pigment suspended in a dielectric fluid medium without dye to estimate the upper limit to reflectivity. He then analyzed experimentally the potential contrast for different dye concentrations and

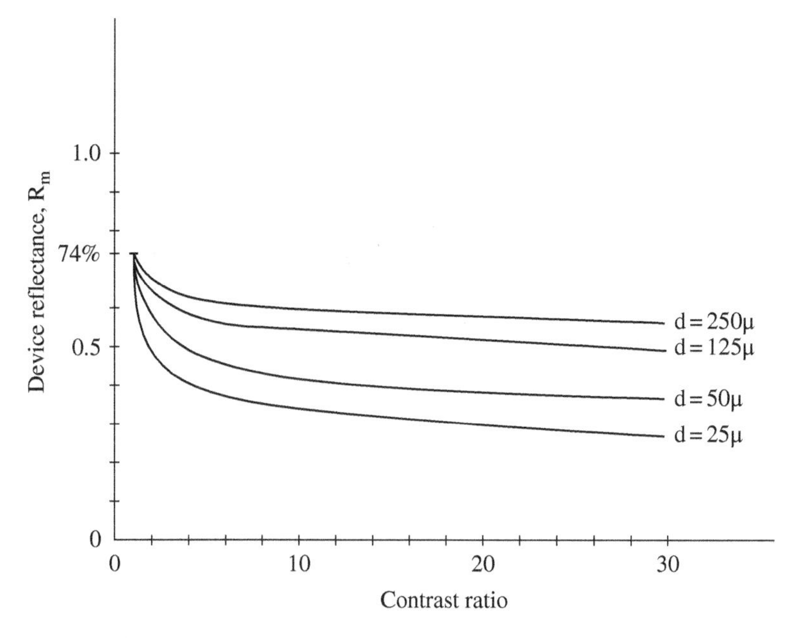

Figure 1.10 Device reflectance vs contrast ratio for EPD at different device thicknesses. From Vance 1977 [66].

Figure 1.11 First EPD product for point-of-purchase (POP) use, and the experimental numeral display EPD [70] / U.S Patents / Public Domain.

cell thickness and the impact on bright state reflectivity. (Figure 1.12) shows the relationship predicted between device reflectance and contrast ratio at different cell thicknesses and dye concentrations. Realizing that significant cell gaps will result in unacceptably high voltages and slow response times, for a 25 μm cell thickness, he estimated that an ideal device could achieve brightness between 35% and40% for a contrast ratio between 5 and 10. However, this estimate assumed that the electrophoretic medium contained a particle concentration of 10% or higher, which was considered unrealistic for a stable device.

Fitzhenry-Ritz published a similar analysis in 1981, examining the combinations of yellow pigment/red dye and green pigment/black dye from both a simulation and experimental perspective [67]. While her results were complicated by using color-contrasting states, the results were similar to Vance in that it did not seem possible to achieve true paper-like brightness and contrast with EPD systems.

Vance states, "*It thus appears unlikely that displays with the contrast (10:1) and reflectivity (>70%) of black and white printed pages can be produced by these systems. Rather, as has been universally observed, practical 'black and*

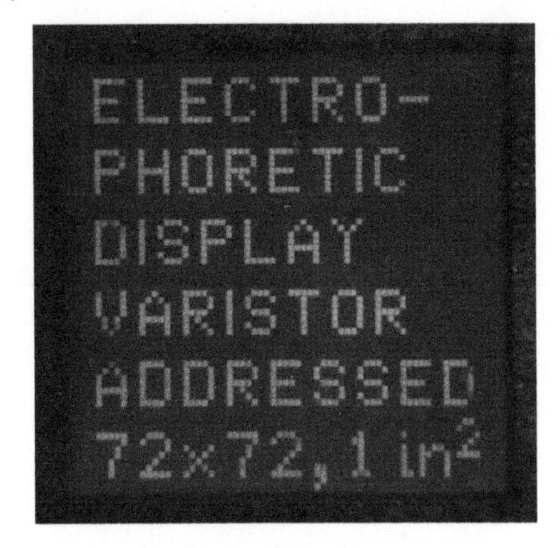

Figure 1.12 Demonstration of the first active-matrix electrophoretic display. *Source:* From A. Chiang. (1980). "Electrophoretic Display Technologies and Characteristics," presented at the Seminar Notes, SID Technical Symposium. pp. 1–32 [78].

white' EP displays will operate between two shades of gray." Interestingly, he notes that *"devices employing undyed suspending fluids state (sic) such as those disclosed in the patent literature will be capable of much higher reflectance than the dye cells described here."* Despite this prediction, the appearance of two-particle black and white EPDs will need to wait until the introduction of the microencapsulated EPDs more than 10 years later.

Analysis by Shu et al. determines that the relative reflectivity of white pigment in a dielectric fluid under glass is limited to <65% due to total internal reflection and other incoming and outcoupling losses [9]. Additional optical elements can be deployed to raise the effective brightness under some lighting conditions, as will be described in Chapter 4.

1.8.4 EPD in Decline

While steady progress was being made in EPDs, progress in LCDs was faster. In a 1986 *IEEE Spectrum* article entitled "*Whatever happened to Electrophoretic Displays?*," Werner and Kmetz essentially announced the demise of EPDs [68]. They pointed out that while great progress was made in developing practical high pixel count EPDs with reasonable switching time, poor reliability still plagued the technology. Over time, the particles would tend to agglomerate, causing image nonuniformities. Moreover, the relatively high switching voltages for EPDs added complexity and cost to the driving electronics. STN LCDs could achieve similar contrast ratios, high pixel counts, and even faster switching speeds, with lower-cost electronics. The future for EPDs seemed grim, with the article ending:" *The two major players in EPDs have weighed these complexities and cast their votes. Philips has cut back its research effort, and the EPID division, which presented the only paper on electrophoretics at the 1986 SID Symposium, has been put on the block by Exxon.*" Additional perspective is provided in the 2005 book *Liquid Gold* by Joseph Castellano [69]. While the book's primary focus is the development of the liquid crystal flat panel display industry, there is mention of competition by EPD displays. For example, Castellano reports that EPID displays (Exxon) had installed equipment to fabricate EPD displays in 1980; expansion was planned to enable 12.5″ displays (7.5″ × 10″) with 70 000 pixels. The EPD displays were able to show white pixels on a darker background, and the contrast was quite good. Nevertheless, the technology was plagued by image sticking, and long-term reliability was unknown.

Philips commissioned a study by Castellano in 1985 to determine if there was still a market segment available for EPDs, given the rise of LCDs. Castellano writes that the only potential entry was to build large screen, high-information content displays without the need for an active matrix backplane. Such products could replace CRT monitors or serve as the displays in portable computers. Unfortunately, the Philips technology suffered from similar reliability and matrix addressing limitations to EPID, Inc. and other companies. Castellano states that Philips ended the project a few years later.

Before closing out the section on the initial wave of EPD displays, it is useful to describe some of the achievements by the teams at that time, recognizing that the technology was moving into a near-dormant phase in the late 1980's and 1990's.

- Matshushita developed EPDs with yellow-red, white-blue, and white-black color pairs around the 1975 time period (Figure 1.11) [70].
- Murau at Philips Laboratories published a paper in 1984 describing an EPD with a control grid structure. The 9 × 16 cm display possessed 230 000 pixels at a pixel size of 250 × 250 μm in 360 rows by 640 column arrangement. Using encapsulated TiO_2, the display achieved a relative brightness of 23% with a contrast ratio of 15 : 1 and an addressing time of 10 ms/line [71].

- Chiang described a 1 in^2 32×32 display addressed using only a passive x-y addressing scheme, enabled by a "packing pulse" followed by complex 2/1 or 3/1 selection Schemes [72].
- Chiang also reported a stylus-addressed EPD in 1979, which finessed the x-y addressing limitations by enabling the user to write an image using a charged stylus [73].
- Shiffman and Parker reported in 1984 an EPD display driven by NMOS circuitry on single-crystal silicon. This first demonstration was limited to 16×16 pixels on a 4×4 mm display area, with peripheral circuitry filling the 6 mm wafer [74].
- Beilin and coworkers reported in 1986 a 9″ diagonal 2000-character (8×8 pixels each) EPD display using control grid technology. The display employed honeycomb grid compartments to segregate the electrophoretic material into single pixels, claiming that five years of a lifetime is feasible. Interestingly, they reported that they improved the particle stability by chemically grafting the stabilizing polymers to the particle surface and several proprietary charging agents and stabilizers. The device was reported to have a reflectance of 13% with a 9:1 contrast ratio [75].
- Chiang reported the first active matrix-driven EPD panels in 1980 [76-78]. The varistor comprised back-to-back ZnO diodes, the panel incorporated pixel-level storage capacitors and enabled a display with 32×32 pixels and an area of 1 in^2. Switching voltages were reported as 70 V, with image switching times of either 16 ms (blue to white) or 11 ms (white to blue). Importantly, the paper also reported that the images were written line-at-a-time, at 20 μs/line without crosstalk, flickering, or refreshing with the incorporation of local storage capacitors. This capability is unique to bistable reflective displays.

Despite these advances, for over 10 years, there was little activity in the development of electrophoretic technology. EPDs had developed the reputation as not viable for primary display application.[i]

The controversial company Copytele, founded in 1983, filed and was granted several patents in EPD technology. While raising millions of dollars from investors and claiming to develop displays for data and graphics, a report in 1992 indicated no evidence of any successful commercialization [79].

Nippon Mektron, in a 1994 paper, reported a 5×7 direct drive display aiming for large-area applications [111]. The drive electronics were bonded directly to the backplane on a printed circuit board. More importantly, this paper demonstrated that EPD switching times could be fast (1.7 ms frame rate) and reliable (with over one 20 million switching cycles with only minor performance loss, and the equivalent of five years exposure to sunlight). Still, the contrast ratio was modest (4 : 1).

1.9 The Revival of EPDs

By the late 1990's several pieces of the puzzle started coming together.

- There was little progress in improving the optical appearance of reflective LCDs. If people wanted high-quality, bright images on a portable electronic device, then backlit LCDs were required, with the usual limitations for emissive displays compared to paper.
- While several other technologies were investigated for reflective displays that could potentially be considered a rival for paper, none met with any commercial success.
- The steady growth of the LCD display marketplace led to a significant investment in TFT panel manufacturing plants. While TFT panels for LCDs were not optimized for EPDs, they potentially could solve the pixel addressing a problem that plagued early versions of EPDs.

EPDs did see a revival in the 1990's, with the introduction of microencapsulated EPDs. In a paper published in the journal *Nature*, Comiskey et al. from MIT demonstrated that it is possible to create an EPD by microencapsulating

---

i The reputation for EPD displays remained poor into the late 1990s. One of this chapter's authors (PSD) was introduced to display technology luminary Sir Cyril Hilsum at an SID event in 1998. When Hilsum learned that he was planning to join a new start-up company (E Ink) as Director of Technology to work on EPDs, Hilsum's response was "I'm so sorry."

Figure 1.13 SID 1998 Digest paper [80] Drzaic P, et.al, 1998 / John Wiley & Sons, Inc.

the electrophoretic medium in urea-formaldehyde microcapsules [22]. The microcapsules were subsequently coated between two electrodes with an *in-situ* polymerized polyurethane binder. Many of the positive features of EPDs were preserved, including near-Lambertian viewing characteristics and bistability. Moreover, the microencapsulation seemed to provide a means of solving the agglomeration and pigment migration problems that plagued conventional EPD displays. Since the pigments and fluid were confined within a small microcapsule, migration of particles across the panel could be avoided.

Since the microcapsules that contained the electrophoretic fluid were dispersed in a polymer film, the display could be flexible as well. A 1998 SID Digest paper by Drzaic et al. demonstrated a fixed (bistable) image microencapsulated EPD display wrapped around a pencil in Figure 1.13 [80]. The display material comprised titanium dioxide particles and a blue dielectric fluid and was reported to exhibit ±90 V switching voltage, 12% luminance, and a 6:1 contrast ratio.

As an interesting footnote, the company Nippon Mektron had filed for several Japanese patents in the area of microencapsulated EPD, the earliest filed in 1987 and published in 1996 [81]. There is no evidence that these patents were filed outside of Japan or publications on this technology until 1998. That year, Nakamura et al. from the Japanese company NOK published a paper on microencapsulated EPDs at the 1998 SID symposium [82]. NOK showed an actual working display images, with a 5×7 matrix design-driven with external switches connected by through-film vias to reflective electrodes. Display characteristics included:

- TiO_2 in blue dye dielectric
- 50 μm capsules made from gelatin/gum arabic
- Mixing of microcapsules into silicone binder, coating into a single layer, and laminating onto flexible film
- +/50 V switching voltage
- 50 nA/cm2 driving current
- Contrast ratio 4:1
- 1000-hour stability (memory function)

Despite this early entry, further activity at NOK on the technology seems limited. There appears to be only one additional publication by that company, in 1999, using a flexible sheet of microencapsulated electrophoretic material with the photoconductive drum of a laser printer to form a rewritable flexible paper.

There are also several Japanese patents on microencapsulated EPDs filed in the late 1980's and early 1990's by Nippon Mektron Corporation, with the earliest appearing in 1987 [83]. Nevertheless, there are no technical publications in this area until the NOK 1998 paper, and no technical papers afterward. For the balance of this chapter, we will focus on developing EPDs at E Ink Corporation and collaborators.

1.10 Developing a Commercial Display

The startup company E Ink Corporation was founded in 1997 to commercialize "electronic ink" based on the MIT technology, and set up shop in Cambridge, MA. Professor Joseph Jacobson first drew attention to the concept in a 1997 publication aptly titled "The Last Book." [84]. In 1998, the company received significant publicity, including being named one of Fortune Magazines 30 Cool Companies of 1998. The company also secured US$15.8 M in venture and industrial funding by mid-1998, backed by the Hearst Corporation, Motorola, and others [85]. The venture capital model expected that E Ink would develop products that would enable the company to scale rapidly, supplying products in high-volume markets such as publishing and advertising.

Much of the interest was driven by the vision of radio paper. In a 2000 interview, Jacobson described radio paper as "*electronic paper, coated with e-ink, with tiny transistors in it to act as electrodes, solar cells, and radio receivers.*" The paper "*would be able to upload pretty much anything, even video, using no more energy than it draws from ambient light.*" "*I fundamentally think five years is the right time scale for the static paper to be phased out and replaced by radio paper*' Jacobson says." [86].

While the microencapsulated technology seemed promising, there was a long way to go to achieve any commercial product, let alone "radio paper." Issues to overcome included:

- Contrast and brightness were not better than reflective LCDs and certainly inferior to backlit LCDs.
- The operating voltage was high. ±90V switching is not compatible with high-resolution devices
- The displays could not be passively addressed. Options were either direct drive (for low pixel count displays) or, in principle, active-matrix backplanes (for high pixel count/high resolution) displays. At that time, TFT panels were expensive, hard to source, and not designed for the switching characteristics needed for EPDs.
- No scalable manufacturing method was demonstrated.
- Reproducibility and reliability not demonstrated

The vision of "radio paper" was compelling but incredibly challenging. Remarkably, the engineers at E Ink and their collaborators managed to make multiple inventions, overcome these limitations, and develop a competitive display technology that, while not quite "radio paper," still met a market need in a compelling way. The following sections describe many of the key achievements to developing a practical display suitable for use in high-volume electronic paper products, up to the introduction of the Amazon Kindle.

1.11 Enhancing Brightness and Contrast

The expectation that electronic paper would look like print on paper requires a display with reasonable brightness and contrast. As mentioned previously in this chapter, for EPDs using titanium dioxide dispersed in a dyed fluid, Fitzhenry-Ritz pointed out a tradeoff in dye concentration, display brightness, and display contrast [67]. Using titanium dioxide in a dyed fluid, the microencapsulated EPD E Ink display exhibited a brightness of 12% and a contrast ratio of 6 : 1, well under the theoretical maxima predicted for EPD displays. While brightness and contrast ratio could be traded off by adjusting dye concentration, there was no expectation that a brightness >40% with good contrast could be achieved in this system.

A way out of this dilemma would be implementing a two-particle system with white and black pigments. If one color of the colloid could be brought to cover the top surface of the microcapsule, with the other color hiding behind this layer, then higher brightness and contrast could be achieved. The challenge is that while previous papers and patents had disclosed the concept of such two-particle systems, there is little indication that such a system was reduced to practice. There are significant challenges to such an implementation:

- There are two types of particles needed to possess opposite charges to respond to the applied field properly. Such opposite charges would attract, so that irreversible particle flocculation was a serious issue.
- The electrophoretic mobilities of the particles would need to be made stable, and any charging agents within the fluid compatible with both types of particles.
- This two-particle system would need to be compatible with the microencapsulation process.

Honeyman and coworkers developed a means of grafting polymers onto the colloid surface, which achieved these goals [87]. They first attached a polymerizable or polymerization-initiating group to the particle, then reacted with one or more polymerizable species. These secondary polymers could comprise branched polymer chains, which would present a barrier retarding the flocculation of oppositely-charged particles. These polymers could also incorporate charged or chargeable groups. Since the polymers were chemically bonded to the particle surface, they would be much more stable over time previous methods of merely physabsorbing polymers onto a surface.

Achieving a two-particle system resulted in a dramatic increase in display brightness compared to the single-particle/dye system. As described in the following section, carefully engineered microcapsules were equally important in improving device optics.

1.12 Microencapsulation Breakthrough

While the microcapsules used by E Ink promised to resolve many of the instabilities found in conventional EPD displays, they introduced another problem regarding reduced contrast and high switching voltages. The following Figures 1.14–1.16 can be used to understand the source of these issues [80]:

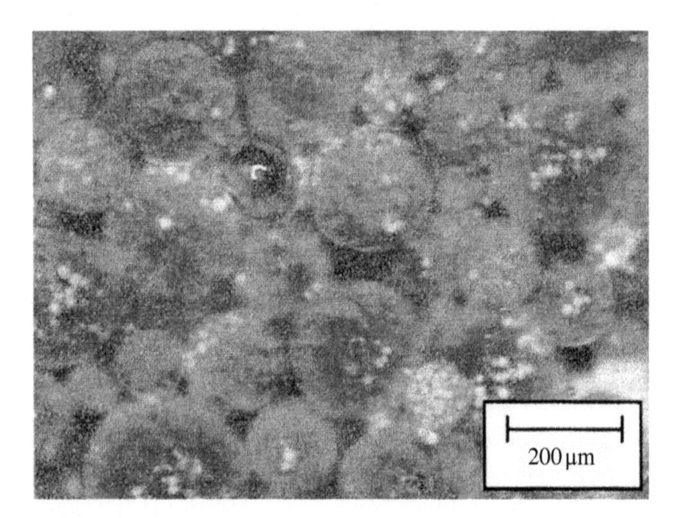

Figure 1.14 Microphotograph of Microcapsules in blue state [112].

Figure 1.15 Microphotograph of Microcapsules in white state. From SID 1997 Digest paper [112].

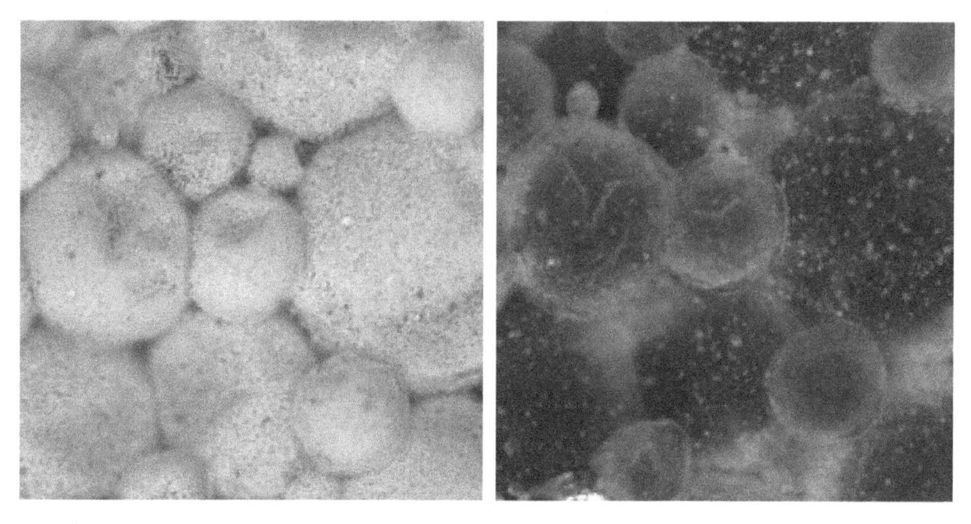

Figure 1.16 Largest capsules are on the order of 100 μm. From 1998 Digest paper [80] Drzaic P, et.al, 1998 / John Wiley & Sons, Inc.

- There is a wide distribution of microcapsule sizes. Since the particles are rigid, they tend to layer on top of each other, increasing the cell gap across the film. This increased cell gap raises the total voltage required to switch.
- The rigid particles do not close pack. There are large gaps between microcapsules in which no switching occurs. These optically "dead" areas do not switch between light or dark states and degrade brightness and contrast.
- The distribution of the electrophoretic particles within the microcapsule is not uniform. Particularly in the dark state, the centers of the microcapsules showed larger optical contrast than the edges, which did not switch to the same degree.

Loxley and colleagues deftly resolved many of these issues by introducing microencapsulation chemistries to make the microcapsule walls compliant and developing a coating formulation to deposit a monolayer of capsules in an aqueous binder [89]. As the coating dried, surface tension pulled the microcapsules together. The compliant sidewalls enable the microcapsules to close pack, with minimal gaps between capsules. Additionally, the sidewalls between neighboring capsules take on a more vertical profile, which minimizes any curvature-induced packing issues in the microcapsules.

Figure 1.17 shows a schematic-view of the close packed capsules in a complete display. Figure 1.18 shows a top view of the microcapsules, with each square spaced at 200 ppi (127 μm on a side). Switching voltages of ±15 V became much more compatible with active-matrix backplanes [90].

The photograph also illustrates another key feature of this formulation. The size of the microcapsules does not directly define the achievable resolution of the display. It is seen that with a given microcapsule, part of the microcapsule can be switched dark, and the remaining part light, depending on the field applied by adjacent electrodes. As such, high-resolution images can be rendered.

1.13 Image Retention

One of the potential advantages of microencapsulated EPD systems is that the display's image can be retained for long periods. While this phenomenon had been noted in the early EPD work, the controlling factors were not fully understood. Sedimentation due to gravity was recognized as a cause of images fading over time, particularly severe in high-density particles such as titanium dioxide. Coating the particles with polymer could reduce their density somewhat.

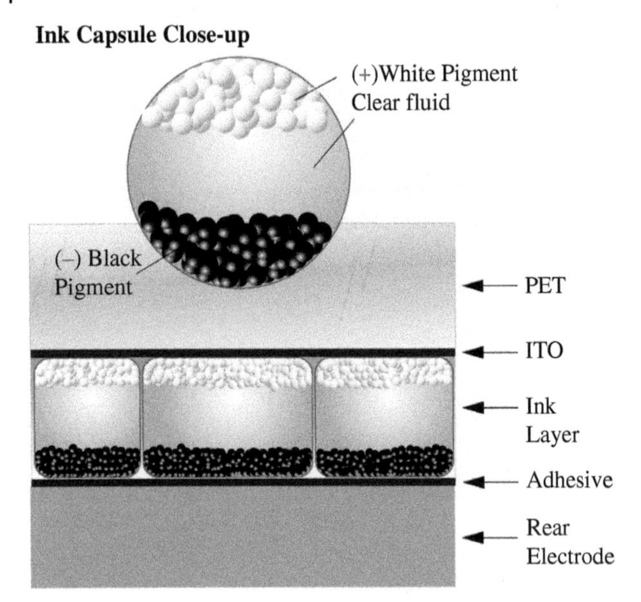

Figure 1.17 Schematic cross-section of microcapsules e-Paper display.

Figure 1.18 Each square is 127 μm on a side. From Bouchard 2004 [88].

As mentioned previously, Ota and Novotny and Hopper's publications attributed memory effects to van der Waals and electrostatic attractive force between the pigment particles and the electrode. In principle, the particle surfaces could be modified to enable sufficient attractive forces to counteract sedimentation forces. Still, the reality is that it is challenging to alter the particles to weakly and reversibly flocculate this way. In a two-particle system, the problems become even more serious.

Webber developed stable and reversible image stability for two particle systems [91]. Webber discovered that adding an oil-soluble high molecular weight polymer (>20 000) to the electrophoretic fluid can dramatically increase image stability without increasing the fluid viscosity. If the polymer is non-absorbing onto the particle surfaces, the well-known phenomenon of depletion flocculation will create osmostic pressure between the particles and induce soft flocculation [92]. Webber demonstrated image stability of 10^6 seconds in a black and white system (Figure 1.19).

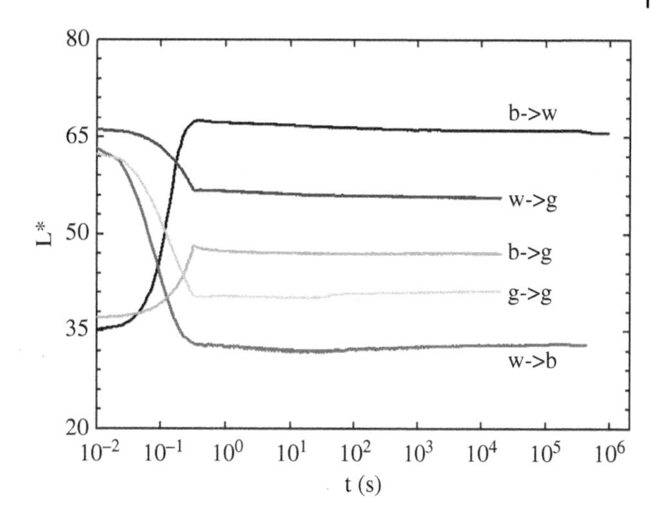

Figure 1.19 Lightness (L*) as a function of time (t). Lightness states were achieved from different starting lightness states as indicated by the notation b→w, w→g, etc. The display was unpowered after completion of the optical change, i.e. t>0.3 s [91] / John Wiley & Sons, Inc.

1.14 Active-Matrix Compatibility

High-resolution images for EPDs require an active matrix backplane for addressing. Moreover, to fulfill the vision of "radio paper," it seemed important that the active matrix backplane was scalable, flexible, and ultimately printable. Fortunately, silicon-based TFT arrays were becoming available, and new flexible and printable technologies were on the horizon.

The first achievement of a silicon TFT, matrix-addressed EPD, was made in late 1999. Bob Wisnieff of IBM provided to E Ink some active-matrix panels designed for use in LCDs that would likely be compatible with the EPD material. This initial demonstration led to a successful collaboration with IBM, resulting in a 2001 report of active-matrix driven EPDs with high pixel counts.[ii] Just as importantly, these prototypes demonstrated compatibility with technology already being deployed in the massive liquid crystal display industry [93].

In 2001 the engineers at E Ink further demonstrated microencapsulated EPD display driven by two other flexible active-matrix panels.

- Using facilities at Princeton University, Chen designed and fabricated an a-Si TFT array on a flexible stainless steel backplane. The resulting display is the first example of a flexible a-Si microencapsulated EPD array [94].
- In collaboration with Lucent/Bell Labs, a 16×16 TFT array was built using organic transistors fabricated using a stamp-transfer methods [95,96]. The resulting array was the first example of using organic semiconductors to fabricate a flexible display. This achievement gained international attention, winning the American Chemical Society Team Innovation award in 2002 and is designated as an "Editor's Choice Best-of-the-Best" R&D100 award from *R&D Magazine* in 2001.

After this basic active-matrix compatibility was demonstrated, custom electronic controllers and backplane designs enabled low power driving of high-resolution panels. Early work leveraged existing AMLCD TFT backplane architectures, using specialized controllers to interface with the panel drivers, and providing specialized pixel-level voltage pulses unique to the E Ink EPD material [97]. Gates et al. described an improved controller for EPD that elegantly repurposed the 16 bit wide LCD interface used for driving color panels to two 8-bit representations of the current and to-be-updated EPD pixel gray levels. The display subsystem was also shut off between updates, allowing for ultra-low power operation due to the EPD image stability [97,98].

ii The E Ink-IBM collaboration was mentioned briefly in a 2000 article in the Wall Street Journal. Interestingly, the reporter noted that IBM had a longstanding interest in newspapers, and was "supplying E Ink with millions of its laptop transistors for testing." A. Klein, "Will the Future Be Written in E-Ink?," *The Wall Street Journal*, pp. 1–4, 04 Jan 2000.

The collaboration with Philips Corporation was critical for the successful evolution of the E Ink displays. Many such improvements are described in the paper by Henzen et al. [98]. EPD-specific backplane modifications included two series transistors per pixel to enable large swing voltages, an enlarged storage capacitor to suppress kickback effects, and the adoption of a thick acrylate coating as an inner-layer dielectric to create a field-shielded pixel design. The Philips collaboration included a team from the Netherlands (Alex Henzen, Jan van de Kamer, Neculai Aileni, Roger Delnoij, Guofu Zhou, Mark Johnson), and Japan (Michael Pitt, Masaru Yasui, Tadao Nakamura, Tomohiro Tsuji).

1.14.1 Reproducible Gray Scale

Relying only on halftones for grayscale limits displaying images at high resolution. Microencapsulated EPD systems can partially switch between black and white states by applying an impulse (voltage*time) less than required to fully switch the pixel state. In principle, a large number of grayscales can be achieved this way. The challenge is that the impulse needed to achieve a specific gray level depends on the prior state of the display. If there is a set of pixels showing white, some showing black, and some showing a certain gray, the impulse to switch to another gray level will be different for each prior state.

Moreover, each pixel can have a somewhat different internal distribution of particles, depending on the history, even at the same nominal gray level. As more and more switching events take place, greater variations are built into each pixel, which appears in varying gray levels to the viewer. The pixels can be reset to a common starting state by applying a series of strong impulses, but the viewer sees these reset pulses as annoying flashes.

Amundson, Zehner, and coworkers demonstrated that it is possible to achieve reliable 4-bit (16 level) grayscale using an algorithm that tracks the history of each pixel's switching for a period of time, applying an impulse based on that history and the target gray level [99]. Through this approach and further application of other pulse waveforms, reasonably accurate 2-bit (4 gray level) images could be achieved without visible flashing. This capability was extended to 16 levels in 2009 [100].

1.14.2 Overcoming "kickback" Phenomena

Since the early days of electrophoretic displays; it was known that charged species besides the pigments within the electrophoretic fluid could affect the performance of the displays. For the microencapsulated EPD, there is the additional complication that the "external phase" (polymer outside of the microcapsules) can also contain mobile ions, leading to deleterious phenomena due to time-dependent polarization that external phase. This polarization can lead to phenomena where the switching characteristics depend on the time between pulses, or where the image slowly erases over time due to this polarization. However, work by Cao, Danner, and others showed that careful selection of the external phase chemistry could prevent these phenomena from occurring, leading to predictable switching and long-term stable images [101,102]. The mechanism why kickback occurs and the waveform for alleviating this phenomenon will be described in more detail in Chapters 2 and 3, respectively.

1.14.3 EPD Panel Manufacturing

Unlike liquid crystal displays, microencapsulated EPDs required the development of several new manufacturing techniques. These include:

- volume production of electrophoretic suspension with all necessary pigment modifications and additives and incorporation of this fluid into a coatable microcapsule suspension
- roll-to-roll coating
- lamination to mechanical protection layer to enable handling

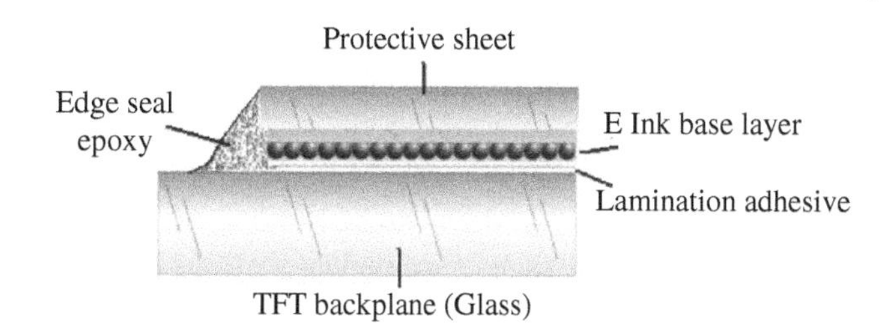

Figure 1.20 Encapsulation construction with [98] / John Wiely & Sons, Inc.

- creating an aperture through the electro-optical material to allow interconnection to a backplane
- mechanical protection layers, and
- edge sealing to protect the microencapsulated EPD from water and humidity

Some engineers responsible for these essential practical advances include Duthaler, Danner, Zehner, Steiner, Doshi, Geramita, Valianatos, LeCain, and others [89,98,103] (Figure 1.20).

1.15 Electronic Book Products, and E Ink Merger

The first EPD-based electronic reader, The Sony Librie EBR 1000, was introduced for sale in Japan in April 2004 [104]. The device was developed through a collaboration between Sony, E Ink, Philips, and Toppan Printing.

The device used a proprietary Sony eBook format. Content could only be purchased from content providers like Publishing Link via a PC and then transferred to the Libre by USB cable or memory stick. Digital rights management built into the system allowed for content to be accessible for 60 days, after which it expired and became unreadable. No support for pdf or other ebook formats was enabled. The device opened to middling reviews, in large part due to these content-based restrictions. While the screen appearance was well-received, many complained about the slow device interface, limited book offering, and the forced erasure of content [105].

Over the next decade, Sony released a series of upgraded readers and started marketing them outside of Japan in 2006. The devices never reached the volumes expected for such a product, and Sony exited the e-reader market entirely in 2014 [106,107].

The Amazon Kindle was released in November 2007 and set the stage for a flourishing ebook ecosystem. This first device was a 6″ diagonal display, 800 × 600 @167 pixels per inch, and four gray levels per pixel. E Ink provided the microencapsulated electrophoretic material, the packaging design, and the driving scheme. Toppan printing mass-produced the coating and conversion into sheet form the moisture-barrier protective sheet. Philips developed the backplane, timing controller, and other ICs, and assembled the final module. Amazon sold the Kindle reader provided a large library of content for sale, and made it easy for customers to access that content. Year-by-year improvements led to greater adoption and increasing sales of e-paper displays [107].

Despite the technical and product success of the Kindle, E Ink, the company was burdened by significant debt. The company had invested in many years of technology and business development. Its large area sign business only achieved modest success, and the Libre did not become a breakout product. Rather than maintain status as a standalone company, E Ink was sold to the Taiwanese Company Prime View International (PVI) in 2009 to enable further commercialization of the EPD technology [108]. Since then, PVI has renamed itself E Ink Holdings and continues to develop electronic ink for ebooks, signage, and other products. Later on, SiPix was merged with E Ink, with the microcup embossing technology enhancing the realization of high-performance color e-paper [109,110].

1.16 Summary

As is true for many technologies, the path for success for the electronic paper was not predictable. Periods of rapid technical progress were followed by periods of little interest due to combinations of application status, competitive technologies, and the commercial and financial landscape. Still, the rise, fall, and rise of electronic ink provides a nice success story for a technology that was envisioned to overcome a limitation of electronic displays and eventually managed to do just that.

The development of electronic paper did not end with the introduction of the Kindle. The remaining chapters of this book describe many exciting areas of development in this important field.

Acknowledgments

PSD is grateful for the assistance of Rob Zehner, Guy Danner, Howie Honeyman, Rich Webber, Glen Crossley, Andrew Loxely, and Gregg Duthaler for useful discussions, and in proofreading the chapter.

BRY is grateful for the assistance of Guangyou Liu, Mingyang Yang, Zheng Zeng, Linyu Song, Hao Shu, Yifan Gu, Yunhe Liu, Jie Liu, Xinzao Wu for manuscript preparation and in proofreading the chapters.

AC would like to acknowledge Joel Pollack for oral accounts of early development activities on Electrophoretic Display Technology at various Xerox R&D Centers up to mid-1970; and Warren Jackson for facilitating fact-checks from Xerox document archives.

References

1 Heikenfeld, J., Drzaic, P., Yeo, J. et al. (2011). Review paper: a critical review of the present and future prospects for electronic paper. *J. Soc. Inf. Disp.* 19(2): 129–156.
2 Eshkalak, S.K., Khatibzadeh, M., Kowsari, E. et al. (2019). Overview of electronic ink and methods of production for use in electronic displays. *Opt. Laser Technol.* 117: 38–51.
3 http://paperproject.org/semgallery/semgallery1a.html#imagetop.
4 Smith, E., Heckaman, R.L., Lang, K. et al. (2020). Evaluating display color capability. *Inf. Disp.* 36(5): 9–15.
5 https://commons.wikimedia.org/wiki/File:Halftoningcolor.svg.
6 https://inkjetinsight.com/knowledge-base/high-speed-inkjet-devices-take-on-commercial-print-quality.
7 https://inkjetinsight.com/knowledge-base/understanding-gray-areas-inkjet.
8 The International Committee for Display Metrology, (2012). *Information Display Measurement Standard*, v 1.03, 477. SID Publishers.
9 Shu, Y., Hagedon, M., and Heikenfeld, J. (2011). Light out-coupling for reflective displays: simple geometrical model, MATLAB simulation, and experimental validation. *J. Disp. Technol.* 7(9): 473–477.
10 Robert, F., Sri, P., Robert, H. et al. (2018). 48-2: Electronic paper 2.0: frustrated eTIR as a path to color and video. In: *SID Symposium Digest of Technical Papers*, 49(1): 630–632.
11 Fleming, R., Kazlas, P., Johansson, T. et al. (2019). 36-3: Tablet-size eTIR display for low-power ePaper applications with color video capability. In: *SID Symposium Digest of Technical Papers*, 50(1): 505–508.
12 Mossman, M.A., Whitehead, L.A., and Rao, S.P. (2002). P-83: Grey scale control of TIR using electrophoresis of sub-optical pigment particles. *SID Symp. Dig. Tech. Pap.* 33(1): 522–525.
13 Satoshi, N., Kodama, Y. et al. (2013). 6.1: Invited paper: electronic paper system using high resolution electrophoretic display. *SID Symp. Dig. Tech. Pap.* 44(1): 34–37.
14 Lee, J.H., Horiuchi, T., and Saito, R. (2007). Confined-error-diffusion algorithm for flat-panel display. *J. Soc. Inf. Disp.* 15: 507–518.

15 Duthaler, G., Au, J., Davis, M. et al. (2002). 53.1: Active-matrix color displays using electrophoretic ink and color filters. In: *SID Symposium Digest of Technical Papers*, 33(1): 1374–1377. Oxford, UK: Wiley Blackwell.

16 Hiji, N., Machida, Y., Yamamoto, Y. et al. (2012). 8.4: Distinguished paper: novel color electrophoretic E-paper using independently movable colored particles. In: *SID Symposium Digest of Technical Papers*, 43(1): 85–87. Oxford, UK: Wiley Blackwell.

17 Telfer, S.J. and Mccreary, M.D. (2016). 42-4: Invited paper: a full-color electrophoretic display. In: *SID Symposium Digest of Technical Papers*, 47(1): 574–577.

18 French, I., Lo, P., Chung, R. et al. (2020). Kaleido Color eReader Displays [J]. *Proceedings of the International Display Workshops*, 27: 754–756.

19 Mukherjee, S., Hsieh, W.L., Smith, N. et al. (2015). Electrokinetic pixels with biprimary inks for color displays and color-temperature-tunable smart windows. *Appl. Opt.* 54(17): 5603–5609.

20 Ivanova, A.G., Khamova, T.V., Gubanova, N.N. et al. (2020). Chemistry and manufacturing technology of electronic ink for electrophoretic displays (a review). *Russ. J. Inorg. Chem.* 65(13): 1985–2005.

21 Werner, K. (2019). LA chapter's one-day conference covers North American display activity. *Inf. Disp.* 35(4): 34–36.

22 Comiskey, B., Albert, J.D., Yoshizawa, H. et al. (1998). An electrophoretic ink for all-printed reflective electronic displays. *Nat.* 394 (6690): 253–255.

23 Verschueren, A., Stofmeel, L., Baesjou, P.J. et al. (2012). Optical performance of in-plane electrophoretic color e-paper. *J. Soc. Inform. Disp.* 18(1): 1–7.

24 Kishi, E., Matsuda, Y., Uno, Y., Ogawa, A., Goden, T., Ukigaya, N., Nakanishi, M., Ikeda, T., Matsuda, H. and Eguchi, K. (2000), 5.1: Development of In-Plane EPD. *SID Sympos. Digest. Tech. Pap.* 31: 24–27.

25 Lenssen, K., Baesjou, P.J., Budzelaar, F. et al. (2009). Novel concept for full-color electronic paper. *J. Soc. Inform. Disp.* 17(4): 383–388.

26 Yeo, J., Emery, T., Combs, G., et al. 2012. 69.4: Novel Flexible Reflective Color Media Integrated with Transparent Oxide TFT Backplane[J]. *SID Sympos. Digest. Tech. Pap.* 41(1): 1041–1044.

27 Koch, T., Yeo, J.S., Zhou, Z.L. et al. (2011). Novel flexible reflective color media with electronic inks[J]. *J. Inform. Disp.* 12(1): 5–10.

28 Sakurai, R., Hattori, R., Asakawa, M. et al. (2008). A Flexible electronic-paper display with an ultra-thin and flexible LSI driver using quick-response liquid-powder technology [J]. *J. Soc. Inform. Disp.* 16(1): 155–160.

29 Hattori, R., Yamada, S., Masuda, Y. et al. (2004). A quick-response liquid-powder display (QR-LPD®) with plastic substrate[J]. *J. Soc. Inform. Disp.* 12(4): 405–409.

30 Yashiro, T., Hirano, S., Naijoh, Y. et al. (2011). 5.3: Novel design for color electrochromic display[C]. *SID Sympos. Digest. Tech. Pap.* 42(1): 42–45. Oxford, UK: Blackwell Publishing Ltd.

31 Ko, I. J., Park, J. H., Kim, Y. C., et al. 2017. P-139: High Optical Contrast Reflective Display with Electrochromism[C]. *SID Sympos. Digest. Tech. Pap.* 48(1): 1781–1784.

32 Noh, C. H., Lee, J. M., Jeon, S. J., et al. (2009). P-105: New Electrochromic Systems having Controllable Color and Bistability[C]. *SID Sympos. Digest. Tech. Pap.* Oxford, UK: Blackwell Publishing Ltd. 40(1): 1508–1511.

33 Cho, S.I., Kwon, W.J., Choi, S.J. et al. (2005). Nanotube-based ultrafast electrochromic display[J]. *Adv. Mater.* 17(2): 171–175.

34 Rawert, J., Jerosch, D., Blankenbach, K. et al. (2010). 15.2: Bistable D3 Electrowetting Display Products and Applications[C]. *SID Sympos. Digest. Tech. Pap.* Oxford, UK: Blackwell Publishing Ltd. 41(1): 199–202.

35 Lo, K.L., Tsai, Y.H., Cheng, W.Y. et al. (2013). 12.4: Recent Development of Transparent Electrowetting Display[C]. *SID Sympos. Digest. Tech. Pap.* Oxford, UK: Blackwell Publishing Ltd. 44(1): 123–126.

36 Blankenbach, K., Schmoll, A., Bitman, A. et al. (2008). Novel highly reflective and bistable electrowetting displays[J]. *J. Soc. Inform. Disp.* 16(2): 237–244.

37 Hagedon, M., Yang, S., Russell, A. et al. (2012). Bright e-Paper by transport of ink through a white electrofluidic imaging film[J]. *Nat. Commun.* 3(1): 1–7.

38 Henzen, A. (2018). 11.1: Invited Paper: Electrofluidic displays: Fast switching, Colorful, low power[C]. *SID Sympos. Digest. Tech. Pap.* 49: 105–107.

39 Bai, P.F., Hayes, R.A., Jin, M. et al. (2014). Review of paper-like display technologies (invited review)[J]. *Prog. Electromagnet. Res.* 147: 95–116.
40 Gally, B., Lewis, A., Aflatooni, K. et al. (2011). 5.1: Invited Paper: A 5.7 "Color Mirasol® XGA Display for High Performance Applications[C]. *SID Sympos. Digest. Tech. Pap.* Oxford, UK: Blackwell Publishing Ltd. 42(1): 36–39.
41 Lee, J.H., Cheng, I.C., Hua, H. et al. (2020). *Introduction to flat panel displays[M].* Wiley.
42 Itoh, Y., Minoura, K., Asaoka, Y. et al. (2007). 40.5: Invited Paper: Super Reflective Color LCD with PDLC Technology[C]. *SID Sympos. Digest. Tech. Pap.* Oxford, UK: Blackwell Publishing Ltd. 38(1): 1362–1365.
43 Wen C J, Ting D L, Chen C Y, et al. (2000). P-1: Optical Properties of Reflective LCD with Diffusive Micro Slant Reflectors (DMSR)[C]. *SID Sympos. Digest. Tech. Pap.* Oxford, UK: Blackwell Publishing Ltd. 31(1): 526–529.
44 Asao (2007). Y. P-126: Advanced Image-Processing Method for Hybrid Color LCDs[C]. *SID Sympos. Digest. Tech. Pap.* Oxford, UK: Blackwell Publishing Ltd. 38(1): 680–683.
45 Akashi, Y., Rea, M.S., and Bullough, J.D. (2007). Driver decision making in response to peripheral moving targets under mesopic light levels[J]. *Light. Res. Technol.* 39(1): 53–67.
46 https://flipdots.com/en/products-services/status-indicators/
47 https://flipdots.com/en/products-services/custom-made-flip-dot-wall/
48 Jones, J.C. (2008). The zenithal bistable display: from concept to consumer[J]. *J. Soc. Inform. Disp.* 16(1): 143–154.
49 Jones, J.C. (2017). Defects, Flexoelectricity and RF communications: the ZBD story[J]. *Liq. Cryst.* 1–28.
50 Kay, A. and Goldberg, A. (1977). Personal dynamic media. *Comput.* 10(3): 31–41.
51 Davey, W. P. 1919. Process for making and applying Japan. US 1294627.
52 Archibald, M. K. 1959. Process for Developing Electrostatic Image with Liquid Developer. US 62936356A.
53 Evans, P. F., Lees, H. D., Maltz, M. S., and Dailey, J. L. 1971. Color Display Device. US 3612758.
54 Ota, I. 1972. Electrophoretic Display Device. US 3668106.
55 Ota, I., Ohnishi, J., and Yoshiyama, M. (1973). Electrophoretic image display (EPID) panel. *Proc. IEEE* 61(7): 832–836.
56 Goodman, L.A. (1975). Passive liquid displays: liquid crystals, electrophoretics, and electrochromics. *IEEE Trans. Consum. Electron.* 3: 247–259.
57 Lewis, J.C., Garner, G.M., Blunt, R.T. et al. (1977). Gravitational, inter-particle and particle-electrode forces in electrophoretic display. In: *Proceedings of the SID*. 18(3–4): 235–242.
58 Mürau, P. and Singer, B. (1978). The understanding and elimination of some suspension instabilities in an electrophoretic display. *J. Appl. Phys.* 49(9): 4820–4829.
59 Croucher, M.D. and Hair, M.L. (1981). Some physicochemical properties of electrophoretic display materials. *Ind. Eng. Chem. Prod. Res. Dev.* 20(2): 324–329.
60 Novotny, V. and Hopper, M.A. (1979). Optical and electrical characterization of electrophoretic displays. *J. Electrochem. Soc.* 126(12): 2211.
61 Dalisa, A.L. (1977). Electrophoretic display technology. *IEEE Trans. Electron Devices* 24(7): 827–834.
62 Dalisa, A.L., Singer, B., and Liebert, R. (1977). Matrix addressed electrophoretic displays. In: *1977 International Electron Devices Meeting*. IEEE, 74–77.
63 Novotny, V. and Hopper, M.A. (1979). Optical and electrical characterization of electrophoretic displays[J]. *J. Electrochem. Soc.* 126(12): 2211.
64 Hopper, M.A. and Novotny, V. (1979). An electrophoretic display, its properties, model, and addressing[J]. *IEEE Trans. Electron Devices* 26(8): 1148–1152.
65 Harbour, J.R. (1979). Subdivided electrophoretic display. *Xer. Disclos. J.* 4(6): 705.
66 Vance, D.W. (1977). Optical Characteristics of Electrophoretic Displays[C]. *Proceed. of the SID.* 18(3): 267–274.
67 Fitzhenry-Ritz B. (1981). Optical properties of electrophoretic image displays[J]. *IEEE Transactions on Electron Devices* 28(6): 726–735.
68 Werner, K. and Kmetz, A.R. (1986). Whatever happened to electrophoretic displays?[J]. *IEEE Spectr.* 23(7): 28–29.
69 Castellano, J.A. (2005). *Liquid gold: the story of liquid crystal displays and the creation of an industry[M].* World Scientific.

70 Ota, I. (2009). History of Electrophoretic Displays and Proposal of a Novel Cell Structure for Lateral Particle Movement Display Devices[C]. *International Display Workshop*. EP1-1, 525.

71 Murau, P. (1984). Characteristics of an X-Y addressed electrophoretic image display SID Technical Digest. 141.

72 Chiang, A. (1981). A matrix addressable electrophoretic display. *Eur. Secur.* 107–110.

73 Chiang, A., Curry, D., Zarzycki, M. (1979). A stylus writable electrophoretic display device. SID. 79: 44–45.

74 Shiffman, R.R. and Parker, R.H. (1984). An electrophoretic image display with internal NMOS address logic and display drivers. *Proc. Soc. Inf. Disp.* 25(2): 105–115.

75 Beilin, S., Zwemer, D., and Kulkarni, R. (1986). 2000-character electrophoretic display. SID Technical Digest. 136.

76 Chiang, A. (1980). Electrophoretic displays: the state of the art, Proceedings of Biennial Display Research Conference. 10–12.

77 Chiang, A. and Fairbairn, D.G. (1980). A high speed electrophoretic matrix display. *SID Tech. Digest.* 1–2.

78 Chiang, A. (1980). Electrophoretic Display Technologies and Characteristics, Seminar Notes, SID Technical Symposium. 1–32.

79 https://www.nytimes.com/1992/09/20/nyregion/a-stock-thrives-without-a-product.html.

80 Drzaic, P., Comiskey, B., Albert, J.D. et al. (1998). *44.3 L: A Printed and Rollable Bistable Electronic Display[C]. SID Sympos. Digest. Tech. Pap.* Oxford, UK: Blackwell Publishing Ltd. 29(1): 1131–1134.

81 Osamu, I., Tadakuma, A., Mori, T., and Naoyuki, M. (1996). Electrophoretic display device[P]. JPS 6486116A.

82 Nakamura, E., Kawai, H., Kanae, N. et al. (1998). *37.3: Development of Electrophoretic Display Using Microcapsulated Suspension[C]. SID Sympos. Digest. Tech. Pap.* Oxford, UK: Blackwell Publishing Ltd. 29(1): 1014–1017.

83 Inoue, O. A., Tadakuma, T., Maida, N. M. (1987). Electrophoretic display device [P]. JP 62244679A.

84 Jacobson, J., Comiskey, B., Turner, C. et al. (1997). The last book[J]. *IBM Syst. J.* 36(3): 457–463.

85 Vinzant, C. (1998). Fortune; New York. 138(1): 76–77.

86 Burdick, A. (2000). The Only Book You'll Ever Need to Read[J]. New York Times Magazine.

87 Honeyman, C. H., Moran, E. A., Zhang, L. et al. (2004). Electrophoretic particles and processes for the production thereof. US 6822782.

88 Bouchard, A., Suzuki, K., and Yamada, H. (2004). P-102: High-resolution microencapsulated electrophoretic display on silicon. *SID Sympos. Digest. Tech. Pap.* Oxford, UK: Wiley Blackwell. 35(1): 651–653.

89 Albert, J. D., Crossley, G., Geramita, K., et al. (2005). Encapsulated electrophoretic displays having a monolayer of capsules and materials and methods for making the same[P]. US 6839158.

90 Pitt, M G. (2002). 53.2: Power consumption of Micro-encapsulated electrophoretic displays for smart handheld applications. SID. 1: 1378–1381

91 Webber, R.M. (2002). *10.4: Image Stability in Active-Matrix Microencapsulated Electrophoretic Displays[C]. SID Sympos. Digest. Tech. Pap.* Oxford, UK: Blackwell Publishing Ltd. 33(1): 126–129.

92 Jenkins, P. and Snowden, M. (1996). Depletion flocculation in colloidal dispersions. *Adv. Colloid Interface Sci.* 68: 57–96.

93 Kazlas, P., Au, J., Geramita, K. et al. (2001). *12.1: 12.1′ SVGA Microencapsulated Electrophoretic Active Matrix Display for Information Appliances[C]. SID Sympos. Digest. Tech. Pap.* Oxford, UK: Blackwell Publishing Ltd. 32(1): 152–155.

94 Chen, Y., Denis, K., Kazlas, P. et al. (2001). 12.2: A conformable electronicilnk display using a foil-based a-Si TFT array. *SID Sympos. Digest. Tech. Pap.* Oxford, UK: Wiley Blackwell. 32(1): 157–159.

95 Amundson, K., Ewing, J., Kazlas, P. et al. (2001). 12.3: Flexible, Active-Matrix Display Constructed Using a Microencapsulated Electrophoretic Material and an Organic-Semiconductor-Based Backplane[C]. *SID Sympos. Digest. Tech. Pap.* Oxford, UK: Blackwell Publishing Ltd. 32(1): 160–163.

96 Rogers, J.A., Bao, Z., Baldwin, K. et al. (2001). Like electronic displays: large-area rubber-stamped plastic sheets of electronics and microencapsulated electrophoretic inks. *Proc. Natl. Acad. Sci.* 98(9): 4835–4840.

97 Gates, H., Ohkami, T., and Au, J. (2006). *37.2: Improved Electronic Controller for Image Stable Displays[C]. SID Sympos. Digest. Tech. Pap.* Oxford, UK: Blackwell Publishing Ltd. 37(1): 1406–1409.

98 Henzen, A., Van de Kamer, J., Nakamura, T. et al. (2004). Development of active-matrix electronic-ink displays for handheld devices. *J. Soc. Inf. Disp.* 12(1): 17–22.
99 Johnson, M.T., Zhou, G., Zehner, R. et al. (2006). High-quality images on electrophoretic displays[J]. *J. Soc. Inform. Disp.* 14(2): 175–180.
100 https://en.wikipedia.org/wiki/Amazon_Kindle.
101 Cao, L., Gates, E., Miller, D. et al. (2007). Electrophoretic media and displays with improved binder. US 20070091417A1.
102 Wilcox, R. J., Whitesides, T. H., Amundson, K. R. et al. (2013). Electro-optic displays with reduced remnant voltage[P]. US 8558783.
103 LeCain, R. D., Knaian, A. N., O'Neil, S. J. et al. (2006). Components and methods for use in electro-optic displays. US 6982178B2.
104 https://goodereader.com/blog/electronic-readers/the-sony-librie-ebr-1000-was-the-first-e-reader-with-e-ink
105 Dvorak, P. (2004). Electronic Readers, Now on Sale in Japan, Still Don't Beat Paper. Wall Street Journal. (Eastern edition). New York, NY: Jul. 15: 13.1.
106 https://goodereader.com/blog/electronic-readers/the-evolution-of-the-sony-e-reader-in-pictures
107 https://techcrunch.com/2014/02/06/sony-closing-reader-store/
108 Chiu, J. (2012). How does financial strategy help to provide a win-win merger and acquisition solution? The case of PVI and E Ink[J]. *Appl. Fin.* 1314.
109 Liang, R.C., Hou, J., Zang, H. et al. (2003). Microcup® displays: Electronic paper by roll-to-roll manufacturing processes[J]. *J. Soc. Inform. Disp.* 11(4): 621–628.
110 Yang, B.-R., Lin, C., and Yih-Ming, K. (2014). Microcup Designs for Electrophoretic Display[P]. US 8891156B2.
111 Toyama, J., Akatsuka, T., Takano, S., Tadakuma, A., Mori, T., Suga, S., Sano, Y., (1994), *An Electrophoretic Matrix Display with External Logic and Driver Directly Assembled to the Panel, SID Sympos. Digest Tech Paper* 25:588–590.
112 Comiskey, B., Albert, J.D., Jacobson, J. (1997) 7.4L: Late-News Paper: Electrophoretic Ink: A Printable Display Material, SID Sympos. Digest Tech Papers, 28:75–76.

2

Fundamental Mechanisms of Electrophoretic Displays

Bo-Ru Yang[1] *and Kristiaan Neyts*[2]

[1]*Professor, School of Electronics and Information Technology, State Key Laboratory of Optoelectronic Materials and Technologies, Guangdong Province Key Laboratory of Display Material and Technology, Sun Yat-Sen University, Guangzhou, China*
[2]*Professor, and Head of Liquid Crystals and Photonics group, ELIS Department, Ghent University, Technologiepark 126, B-9052 Ghent, Belgium*

2.1 General View of Electronic Ink Operation

Electronic ink contains pigment particles, charge controlling agents (CCAs), non-polar solvents, and various additives. The particles are charged by the CCAs to be transported in the solvent by an electric field. The interaction between these ingredients in ink is complicated. Therefore, to explain the whole operation in a more easily understandable way, we will start from the general concepts of the driving waveform, the charging mechanism of inverse micelles (IMs), the drift and diffusion of inverse micelles, the charging of particles with additives, the rheological behavior of charged particles, and then the driving method for realizing full-color e-paper.

The driving voltage waveform, which updates an electronic ink display, typically includes several stages [1]. The particles' track during the update process is illustrated in Figure 2.1a for a pixel with top and bottom electrodes. In a typical driving waveform, particles are first agitated by an alternating voltage ("shaking pulses") to activate the particles, then driven to the upper and bottom boundaries to erase the image history, and finally driven to the desired gray level with a defined particle distribution. In this stage, the distribution of pigment particles can be controlled by applying specific voltage waveforms over electrodes. Different particle distributions will yield different reflectivities specified by the term "grayscale" or "color scale" when colored particles are used.

Information written to the display during previous addressing cycles can lead to a residual image (also known as a "ghost image") [2]. The ghost image results from a difference between the target grayscale and the actual grayscale shown in Figure 2.1b. Suppose the aim is to reach e.g. the 13th grayscale (G13). In that case, the final reflectivity may be slightly different (G13*) because it has a dependence on the previous gray level (i.e., previous particle distribution). The commonly used solutions to solve ghost image are multi-flash transition, weakly indirect transition, local DC balance, and global DC balance methods. With the multi-flash transition, electrophoretic particles are pushed several times to the capsule boundaries to erase the memory of previously applied gray scales. However, this introduces a rapid bright/dark flickering which is annoying for the viewers, and therefore the weakly indirect transition [3] was adopted.

The driving waveform is usually composed of a history erasing phase, an activation phase, and a display phase which reduces grayscale deviations [4]. Figure 2.2 shows a modified driving waveform, including first shaking pulses, a help reset pulse of opposite polarity to the reset pulse, a reset pulse, second shaking pulses, and a driving pulse [5]. Furthermore, in the case of an area of a static displayed image, the electrophoretic particles are driven in the same direction for a long time, which leads to the accumulation of charge in the direction of the built-in field, which affects the accuracy of the grayscale, and reduces the lifetime of the EPD. This charge accumulation-induced phenomenon is termed "DC imbalance."

E-Paper Displays, First Edition. Edited by Bo-Ru Yang.

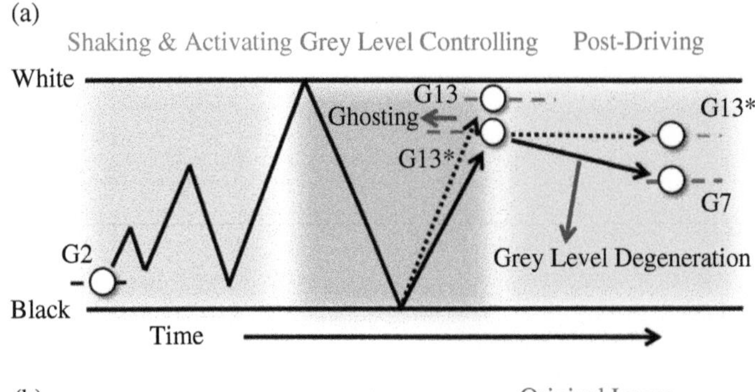

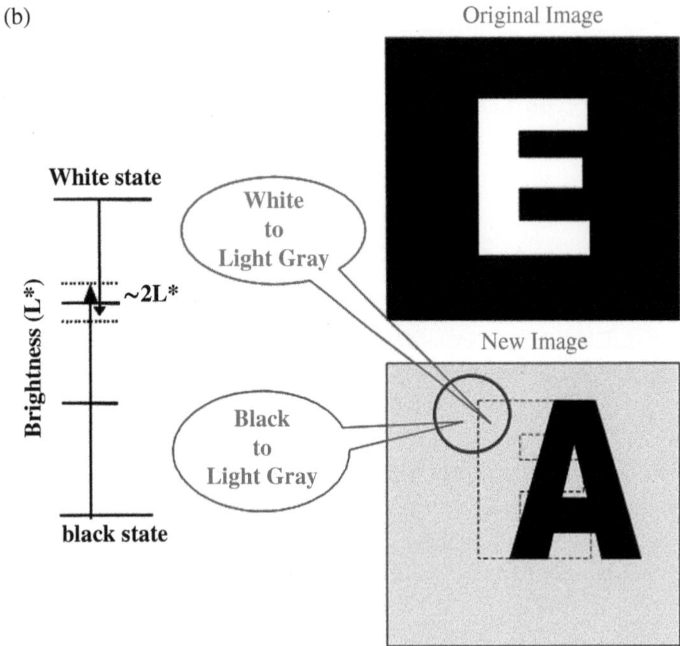

Figure 2.1 The ghosting phenomenon. (a) schematic optical trace of particles under the influence of the driving waveform [1] and (b) illustrative diagram of ghost image [2].

To fix this problem and obtain DC-balance in the driving waveform, it is important that the integral of the applied voltage waveform over time should be approximately zero for each pixel. To switch from one state to another, a series of voltage pulses can be used in which the negative voltage pulses of appropriate duration compensate the positive voltage pulses to obtain DC balance. The approach to achieve DC balance in a single driving cycle is called "Local DC Balance" [6]. The "Global DC Balance" approach [7] was further adopted to shorten the image updating time and flashing times. These two methods will be elaborated in Chapter 3 of this volume.

After removing the external driving voltage, the distribution of particles is not supposed to change due to the bistability characteristics of the E-paper. However, the distribution of particles is typically influenced by factors such as the built-in electric field of each charged species, which causes a slow variation of the grayscale of E-paper over time. This phenomenon is called degeneration of the gray scale.

The multi-gray scales of the E-paper can be achieved by controlling the polarity and the amplitude of the voltage, which is called pulse amplitude modulation (PAM). Alternatively, it can be achieved by controlling the polarity and the duration of the voltage applied between the two electrodes of the E-paper, which is called pulse width modulation (PWM). As shown in Figure 2.3, both PWM and PAM can be used to control the movement of

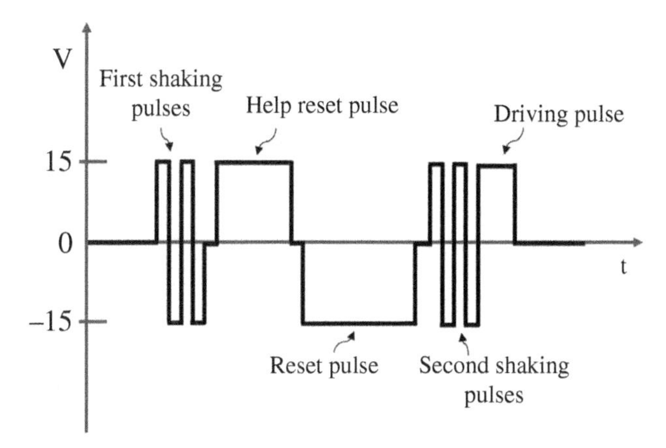

Figure 2.2 The general structure of a waveform for EPD driving.

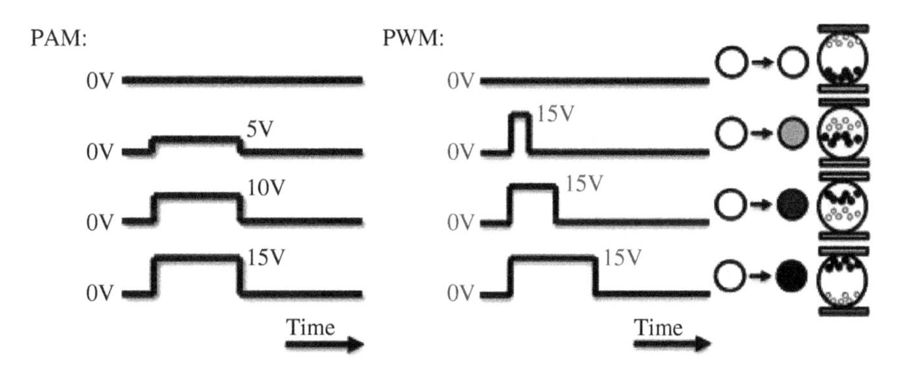

Figure 2.3 Multi-gray scale of E-paper driven by PAM and PWM [1] / John Wiley & Sons.

electrophoretic particles so that the spatial distribution of electrophoretic particles can be differentiated. The grayscale is determined by the final spatial distribution of the particles [8–10].

The electronic ink device is a complicated stack, as the properties of the particle dispersion and the dielectrics are image history-dependent, time-dependent, and sensitive to slight changes in the materials due to batch-to-batch variations. Different sets of electrophoretic particles may show different performances even with the same driving waveform. Slight differences in particle size distribution, zeta potential, amount of charging agent, or the dielectrics' condition can significantly influence the electronic ink performance. Moreover, even in the same batch of electronic ink devices, different driving histories may lead to differences in the electronic ink dispersion system and cause other particle behavior. Therefore, a waveform with higher tolerance that can be applied under different conditions still leads to the same performance is critical. In this regard, summarizing the particles' behaviors with a look-up table (LUT) is an economical way to simplify the complexity for realizing electronic ink display mass-production [11–13]. The details of this approach will be elaborated in Chapter 3.

2.2 Charging Mechanism with Inverse Micelle Dynamics

In EPD applications, electrophoretic particles are generally dispersed in non-polar solvents. In non-polar solvents, the separation of opposite charges requires more energy, leading to a low concentration of charges and a low conductivity. This is beneficial for EPDs because it reduces the leakage current. As a result, non-polar solvents contain few electrical charges, the double layers near the electrodes have larger dimensions compared with those of polar solvents, and the electric fields extend over a large distance without screening. Polar solvents like water with a high dielectric constant and high conductivity, are considered inappropriate for EPDs.

However, the electrical properties of non-polar solvents also limit the charging of particles. The Bjerrum length is the distance at which the electrostatic energy between two elementary charges is comparable to the thermal energy. It is expressed as:

$$\lambda_B = \frac{e^2}{4\pi\varepsilon_0\varepsilon_r k_B T} \tag{2.1}$$

where e is the elementary charge, ε_0 is the vacuum permittivity; k_B is the Boltzmann constant, and T is the absolute temperature in Kelvin. The large Bjerrum length due to the low dielectric constant of non-polar solvents prevents the separation of charges. Therefore, the pigment particles with a large charge are unlikely to exist, which impedes the movement of pigment particles under the electric field.

To charge the pigment particles in a non-polar solvent, it is necessary to provide counter-charges in the solvent with lower electrostatic energy to maintain charge neutrality. Charging of particles in an EPD is realized by adding amphiphilic CCAs or surfactant molecules. These molecules self-organize into reverse or IMs [14–16] with a diameter of a few nanometers when the concentration of the CCAs is above the critical micelle concentration (CMC). This process is driven by electrostatic attraction between polar heads of the CCA molecules that associate together in the center of the inverse micelle. In contrast, the non-polar tails of the CCA molecules point outward into the solvent. Most IMs are uncharged, but a small fraction has a charge of +e or −e, with the elementary charge, which we term Charged Inverse Micelles (CIM) [17]. The net charge is expected to be present in the core of the IM in the neighborhood of the polar heads of the CCA molecules. Two IMs with opposite charges attract each other, but close contact is avoided by the tails of the CCA molecules, which gives the charge in the CIMs high stability. As shown in Figure 2.4, two neutral IMs can exchange an ion and obtain opposite charges. The generation rate of CIMs is proportional to the square of the concentration of uncharged IMs, as can be expected from a process determined by collisions of uncharged IMs [18]. In a steady-state in the absence of a voltage, there

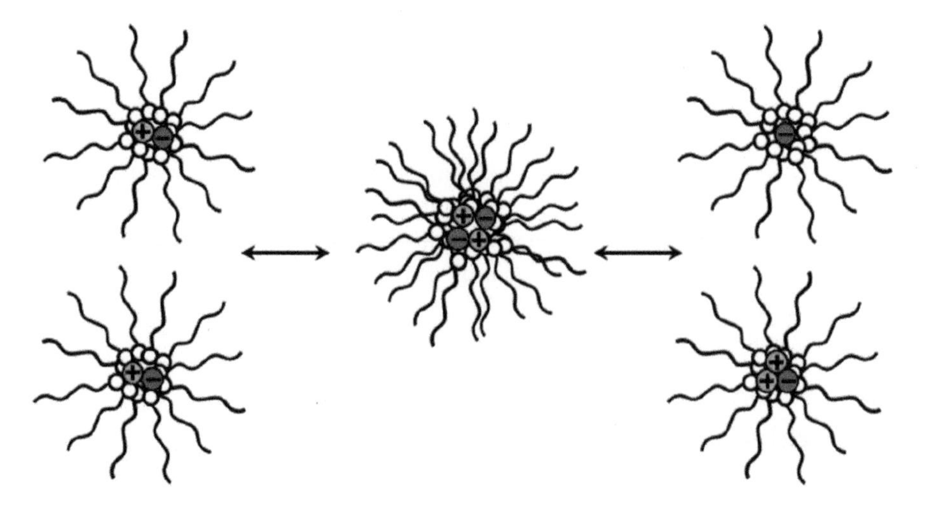

Figure 2.4 Schematic picture of the disproportionation mechanism of inverse micelles, in which two uncharged inverse micelles (left) form a positively and a negatively charged inverse micelle (right) through an intermediate merged state, and vice versa. Figure from [19] / IOP Publishing.

is an equilibrium between the generation of CIMs and the recombination of CIMs with an opposite charge into two uncharged IMs [19]:

$$M_0 + M_0 \rightleftarrows M_+ + M_- \tag{2.2}$$

This process is described by a generation rate constant β and a recombination rate constant α. When we describe the uncharged IMs, positive and negative CIMs with concentrations n_0, n_+ and n_- respectively, we obtain for the generation and recombination processes:

$$\left(\frac{\partial n_\pm}{\partial t}\right)_{GR} = \beta n_0^2 - \alpha n_+ n_- \tag{2.3}$$

$$\left(\frac{\partial n_0}{\partial t}\right)_{GR} = -2\beta n_0^2 + 2\alpha n_+ n_- \tag{2.4}$$

In equilibrium, we find $n_+ = n_- = \sqrt{\beta / \alpha}.n_0$. The free energy in the neutral state is lower than that in the charged state, and the generation constant β is lower than α, which means that most of the IMs are uncharged in equilibrium. The electrostatic energy required to separate two touching CIMs with opposite charges is larger for smaller IMs. Therefore, smaller IMs, like those formed by AOT (also called sodium docusate or sodium bis [2-ethylhexyl] sulfosuccinate), have a smaller β/α ratio, and a smaller fraction of the IMs is charged [20]. On the other hand, for the same vol% of CCA, smaller IMs (like AOT) lead to a higher concentration n_0 and the generation term βn_0^2 may be higher [21]. For this reason, small IMs like AOT are usually not appropriate for EPDs.

2.3 Drift and Diffusion of Charged Inverse Micelles

When a potential difference is applied between two electrodes, the CIMs will move under the influence of the electric field toward the electrode with the opposite polarity. A CIM's drift velocity is proportional to the electric field and its mobility. The electrical mobility of single-charged positive and negative CIMs is given by Stokes' law:

$$\pm\mu = \frac{\pm e}{6\pi\eta r} \tag{2.5}$$

with r the hydrodynamic radius of the particle and η the viscosity of the liquid. When the concentration of CIMs is inhomogeneous, diffusion becomes important. The diffusion constant is related to the electrical mobility by the Einstein relation:

$$D = \frac{\mu k_B T}{e} \tag{2.6}$$

with k_B being Boltzmann's constant and T the absolute temperature. The continuity equation relates the change in concentration with the flux of particles due to drift and diffusion and generation and recombination (in the absence of hydrodynamic flow):

$$\frac{\partial n_\pm}{\partial t} = \frac{\mu kT}{e}\nabla^2 n_\pm \mp \nabla \cdot \left(\mu n_\pm E\right) + \beta n_0^2 - \alpha n_+ n_- \tag{2.7}$$

When the local concentration of positive and negative CIMs is not the same, this leads to a charge density in the non-polar medium. From Poisson's equation, we know that the charge density is equal to the divergence of the displacement field:

$$\rho = e(n_+ - n_-) = \nabla \cdot (\varepsilon E) \tag{2.8}$$

The continuity equation above, and the boundary condition for the electrode potential enables us to model the evolution of the CIM concentration as a function of time. When the concentrations are known as a function of time, the electric transient current supplied to the electrodes can be determined. This kind of modeling has been used to study experimental transient current measurements for different materials [20, 22].

The generation and recombination terms are often small, at least for low CCAs concentrations, IMs with a large radius, and small time scales. When these two terms are neglected, the continuity equation can be used to link the equilibrium distribution of CIMs with the local potential, according to a Boltzmann distribution:

$$n_\pm = n_\pm^0 \exp\left(\mp \frac{eV}{kT}\right) \tag{2.9}$$

with $n_\pm^0$ being reference values that depend on the boundary conditions. The Poisson-Boltzmann equation is then obtained by inserting this solution in the Poisson Eq. (2.8):

$$en_+^0 \exp\left(-\frac{eV}{kT}\right) - en_-^0 \exp\left(\frac{eV}{kT}\right) = \nabla \cdot (\varepsilon E). \tag{2.10}$$

This equation is the basis for the Gouy-Chapman theory [23] or Debye-Huckel theory for calculating the charge distribution between two planar electrodes [17].

The steady-state charge distribution in a one-dimensional EPD between two electrodes can have different forms. Two dimensionless parameters have been identified to characterize the behavior: the dimensionless voltage φ and the dimensionless concentration λ, for the case of unit charges ($z = 1$):

$$\varphi = \frac{eV_A}{kT}; \quad \lambda = \frac{2e^2d^2}{\varepsilon kT}\bar{n}, \tag{2.11}$$

with V_A the potential difference between the two electrodes, $\bar{n}$ the initial concentration of positive and negative CIMs in the absence of a voltage, and d the distance between the electrodes. As shown in Figure 2.5, the parameter space of φ and λ can be divided into five regions, which are termed respectively, limiting case I: uniform concentration n and field E; limiting case II: uniform field E; limiting case III: uniform concentration n; limiting case IV: screened field E; limiting case V: separated charges n.

Considering both drift and diffusion of the positive and negative CIMs, the electric field and the concentration in the middle of the layer of electronic ink can be determined numerically, with the result shown in Figure 2.6.

The figures illustrate that when the applied potential difference is high ($\varphi > 10$), practically all CIMs move toward the regions close to the electrodes, the concentration of CIMs in the middle of the EPD is low (charges depleted), and the field remains similar to the original field (no screening). This is the preferred situation to have reliable transfer of pigment particles. On the other hand, when the potential difference is small ($\varphi < 10$), and the initial concentration of CIMs high ($\lambda > 100$), then the field in the middle of the EPD becomes practically zero, because CIMs near the electrodes form a double layer that screens the electric field in the bulk. For low voltages and low concentrations, the concentration of CIMs in the bulk remains practically homogeneous, and the field in the bulk is given by V_A/d. According to the above discussions, four conditions are schematically illustrated in Figure 2.7: (a) $\varphi < 10, \lambda > 100$; (b) $\varphi > 10, \lambda > 100$; (c) $\varphi < 10, \lambda < 100$; (d) $\varphi > 10, \lambda < 100$.

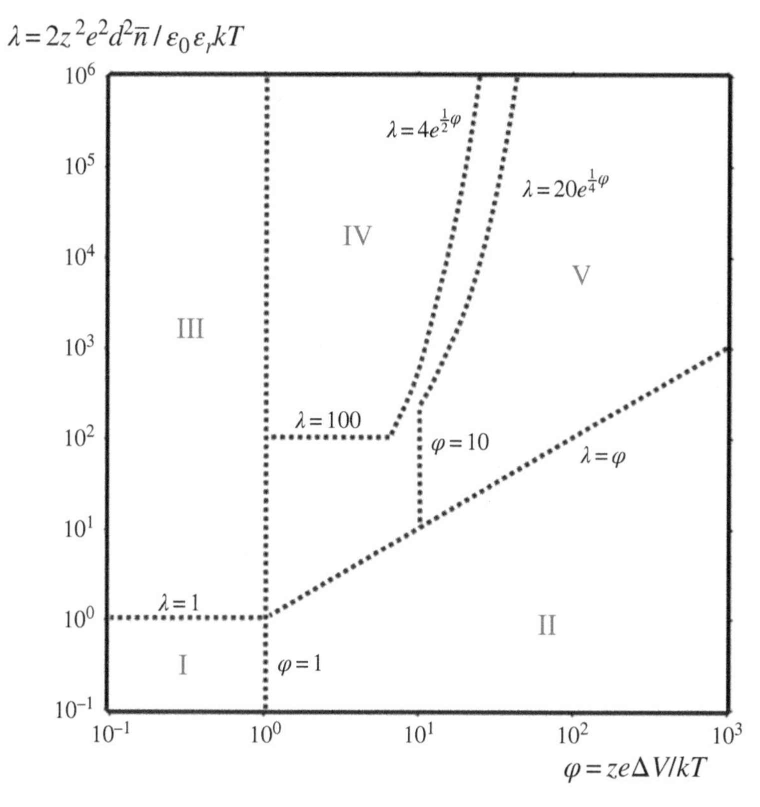

Figure 2.5 Five analytical regimes in φ and λ parameters space, respectively, limiting case I: uniform n, E; limiting case II: uniform E; limiting case III: uniform n; limiting case IV: screened E; limiting case V: separated n [17] American Chemical Society.

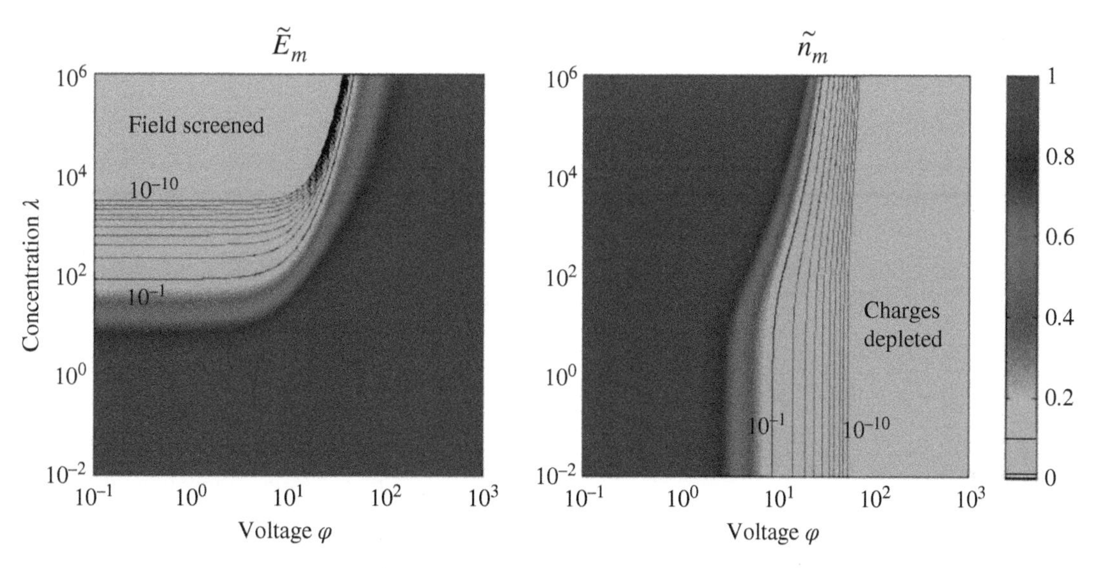

Figure 2.6 Normalized electric field $\tilde{E}_m = E_m d / V_A$ (left) and normalized concentration $\tilde{n}_m = n_m/\bar{n}$ (right) in the middle of the layer in steady state, as a function of the normalized applied voltage φ and normalized CIM concentration λ in absence of generation and recombination. Figure from [24].

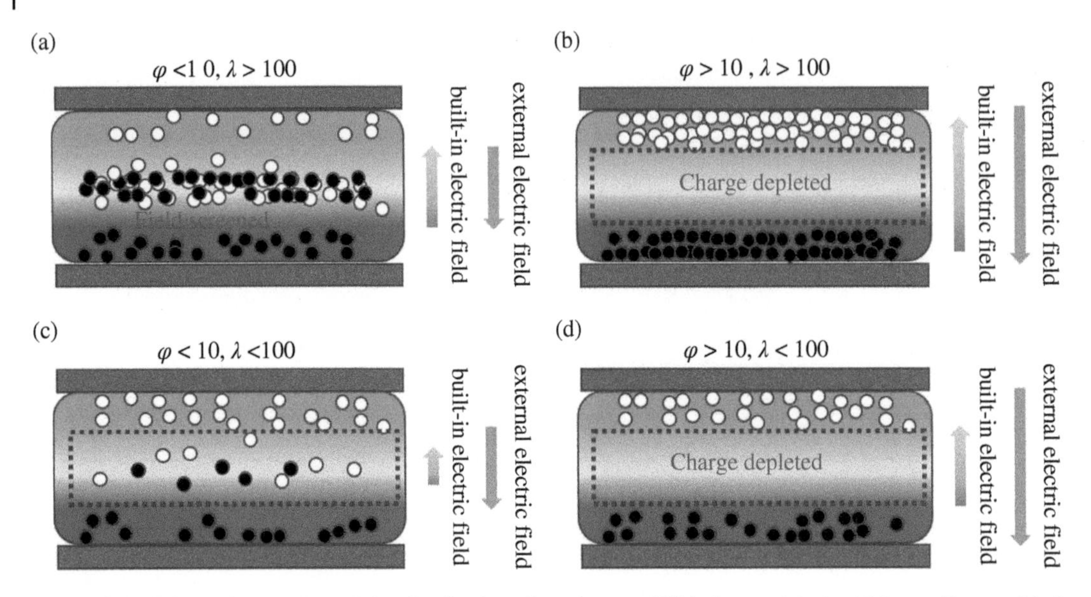

Figure 2.7 Schematic overview of the distribution of two types of CIMs (or particles), which are discussed in four conditions, with φ the dimensionless voltage and λ the dimensionless concentration: (a) low φ, high λ; (b) high φ, high λ; (c) low φ, low λ; (d) high φ, low λ.

2.4 Motion of Charged Inverse Micelles Under External Field Driving

When a voltage step is applied between the two electrodes at $t = 0$, CIMs that are initially homogeneously distributed start to move under the influence of the electric field, which is initially homogenous and equal to V_A/d, with d the distance between the two electrodes. Based on this, we can estimate the initial current density J (per unit area of the EPD) that will flow to the electrodes immediately after switching:

$$J_0 = \frac{2\bar{n}e\mu V_A}{d^2} \tag{2.12}$$

Due to the motion of the CIMs, their distribution in the liquid changes, which may influence the distribution of the electric field. As expected, the interaction between the electric field and the distribution of CIMs depends strongly on the initial concentration of CIMs and the amplitude of the voltage. Different regimes of transient currents have been observed based on the dimensionless voltage and concentration. When the applied voltage is high, and the concentration of CIMs is low (case (d) in Figure 2.7), the field is not influenced a lot and the transient current density per unit area is limited by the geometry of the structure:

$$J_{geometry} \cong J_0\left(1 - \frac{\mu V_A}{d^2}t\right) \quad t < \frac{d^2}{\mu V_A}. \tag{2.13}$$

When the applied voltage is high, and the concentration of CIMs is also high (case (b) in Figure 2.7), the transient current is limited by the presence of space charge. The decrease in the current (in a limited time interval) in this case is given by the Mott-Gurney law:

$$J_{space\ charge} \cong \frac{\sqrt{3}}{4}\left(\frac{\varepsilon\bar{n}^3e^3\mu V_A^2}{t^3}\right)^{1/4} \tag{2.14}$$

When the applied voltage is low, and the concentration of CIMs is high (case (a) in Figure 2.7), the transient current is determined by the formation of two double layers close to the electrodes, and the associated screening of the electric field:

$$J_{double\ layer} \cong J_0 \exp\left(-\sqrt{\frac{2\overline{n}e}{\varepsilon V_A}\frac{2Dt}{d}}\right) \quad t < \frac{d^2}{\mu V_A}. \tag{2.15}$$

When the applied voltage is low, and the concentration of CIMs is also low (case (c) in Figure 2.7), the transient current after some time is determined by diffusion:

$$J_{diffusion} \cong J_0 \frac{8}{\pi^2} \exp\left(-\frac{\pi^2 Dt}{d^2}\right) \quad t > \frac{d^2}{9\pi^2 D} \tag{2.16}$$

For practical EPDs, the applied voltages are relatively high, and therefore only the first and the second cases are relevant. After the depletion of charges in bulk due to the application of a DC voltage, CIMs are only present near the electrodes. However, there are still many uncharged inverse micellesIMs in the bulk. The generation process discussed in the previous section will create pairs of oppositely charged CIMs that are immediately transported in opposite directions. The current density associated with this process is called the generation current and can be measured in the external circuit [18, 19]:

$$J_G = e\beta n_0^2 d \tag{2.17}$$

Experimentally the electrode polarization charge can be determined from the transient current that flows externally between the electrodes after a voltage has been applied over the electrolyte [17]. The measurement circuit is shown in Figure 2.8. The dispersion is placed between two parallel ITO electrodes from 20 to 100 μm.

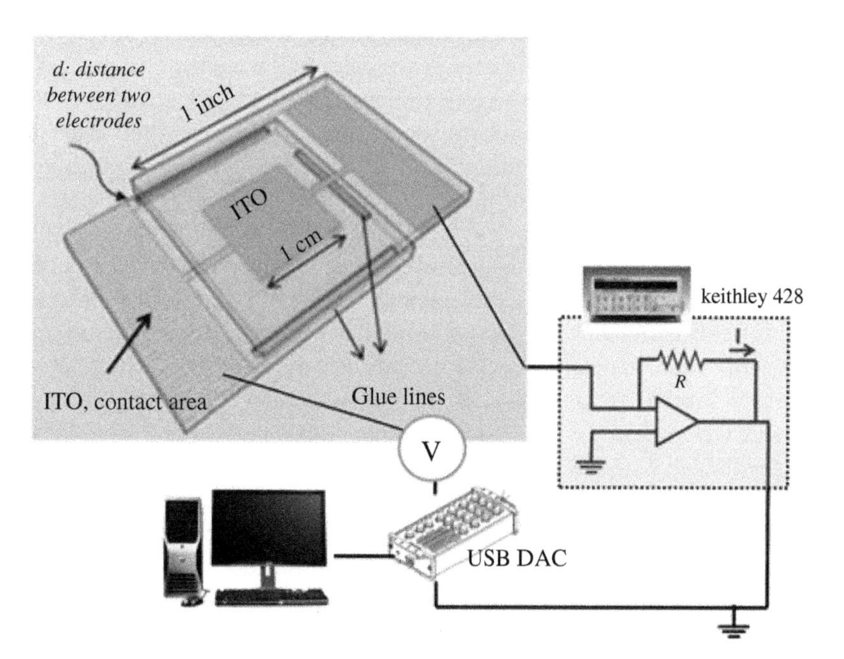

Figure 2.8 Transient current measurement setup.

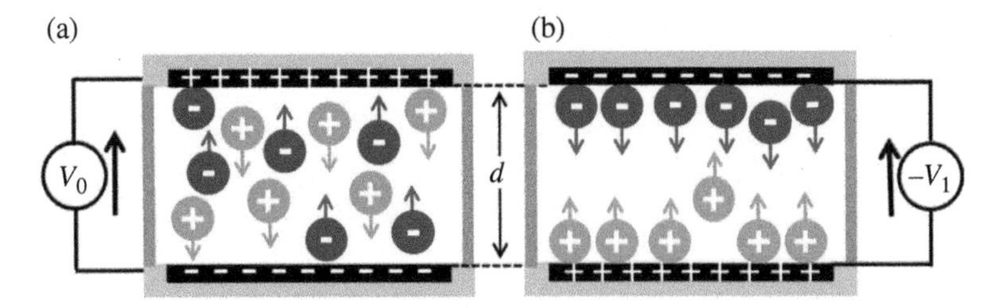

Figure 2.9 The schematic of the transient current measurements. (a) Measuring the transient current from the homogeneous equilibrium situation. (b) Measuring the transient current from the opposite polarity situation [25] with permission of Elsevier.

Table 2.1 Properties of the five devices used for measurements.

Device number	Surfactant concentration (wt %)	Expected $\bar{n}$ (m^{-3})	Measured d(μm)
1	0.01	5.0×10^{17}	2.2
2	0.03	1.5×10^{18}	4.5
3	0.10	5.0×10^{18}	9.4
4	0.30	1.5×10^{19}	14.8
5	1.00	5.0×10^{19}	26.8

A source-measure unit (Keithley 428) is used to apply a voltage across the electrodes and simultaneously measure the corresponding current.

The transient current refers to the current as a function of time when a voltage pulse is applied. For example, a polarization voltage V_0 is applied for a long duration (30 seconds) to drive all CIMs toward the side of the electrodes, and the transient current starting from the homogenous equilibrium situation is measured, as shown in Figure 2.9a. Then, a reverse voltage pulse ($-V_1$) is applied for a duration, and the transient current from the opposite polarity situation is measured, as shown in Figure 2.9b.

Several transient currents of devices with different thicknesses, surfactant concentrations (see Table 2.1), and driving voltages were measured, as shown in Figure 2.10 [24]. Before each measurement, the devices were short-circuited for a sufficient duration to ensure a homogeneous distribution of charges. Then the voltage was abruptly changed, and the transient current was measured. Besides, Beunis et al. derived approximate analytical expressions, which are a solution of the Poisson–Nernst–Planck (PNP) equation in four limiting cases. Figure 2.10a shows the current measurements and numerical fit for the five devices for a low voltage step (25 mV), while Figure 2.10b shows measurements for a high voltage step (2.5 V). Figure 2.10c shows the current measurements and numerical fit for device 1 with a low charge content (different voltages), and Figure 2.10d shows the current measurements and numerical fit for device five with high charge content (different voltages).

Electrohydrodynamic (EHD) flow occurs when the voltage (V_0) of a device with high charge content is switched to a high reverse voltage ($-V_1$) [26]. As shown in Figure 2.11, the transient reversal currents of the device with 0.3 wt% OLOA 11 K, d = 48.4 μm and overlapping area S = 1 cm^2 were measured and simulated. The measured data with high V_1 shows two current peaks, while each simulated curve only has one peak. Besides, the first peak in the measurements is observed earlier than in the simulation. The discrepancy between measurement and simulation is caused by EHD flow, which is not considered in the simulations. EHD instabilities occur when space charge layers are inverted quickly, which appears above a threshold voltage. The first peak is due to charged IMs transported faster than predicted by the PNP equations, and the second peak is due to the charged micelles induced by a large polarization voltage.

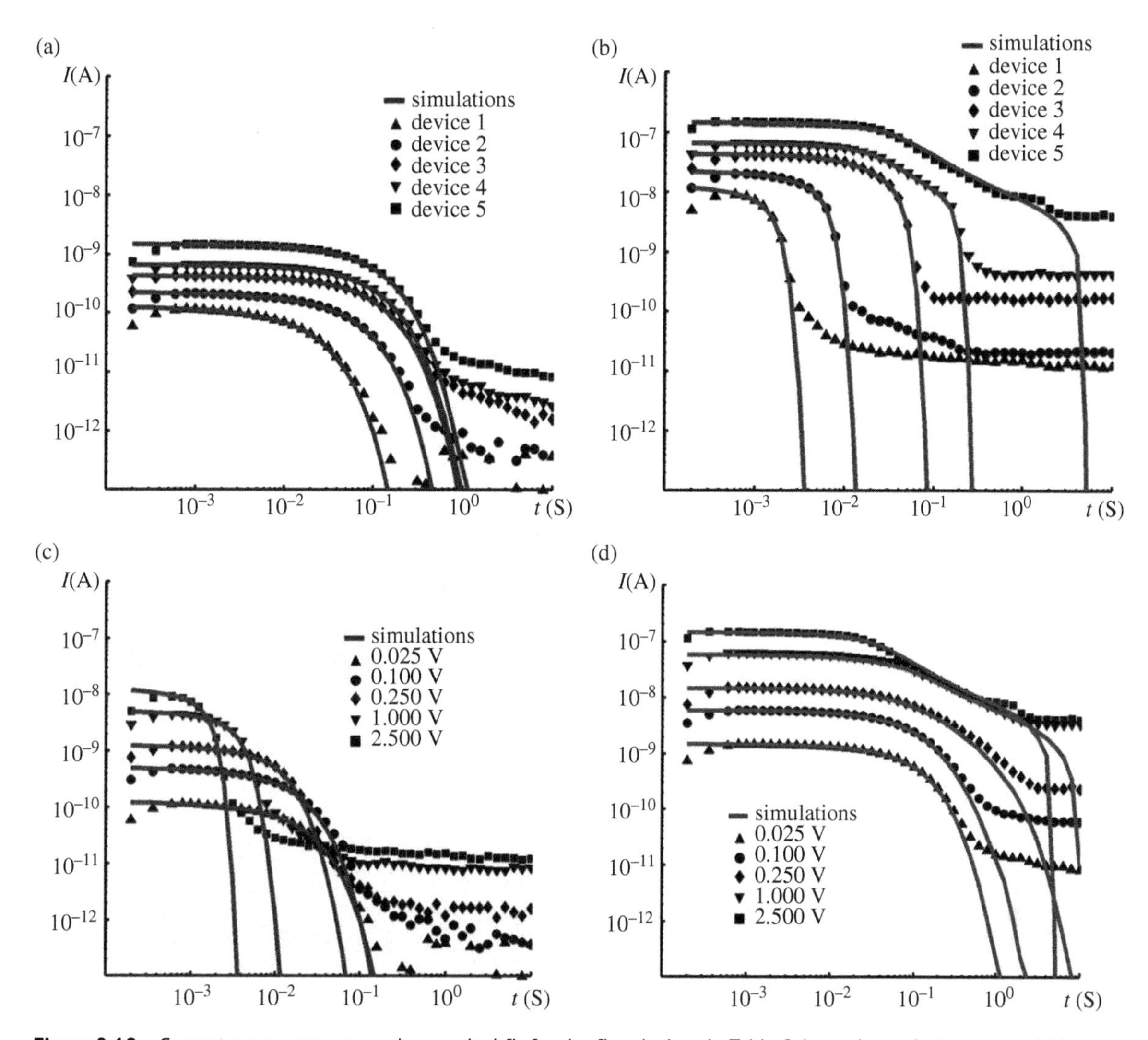

Figure 2.10 Current measurements and numerical fit for the five devices in Table 2.1 are shown in two cases: (a) low voltage; (b) high voltage; (c) Current measurements and numerical fit for device 1 with low charge content; (d) Current measurements and numerical fit for device 5 with high charge content [24].

The transient currents of five surfactants under different conditions were measured, as shown in Figure 2.12. The five surfactants are Solsperse 13 940, OLOA 11K, Span 80, Span 85, and AOT, respectively. Micelles formed by different surfactants have different radii and different current characterizations. Figure 2.12 indicates tha the initial current is larger for CIMs with a larger radius (because more CIMs are charged), and the generation is lower (lower current for larger times).

2.5 Stern Layer Formation

CCA molecules typically attach to all surfaces, forming a monolayer with the apolar tails of the CCA molecules pointing into the non-polar liquid. When a voltage is applied between the two electrodes, CIMs are attracted to the electrode with the opposite polarity. In most cases, the CIMs form a diffuse double layer with a charge opposite to

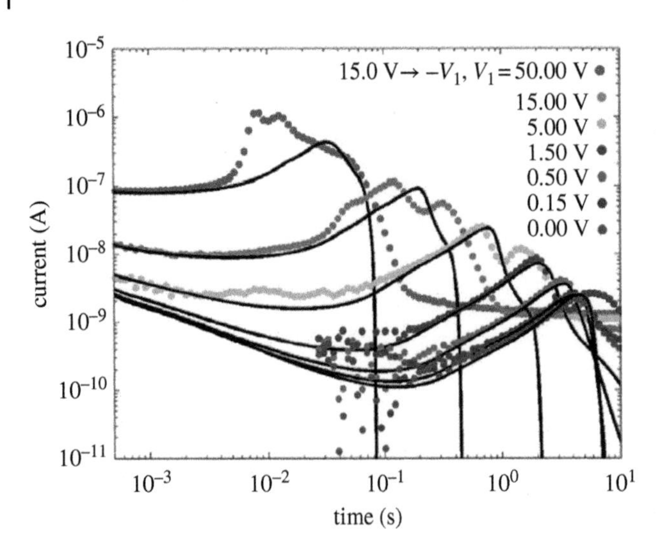

Figure 2.11 Measured (colored dots) and simulated (black lines) transient reversal currents for a device with 0.3 wt% OLOA 11 K, d = 48.4 μm and S = 1 cm^2 [26] with permission of Elsevier.

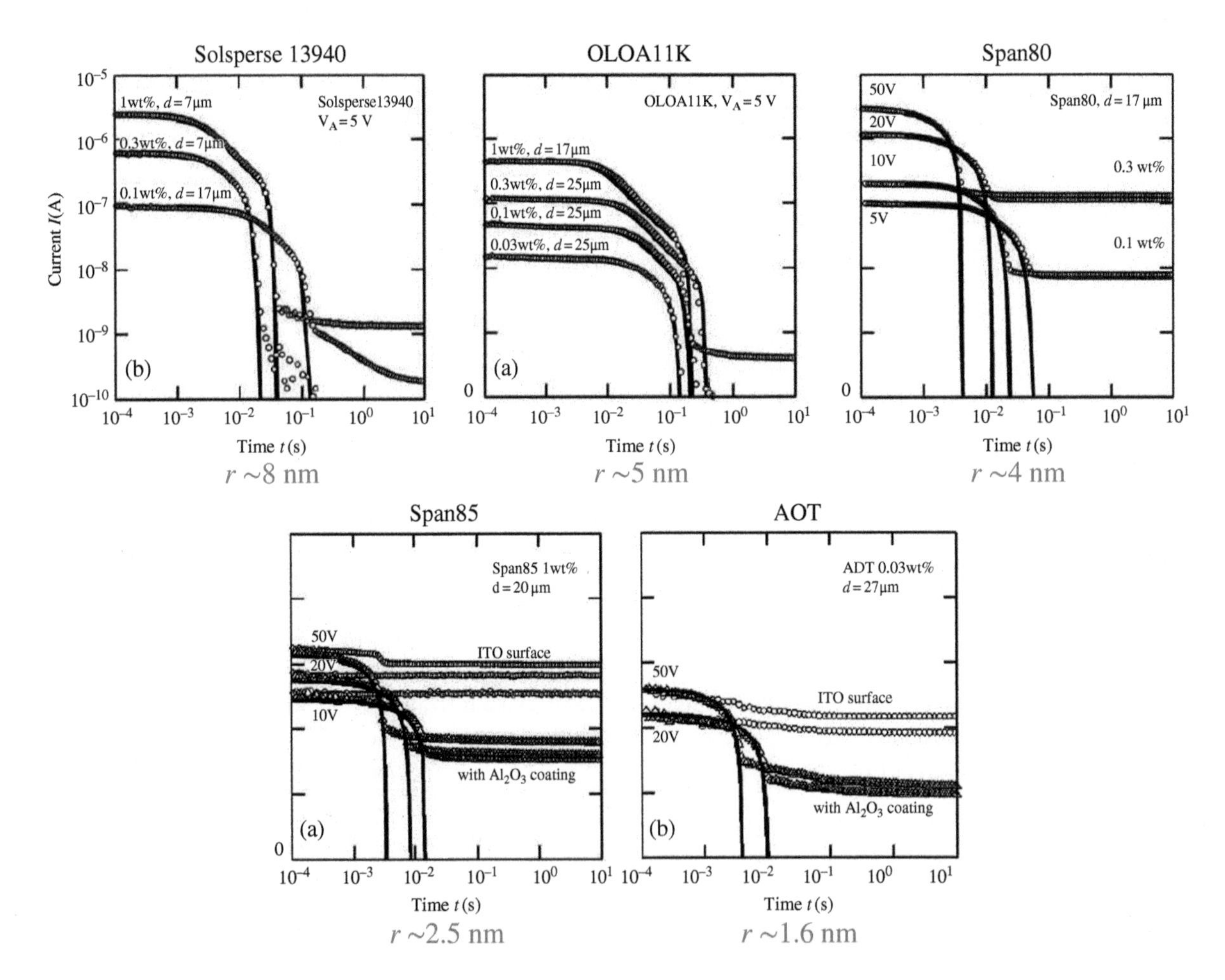

Figure 2.12 Characterization of five surfactants in dodecane by current measurements, respectively, Solsperse 13940, OLOA 11 K, Span 80, Span 85 and AOT. Micelles formed by different surfactants have different radius (r).

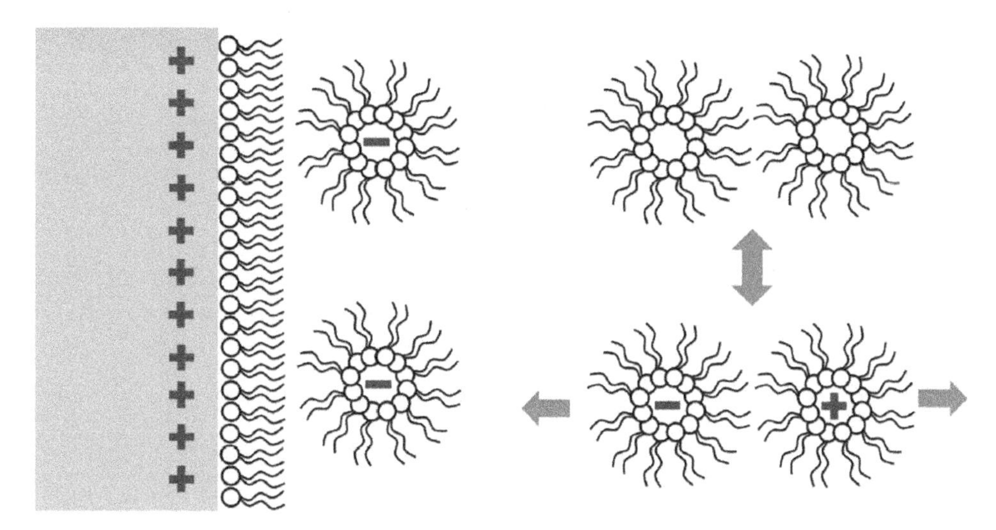

Figure 2.13 Electrical double layer in non-polar suspension [27] American Chemical Society. CCA molecules cover the positively charged electrode. Negatively charged inverse micelles are attractred and can form a diffuse double layer of mobile CIMs or can attach to the surface and form a Stern layer.

the charge on the electrode. The Gouy Chapman theory describes the charge density in the double layer, and for relatively high voltages applied, the characteristic distance is smaller than the Debye length (Figure 2.13).

It has been observed that CIMs can transfer from the diffuse double layer to form a charged Stern layer of CIMs attached to the surface. Sometimes this layer is referred to as the Helmholtz layer, with the charges present in the inner Helmholtz plane [28]. The typical distance between the charge of the CIM and the metallic surface is then roughly the radius of the CIM or the length of the CCA molecules, leading to a large Stern capacitance. The voltage over this thin Stern layer is usually small. However, when a voltage is applied over a device for a long time, many CIMs can be formed by generation in the bulk and may add to the Stern layer. In this case, there may be an appreciable voltage drop over the Stern layer, which can influence the driving of the electrophoretic display. The CIMs in the Stern layer is adsorbed at the solid/liquid interface, and (usually) they remain there, even after the polarity of the voltage is reversed. The buildup of a Stern layer has been observed in dodecane with AOT as CCA [27]. In this case, the CIMs are small, and the attraction between the charge and its mirror image in the electrode leads to effective trapping of the charge in the Stern layer.

When a high voltage is applied over the electrodes of a dodecane cell with CCA for a long time (>1000 s), a lot of charge is added to the system, corresponding to the time integral of the generation current. When the voltage polarity is reversed, that charge does not return, indicating that the CIMs that have been generated during the generation process are adsorbed in the Stern layer. The CIMs generated in the high field have different properties from the CIMs that were initially present in the solvent and form a diffuse double layer.

It was shown that a small fraction of the generated negative CIMs, that are in a Stern layer during a long DC voltage pulse (5 V) is released when the field at the electrode reverses sign [29]. Apparently, this particular kind of negative CIMs is not bound to the interface but remains there under the influence of the applied electric field. The release of a relatively large amount of charge during field reversal leads to a well described current the Mott-Gurney theory [30].

Another set of experiments was carried out for a fluorocarbon ink with fluorosurfactants and low voltage DC pulses (0.5 V) [31]. Under these conditions, many CIMs remain in the bulk, and the electric field is screened by the charges near the surfaces. The experiments can be well explained by assuming a surface- and polarity-dependent probability for adsorption of diffuse double layer CIMs into the Stern layer. In these experiments,

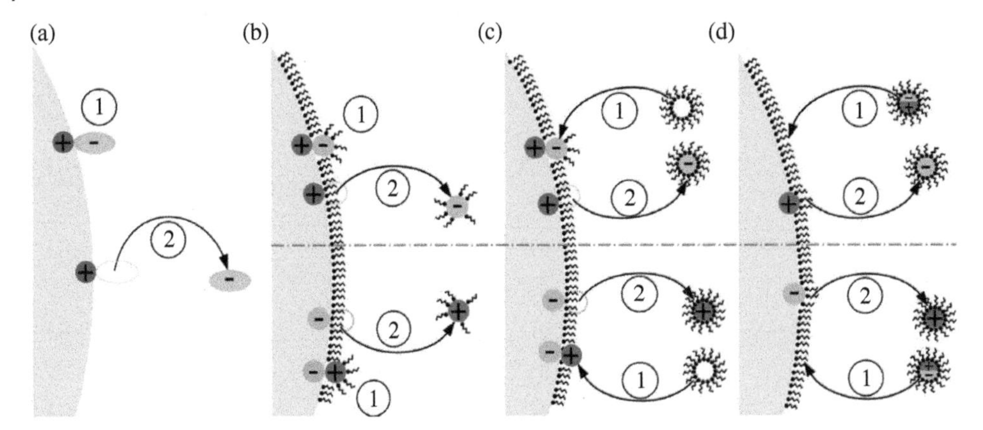

Figure 2.14 Three charging mechanisms reported by Schreuer et al. (a) field induced charging; (b) surfactant-mediated charge desorption; (c) and (d) charging mediated by inverse micelles [33] American Institute of Physics / CC BY 4.0.

adsorption continues until a thermodynamic equilibrium is reached. Typically, this results in a large voltage drop over the Stern layer. In these experiments, the release of charges from the Stern layer after the voltage reversal is not observed.

2.6 Charging Mechanism with Particles and Additives

When the CCAs are added into an electrophoretic dispersion containing pigment particles, the hydrophilic side of the surfactant molecules rapidly coats the particles, with the hydrophobic tail turned toward the non-polar solvent. In this way, CCAs help avoid aggregation of particles, leading to better dispersion and a stable suspension. The charged IMs (CIMs) can transport chs between speciesispersed in the non-polar solvent [25]. In thermal equilibrium, after many collisions between particles and CIMs, the pigment particles typically obtain a net charge [32, 33]. The average charge of the particles depends on the surface treatment of the particles and the functional groups of the CCAs.

In non-polar solvents, charges can only be transported with IMs as the carrier. Some particles modified by quaternization [34] or polymerization [35] of functional monomers can be charged by the physical desorption of ions to IMs [36]. Schreuer et al. reported three mechanisms to explain the charging of a particle both below and above the CMC, as illustrated in Figure 2.14 [33]: field-induced charging, surfactant-mediated charge desorption, and charging mediated by inversed micelles. It is also possible to have Lewis acidbase reactions between IMs and particles [37]. Particles can be charged positively or negatively, while a majority of the IMs obtain the opposite charge.

2.7 Observations on a Single Particle

Microscope observation of individual charged colloidal particles allows the study of fundamental particle properties in great detail. Typically, a sinusoidal electric field is applied to the liquid and the position of the particle as a function of time is measured by a camera attached to the microscope. By calculating the ratio of the particle velocity to the electric field in the medium, the mobility and the charge of the particle can be estimated.

In other experiments [38], the particles are trapped with optical tweezers, which allows them to be studied over an extended period, as shown in Figure 2.15. The oscillating electric field has a frequency in the kHz range, while

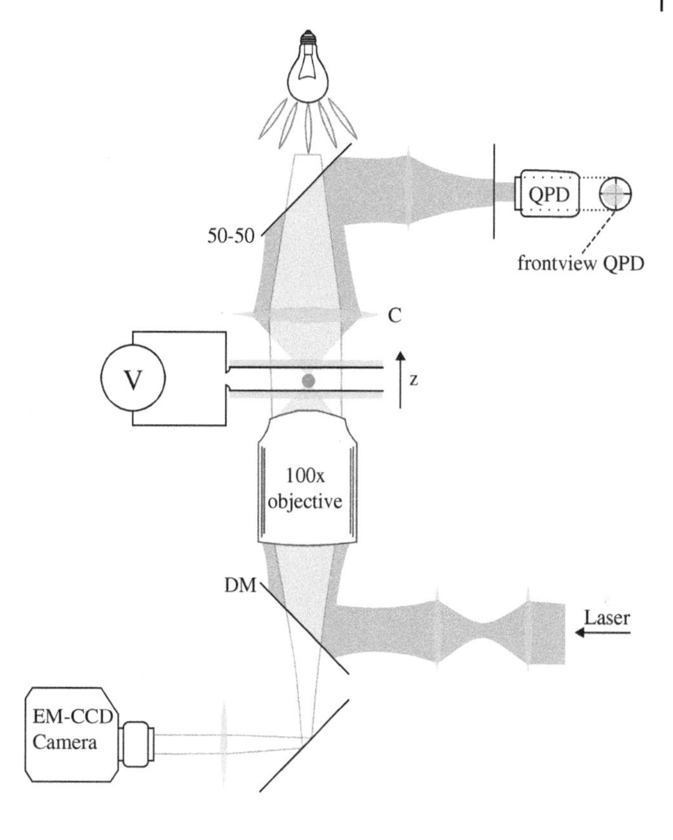

Figure 2.15 Schematic overview of the optical trapping electrophoresis (OTE) setup, showing the path of the infrared ray (IR) laser beam (pink) that traps the particle and arrives at the quadrant photodiode (QPD) and the path of the white illumination (yellow) from the lamp to the EM-CCD camera [33] American Institute of Physics / CC BY 4.0.

the force due to the optical tweezers is only relevant at much lower frequencies. In the absence of CCA, the charge of the particles is relatively stable, and there is no double layer of counter-charges around the particle.

By applying a high electric field, it is possible to measure the charge of a particle with the resolution of one elementary charge q, which corresponds to the experiment of Millikan but is carried out in a non-polar liquid instead of air [38]. Over 3000 seconds, the charge on a particular particle varies between 0 and −21q, with a typical time of 1 second between variations (Figure 2.16).

Adding CCA to the mixture makes the measurement more challenging because also CIMs move in an oscillating electric field. Superimposing a small DC voltage makes it possible to deplete CIMs in the region near the optically trapped particle and measure with the resolution of a single electron charge [38]. In some experiments, the typical time between elementary charging events was increased to more than 50s, indicating a very stable charge.

The influence of CIMs in the neighborhood of a particle on its mobility has been investigated [39]. In this

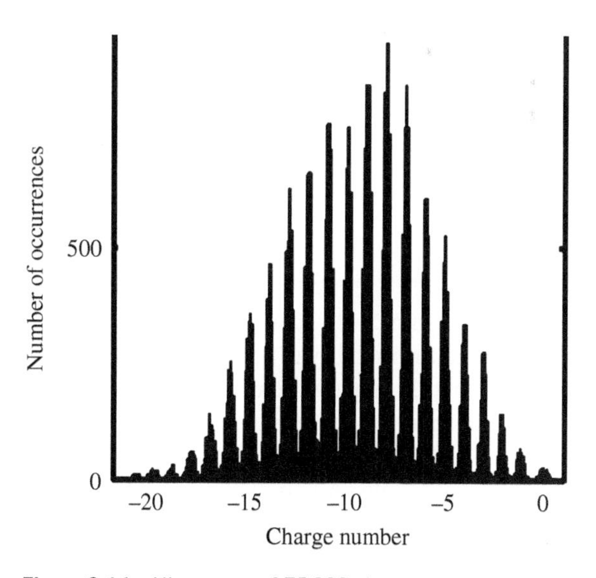

Figure 2.16 Histogram of 75 000 charge measurements in units of the elementary charge, for an individual optically trapped PMMA particle with radius 0.5 ϕm in dodecane, measured over 3000 s (ac field with amplitude 3.3 MV/m and f = 1 kHz) [38] / American Physical Society.

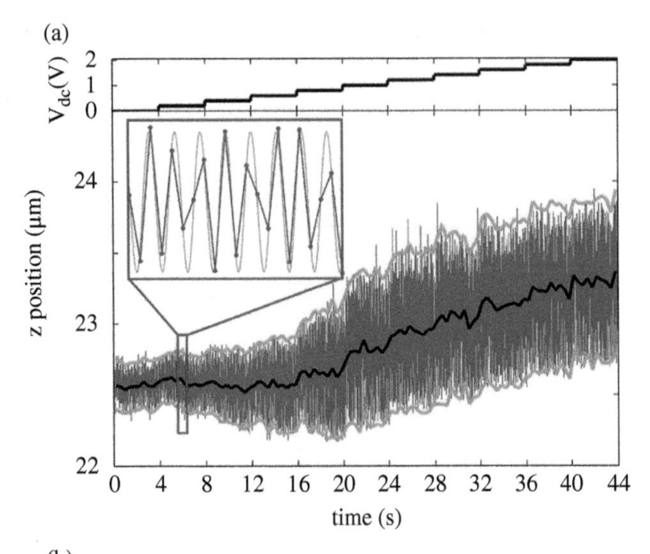

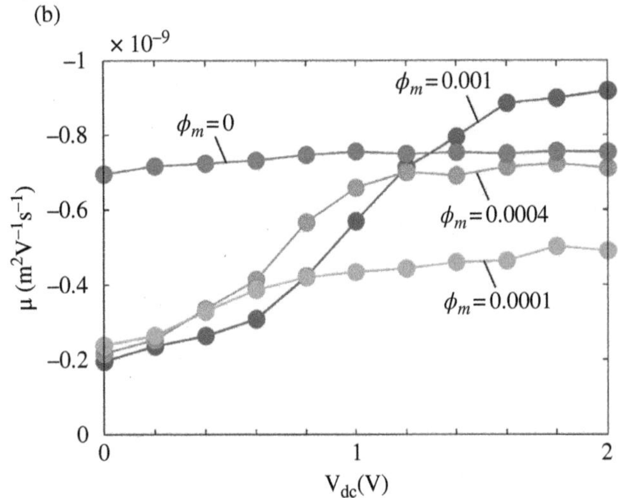

Figure 2.17 (a) Measurement of the z-position of a single optically trapped polystyrene divinylbenzene particle with radius 3.38 μm in dodecane with PIBS (mass fraction $\phi_m = 0.0004$), with $V_{ac} = 10\ V_{pp}$, $f_{ac} = 21$ Hz (sampling at 50 Hz) with stepwise increase of V_{dc}. (b) Resulting electrophoretic mobility of four optically trapped particles in dodecane with PIBS (different mass fractions ϕ_m) as a function of an applied dc voltage. In the absence of V_{dc} the mobility is lower due to counter charges in the double layer. For increasing concentrations, a higher V_{dc} is needed to reduce the concentration of CIMs in the bulk, and to obtain the mobility of the particle that corresponds to its charge [39] / American Physical Society / CC BY 3.0.

experiment, an AC voltage with a frequency of 21 Hz is used, the axial position is determined from microscope images, and the mobility of an individual, optically trapped particle is estimated, as shown in Figure 2.17a. It is found that the apparent mobility increases by a factor up to four when a DC voltage is applied simultaneously. The small DC voltage is enough to completely deplete the bulk of the layer of CIMs, as illustrated in a previous section. In the absence of a DC voltage, the particle has an extended double layer with excess counterions over co-ions. A slip-plane can be identified by the separating inner parts of the double layer, which move together with the particle, from the outer part of the double layer, which flows around the particle [28]. As the counter-charges near the particle move together with the particle, the effective charge of the moving unit is reduced. In addition, the dynamic charging and decharging of the double layer outside of the slip plane leads to an electric field that slows down the particle. As a result of these effects, the adequate mobility of the particle is lower than would be expected from the charge on the particle.

When a a small dc-field depletes the CIMs, higher mobility of the particle is measured, corresponding to the actual charge of the particle.

2.8 Rheological Effects During Driving

The rheology of the electronic ink dispersion within the micro-chamber is complicated. It involves many factors, such as the electrical field, the charge of the pigment particles, the size of particles, the solvent's viscosity, and the concentration of CCA and IMs. Carrique et al. reported that the mobility is linearly connected to the charging number of the particles when the charge is below a critical point and that the mobility reaches a plateau when the charge is higher than the critical point [40]. Morrison and Tarnawskyj proposed an equation that relates the mobility μ of the particle with the average particle diameter *d*: [41]

$$\frac{\frac{Q}{M}d^2}{\mu} = \frac{18\eta}{\rho_2} \tag{2.18}$$

with *Q/M* the charge to mass ratio of the particle, η the viscosity of the solvent, and ρ_2 the density of the particle. For a spherical particle, *M* is proportional to the third power of *d*. In conclusion, Eq. (2.18) illustrates that for a given charge, the mobility is inversely proportional to the size of the particle, as in Eq. (2.5).

A set of iso-alkane aliphatic hydrocarbons has been investigated to further investigate the relationship between the mobility of the particles and the solvent's viscosity. In Figure 2.18, the solvents from A to D have increasing viscosities. Furthermore, an opto-electric setup for a reflective test has been designed, which gradually increases the amplitude of voltage pulses applied to the EPDs to study the optical response. Figure 2.18 illustrates that for a solvent with a lower viscosity (solvent A), the optical response of the EPD for the same voltage pulse is faster, which means that the particles have a higher mobility and encounter less resistance [42, 43]. This finding is significant for the realization of full-color ink. As the full-color electronic ink accommodates several kinds of color particles to realize color by micro-scale dithering, the ink may become too dense to be driven with sufficient speed [41]. Thus, lowering the viscosity is essential.

An effective method to increase electrophoretic mobility is to reduce viscosity. It may also reduce the EPD's bistability. A more feasible strategy is to realize a viscosity reduction of the suspension when a voltage is applied and recover the original viscosity after removing the voltage, an effect known as the inversed electrorheological (IER) effect [44, 45].

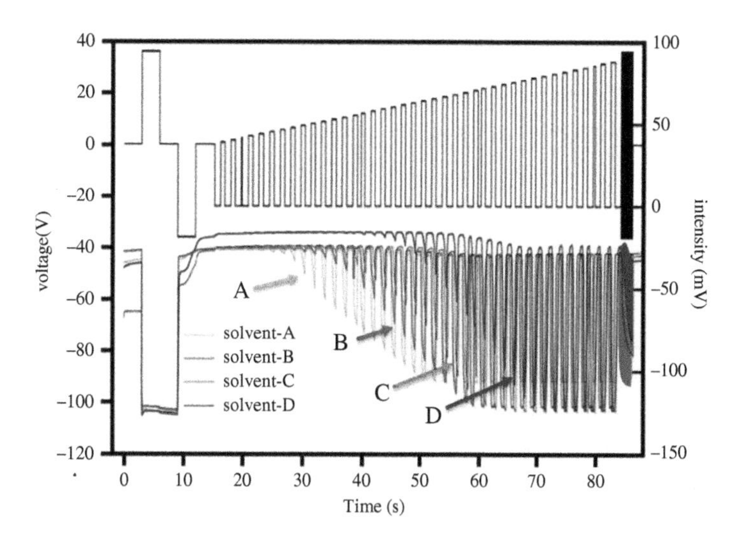

Figure 2.18 Optical response of EPDs with different solvent viscosity [1] / John Wiely & Sons.

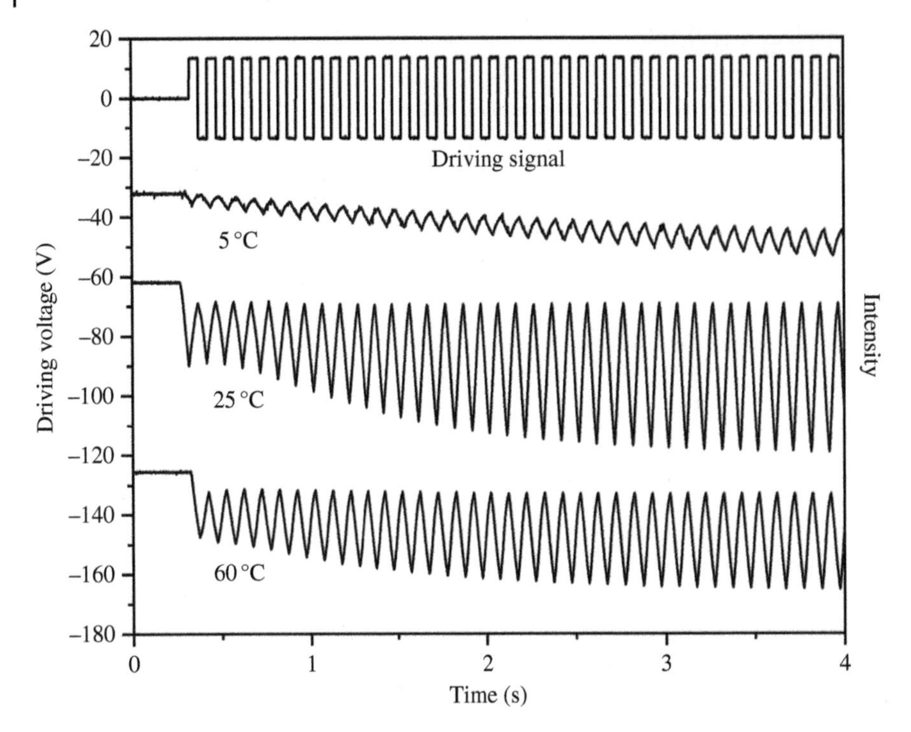

Figure 2.19 Activation of particles with various temperatures in EPD [1] / John Wiley & Sons.

The switching rate increase suggests that the electronic ink is a pseudo-plastic fluid (non-Newtonian fluid). The behavior of pseudo-plastic fluids can be described by the Herschel Bulkley model [44, 46]:

$$\eta = \frac{\tau_{el}}{D} = \frac{\tau_{el}\eta_H}{\left(\tau_{el} - \tau_H\right)^p} \tag{2.19}$$

here τ_{el} and τ_H denote the shear force applied by the electric field to the particles and the yield stress of the fluid, η is the liquid viscosity, D is the shear rate, and p is the Herschel Bulkley index with a typical value of 0.8 [44].

The fluid behavior also varies with temperature. As shown in Figure 2.19, owing to the variable viscosity of the solvent, the optical response at 5 °C is smaller than that at 25 °C. On further increasing the temperature to 60 °C, the optical response speed drops again, which is believed to result from the desorption of IMs or a lower charge exchange rate between the particles and CCAs under the effect of high-temperature Brownian motion [1].

2.9 Bistability After Removing External Fields

After removing the external fields, the particles tend to stay still under the balance of forces. As shown in Figure 2.20, many forces acting on the particles, such as buoyancy, Vander Waals forces, coulombic forces, frictional forces, gravitational forces, and solubility forces. The ability to sustain this balance is called "bistability." This force balance is why particles can remain at the same place after removing the fields, and why EPDs have low power consumption. One feasible way to further increase the bistability is to utilize the depletion flocculation effect. This effect was first predicted theoretically by Oosawa et al. proposing that colloidal particles in a solution

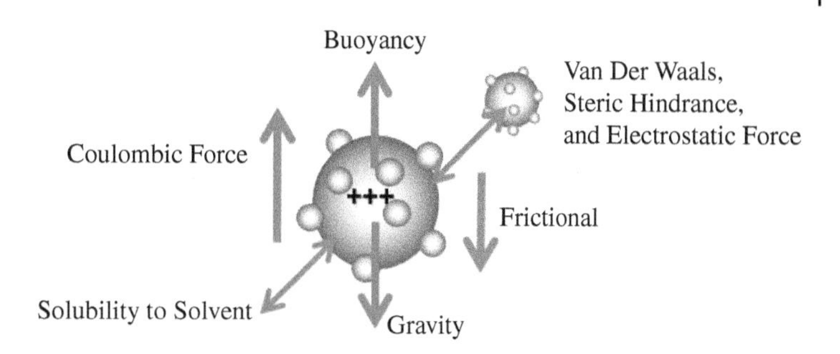

Figure 2.20 Forces encountered by electronic ink particles.

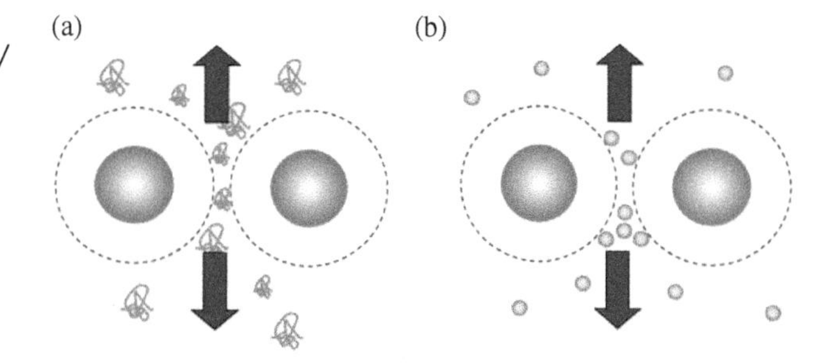

Figure 2.21 Depletion flocculation for bistability. (a) with free polymers and (b) with nano-particles [1] / John Wiley & Sons.

would tend to be stable by flocculation, caused by the Vander Waals force and electrostatic force [47]. The depletion flocculation effect was later observed by other researchers [48, 49]. Furthermore, some researchers found that this effect could be enhanced by adding polymers and nano-particles to the solution [50–52].

Since electronic ink is a colloidal particle system, the depletion flocculation effect could be utilized to optimize the bistability of EPD. By adding some nanoparticles or free polymers to the solvent, the pigment particles would be more stable, while they could still be separated by an externally applied field, as shown in Figure 2.21.

2.10 Full Color E-Paper

There are many approaches to realizing full-color electrophoretic e-paper, including the use of a color filter, shutter mode, and multiple particles with different mobility and voltage sensitivity. In Chapter 1 of this volume, these approaches are introduced in detail. Among these approaches, the color e-paper demonstrated by Fuji Xerox corporation with shuffling CMY particles was the earliest demonstration providing outstanding color performance [53]. The structure was termed "Independently Movable Colored Particles" (IMCP) and comprised a transparent fluid, floating white particles, and three types of particles with different colors, cyan, magenta, and yellow, respectively, in one electrophoretic layer, as shown in Figure 2.22a. The cyan, magenta, and yellow particles were assumed to have a positive charge and different threshold fields Etc, Etm, and Ety, respectively, where $Etc > Etm > Ety$. The driving method of IMCP is based on the different threshold fields of CMY particles and the combination of varying levels of voltage to drive the CMY particles independently. As previously mentioned, the particles' behaviors can be differentiated by modifying the charging additives and particle surface conditions. In Figure 2.22b, fields E1, E2, E3, and E4 are defined as $E1 < Ety < E2 < Etm < E3 < Etc < E4$. A driving waveform for IMCP consists of four phases reset, C writing, M writing, and Y writing, respectively. In the reset phase, field -E4 is applied to move all the particles to the front substrate. In each color state, the primary-colored particles can be driven independently. By controlling the mixing extent of the CMY particles, the designated color coordinate can be realized z, and thus the full color can be shown.

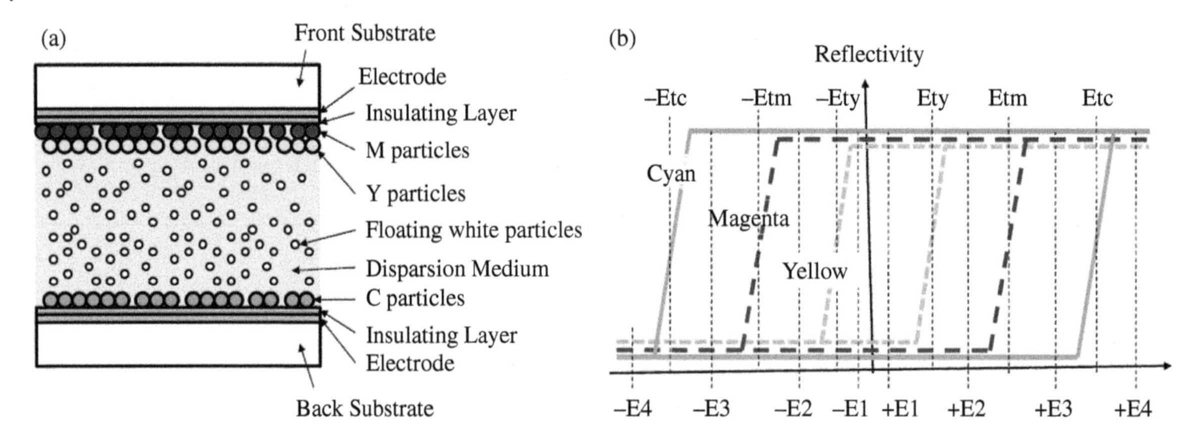

Figure 2.22 (a) Cross section of IMCP EPD. (b) Schematic diagram of field-reflectivity characteristics of CMY particles [53] / John Wiley & Sons.

2.11 Conclusion

This chapter reviewed and summarized the operational principles and fundamental mechanisms of the electrophoretic electronic ink display, including the EPD driving scheme, materials design, CIMs dynamics, charged particle behaviors, and color EPD. However, EPDs still have several challenges to be overcome, such as high switching speed with good bistability, better color performance without waveform flickering, hand-writing capability with a more intuitive experience, and environmental endurance for IoT applications. E-paper displays will undoubtedly become even more common when some of these challenges can be resolved.

References

1 Yang, B.-R., Wen-Jie, H., Zeng, Z. et al. (2020). Understanding the mechanisms of electronic ink operation. *J. Soc. Inf. Disp.* 29: 38–46.

2 Johnson, M.T. and Zhou, G. (2005). 56.1: Invited paper: high quality images on electronic paper displays. *SID Symp. Dig. Tech. Pap.* 36 (1): 1666–1669.

3 Amundson, K. and Sjodin, T. (2006). Achieving Graytone images in a microencapsulated electrophoretic display. *SID Symp. Dig. Tech. Pap.* 37 (1): 1918–1921.

4 Kao, W.-C., Jia-An, Y., Fong-Shou, L. et al. (2009). Configurable timing controller Design for Active Matrix Electrophoretic Display. *IEEE Trans. Consum. Electron.* 55 (1): 1–5.

5 Zhou, G., Cortie, R.H.M., Johnson, M.T. et al. (2009). Method of compensating temperature dependence of driving schemes for electrophoretic displays. US 7623113 B2.

6 Kao, W.C., Chang, W.T., and Ye, J.A. (2012). Driving waveform design based on response latency analysis of electrophoretic displays. *J. Disp. Technol.* 8 (10): 596–601.

7 Lin, C. and Yang, B.R. (2016). Driving method for electrophoretic displays with different color states. US9299294 B2.

8 Wang, L., Yi, Z.C., Jin, M.L. et al. (2007). Improvement of video playback performance of electrophoretic displays by optimized waveforms with shortened refresh time. *Displays* 49: 95–100.

9 Shen, S.T., Gong, Y.X., Jin, M.L. et al. (2018). Improving electrophoretic particle motion control in electrophoretic displays by eliminating the fringing effect via driving waveform design. *Micromachines* 9 (4): 143.

10 Kao, W.C. and Tsai, J.C. (2018). Driving method of three-particle electrophoretic displays. *IEEE Trans. Electron Devices* 65 (3): 1023–1028.

11 Zehner, R., Amundson, K., Knaian, A. et al. (2003). Drive waveforms for active matrix electrophoretic displays. *SID Symp. Dig. Tech. Pap.* 842–845.

12 Chi-Ming, L. and Wey, C.-L. (2011). A controller design for micro-capsule active matrix electrophoretic displays. *J. Disp. Technol.* 7 (8): 434–442.

13 Kao, W.C., Liu, J.J., Chu, M.I. et al. (2010). Photometric calibration for image enhancement of electrophoretic displays. In: *IEEE International Symposium on Consumer electronics* (ISCE 2010), 1–2.

14 Parent, M.E., Yang, J., Jeon, Y. et al. (2011). Influence of surfactant structure on reverse micelle size and charge for non-polar electrophoretic inks. *Langmuir* 27 (19): 11845–11851.

15 Smitha, G.N., Brown, P., James, C. et al. (2016). The effect of solvent and Counterion variation on inverse micelle CMCs in hydrocarbon solvents. *Colloids Surf., A* 494: 194–200.

16 Eicke, H.F. (1980). Surfactants in non-polar solvents. Aggregation and micellization. *Top. Curr. Chem.* 87: 85–145.

17 Verschueren, A.R.M., Notten, P.H.L., Schlangen, L.J.M. et al. (2008). Screening and separation of charges in microscale devices: complete planar solution of the Poisson–Boltzmann equation. *J. Phys. Chem. B* 112 (41): 13038–13050.

18 Strubbe, F., Verschueren, A.R.M., Schlangen, L.J.M. et al. (2006). Generation current of charged micelles in nonaqueous liquids: measurements and simulations. *J. Colloid Interface Sci.* 300 (1): 396–403.

19 Strubbe, F. and Neyts, K. (2017). Charge transport by inverse micelles in non-polar media. *J. Phys.-Condens. Matter* 29 (45): 453003.

20 Karvar, M., Strubbe, F., Beunis, F. et al. (2014). Investigation of various types of inverse micelles in nonpolar liquids using transient current measurements. *Langmuir* 30 (41): 12138–12143.

21 Karvar, M., Strubbe, F., Beunis, F. et al. (2015). Charging dynamics of aerosol OT inverse micelles. *Langmuir* 31 (40): 10939–10945.

22 Neyts, K., Karvar, M., Drobchak, O. et al. (2014). Simulation of charge transport and steady state in non-polar media between planar electrodes with insulating layers. *Colloids Surf. A Physicochem. Eng. Asp.* 440: 101–109.

23 Gouy, M. (1910). Sur la constitution de la charge électrique à la surface d'un electrolyte. *J. Phys. Theor. Appl.* 9: 457–468. (in French).

24 Beunis, F., Strubbe, F., Marescaux, M. et al. (2008). Dynamics of charge transport in planar devices. *Physics Review E* 78: 011502.

25 Bert, T. and Smet, H.D. (2003). Dielectrophoresis in electronic paper. *Displays* 24: 223–230.

26 Prasad, M., Beunis, F., Neyts, K. et al. (2015). Switching of charged inverse micelles in non-polar liquids. *J. Colloid Interface Sci.* 458: 39–44.

27 Karvar, M., Strubbe, F., Beunis, F. et al. (2011). Transport of charged aerosol OT inverse micelles in nonpolar liquids. *Langmuir* 27 (17): 10386–10391.

28 Delgado, A.V., Gonzalez-Caballero, E., Hunter, R.J. et al. (2005). Measurement and interpretation of electrokinetic phenomena – (IUPAC technical report). *Pure Appl. Chem.* 77 (10): 1753–1805.

29 Prasad, M., Strubbe, F., Beunis, F., and e.a. Different types of charged-inverse micelles in nonpolar media. *Langmuir* 32 (23): 5796–5801.

30 Petrenko, V.F. and Ryzhkin, I.A. (1984). Space-charge-limited currents in ice. *Soviet Physics - JETP* 60 (2): 320–326.

31 Robben, B., Beunis, F., Neyts, K. et al. (2020). Optical evidence for adsorption of charged inverse micelles in a stern layer. *Colloids Surf. A Physicochem. Eng. Asp.* 589: 124451.

32 Schreuer, C., Vandewiele, S., Strubbe, F. et al. (2018). Electric field induced charging of colloidal particles in a non-polar liquid. *J. Colloid Interface Sci.* 515: 248–254.

33 Schreuer, C., Vandewiele, S., Brans, T. et al. (2018). Single charging events on colloidal particles in a non-polar liquid with surfactant. *J. Appl. Phys.* 123 (1).

34 Noël, A., Mirbel, D., Charbonnier, A., and e.a. (2017). Synthesis of charged hybrid particles via dispersion polymerization in non-polar media for color electrophoretic display application. *J. Polym. Sci., Part A: Polym. Chem.* 2016 (55): 338–348.

35 Kim, M.K., Kim, C.A., Ahn, S.D. et al. (2005). Density compatibility of encapsulation of white inorganic TiO2 particles using dispersion polymerization technique for electrophoretic display. *Synth. Met.* 91: 370–374.

36 Morrison, I.D. (1993). Electrical charges in nonaqueous media. *Colloids Surf., A* 71: 1–37.

37 Gacek, M.M. and Berg, J.C. (2015). The role of acid–base effects on particle charging in apolar media. *Adv. Colloid Interf. Sci.* 220: 108–123.

38 Beunis, F., Strubbe, F., Neyts, K., and e.a. (2012). Beyond Millikan: the dynamics of charging events on individual colloidal par. *Phys. Rev. Lett.* 108 (1): 016101.

39 Strubbe, F., Beunis, F., Brans, T. et al. (2013). Electrophoretic retardation of colloidal particles in nonpolar liquids. *Phys. Rev. X* 3 (2): 021001.

40 Carrique, F., Ruiz-Reina, E., Arroyo, F.J. et al. (2008). Dynamic electrophoretic mobility of spherical colloidal particles in salt-free concentrated suspensions. *Langmuir* 24: 2395–2406.

41 Morrison, I.D. and Tarnawskyj, C.J. (1991). Toward self-consistent characterizations of low conductivity dispersions. *Langmuir* 7: 2358–2361.

42 Yang, B.R., Hsieh, Y.J., Zang, H.M. et al. (2011). Electrophoretic display fluid. US9341915.

43 Sprague, R.A., Yang, B.R., and Laxton, P. (2015). Electrophoretic dispersion. US9052564.

44 Herb, C.A., Morrison, I.D., and Loxley, A.L. (2004). Threshold addressing of electrophoretic displays. US6693620.

45 Zhang, Y.D., Hu, W.J., Qiu, Z.G. et al. (2019). Backflow effect enabling fast response and low driving voltage of electrophoretic electronic ink dispersion by liquid crystal additives. *Sci. Rep.* 9 (1): 1–8.

46 Bert, T. and Smet, H.D. (2005). How to introduce a threshold in EPIDs. In: *Proceedings of the Twenty-Fifth International Display Research Conference - Eurodisplay*, 337–339.

47 Asakura, S. and Oosawa, F. (1958). Interaction between particles suspended in solutions of macromolecules. *J. Polym. Sci.* 31: 183–192.

48 Royall, C.P., Louis, A.A., and Tanaka, H. (2007). Measuring colloidal interactions with confocal microscopy. *J. Chem. Phys.* 127: 044507.

49 Ji, S. and Walz, J.Y. (2013). Synergistic effects of nanoparticles and polymers on depletion and structural interactions. *Langmuir* 29: 15159–15167.

50 Edwards, T.D. and Bevan, M.A. (2012). Polymer mediated depletion attraction and interfacial colloidal phase behavior. *Macromolecules* 45: 585–594.

51 Tulpar, A., Tilton, R.D., and Walz, J.Y. (2007). Synergistic effects of polymers and surfactants on depletion forces. *Langmuir* 23: 4351–4357.

52 Sprague, R.A., Yang, B.R., and Laxton, P.B. (2019). Electrophoretic dispersion including charged pigment particles, uncharged additive nano-particles, and uncharged neutral density particles. US10288975.

53 Hiji, N., Machida, Y., Yamamoto, Y. et al. (2012). Novel color electrophoretic E-paper using independently movable colored particles. *SID Digest* 8 (4): 85–87.

3

Driving Waveforms and Image Processing for Electrophoretic Displays

Zong Qin[1] *and Bo-Ru Yang*[2]

[1] *School of Electronics and Information Technology, Sun Yat-Sen University, Guangzhou, China*
[2] *Professor, School of Electronics and Information Technology, State Key Laboratory of Optoelectronic Materials and Technologies, Guangdong Province Key Laboratory of Display Material and Technology, Sun Yat-Sen University, Guangzhou, China*

Overview

Driving waveforms are composed of voltage signals and timing sequences, which are essential to control the grayscale and color of a specific pixel accurately. The design concept of producing a waveform is to derive a quantitative relationship between optical responses and the electric field applied. Also, image processing technology is used to manage the optical response further because a TFT backplane's temporal resolution is limited, and software-based image processing is more convenient and efficient. This Chapter will introduce the waveform technology for electrophoretic displays (EPDs) in the first Section. Next, the second section will discuss how image processing technology facilitates high image quality in EPDs. The third section will introduce several recent advances in driving technologies for future EPDs.

3.1 Driving Waveforms of EPDs

3.1.1 Fundamental Concepts of Waveforms

The movement of charged particles determines an EPD's optical response. The velocity is given by the Stokes equation, as Eq. (3.1) [1, 2].

$$v(t) = \frac{qE(t)}{6\pi\eta r} \tag{3.1}$$

Equation (3.1), where v(t) is a particle's velocity, q is the particle's electric charge, E(t) is the effective electric field varying with time t, η is the solvent's viscosity, and r is the particle's radius.

The driving voltage produces the electric field, and the corresponding signal duration determines reflectance, contrast ratio, response time, etc. In particular, grayscale images are theoretically feasible by applying different combinations of driving voltage and signal duration on each pixel. However, it is challenging to design driving signals by directly adopting the Stokes equation because many particles of different sizes are encapsulated in a single micro-cup or micro-capsule. The electric field E(t) on a single particle is affected by other particles – for example, particles moving ahead of others produce resistance to those moving behind with the same charge

E-Paper Displays, First Edition. Edited by Bo-Ru Yang.

polarity. Moreover, mechanical interactions between particles also affect their movement. As a result, the particles within one pixel have a distribution of velocities, and the velocity is time-variant while the particles move to change the gray level states.

The optical response accounting for particles' differing and time-variant velocity can be theoretically estimated by assuming the particles' velocities have a Gaussian distribution at time t, as given by Eq. (3.2) [1, 2].

$$n(x,t,t_i) = \frac{N}{\sqrt{2\pi} v_s t} \exp\left\{ -\frac{\left[x - v_a(t - t_i)\right]^2}{2\left[v_s(t - t_i)\right]^2} \right\} \tag{3.2}$$

Equation (3.2), where x is the positional coordinate along the direction of particle movement (a simplified 1-dimensional case here), t_i is a time delay associated with removing particles from an electrode, v_a and v_s are respectively the average velocity and the standard deviation.

Furthermore, assuming the overall optical response is an integral of reflected light from multiple layers of particles with different time delays t_i (t_i can also be assumed to have a Gaussian distribution), the optical response can be derived.

With the above estimate, it is still difficult to control gray scales accurately and precisely because the estimation model incorporates several assumptions of ideal conditions of an EPD device. It is a highly nonlinear electro-optic response model, which is a challenge for the driving circuit. Therefore, in practice, an EPD's gray levels are usually implemented by pulse width modulation (PWM) of constant-voltage signals, as shown in Figure 3.1a. Modern TFT-based EPDs can conveniently implement PWM of constant-voltage signals. The sequence of voltage pulses applied to a pixel electrode to transform one gray level to another is called waveforms. The correct electric-optic response is further guaranteed by mapping between optical responses and the voltage pulses with the help of optical measurement.

For acquiring more precise gray level control and better color saturation, Kim et al. [3] adopted driving waveforms with multilevel voltages at the cost of backplane complexity. By adding two voltage levels, the number of gray levels was expanded from 8 to 165, and the average response time was reduced from 90 to 61 ms. Also, by varying the driving voltage, Yi et al. [4] proposed a driving waveform based on a damping oscillation to optimize the red saturation of three-color EPDs. Their experimental results showed that the maximum red saturation could reach 0.583, which was increased by 27.57% compared with the traditional driving waveform.

Alternatively, the multi-gray scales of the E-paper can be achieved by controlling the polarity and the amplitude of the voltage, which is called pulse amplitude modulation (PAM) [5]. As shown in Figure 3.1b, people utilize PAM to control the movement of electrophoretic particles so that the spatial distribution of electrophoretic particles can be controlled. The spatial distribution of the particles determines the grayscale. It is challenging to adopt

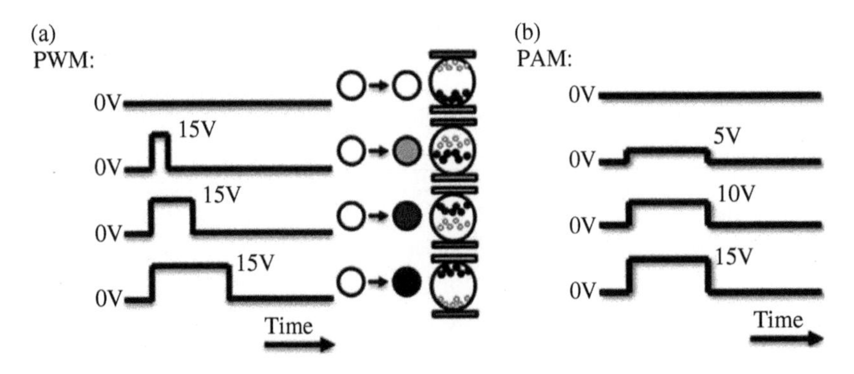

Figure 3.1 Multi-gray scale of E-paper driven by (a) PWM and (b) PAM.

a PAM waveform for TFT circuit design, so it is rarely used in practical applications, and the waveforms we discuss next are all based on PWM unless it is explicitly stated to the contrary.

By defining how the voltage is applied between common electrodes and pixel electrodes (i.e. voltages on source lines), the EPD's driving waveform can be classified into bipolar and unipolar waveforms [2]. The bipolar waveforms apply +15V or −15V on the pixel electrodes while keeping the common electrodes at 0V, as shown in Figure 3.2a. Alternatively, the unipolar waveforms always apply opposite voltages on common and pixel electrodes to achieve a doubled voltage difference, as shown in Figure 3.2b. The unipolar waveform produces a faster response because a larger electric field speeds up the charged particles.

A waveform composed of multiple stages is usually used for showing a desired gray level. Specifically, EPD's waveforms contain three stages. (i) Erasing the previous particles' state of a pixel before a new image is displayed. (ii) Activating the particles by quickly pushing them several times. (iii) Displaying a new image. Figure 3.3 illustrates such waveforms containing such three stages.

The erasing stage is critical in the waveform design since the "memory" of EPDs will cause so-called "ghost images" if the erasing is not appropriately performed. The memory effect was demonstrated in [6], as shown in Figure 3.4, where the second image (State 2) contains visible residual contrast from the first image (State 1). The memory effect also suggests that waveform design should consider the target grayscale and the previous grayscale.

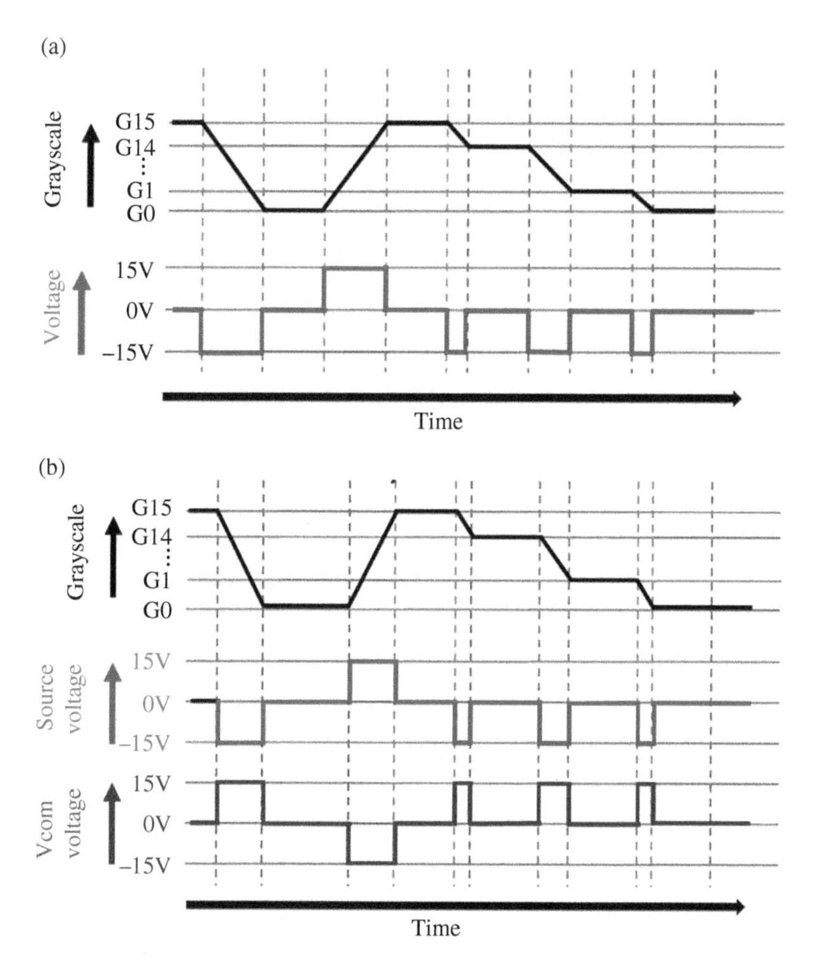

Figure 3.2 (a) Bipolar and (b) unipolar driving waveforms.

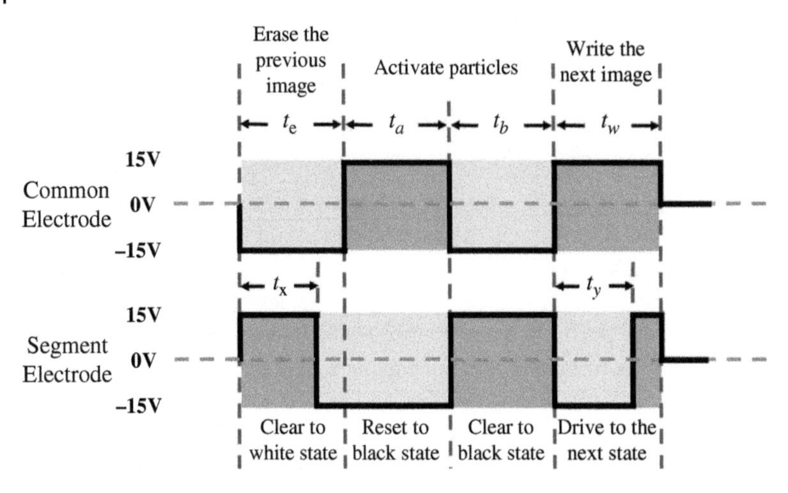

Figure 3.3 Typical EPD waveform containing three stages: erasing, activating, and display.

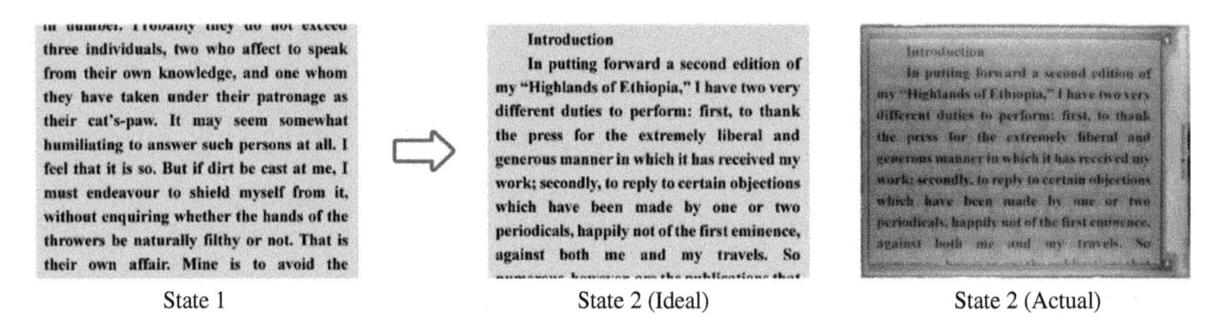
three individuals, two who affect to speak from their own knowledge, and one whom they have taken under their patronage as their cat's-paw. It may seem somewhat humiliating to answer such persons at all. I feel that it is so. But if dirt be cast at me, I must endeavour to shield myself from it, without enquiring whether the hands of the throwers be naturally filthy or not. That is their own affair. Mine is to avoid the

State 1

Introduction

In putting forward a second edition of my "Highlands of Ethiopia," I have two very different duties to perform: first, to thank the press for the extremely liberal and generous manner in which it has received my work; secondly, to reply to certain objections which have been made by one or two periodicals, happily not of the first eminence, against both me and my travels. So

State 2 (Ideal)

State 2 (Actual)

Figure 3.4 Demonstration of historical dependence of EPDs.

3.1.2 DC-Balanced Waveforms

As mentioned above, the common practice solution to address the memory effect and guarantee accurate gray levels is to push charged particles to the very boundaries of a micro-capsule or a micro-cup in the erasing and the activating stages so that the particles' history can be eliminated. Nevertheless, it is necessary that for each pixel, the integral of the applied voltage waveform overtime should be zero; that is, the charge used to push particles in these stages must be balanced out during the driving period, a condition called "DC balance" [5–7]. As shown in Figure 3.5, in the activating stage (Phase 2), the time t_w is used to activate particles and reset the image history by forcing a pixel to change from a black state to a white state or vice-versa. In the display stage (Phase 3), the time t_b is used for a pixel to change from the black state to a specific gray level G_b. In Phase 2, DC balance is satisfied because the durations of the positive and negative pulses are equal. Thus, to balance out the remaining DC charge, the pulse width in the erasing phase (Phase 1) should equal t_b so that the internal residual charge could be canceled. This method achieves DC balance in a single driving period (i.e. a single grayscale conversion from G_a to G_b), also known as the "Local DC Balance" method.

Alternatively, a "Global DC Balance" method can be adopted to shorten the image update time and alleviate the flickering that occurs while resetting the image history [8, 9]. The Global DC Balance method comprises applying a series of driving voltages to a pixel, and the accumulated voltage integrated over a period of time from the first image to the last image is 0 (zero) or substantially 0 (zero), as shown in Figure 3.6a, where DC balance is achieved across two consecutive image frames. Figure 3.6b provides a further example of Global DC Balance, where the numbers 0, +50, +100, +150, −50, −100, or −150, with a unit of volt·ms, represent the accumulated voltage

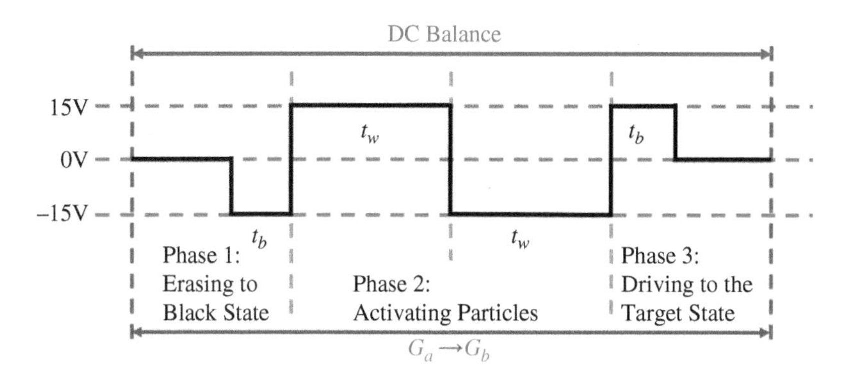

Figure 3.5 Illustration of the local DC balance method.

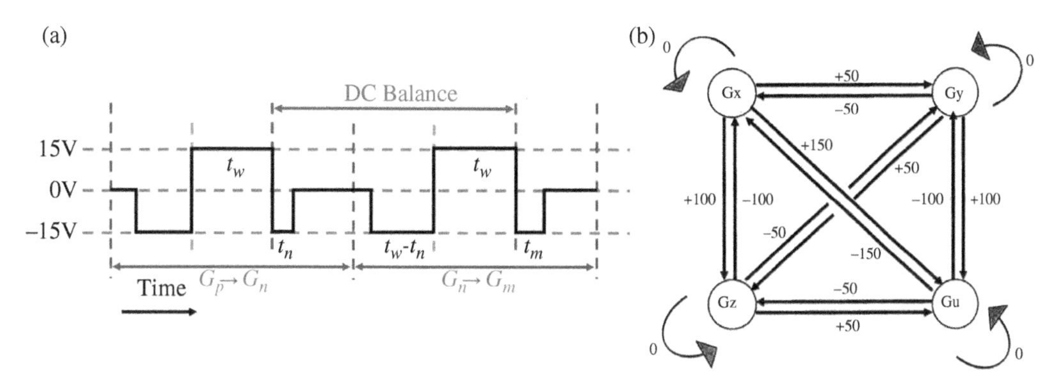

Figure 3.6 The driving method of Global DC balance: (a) general concept; (b) an example.

integrated over time. Gx, Gy, Gz, and Gu indicate gray levels x, y, z, and u, respectively. When a pixel is driven from Gx to Gy, the accumulated voltage integrated over time is +50 V·msec. In the case of driving a pixel from Gx, Gy, Gz, then to Gx, the image undergoes three driving waveforms, and the accumulated voltage integrated over time is $(+50)+(+50)+(-100) = 0$ V msec. The value of zero could result from several possibilities.

3.1.3 Waveform Design with Temperature Considerations

The device parameters cannot simply determine the appropriate driving waveforms [10] because the EPD's response curve is nonlinear [11, 12]. Different batches of electrophoretic particles may show various performances even with the same driving waveform. Differences in particle size distribution, zeta potential, charging agent amount, or the dielectrics' condition can significantly influence the electronic ink performance. Moreover, even in the same batch of electronic ink devices, different driving histories may lead to differences in the electronic ink dispersion system and cause other particle behaviors. Therefore, a waveform with higher tolerance that can lead to the same performance under different conditions is critical. In this regard, summarizing the particles' behaviors with a lookup table (LUT) is an economical way to simplify the complexity for realizing electronic ink display mass-production [13].

Of the factors affecting electronic ink's behavior, the temperature is one of the most important, which directly influences the performance of EPDs [13]. For commercial EPDs, there are usually some driving methods that allow for temperature compensation [13–17]. Designing several driving waveforms for different ambient temperatures, stored as a LUT, is a facile method.

Especially at a low temperature, a bounce-back effect may occur, decreasing an EPD dynamic range when power is off [18]. In electrophoresis, charged reverse micelles are bounded around charged particles due to the electrostatic force. The bounce-back effect occurs when the charged particles and charged inverse micelles move in the same direction toward the electrode of opposite polarity under an applied driving voltage. When the charged particles reach the electrode, the charged inverse micelles are desorbed. The freed charged inverse micelles and charged particles create an electric space charge and consequently a built-in electric field. When the driving voltage is turned off, the external electric field disappears. However, the built-in electric field remains for some time, and the particles move under its influence, resulting in a grayscale rebound.

A method called "invisible shaking" was proposed to alleviate the bounce-back effect [18]. Invisible shaking applies a pulsed voltage smaller than the driving voltage after the pixel is addressed and before power is turned off. It is used to eliminate the influence of the built-in electric field on the particles. It resets the position of the charged inverse micelles on the premise of no displacement of the particles. In this way, the particles will not bounce back after power off, and there will be no grayscale mutation, as shown in Figure 3.7.

3.1.4 Driving System and Lookup Table

A complete EPD driving system contains an EPD film, a TFT-based backplane, a timing controller (TCON), driver chips, a power circuit, storage, etc., as shown in Figure 3.8. Similar to TFT-LCDs, the TFT technology in an EPD plays the role of providing particular voltages on source electrodes, applying the voltages to a line of pixels, keeping the biased voltages on the pixels through storage capacitors, and fast scanning line by line. A TCON meantime reads image data and a waveform LUT stored in random-access memory (RAM), translates gray levels in the image data to waveforms according to the waveform LUT, and then sequentially sends the waveforms to the driver chips. Here, the image data and the LUT may be loaded from external memory (e.g. flash memory) into the RAM in advance. Finally, since the driving voltage of an EPD is typically +15 V to −15 V, which is higher than TFT-LCDs, a power circuit for boosting the voltage is needed.

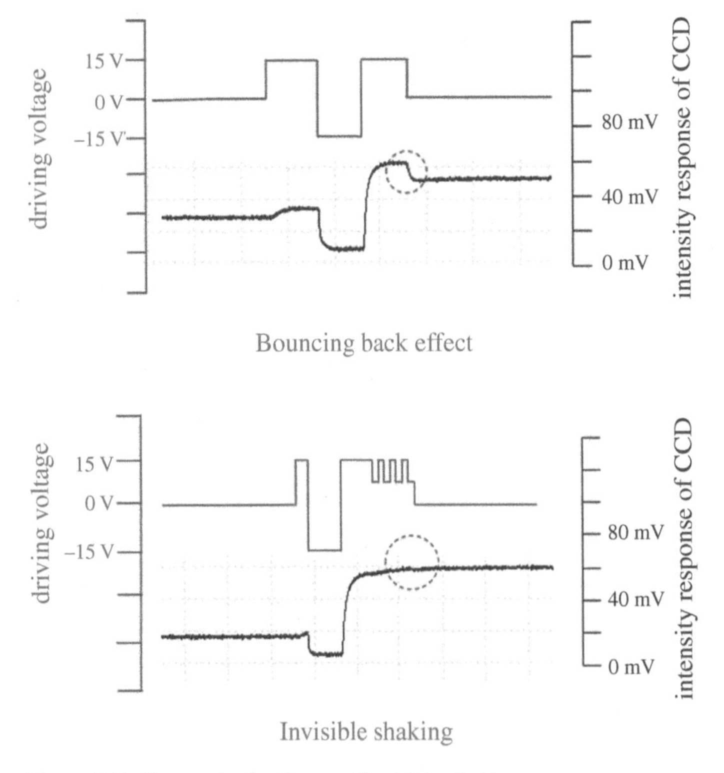

Figure 3.7 Bounce back effect and invisible shaking.

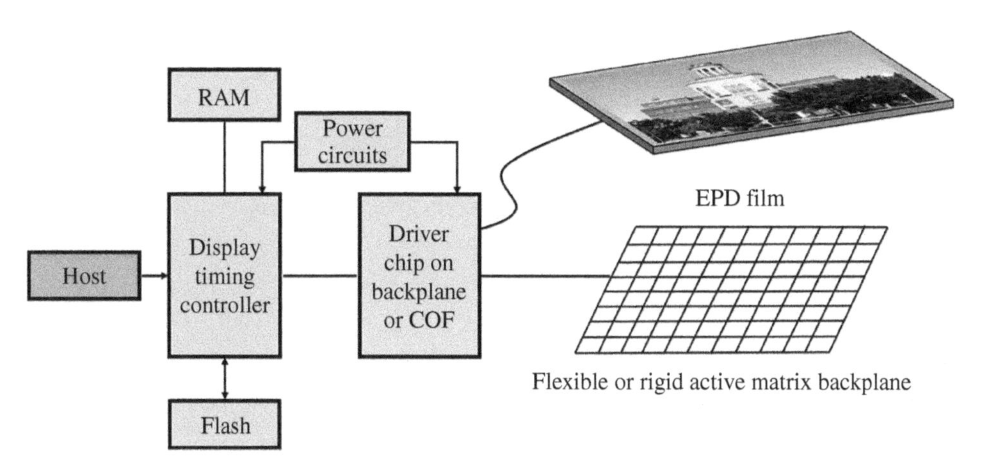

Figure 3.8 A complete EPD driving system.

As discussed in the previous Section, electrophoresis is sensitive to working conditions such as temperature, and device characteristics vary. Thus, driving waveforms should be configurable to adapt to different working conditions and device characteristics. That is, configurable LUTs linking image data and waveforms are needed. However, creating configurable LUTs for different devices and working conditions is very time-consuming; hence, recent studies considered artificial intelligence for automatic waveform configuration.

For example, to suppress ghost images caused by the memory effect mentioned above, the conventional solution would design multiple refresh waveforms to reset the EPD state, manually adjust different waveforms to specific gray levels, and finally establish LUTs. As Figure 3.9a shows, the manual adjustment requires five hours for one set of LUTs with 256 waveforms. Therefore, an automatic system for waveform adjustment is required to reduce the cost.

To overcome the shortcomings of manual waveform adjustment, ghost image recognition based on a convolutional neural network (CNN) and an automatic waveform system were proposed in [6], as shown in Figure 3.9b. A system that automatically generated several LUTs based on the initial LUT was designed. These waveforms were input to an EPD to display images. A camera captured each display image driven by every input waveform. CNN is a

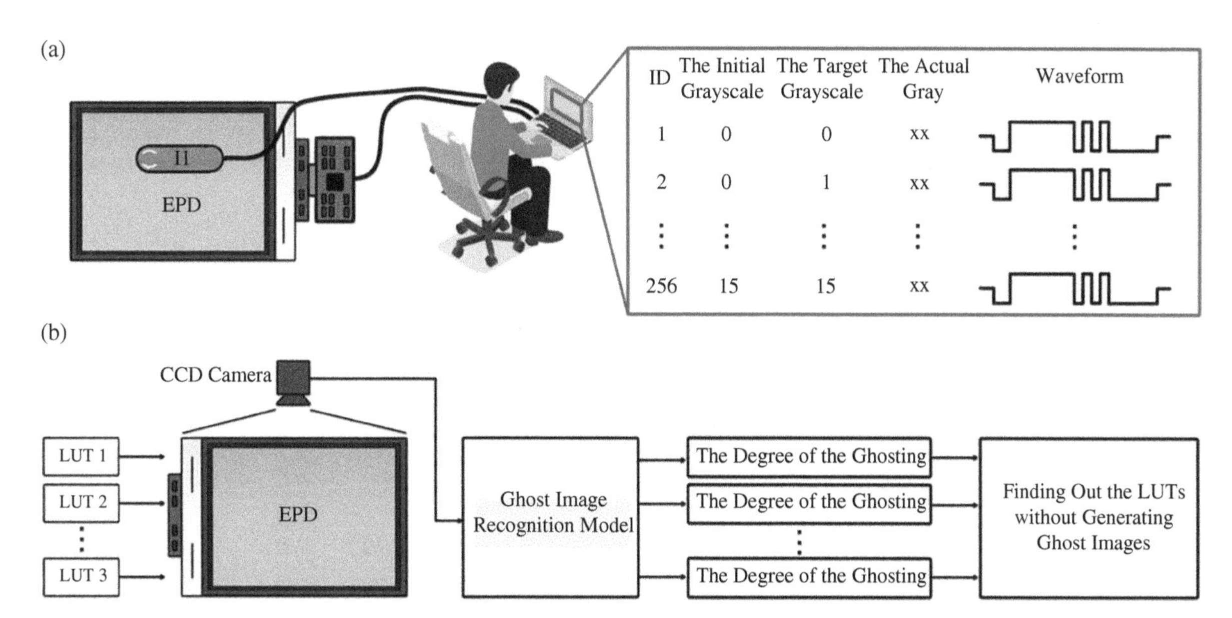

Figure 3.9 (a) Manual waveform adjustment for an EPD with 16 grayscales and 256 specific waveforms. (b) The schematic of ghost image recognition and automatic waveform system.

standard method for image classification, which can be trained to recognize whether a photograph contains ghost images. The image data obtained from the camera were transmitted to the ghost image recognition model – a trained CNN. The amount of ghosting corresponding to every input LUT was calculated through the above model. Finally, the severity of the ghost images was compared, and a set of LUTs without ghost images was selected.

A computer processed the ghost image recognition and the automatic waveform system entirely without manual adjustment, which is time-saving and low-cost. The automatic waveform generation system spent approximately 30 minutes to generate 3 sets of LUTs and selected a set of LUTs with the slightest ghost image within 1 minute [6].

3.1.5 Driving Schemes for Color E-Papers

Recently, color EPDs have gained increasing attention, and their waveform design was investigated in the literature [19]. The basic concept of waveform design for color EPDs is similar to B&W EPDs, but the color gamut, in addition to gray levels, should be considered to be affected by waveforms.

For instance, Figure 3.10 shows the architecture of the 3-color EPDs from E Ink Corp. [20]. Positively charged white particles and negatively charged black and red particles are encapsulated in a single microcup. At any time, only one of the three colors is driven to the top electrode to form a color. Although black and red particles carry the same polarity, the size and the mass of the red particles are much larger than those of the black particles, which makes it possible to drive the two particles separately with the same polarity of the electric field. This is achieved by driving the red particles with a lower voltage for a longer duration. In this manner, the driving waveform design needs to consider driving stages, voltages, and durations.

In contrast to driving waveforms for B&W EPDs that usually contain three stages (i.e. erasing, activating, and display), the waveforms for 3-color EPDs include one further stage for driving the red particles with a lower voltage, as shown in Figure 3.11. The DC balance rule should still be obeyed in the first erasing stage to balance the charge

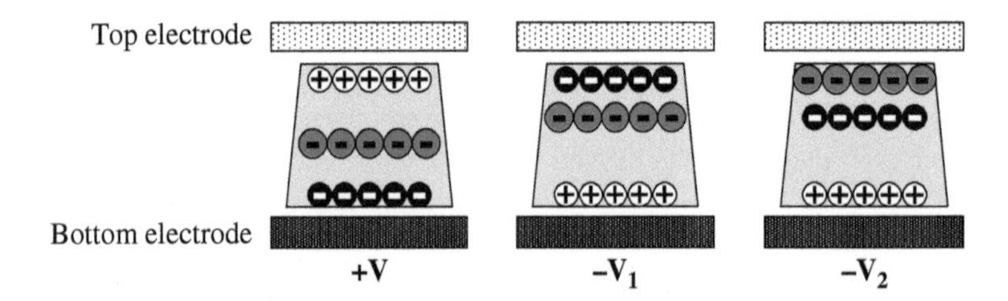

Figure 3.10 Architecture of 3-color EPDs. Black dots with negative signs: black particles; Dark gray dots with negative signs: red particles.

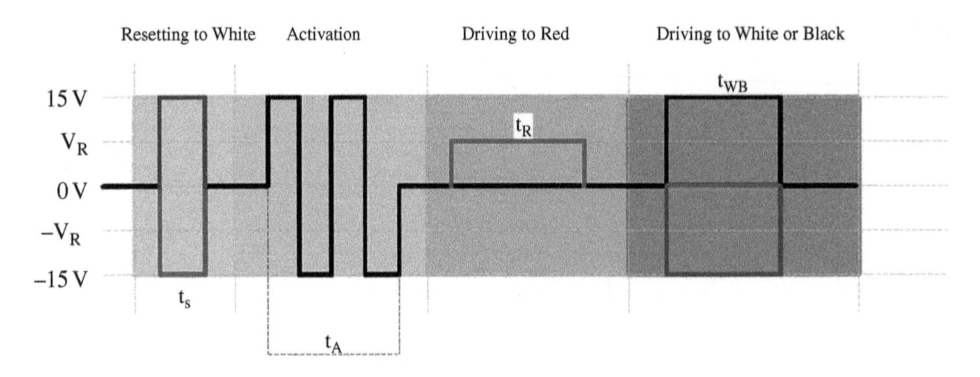

Figure 3.11 Driving waveforms for 3-color EPDs with four stages.

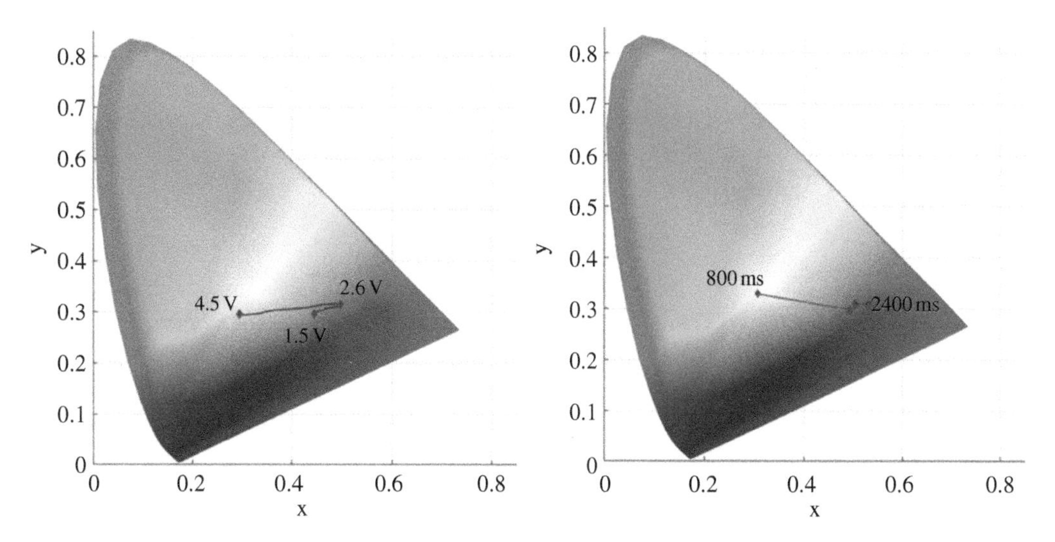

Figure 3.12 Red color's chromaticity coordinate versus the voltage (left) and the duration time (right) of waveforms. The chromaticity coordinate (0.735, 0.265) is the ideally red the waveform optimization pursued.

Figure 3.13 Three-color EPD. From left to right: black-white, red-white, and red-black tones, where the continuous tones are achieved through digital dithering. (In the printed B/W format, the rightmost picture looks dark because the red-black tones intrinsically have low lightness.)

in the following display stages. Since the red stage adopts a lower voltage with a longer duration, the erasing time (t_s in Figure 3.11) should use a higher voltage with a shorter duration to guarantee equal energy. The third stage directly determines the image quality of red, thus requiring careful design optimization.

Kao et al. [19] optimized the red stage by tuning the voltage and the pulse duration. As a result, by investigating the red color's chromaticity coordinate versus the voltage and the duration, as shown in Figure 3.12, they determined the optimal voltage and duration time (2.6 V and 2400 ms in their study). Besides, the fourth driving stage for black or white with a normal voltage was the same as B&W EPDs. Figure 3.13 shows photographs of a three-color EPD with optimized red color, where the digital dithering technology discussed later in this Chapter is used for continuous tones.

3.2 Image Processing

Theoretically, an EPD's image quality can be fully optimized through waveform design. However, the time resolution of a TFT backplane is limited, and the voltage applied is usually constant. Therefore, image processing is a complementary method to achieve high-quality EPD images in addition to waveform design. Typical EPD image processing methods include digital dithering for richer grayscales and image calibration for faithful image reproduction.

3.2.1 Digital Dithering and Halftoning

As mentioned in the previous section, under the limited scanning rate of a backplane, it is challenging to precisely control a portion of charged particles through the driving waveform to achieve as many grayscales as conventional displays (e.g. usually 8-bit for TFT-LCDs). For example, the B&W microcapsule EPD usually has 4-bit grayscales [21, 22]. The commercialized microcup 3-color EPD only has 1-bit grayscales, namely only 2 states for each color [20]. More grayscales must be achieved to deliver rich information to users, based on which more colors can be displayed by mixing multiple color channels.

The typical approach to increasing the available grayscales from a low intrinsic grayscale number is digital dithering, which operates based on the human eye's characteristics. From a signal processing point of view, the human eye is approximately a low-pass spatial filter with the spatial frequency set by diffraction, aberration, etc. Figure 3.14 shows the human eye's typical modulation transfer function (MTF) acquired by ophthalmic studies [23]. By applying the inverse Fourier transform to the MTF, the point spread function (PSF) depicting how the human eye perceives a point can be obtained. Next, the way an EPD image is perceived is formulaically given by Eq. 3.3 [24].

$$I(x,y) = I_0(X,Y) \otimes W(x,y) \otimes P(x,y) \tag{3.3}$$

Equation (3.3), where $I_0(X, Y)$ is an EPD's pixel values, W(x,y) is a square window function whose size equals that of the pixel, P(x,y) is the human eye's PSF, and I(x,y) is the perceptual EPD image. Note that X and Y denote discrete positions of pixels, and x and y denote continuous positions on the EPD panel.

When multiple-tonal images are constructed using only binary values (i.e. only black and white), digital dithering is known as digital halftoning. Figure 3.15 shows an example where a halftone image formed by binary pixels is perceived as a multi-tonal image in accordance with Eq. (3.3).

There are three main methods of obtaining a halftone image from a grayscale image, i.e. pixel-wise (screening), error-diffusion, and model-based iterative methods [25]. The pixel-wise method compares a threshold matrix with the original grayscale image via Boolean operations. For example, the classic Bayer matrix proposed in the 1970s [26] devises the matrix so that two successive thresholds are as far as possible, as shown in Figure 3.16a. The error-diffusion method compares a pixel in the grayscale image with a threshold (i.e. 50% grayscale) to generate a binary value and then calculates the quantization error concerning the original grayscale. Next, the error is diffused and accumulated to adjacent pixels by following a diffusion matrix. The entire halftone image is generated by parsing the original image along a specific path (e.g. raster or serpentine paths). Figure 3.16b illustrates the standard Floyd–Steinbergs error-diffusion method [27] with a serpentine parsing path, a representative in the error-diffusion family. The model-based iterative method, also known as the direct binary search (DBS) method [28], starts with an initial

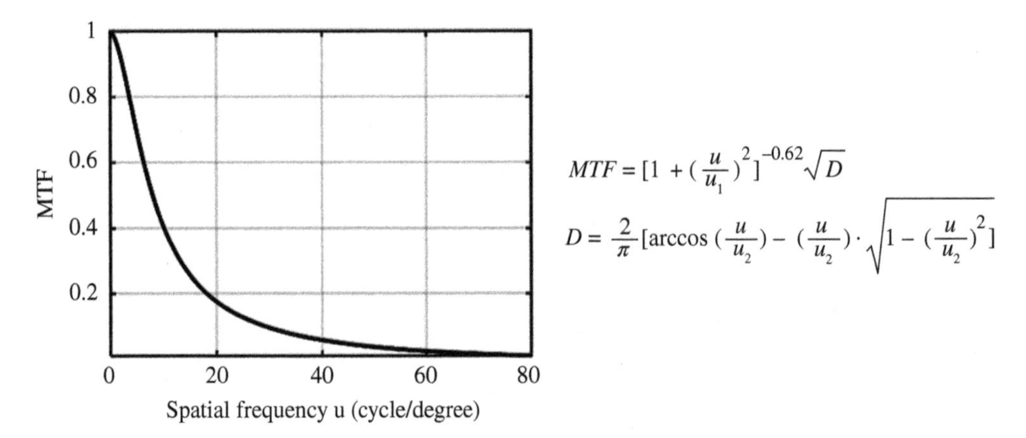

Figure 3.14 Human eye's MTF, given by the equations in the right, where u_1 and u_2 are pupil diameter-dependent terms that respectively equal 6.18 and 125.78 for a typical pupil diameter of 4 mm.

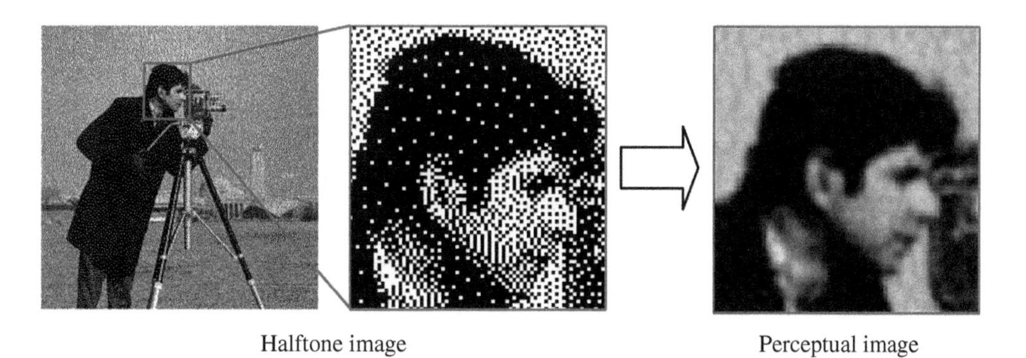

Figure 3.15 A halftone image formed by binary pixels and its perceptual image [54] Zhou, X., et.al, 2018 / MDPI / CC BY 4.0.

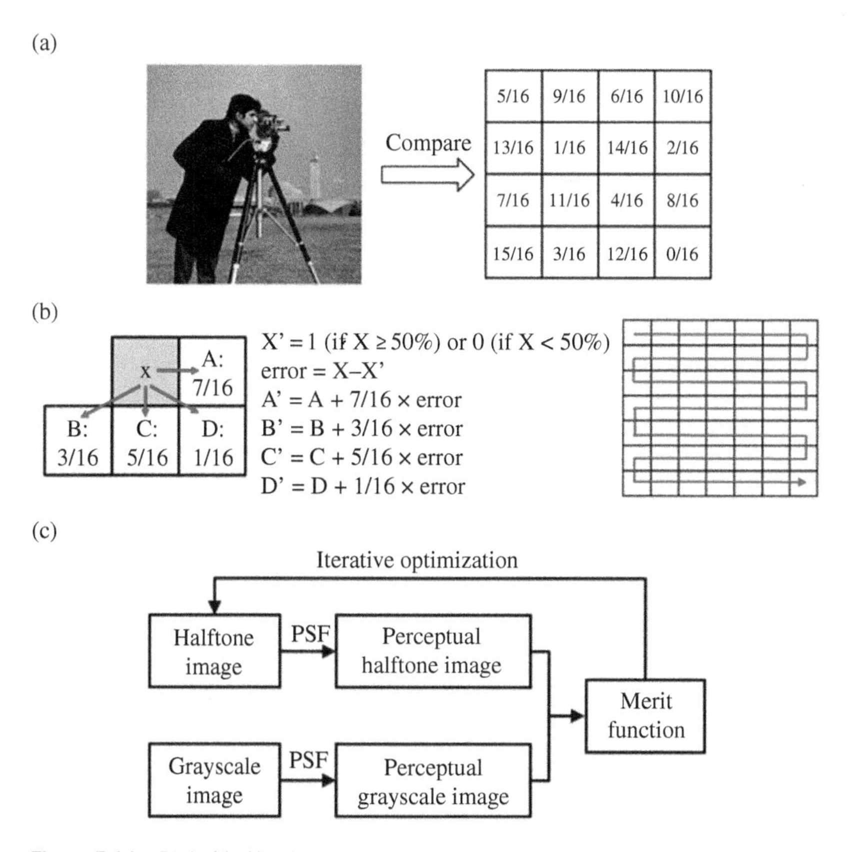

Figure 3.16 Digital halftoning methods: (a) the 4-by-4 Bayer matrix, a pixel-wise method; (b) the standard Floyd–Steinberg's error-diffusion method [54] Zhou, X., et.al, 2018 / MDPI / CC BY 4.0; (c) flow chart of the DBS method. The three methods have different pros and cons, and accordingly, are applied for EPDs in different scenarios.

halftone image and minimizes the difference between the perceptual halftone image and the original grayscale image through iterative optimization. In every iteration, heuristic search, e.g. swapping a binary pixel with its eight neighbor pixels or toggling the pixel, is adopted to modify the current halftone image. Its perceptual form is then calculated and compared with the original grayscale image to compute a merit function, usually root mean square error (RMSE). The merit function is minimized after several iterations across the whole image. Figure 3.16c shows the DBS method's flow chart. The above three methods can be similarly performed for a digital dithering problem, such as displaying 8-bit grayscales on a 4-bit display, where quantization operations replace Boolean operations.

3.2.1.1 Pixel-Wise Method

The pixel-wise method has the fastest speed as only Boolean, or quantization operations are required. Kao et al. [14, 29] utilized the pixel-wise method for displaying 8-bit grayscale images on a 4-bit B&W EPD that features a real-time driving controller. Instead of directly quantizing an 8-bit grayscale image to a 4-bit one, they adopted the Nasik pattern matrix **K** given in Eq. (3.4) as the screening matrix. The Nasik pattern matrix distributes consecutive numbers as far as possible and makes the summation of each row, each column, either diagonal, and each 2-by-2 submatrix identical so that any 4-by-4 matrix cropped from titled matrices is a new Nasik pattern matrix. By doing so, the spatial periodicity of tiled screening matrices, which may induce the cross-hatch artifact, can be largely suppressed. Figure 3.17 compares two images on the 4-bit EPD, acquired through direct quantization and digital dithering. The dithered one exhibited richer grayscales and much slighter quantization-induced contour artifacts. Kao et al. also adopted the Nasik pattern matrix to display grayscale paintbrushes on a binary EPD [30]. Later, they extended the dithering-based paintbrush function to a 2-bit EPD with 6-bit grayscales inputted [31].

$$\mathbf{K} = \begin{bmatrix} k_{00} & k_{01} & k_{02} & k_{03} \\ k_{10} & k_{11} & k_{12} & k_{13} \\ k_{20} & k_{21} & k_{22} & k_{23} \\ k_{30} & k_{31} & k_{32} & k_{33} \end{bmatrix} = \begin{bmatrix} 0 & 14 & 3 & 13 \\ 11 & 5 & 8 & 6 \\ 12 & 2 & 15 & 1 \\ 7 & 9 & 4 & 10 \end{bmatrix} \tag{3.4}$$

3.2.1.2 Error-Diffusion Method

The error-diffusion method is also frequently used in EPDs with better image quality [32], but its computational complexity is higher because it involves neighboring pixels while quantizing a pixel. Figure 3.16b illustrates the widely used standard Floyd–Steinberg's error diffusion and its diffusion matrix. Using this method, a seven-color EDP with binary values for each color can present continuous tones between every two colors, as shown in Figure 3.18. Similarly, the error-diffusion method can be used for full-color EPDs by diffusing the color error instead of the grayscale error. The color error should be quantified using a specific color difference model such as CIE76 or CIEDE2000. For instance, the seven-color EPD used above presents a full-color halftone image derived from an original image with continuous colors, as the last photograph in Figure 3.18 shows. Nevertheless, the color gamut of current full-color EPDs is still poor compared with mainstream LCDs and OLED displays, which needs further efforts in device and driving technologies.

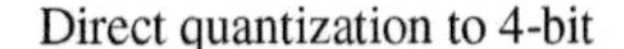

Figure 3.17 Photographs on a 4-bit EPD: direct quantization from an 8-bit input image (left) and digital dithering (right).

Figure 3.18 Using the standard Floyd–Steinberg's error-diffusion method, a 7-color EPD presents two continuous-tone images between two specific colors and a full-color image: (a) campus; (b) baboon. (Color information may be missing from the printed B/W format)

Kao et al. [13, 15, 33] also adopted the standard Floyd–Steinberg's diffusion matrix and modified it to be temperature adaptive. Considering an EPD's electric-optic response is dependent on temperature and thus varying the number of available grayscales, they modified the standard Floyd–Steinbergs error-diffusion method by changing the target grayscale number from binary to temperature-dependent. First, the reflectance function curves that provided the time for the white state to transit to the black state were measured, as shown in Figure 3.19. The target grayscale number N was then determined to be $N = 1 + \lfloor t/t_0 \rfloor$, where t is the transition time, and t_0 is the time the backplane uses to scan for a frame. For example, under the room temperature, the transition time was 220 ms, and the frame rate was 50 FPS, so there were 12 target grayscales for quantization while performing

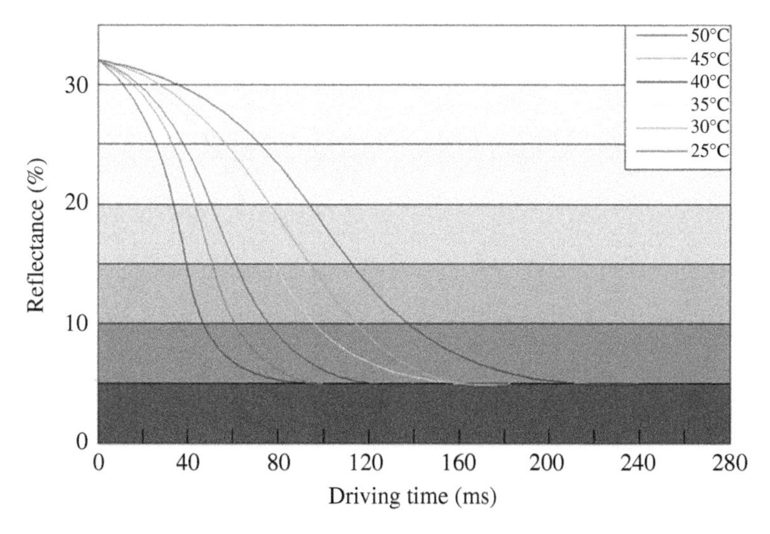

Figure 3.19 Reflectance function curve depicting how fast the white state transitions to the black state at different temperatures.

the error-diffusion algorithm. With a higher temperature detected by a sensor, a shorter transition time brought about fewer target grayscales. In their studies, images acquired through the modified error-diffusion method exhibited much better grayscale reproduction than the standard error-diffusion method under different temperatures.

Kao et al. [15] implemented the above error-diffusion algorithm with an efficient image processing pipeline targetted toward real-time applications. According to the basic principle of the error diffusion method, the first pixel in an input grayscale image is quantized to a binary value, and the quantization error is recorded. Next, the error is diffused to four new pixels (see the outward arrows in Figure 3.20a). In this manner, each pixel's original value is added with quantization errors diffused from four previously processed pixels [see the inward arrows in Figure 3.20a] and then quantized. The quantization error is further diffused to four unprocessed pixels. If there are no pixels at the inward arrows' origins or the outward arrows' destinations, directly omit the operations. Kao et al. [15] considered that three error values were diffused from the previous line, so they adopted a circular queue-shaped line buffer storing N-3 pixels (N: pixel number of a row) to store the error values for later use. In addition, the multiplication and division operations for the constant diffusion weights (i.e. 1/16, 3/16, 5/16, and 7/16) were replaced by faster addition and shift operations. For example, the weight 7/16 was regarded as (8–1)/16. Multiplication by eight was implemented as a left shift of three bits, and division by 16 was executed by excluding the four least significant bits. Figure 3.20b illustrates the alternative implementation of multiplication and division.

In later studies [34–36] aimed at video applications, Kao et al. considered it difficult to use a simple waveform for multi-tonal images because a pixel's response time strongly depended on its image retention time and the electric field on a pixel was affected by its neighbor pixels, they used an EPD in two-level B&W mode. They then adopted the Jarvis error-diffusion matrix [37] shown in Figure 3.20c for grayscales. They further optimized the image processing pipeline; e.g. the difference between two successive frames was encoded because not too much content would be changed in a video's consecutive frames.

3.2.1.3 DBS Method

The DBS method provides the best image quality because it essentially seeks the halftone image most approaching the reference grayscale image [38, 39]. Still, its high computational complexity prevents it from being widely used in EPDs. Nevertheless, its unique feature that a merit function can be constructed as needed means it has been considered in a few studies

The pixel value distribution along source lines affects the drive current in a binary EPD (e.g. a B&W or three-color EPD). A black-to-white or white-to-black reversal along a source line introduces an extra current compared with constant values [40]. Thus, a halftone image is preferred to contain continuous black or white pixels along source lines; whereas, a halftone image tends to use frequently reversed black and white pixels to avoid visible

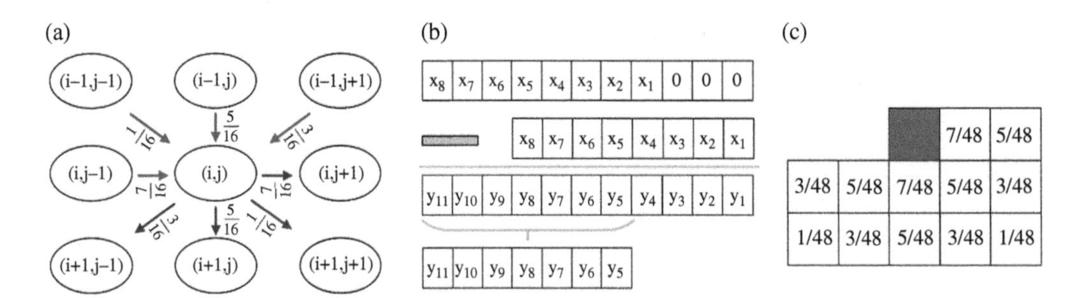

Figure 3.20 (a) Quantization errors diffused from previously processed pixels (inward arrows) and diffused to unprocessed pixels (outward arrows); (b) Implementing the diffusion weight of 7/16 with shift and subtraction operations instead of multiplication and division; (c) Jarvis error-diffusion matrix (gray pixel: the target pixel to be quantized).

patterns. Qin et al. [24] adopted the DBS method to address this conflict. They modified the conventional DBS method by changing the merit function from RMSE to a combination of RMSE and the drive current, as Figure 3.21a shows. Here, they proved that the drive current was near-perfectly linear with the total pixel reversals along source lines, and consequently, a current prediction model was conducted. The weighting of image quality versus drive current was optimized to guarantee that the RMSE was below 5%, which ensured that artifacts were invisible. Figure 3.21b compares EPD images acquired through the standard Floyd–Steinberg's error-diffusion method and the modified DBS method. Little image degradation concerning the error-diffusion method was observed, and the drive current was significantly reduced. Although the DBS method is currently challenging to implement in video applications, Qin et al. [24] discussed the possibility of introducing deep learning to replace iterative optimization.

3.2.2 Image Calibration

The essential goal is to create a correct mapping between device-dependent quantities and device-independent quantities, e.g. luminance versus grayscale and tristimulus versus RGB values. Considering the most widely used sRGB color space, Figure 3.22a shows luminance as a function of grayscale, also known as the "gamma curve," which each color channel in a color display should obey. Figure 3.22b shows chroma ΔC^* as a function of saturation (S) at different values of brightness (B), where ΔC^* is the root sum square of a* and b* in the device-independent CIELAB color space, and S and B are from the device-dependent HSB color space. The curves demonstrate a color display's relationship between perceptual chroma and input grayscales.

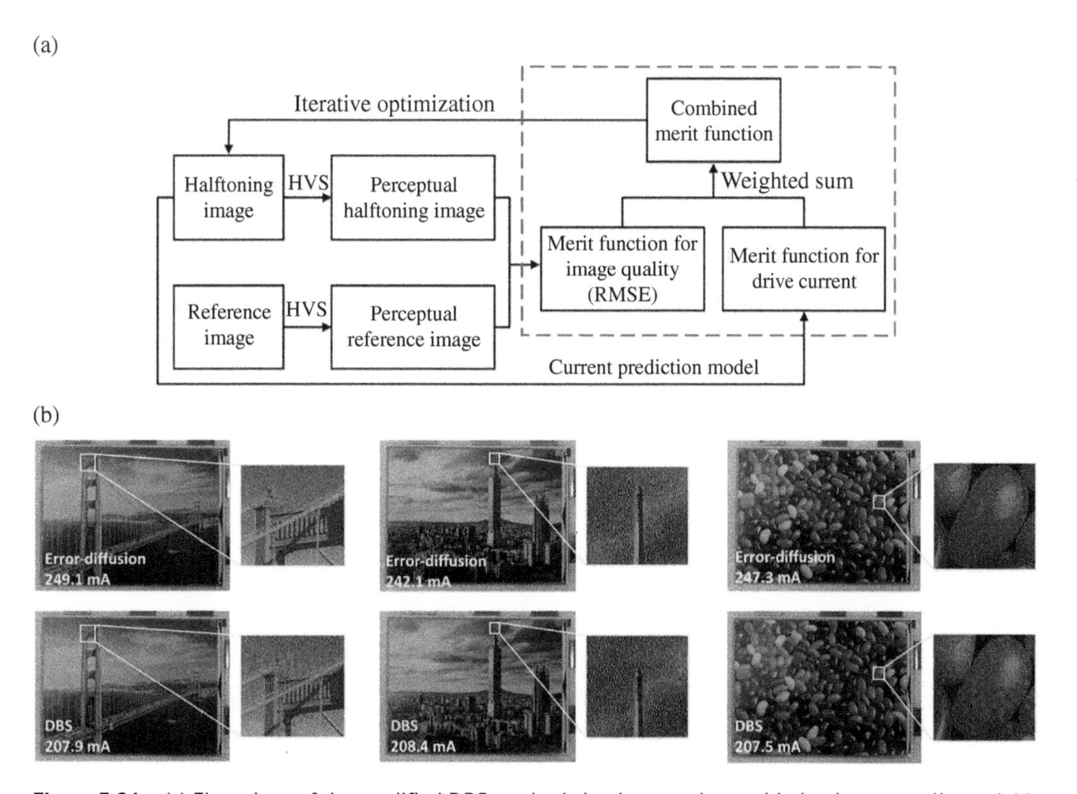

Figure 3.21 (a) Flow chart of the modified DBS method simultaneously considering image quality and drive current; (b) EPD images acquired through the standard Floyd–Steinberg's error-diffusion method and the modified DBS method, with their respective drive currents, [24] Zong Qin, et al. 2020 / Optica Publishing Group.

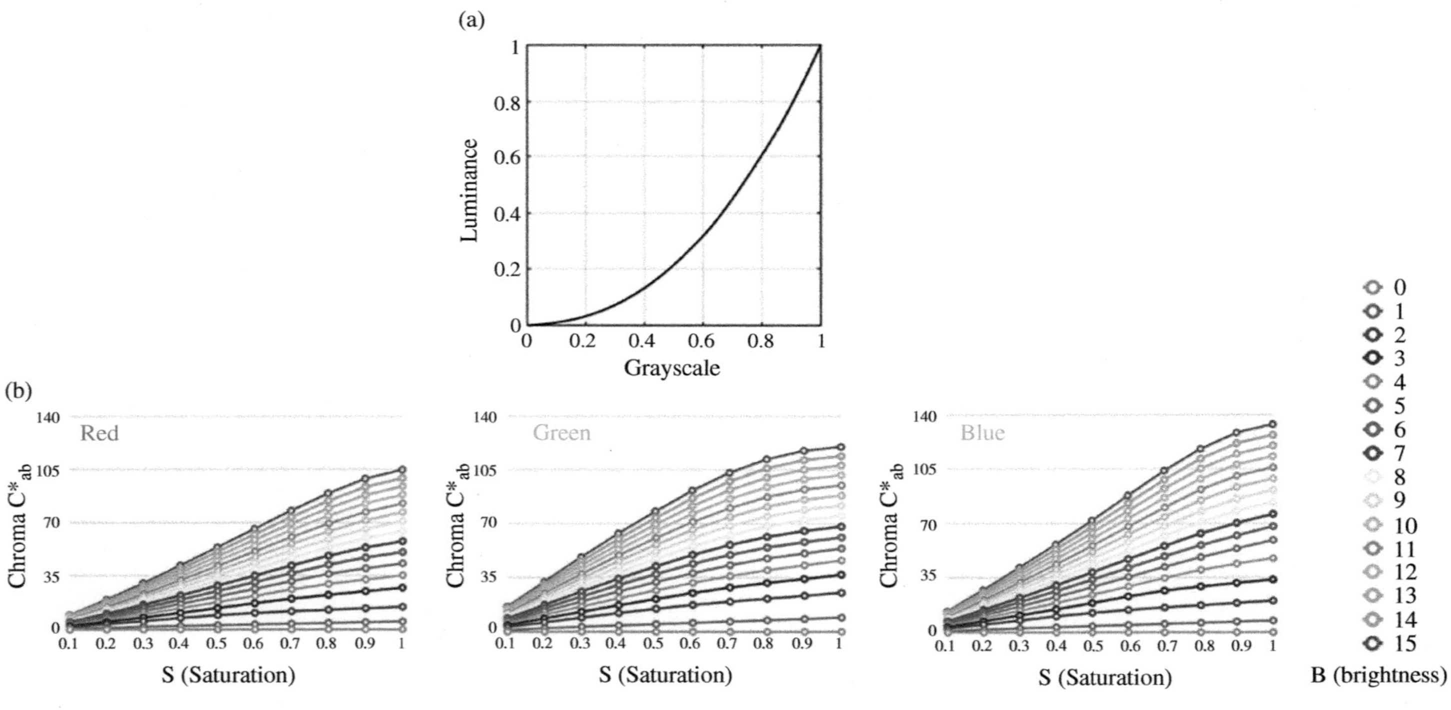

Figure 3.22 Ideal electric-optic response of a display in the sRGB color space: (a) luminance versus grayscale; (b) chroma ΔC^*in the CIELAB color space versus saturation (S) at different values of brightness (B) in the HSB color space.

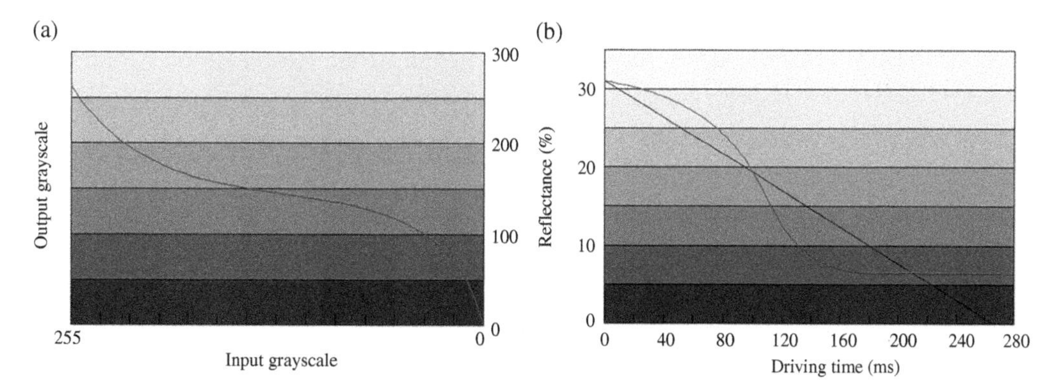

Figure 3.23 (a) Output optical response versus input grayscale; (b) tone-mapping calibration curve for input grayscales (the thick line) and the ideal response curve after the calibration (the thin line).

In practice, it isn't easy to perfectly match the electro-optic response of a display to the above curves. For this reason, the tone-mapping method is frequently adopted [41, 42], which pre-calibrates input grayscales to compensate for the non-ideal electric-optic response. The non-ideal electric-optic response of an EPD can be improved through waveform design; however, limited by the backplane's scanning rate (i.e. limited time resolution), waveform-based calibration can sometimes not produce an ideal electric-optic response. Thus, tone-mapping calibration is further used for EPDs.

As discussed in the previous Section on digital dithering, Kao et al. [15, 33] developed a temperature-dependent dithering method whose grayscale number was determined according to a sensed temperature. After determining the grayscale number, each grayscale's specific driving time was allocated by assuming the reflectance (i.e. output luminance) was ideally linear with the driving time. However, the electronic-optic response was nonlinear in reality, as Figure 3.23a shows; therefore, the inverse function of the electric-optic response was adopted as a tone-mapping curve for input grayscales, as Figure 3.23b shows. By doing so, the luminance recorded in an image could be correctly reproduced.

Image calibration can also be used in color EPDs for faithful color reproduction. A convenient approach to full-color EPDs is laminating a color filter array on a B&W EPD [43]. However, irregular electric-optic responses introduced by the color filter array caused contour-like tone distortions, as shown in Figures 3.24a and 3.25. Qin et al. [44, 45] performed photometric measurement on a full-color EPD between the device-independent CIELAB color space and the device-dependent HSB color space. They found the root cause of the contours was an abnormal chroma-saturation relationship under high saturation; i.e., the chrome (ΔC^*) started to decline when the input saturation (S) was higher than around 0.8, as Figure 3.24b shows. Qin et al. devised saturation-based tone-mapping curves to address such an abnormal electric-optic response as shown in Figure 3.24b. The input saturation ranging from zero to one was mapped to a new range excluding the abnormal segment. As a result, contours occurring in high-saturation regions were suppressed, as photographs in Figures 3.24a and 3.25 show.

The above-discussed digital dithering and image calibration methods for EPDs have unique features compared with image processing methods for conventional displays. In addition, regular image processing techniques, such as edge-based contrast enhancement, can be applied to EPDs [29].

3.3 Advanced Driving Methods for Future E-Papers

In addition to the driving methods introduced above (including the image processing methods), novel driving technologies are being rapidly developed for EPDs for promising applications such as IoT, public information

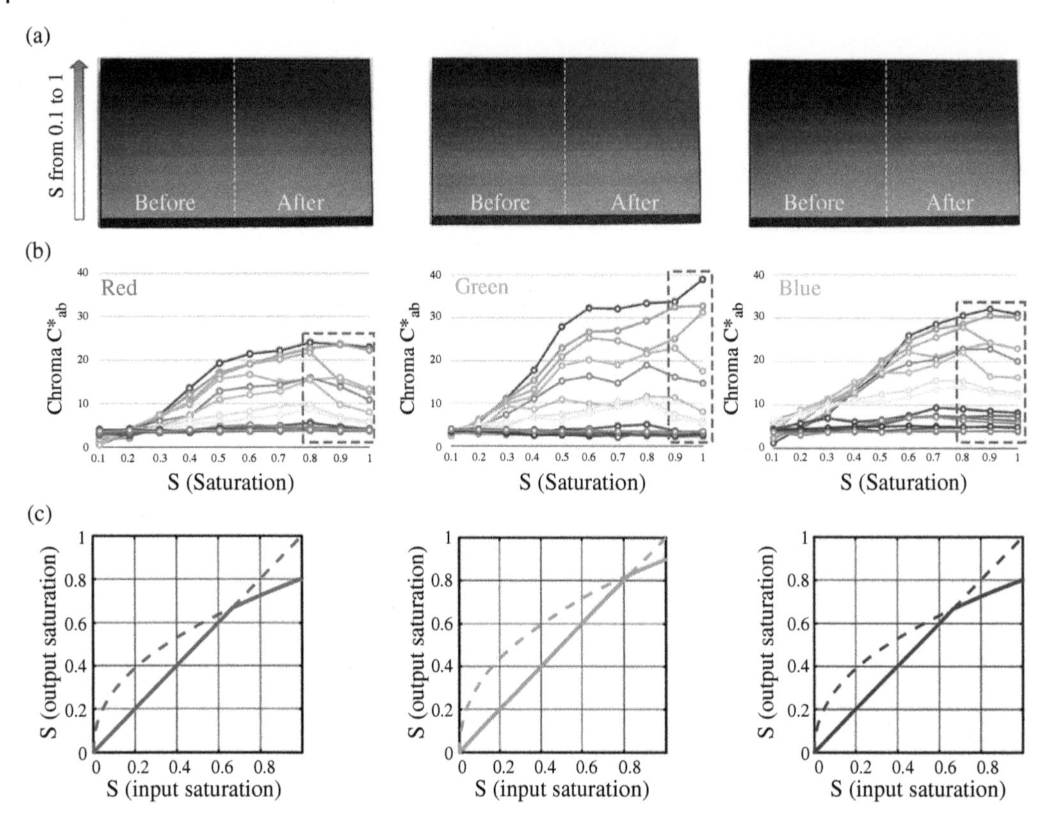

Figure 3.24 Image calibration for a color filter-based full-color EPD. (a) Photographs corresponding to gradient input saturation (S) under a certain brightness (B), where "before" and "after" denote before and after applying the saturation-based tone-mapping curves. (b) Measured chroma-saturation relationships under different brightness, highlighted abnormal chroma declinations (see Figure 3.22b for normal curves). Different curves denote different brightness (B) (c) Saturation-based tone-mapping curves for suppressing the contour-like artifacts. All subfigures are respectively drawn for R (left), G (middle), and B (right).

Figure 3.25 Photographs of two images before and after applying the saturation-based tone-mapping curves, [44] Qin, Zong, et al, 2018 / Reproduced from JOHN WILEY & SONS, INC.

display, commercial display, and so forth. In this Section, two recently proposed driving methods, i.e. self-powered EPDs enabled by a triboelectric nanogenerator (TENG) and three-dimensional driving for more precise grayscale and saturation control, will be introduced.

3.3.1 Self-Powered EPDs

Considering IoT applications, which call for ultra-low power consumption, self-powered EPDs that harvest energy in-situ from the working environment were studied using an energy-generating device called the triboelectric nanogenerator (TENG) [46].

The TENG is an emerging technology, which converts mechanical energy into electricity by coupling the triboelectric effect and electrostatic induction [47, 48]. High conversion efficiency provides TENG with various application opportunities as a power supply device and even a standalone input device triggered by low-frequency mechanical motions, such as touching, sliding, and stretching [49–51]. These features make TENG ideal for integrating with devices with low power consumption, especially EPDs.

As shown in Figure 3.26, when friction between a tribo-positive and a tribo-negative layer occurred, alternating current (AC) was generated through the triboelectric effect and electrostatic induction. After rectification, the AC was applied to the driving electrodes of the EPD. When the friction stopped, the grayscale of E-paper could be maintained due to the EPD's bistability. A high-transparency TENG with high output current was prepared for the integrated device to optimize the performance. The driving current density directly influences the EPD's performance, such as contrast ratio and response time. After rectification, the output current could reach approximately seven μA, which was enough for driving the self-powered electronic paper (SPEP) device. In the research, the SPEP was still a single-pixel device without an external power management module, which limited its application. Yang et al. [46] supposed that a multi-pixel SPEP device with power storage and a

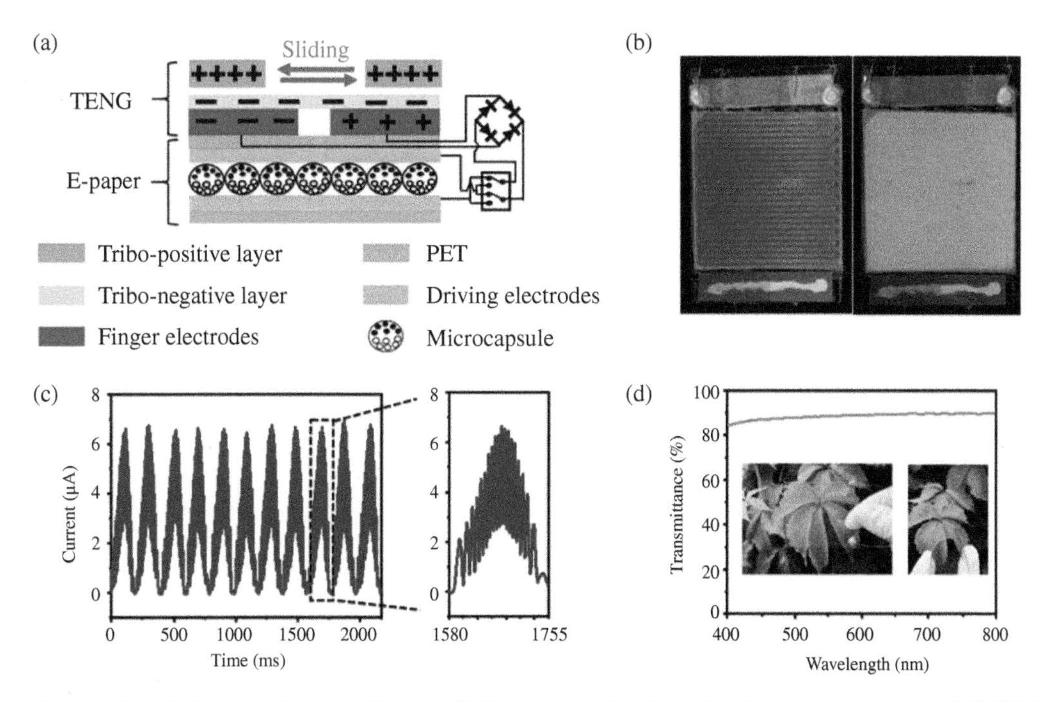

Figure 3.26 Self-powered electronic paper (SPEP) integrated with a triboelectric nanogenerator (TENG) [46] Gu, Yifan, et.al, 2020 / Optica Publishing Group / Public Domain. (a) The structure of the SPEP. (b) Chromatic-type SPEP showing different grayscales. (c) Rectified short-circuit current of the transparent TENG. (d) Transmittance versus wavelength.

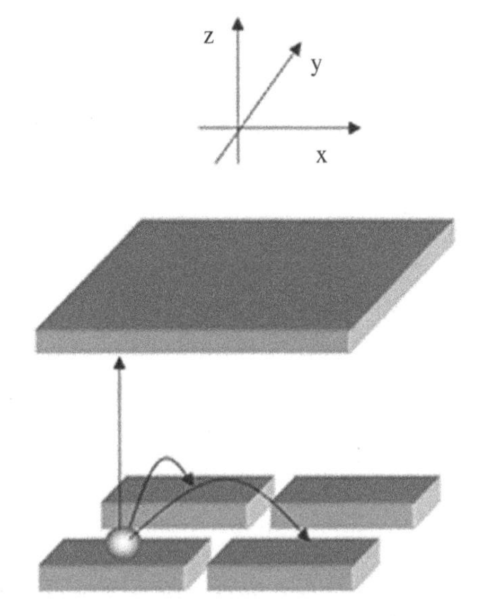

Figure 3.27 Illustration of the three-dimensional driving method With Adapted from [53].

management module would have great application value in the future while admitting that a mature power management module was necessary for it. As a possible solution, Mendis et al. [52] provided an accumulative display updating (ADU) design for intermittently powered systems, which could be used in the management module for SPEP devices.

3.3.2 Three-Dimensional Driving

Flickering is always the most annoying phenomenon while using e-readers. However, it is inevitable as flicker is a consequence of the addressing steps that activate and reset the particle conditions, which alleviates ghost images. However, we can shuffle the particles to reset the image underneath without affecting the upper layer particle distribution if we adopt three-dimensional driving. Recently, the CLEARink company proposed separated chambers (upper and lower) to demonstrate fast-switching response EPDs, which is a good practice to realize such a new driving scheme. This driving scheme can also incorporate high-frequency di-electrophoresis to realize advanced applications showing transparent, B/W, and color states Figure 3.27.

References

1 Hopper, M.A. and Novotny, V. (1979). An electrophoretic display, its properties, model, and addressing. *IEEE Trans. Elect. Dev.* 26 (8): 1148–1152.

2 Kao, W.-C., Ye, J.-A., Lin, F.-S. et al. (2009). Configurable timing controller design for active matrix electrophoretic display. *IEEE Trans. Consum. Electron.* 55 (1): 1–5.

3 Kim, J.M., Kim, K., and Lee, S.W. (2014). Multilevel driving waveform for electrophoretic displays to improve grey levels and response characteristics. *Electron. Lett.* 50 (25): 1925–1927.

4 Yi, Z., Zeng, W., Ma, S. et al. (2021). Design of driving waveform based on a damping oscillation for optimizing red saturation in three-color electrophoretic displays. *Micromachines* 12 (2): 162.

5 Yang, B.-R., Wen-Jie, H., Zeng, Z. et al. (2021). Understanding the mechanisms of electronic ink operation. *J. Soc. Inf. Disp.* 29 (1): 38–46.

6 Jin-Xin, C., Qin, Z., Zeng, Z. et al. (2020). A convolutional neural network for ghost image recognition and waveform design of electrophoretic displays. *IEEE Trans. Consum. Electron.* 66 (4): 356–365.

7 Kao, W.-C., Chang, W.-T., and Ye, J.-A. (2012). Driving waveform design based on response latency analysis of electrophoretic displays. *J. Disp. Technol.* 8 (10): 596–601.

8 Wong, J., Chen, Y., Sprague, R., and Zang, H. (2013). Driving method for bistable display. US Patent 8,46,210,2B2, June 11 2013.

9 Lin, C. and Yang, B. (2016). Driving method for electrophoretic displays with different color states. US Patent 9,299,294, 29 March 2016.

10 Zehner, R., Amundson, K., Knaian, A. et al. (2003). Drive waveforms for active matrix electrophoretic displays. *SID Symp. Dig. Tech.* 34 (1): 842–845.

11 Bert, T. and Smet, H.D. (2003). The microscopic physics of electronic paper revealed. *Displays* 24 (3): 103–110.

12 Bert, T., Smet, H.D., Beunis, F., and Neyts, K. (2006). Complete electrical and optical simulation of electronic paper. *Displays* 27 (2): 50–55.

13 Kao, W., Liu, J., Chu, M. et al. (2010). Photometric calibration for image enhancement of electrophoretic displays. IEEE International Symposium on Consumer Electronics (ISCE 2010). IEEE.

14 Kao, W.-C. (2010). Electrophoretic display controller integrated with real-time halftoning and partial region update. *J. Disp. Technol.* 6 (1): 36–44.

15 Kao, W.-C., Guan-Fan, W., Shih, Y.-L. et al. (2011). Design of real-time image processing engine for electrophoretic displays. *J. Disp. Technol.* 7 (10): 556–561.

16 Zehner, R. W., Gates, H. G., Amundson, K. R., et al. (2007). Methods for driving bistable electro-optic displays, and apparatus for use therein. US7312794.

17 Zehner, R. W., Gates, H. G., Amundson, K. R., et al. (2013). Methods for driving bistable electro-optic displays, and apparatus for use therein. US8558785.

18 Yang, B., Cheng, P., Hung, C, Wei, C., and Hsieh, Y. (2011). Driving method of electrophoretic display. US Patent Application 13/042,467, filed 15 September 2011.

19 Kao, W.-C. and Tsai, J.-C. (2018). Driving method of three-particle electrophoretic displays. *IEEE Trans. Electron Devices* 65 (3): 1023–1028.

20 Wang, M., Lin, C., Du, H. et al. (2014). 59.1: Invited paper: electrophoretic display platform comprising B, W, R particles. *SID Symp. Dig. Tech.* 45 (1): 857–860.

21 Inoue, S., Kawai, H., Kanbe, S. et al. (2002). High-resolution microencapsulated electrophoretic display (EPD) driven by poly-Si TFTs with four-level grayscale. *IEEE Trans. Electron Devices* 49 (9): 1532–1539.

22 Telfer, S.J. and McCreary, M.D. (2016). 42-4: Invited paper: a full-color electrophoretic display. *SID Symp. Dig. Tech. Pap.* 47 (1): 574–577.

23 Watson, A.B. (2013). A formula for the mean human optical modulation transfer function as a function of pupil size. *J. Vis.* 13 (6): 18–18.

24 Qin, Z., Wang, H.-I., Chen, Z.-Y. et al. (2020). Digital halftoning method with simultaneously optimized perceptual image quality and drive current for multi-tonal electrophoretic displays. *Appl. Opt.* 59 (1): 201–209.

25 Baqai, F.A., Lee, J.-H., Ufuk Agar, A., and Allebach, J.P. (2005). Digital color halftoning. *IEEE Signal Process. Mag.* 22 (1): 87–96.

26 Bayer, B.E. (1973). An optimum method for two-level rendition of continuous tone pictures. *IEEE Int. Conf. Commun.* 26: 11–15.

27 Floyd, R.W. (1976). An adaptive algorithm for spatial grayscale. *Proc. Soc. Inf. Disp.* 17: 75–77.

28 Analoui, M. and Allebach, J.P. (1992). Model-based halftoning using direct binary search. In: *Human Vision, Visual Processing, and Digital Display III*, vol. 1666, 96–108. International Society for Optics and Photonics.

29 Kao, W.-C., Ye, J.-A., Chu, M.-I., and Chung-Yen, S. (2009). Image quality improvement for electrophoretic displays by combining contrast enhancement and halftoning techniques. *IEEE Trans. Consum. Electron.* 55 (1): 15–19.

30 Kao, W., Chen, H., Liu, Y. et al. (2012). Real-time engine for paintbrush function on electronic papers. The 1st IEEE Global Conference on Consumer Electronics 2012. IEEE.

31 Kao, W.-C., Chen, H.-Y., Liu, Y.-H., and Liou, S.-C. (2014). Hardware engine for supporting gray-tone paintbrush function on electrophoretic papers. *J. Disp. Technol.* 10 (2): 138–145.

32 Kite, T.D., Evans, B.L., and Bovik, A.C. (2000). Modeling and quality assessment of halftoning by error diffusion. *IEEE Trans. Image Proc.* 9 (5): 909–922.

33 Kao, W.-C., Liu, J.-J., and Ming-I. Chu. (2010). Integrating photometric calibration with adaptive image halftoning for electrophoretic displays. *J. Disp. Technol.* 6 (12): 625–632.

34 Kao, W., and Liu, C. (2014). Real-time video signal processor for electrophoretic displays. 2014 IEEE International Conference on Consumer Electronics-Taiwan. IEEE.

35 Kao, W. and Liu, C. (2015). Driving waveform design for playing animations on electronic papers. 2015 IEEE International Conference on Consumer Electronics (ICCE). IEEE.

36 Kao, W.-C., Liu, C.-H., Liou, S.-C. et al. (2016). Towards video display on electronic papers. *J. Disp. Technol.* 12 (2): 129–135.

37 Jarvis, J.F., Ni Judice, C., and Ninke, W.H. (1976). A survey of techniques for the display of continuous tone pictures on bilevel displays. *Comput. Graph. Image Proc.* 5 (1): 13–40.

38 Liao, J.-R. (2015). Theoretical bounds of direct binary search halftoning. *IEEE Trans. Image Proc.* 24 (11): 3478–3487.

39 Kim, S.H. and Allebach, J.P. (2002). Impact of HVS models on model-based halftoning. *IEEE Trans. Image Proc.* 11 (3): 258–269.

40 Wang, H.-I., Qin, Z., Tien, P.-L. et al. (2019). 29-2: simultaneous optimization of Halftoning image quality and TFT power consumption for multi-tonal E-papers. *SID Symp. Dig. Tech.* 50 (1): 402–405.

41 Mantiuk, R., Daly, S., and Kerofsky, L. (2008). Display adaptive tone mapping. ACM SIGGRAPH 2008 papers. 1–10.

42 Kerofsky, L. and Daly, S. (2006). Brightness preservation for LCD backlight dimming. *J. Soc. Inf. Disp.* 14 (12): 1111–1118.

43 Yang, B.-R., Wang, Y.-C., and Wang, L. (2017). The design considerations for full-color e-paper. In: *Advances in Display Technologies VII*, vol. 1012602 (ed. L.-C. Chien, T.-H. Yoon and S.-D. Lee), 1–11. International Society for Optics and Photonics.

44 Qin, Z. et al. (2018). Ambient-light-adaptive image quality enhancement for full-color e-paper displays using a saturation-based tone-mapping method. *J. Soc. Inf. Disp.* 26 (3): 153–163.

45 Chen, Y.-W. et al. (2018). 48-3: distinguished student paper: ambient-light-adaptive image quality enhancement for full-color E-paper displays using saturation-based tone-mapping method. *SID Symp. Dig. Tech. Pap.* 49 (1): 153–163.

46 Gu, Y., Hou, T., Chen, P. et al. (2020). Self-powered electronic paper with energy supplies and information inputs solely from mechanical motions. *Photon. Res.* 8 (9): 1496–1505.

47 Fan, F.-R., Tian, Z.-Q., and Wang, Z.L. (2012). Flexible triboelectric generator. *Nano Ener.* 1 (2): 328–334.

48 Wang, Z.L. (2013). Triboelectric nanogenerators as new energy technology for self-powered systems and as active mechanical and chemical sensors. *ACS Nano* 7 (11): 9533–9557.

49 Zi, Y., Guo, H., Wen, Z. et al. (2016). Harvesting low-frequency (< 5 Hz) irregular mechanical energy: a possible killer application of triboelectric nanogenerator. *ACS Nano* 10 (4): 4797–4805.

50 Xiong, J., Cui, P., Chen, X. et al. (2018). Skin-touch-actuated textile-based triboelectric nanogenerator with black phosphorus for durable biomechanical energy harvesting. *Nat. Commun.* 9 (1): 1–9.

51 Liu, W., Wang, Z., Gao, W. et al. (2019). Integrated charge excitation triboelectric nanogenerator. *Nat. Commun.* 10 (1): 1–9.

52 Mendis, H.R. and Hsiu, P.-C. (2019). Accumulative display updating for intermittent systems. *ACM Trans. Embed. Comput. Syst. (TECS)* 18 (5s): 1–22.

53 Yang, B. R., Liu, Y. T., Wei, C. A., and Lin, C. (2014). Three dimensional driving scheme for electrophoretic display devices. US8681191 B2.

54 Zhou X, et. al., Lossless and Efficient Polynomial-Based Secret Image Sharing with Reduced Shadow Size. Symmetry. 2018; 10(7):249. MDPI. Licensed under CC BY 4.0.

4

Fast-Switching Mode with CLEARInk Structure

Robert J. Fleming, PhD

CTO CLEARink Displays, San Jose, CA, USA

4.1 Introduction

Ideally, reflective displays offer the same features as traditional emissive displays but with lower power usage, excellent outdoor readability, better viewing angle performance, and a paper-like experience that reduces eye strain. Researchers for decades have been developing reflective display technologies to meet as many of these features as possible. Reflective displays modulate the light that impacts the surface of the display from the surrounding environment. One way to broadly categorize reflective display technologies is through the optical mechanism by which the display returns the ambient light to the user. This classification leads to three general categories of reflective displays, diffuse reflective, specular reflective, and retroreflective (Figure 4.1).

White paper is a Lambertian or diffuse reflector, meaning light that impacts the paper's surface is reflected uniformly in all directions. E Ink's (E Ink Corporation/E Ink Holdings – Boston/Taiwan) electrophoretic display technology is also a diffuse reflector and can exhibit a paper-like viewing experience (Figure 4.1a. In a clear fluid, the E Ink technology utilizes negatively charged white particles and positively charged black particles. [1]. The light is modulated by applying an electric field and preferentially moving either the white particles or the black particles to the surface to selectively reflect or absorb the ambient light impinging the front surface of the display [1]. The E Ink technology can provide a very nice paper-like viewing experience (as seen in the Amazon Kindle), but the limited reflected brightness brings challenges we will discuss below in the section about color [2].

Specular reflection is when modulated light returns at the same angle as the light source, Figure 4.1b. Multiple display technologies can be classified as "specular reflective" type displays, and examples are reflective LCD [3], electro-wetting type, MEMs type [4, 5], and phase-change type [6]. These types of displays generally employ a metallic reflector. What is important to note about specular reflective-type displays is that light is not returned uniformly in all directions but preferentially at the same angle as the incoming light source. This results in "optical gain" meaning, more light can be returned at certain lighting and viewing directions than the Lambertian type. However, there are disadvantage, the optical gain peak can be very close to the glare angle (reflection off the front surface, which can be particularly problematic in direct sunlight), and viewing angle performance can be significantly compromised. In the retro-reflective type display [7, 8] (Figure 4.1c), light is preferentially returned to its source by a principle referred to as retroreflection, to be discussed in detail below. Retroreflective displays also result in optical gain but the opposite direction as specular reflective-type displays; this can be advantageous because the optical gain peak is in the opposite direction as the glare peak.

E-Paper Displays, First Edition. Edited by Bo-Ru Yang.

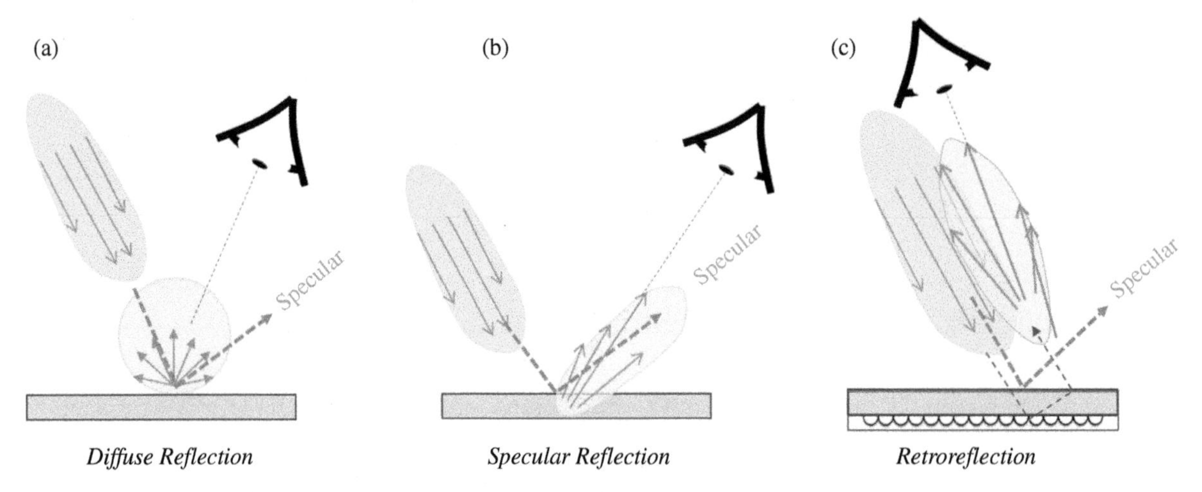

Figure 4.1 Types of reflective displays, (a) diffuse reflective, (b) specular reflective, and (c) retroreflective. The incoming light is shown coming in from the top left for all cases for reference. The direction of light returned by the display is shown by the light-yellow bubble. The ideal viewing condition is shown by the location of the eye cross-section.

The core of the CLEARink Displays' technology is based on two fundamental concepts in optics. The first is the phenomenon known as Total Internal Reflection (TIR) [9, 10], the second is a clever use of geometrical optics known as *retroreflection.*

When a light ray impacts the surface of a substantially uniform medium, the light ray is refracted (transmitted) and/or reflected (given the optical structure is much larger than the wavelength of interest) [9]. In the 1920s it was discovered (likely by observation, hence the nickname "cat's eyes") that glass spheres, when attached to information signs, improved their visibility at night. In the 1930s the 3M Corporation developed the first glass bead type retroreflectors used in traffic signs [11]. Examples of two types of simple retroreflectors are shown in Figure 4.2. These simple optical engines return a light ray substantially back to its source, i.e., *retroreflected.* This is a highly advantageous property for viewing traffic signs at night because the light from the car's headlights hits the surface of the traffic sign and is preferentially returned to the eyes of the car's driver. Even in 2020, 3M has substantial businesses, offering more than 30 products based on retroreflection. Examples of applications are traffic signage, personal safety (worn on the person for visibility), commercial graphics, highway pavement markings, etc.

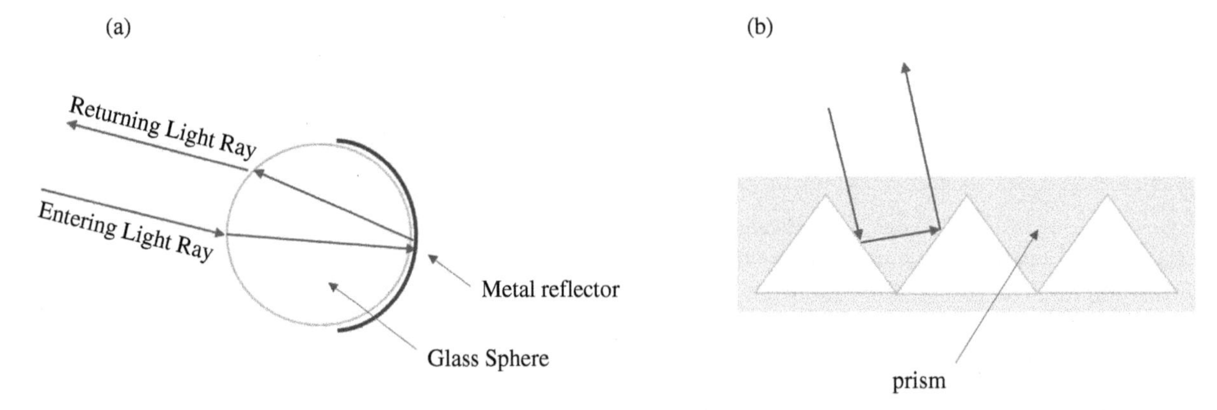

Figure 4.2 Two types of simple retroreflectors, (a) spherical glass bead type, (b) prism type. In (a) a light ray enters a high refractive index glass sphere with a curved metalized reflector placed at the focal point behind the glass sphere. In (b) a light ray enters the prism of high refractive index and reflects at the low refractive index boundaries (white triangle region).

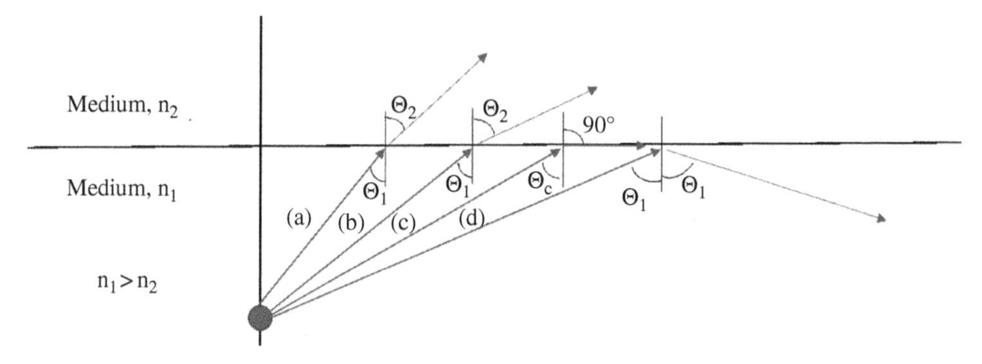

Figure 4.3 Description of Snell's law from rays (a)–(d) emitting from a light source (blue dot) in high refractive index medium n_1 and impacting the surface of lower refractive index medium n_2 with adapted from [9].

Retroreflectors developed by 3M can return more than 60% of the light to its source [12]. In a display application, particularly of the reflective display type, it would be very beneficial if a significant fraction of the available light could be preferentially re-directed to the viewer's eyes. This was the original idea behind the CLEARink Display approach to making a reflective display brighter [13]. However, there was a problem, how could a retroreflector be selectively "turned off" to make a digitally addressable display? The answer to this question lies within nuanced features of what is known as TIR.

When light passes from a high refractive index medium to a lower and the boundary between the two media is sharp (substantially discontinuous), the speed of light increases, and the light ray is refracted (changes angle). Snell's Law conveniently describes the relationship between the refractive indices of the two media and their angles of refraction. Snell's law is commonly written in the form $n_1 \sin Q_1 = n_2 \sin Q_2$ [9]. Figure 4.3 below shows light rays; (a), (b), (c), and (d) emanating from a point in a high refractive index medium, n_1, and impacting a low refractive index medium, n_2. Because n_1 is larger than n_2, Snell's law requires that rays (a) and (b) bend or refract away from the normal. A critical angle of incidence is reached when the angle of refraction reaches 90°. This is designated at Q_c. The critical angle is critical for two reasons; (i) because all incident angles $Q_1 > Q_c$ undergo *TIR* (i.e., the light ray is reflected) as shown by ray (d), and (ii) because many interesting phenomena happen at and very near the critical angle (this will be discussed in more detail below).

But how can light traveling through a high refractive index medium impact a surface of low refractive index at an angle greater than the critical angle and *not* undergo TIR? How can TIR be inhibited or "frustrated"? A closer investigation of phenomena near the critical angle reveals very complex behavior. Returning to Figure 4.3, a complete description would show the rays as plane waves in the form of $\mathbf{E}_i = \mathbf{E}_{0i} \exp [i(\mathbf{k}_i \cdot \mathbf{r} - \omega_i t)]$. When the transmitted and reflected waves are derived using the Fresnel Equations, a surface wave propagates along the plane between the two media [14]. This surface wave is known as the *evanescent wave*. It decays rapidly in the direction of the low index medium, the rate of decay is determined by the wavelength of light, refractive indices of the two media, and the angle of incidence. What is most important to note is that this surface wave *does* penetrate the low index medium, making it possible for energy to be transferred across a gap in what is known as *frustrated total internal reflection* (fTIR) [14]. The above description is based on the assumption of a system as idealized boundaries and infinite plane waves, whereas natural systems are much more complicated. In natural systems, the boundary between two interfaces is not perfectly sharp, "boundaries" are made of materials comprising molecules and atoms. The real interface between the low and high index materials is a medium with different optical properties than either layer. Consequently, the critical angle in most natural systems is far from perfectly sharp [10]. However, with a proper fundamental understanding of the material and interfacial properties, the critical angle can be "engineered" to be sharp or broad depending on the desired result of the particular system.

4.2 CLEARink Display Optics

The original concept for a retroreflective type display was first developed by the University of British Columbia, Professor Lorne Whitehead. The principles of TIR and retroreflection were well known because of his previous work with 3M on Brightness Enhancement Films for liquid crystalline displays and prismatic type retroreflectors for traffic safety signs. Combining these observations, knowledge, and experience, he theorized that by selectively frustrating the TIR of a retroreflector, a "retro-reflective type" reflective display could be made. He filed a patent on this invention in 1997 [13]. Coincidentally, 1997 was the same year the E Ink Corporation was spun out of the MIT Media Lab.

The original invention for a retroreflective type display was to frustrate TIR in a prismatic type retroreflector by selectively contacting the surface undergoing reflection with a silicone rubber filled with light-absorbing particles [13]; thereby "frustrating" or inhibiting TIR. In this case, the low refractive index medium would be air. Light modulation were by the physical application of force to contact the filled silicone rubber to the reflecting surface. The next iteration of the original invention was to suspend charged absorptive particles in a low refractive index fluid. By applying an electrostatic force, the particles could be moved, by electrophoresis, into and away from the reflecting surface [15]. High reflectivity with good contrast ratios was achieved using the prism type retroreflectors, but the angular performance remained poor.

The next iteration on the retroreflective type display was to employ hemispherical shaped micro-lenses in an array. This incarnation became the basis for the modern CLEARink type display. This invention by Professor Lorne Whitehead developed from his observation of hemispherical water droplets on a transparent polyethylene bag one wet Vancouver day. Prof. Whitehead and then Ph.D. student Dr. Michele Mossman, were working on the prismatic retroreflective type display and together built the first prototype demonstrating this display concept. The micro-lens array approach dramatically improved the angular performance. However, at angles near the retroreflection peak, the reflected brightness was lower than what could be achieved with prism-type retroreflectors. Monte Carlo ray-trace simulations of different light paths entering a single hemispherical microstructured lens (Figure 4.4) yielded several insights on the optics of lens-type retroreflective display. The first insight was that the light was not truly retroreflected. The light rays rarely returned directly to the source. Instead, the retroreflected light returned broadly back to the source. As a result, it is more accurate to classify the CLEARink type display as a *semi-retroreflective* display. The *semi-retroreflective* properties of the optical structures result in a paper-like viewing experience with good angular performance. The second major insight was that light entering the center portion of the lens impacts the back surface of the lens at an angle less than the critical angle, Q_c, and therefore does not reflect but passes straight through; these paths can be observed in Figure 4.4. The light passing through the lens at angles less than the critical angle became referred to as "the dark pupil" effect. The dark pupil effect was initially a major obstacle to achieving high brightness..

The semi-retroreflective properties of the microlens array provide a significant advantage over current reflective display technologies. When using a reflective display, the light source is generally fixed over the viewer, shining

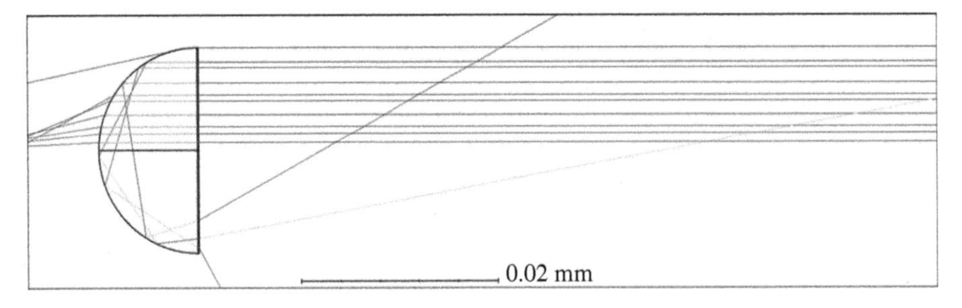

Figure 4.4 Monte Carlo simulation of 10 light rays entering a hemisphere type lens structure.

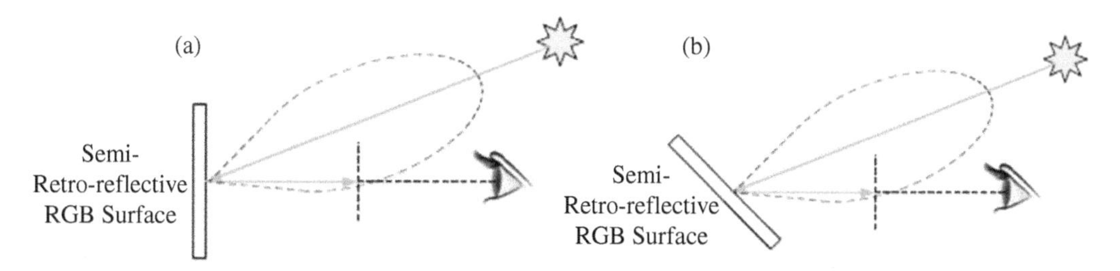

Figure 4.5 Simple diagram showing optical gain of a semi-retroreflective display at tilt angles (a) perpendicular to the viewer, and (b) 45° tilt away from the viewer.

down on display. A diffuse white surface appears white when viewed from straight on, at an angle, and moved around the display. However, the white appearance is reduced to gray when colored filters are added because of the loss in brightness. The apparent brightness can be regained by introducing optical gain. The disadvantage of semi-specular gain is that the gain is lost if the angle of the display is changed. The semi-retro-reflective gain approach works well for many display angles, as shown in Figure 4.5. The dotted oval in Figure 4.5 represents the semi-retroreflection region of high optical gain, which changes very little from Figure 4.5a,b as the sample is rotated. The viewing brightness is similar between Figure 4.5a,b. Detailed analysis on viewing angle performance will be provided in the following sections.

CLEARink Displays was spun out of the University of British Columbia in 2012 by the co-founders, Professor Lorne Whitehead, the original inventor, and Frank Christiaens. A small team in Vancouver, Canada, was formed to continue the development of the CLEARink technology. In parallel, Merck KGaA developed electrophoretic ink systems [16, 17], and in 2012 CLEARink began testing Merck electrophoretic ink formulations. The first active-matrix prototype using a Merck developed single-particle ink and a CLEARink hemisphere microstructured array demonstrating fTIR switching was reported in 2015 [18]. Then in 2015, the CLEARink Displays headquarters was moved to Silicon Valley, first in Sunnyvale, then to Fremont, California. In 2016 a materials development R&D lab, cleanroom, microreplication lab, and display fabrication developmental processes were brought online. In 2017 CLEARink entered a materials development agreement with Merck KGaA to commercialize the CLEARink Display technology. In May 2017, the small CLEARink team led by Dr. Robert Fleming, CTO, exhibited its prototype displays at the Society for Information Displays annual DisplayWeek trade show and won the Best in Show Award. The company won several other awards, including the IDTechEx award (Figure 4.6).

Figure 4.6 CLEARink Displays' Dr. Robert Fleming, CTO, and Dr. Joel Pollack, Board Member, accepting the 2017 IDTechEx award. *Source:* Photo courtesy of Sri Peruvemba.

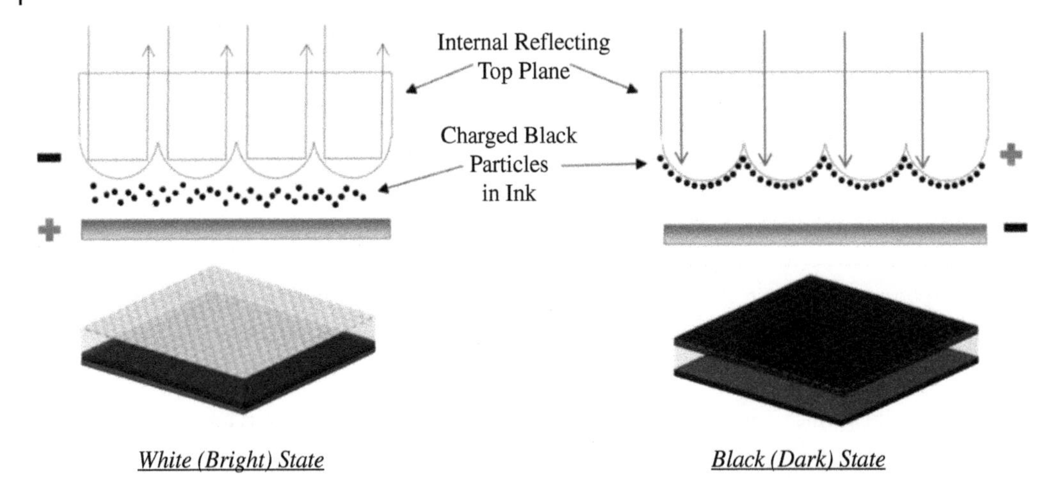

Figure 4.7 General description of semi-retroreflective display using sub-micron black particles to frustrate total internal reflection.

The first reported display brightness by CLEARink, using a hemisphere microstructured array with a refractive index of approximately 1.6 and a non-aqueous colloidal dispersion ink system, gave a reflected brightness of 63% relative to a Lambertian white standard [8, 18]. The hemisphere microstructure is the CLEARink Gen 1 lens design. Monte Carlo ray-trace simulations, found that the resulting optical gain could be modified by changing the shape of the lens (Figure 4.7).

The shape of the lens modifies the light path through the lens, the number of bounces inside the lens before the light is returned, and the overall angular distribution of light returned. Lenses could be designed to return a very narrow angular distribution of light but with higher brightness or broader angular distributions with reflected brightness approaching that of a Lambertian reflector. CLEARink began experimenting with different lens designs, first in ray-trace simulations, then in real fabricated microstructured lens array prototypes. CLEARink is currently on its third-generation microlens array design. The first-generation design lens was a 20 um diameter hemisphere. The second-generation design was a 20 um diameter round, non-hemispherical design exhibited at the 2017 Society for Information Displays annual DisplayWeek trade show and won CLEARink the Best in Show Award. The third-generation lens design is significantly smaller than the Gen 2 lens and is non-hemispherical. The Gen 3 design was debuted in the 2019 Society for Information Displays annual DisplayWeek trade show and won CLEARink the "People's Choice Award.." In the following sections on optical performance, the different lens designs will be referred to as Gen 1, Gen 2, and Gen 3.

There are nearly an infinite number of lenses that could be designed and built. The difficult problem was efficiently narrowing down the infinite possibilities to potential lens designs that would give the overall best user experience in a functioning display. To meet this objective, CLEARink first developed a simple ring light test for measuring reflected brightness. This test was an attempt to mimic the user experience [8]. Then CLEARink conducted laser light and collimated white light goniometer measurements to map the reflected light [8, 19] finally designs were validated with real-life user conditions [19].

To optimize the best lens design, CLEARink needed to develop a brightness measurement test that best represented the use case of the intended consumer. If we assume the display is a portable device being held or sitting on a table, the viewer would be looking approximately normal to the surface of the display. Then we assume the ambient light is semi-diffuse/semi-directional, shining down from behind the viewer's head. As a result, the viewer's head would be blocking approximately ±5° of the light impinging the surface of the display. The results of these assumptions became the foundation of the test, which CLEARink refers to as the "5° to 30° Ring Light Test" [8]. The resulting measured reflected brightness became referenced as the "5–30 brightness.." Figure 4.8 below shows the test setup. The sample is placed at a controlled distance from an annular Lambertian light source

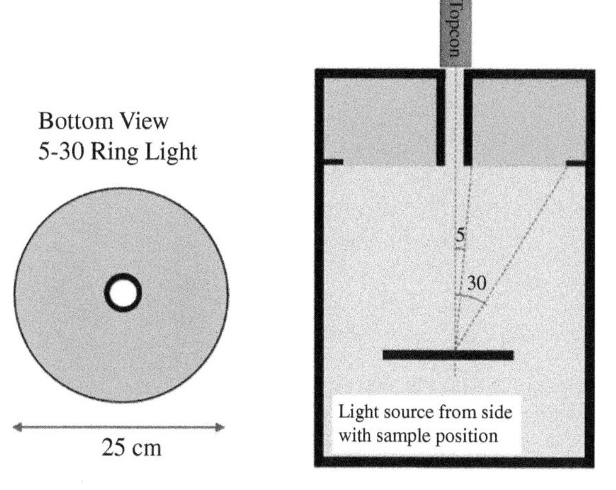

Figure 4.8 Description of the 5–30 reflected brightness test method.

Table 4.1 White and dark state reflectance measured using the 5° to 30° Ring Light Test referenced in Figure 4.8.

Sample	Reflectance white state	Reflectance black state	Contrast
First generation CLEARink Monochrome	60%	4%	1 : 15
Second generation CLEARink Monochrome	83%	4.2%	1 : 20
Epaper 1.0	48%	4%	1 : 12

White state reflectance values are referenced to a 98% Lambertian white standard and dark state values are referenced to a 2% black standard. Contrast is calculated by dividing White/Black reflectance.

with a viewing port through its center. The sample is illuminated with rays at angles ranging from 5° to 30° to the sample's normal. The reflection from the sample is measured through the central hole with a calibrated Topcon BM9 luminance meter equipped with a 2° acceptance angle optic. Specular reflection is masked by the ±5° angular selective tube. Light intensity is normalized to a 98% reflective Lambertian white standard from Labsphere: SRS-99-020. If the sample is more reflective than the standard in the direction of the detector (due to optical gain), then the relative reflectivity may exceed 100%.

Using this 5–30 brightness test, we could compare first-generation lens designs to second-generation lens designs, and compare to a Lambertian type reflective display [8], The results are shown in Table 4.1. The Gen 2 lens design s significantly improves measured reflective brightness without sacrificing dark state, thereby improving both brightness and contrast ratio. This is a very significant result because it indicates frustration of TIR is not significantly impacted by the shape of the lens. The brightness improvement with the Gen 2 lens design was significant enough that a color reflective display using a printed color filter array (CFA) became possible [7, 8]. A detailed discussion on color will follow. These results were encouraging but lacked details on the complete viewing angle performance picture.

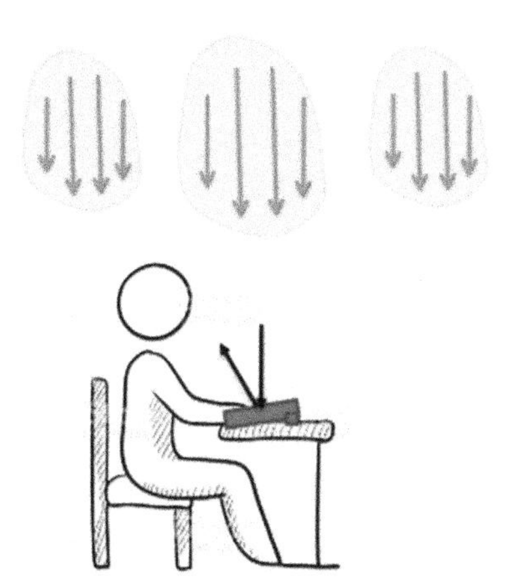

Figure 4.9 Conference room, office, or classroom use case study. Overhead semi-diffuse, low glare office lighting measuring 580 lux.

The team also validated the 5–30 test by measuring the luminance of printer paper, the CLEARink Gen 2 type display, and an E Ink type display. They approximated a "standard office" type lighting condition and viewing condition to mimic a tablet or eReader type display. The test conditions are shown in Figure 4.9 below, and the results are in Table 4.2. Please note the

Table 4.2 Measured white state luminance and dark state luminance from standard printer paper, a CLEARink Gen 2 type display with an unoptimized diffuser and an E Ink type eReader.

Test Conditions: Open Office, low glare, 580 lux	**Display Brightness (nits), ±15 deg optimal viewing**		
Display	White state	Dark state	CR
White printer paper/Black print	145	9.98	14.5
CLEARink Monochrome w/Diffuser	170	16	10.6
eReader	94	9	10.4

The luminance measurement at the location of the "viewer" as shown in Figure 4.9.

dark state in the CLEARink display is very poor because, at this time, a diffuser not optimized for the CLEARink technology was used (resulting in high surface scatter comprising the dark state).

The results in Table 4.2 generally correlate with the results from the 5° to 30° Ring Light Test and further increased confidence in this test. The CLEARink Gen 2 brightness is almost 2× that of the Lambertian E Ink type display. The CLEARink display is also brighter than the standard printer paper, not a surprising result because printer paper does not reflect as much as the 98% white reflectance standard used as the reference in Table 4.1.

The 5° to 30° Ring Light Test is simple and reduces the brightness metric to a single value but clearly oversimplifies the total display performance. A goniometric test was devised to map the light returned across a wide range of discreet angles. A sample on a rotating stage was illuminated by a green laser that was fixed in place as shown in Figure 4.10a. The photodetector was fixed in the same plane as the sample but could be rotated independent of the rotation of the sample, also shown in Figure 4.10a. The angle of the sample to the laser is the incidence angle, 0° being where the sample is normal to the laser. The viewing angle is the angle of the detector to the sample, 0° being where the sample is normal to the detector (Figure 4.10b,c).

It can be observed from (Figure 4.10b) the optical gain at certain angles derived from TIR and semi-retroreflection off the Gen 2 type microlens array. From 0° to 20° viewing and incidence angles, the reflection is almost a factor of 2 greater than the E Ink Lambertian reflection type display. The optical gain is still present up to 60° viewing and incidence angles when rotated together, maintaining the semi-retroreflection viewing condition. When the incidence angle is opposite the viewing angle, the reflection drops off. Ergonomically, this may not be as bad as it appears because the viewing conditions where the reflection drops off the fastest is also the viewing condition near the glare peak. The glare peak is the reflection off the top surface of the display.

In addition to measuring bright state reflection of the displays, we also measured the dark state reflection. Using the same goniometric setup as described in Figure 4.10a, we measured the white state reflection and the dark state reflection and plotted the ratio of white state and dark state reflection versus incidence and viewing angle. A CLEARink Gen 2 lens type display was measured and compared to reflective LCD type Memory in Pixel (MIP) display (Figure 4.11). An E Ink type display was also measured but is not reported here. As observed from (Figure 4.10c), the performance was pretty uniform across a broad range of angles (except for the glare angle).

As would be expected from (Figure 4.10b), the CLEARink Display has very good contrast along the retroreflection direction to 80°. What is important to observe, and was not expected, is that the contrast ratio even well outside of the retroreflection conditions and does not fall to 1 as in the case of the reflective LCD. The black regions in Figure 4.11 are where the contrast ratio is less than two as measured using the green laser (but corresponds to where the displays lose contrast under direct white light viewing). This phenomenon was observed in live demos and images of viewing angle performance of the content on active matrix displays [7, 8] as shown in Figure 4.12 below. The results are shown in Figure 4.12b validate the superior optical performance of a semi-retroreflective type display.

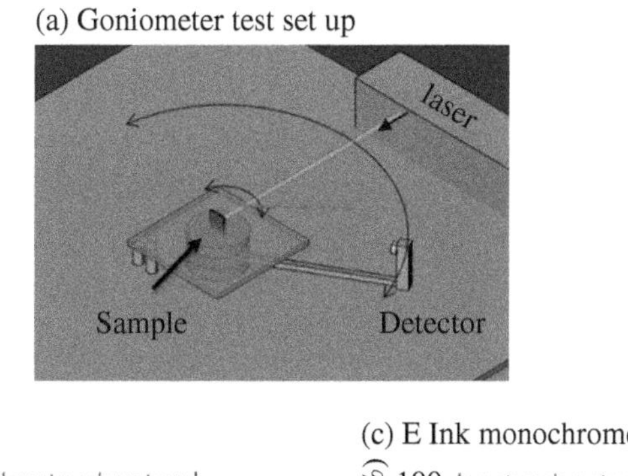

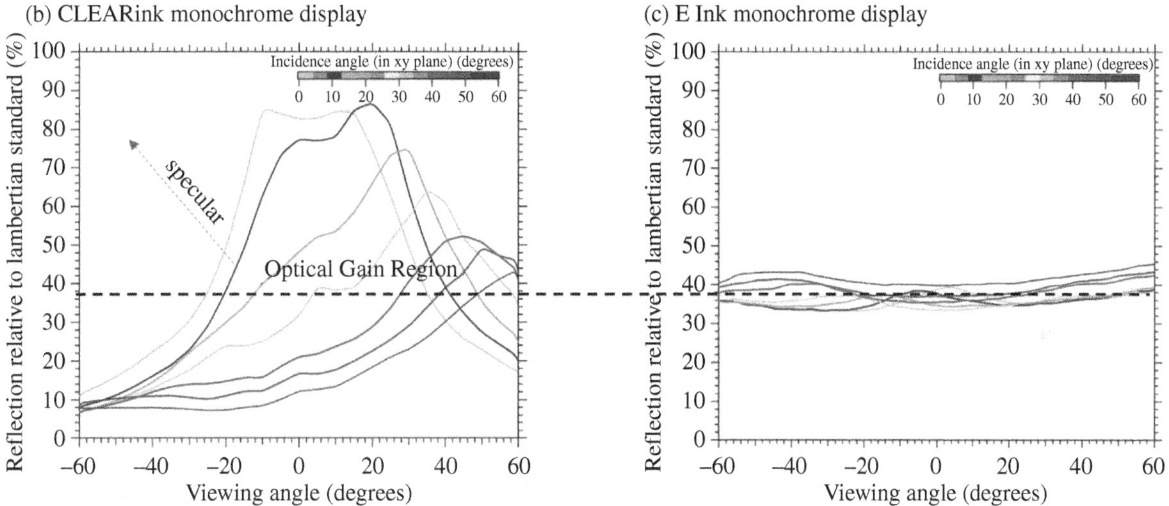

Figure 4.10 The goniometer measurement set up (a), the measured normalized reflection versus incidence angle and viewing angle of a Gen 2 type microlens array device (b). For reference an E Ink type Lambertian monochrome display (c).

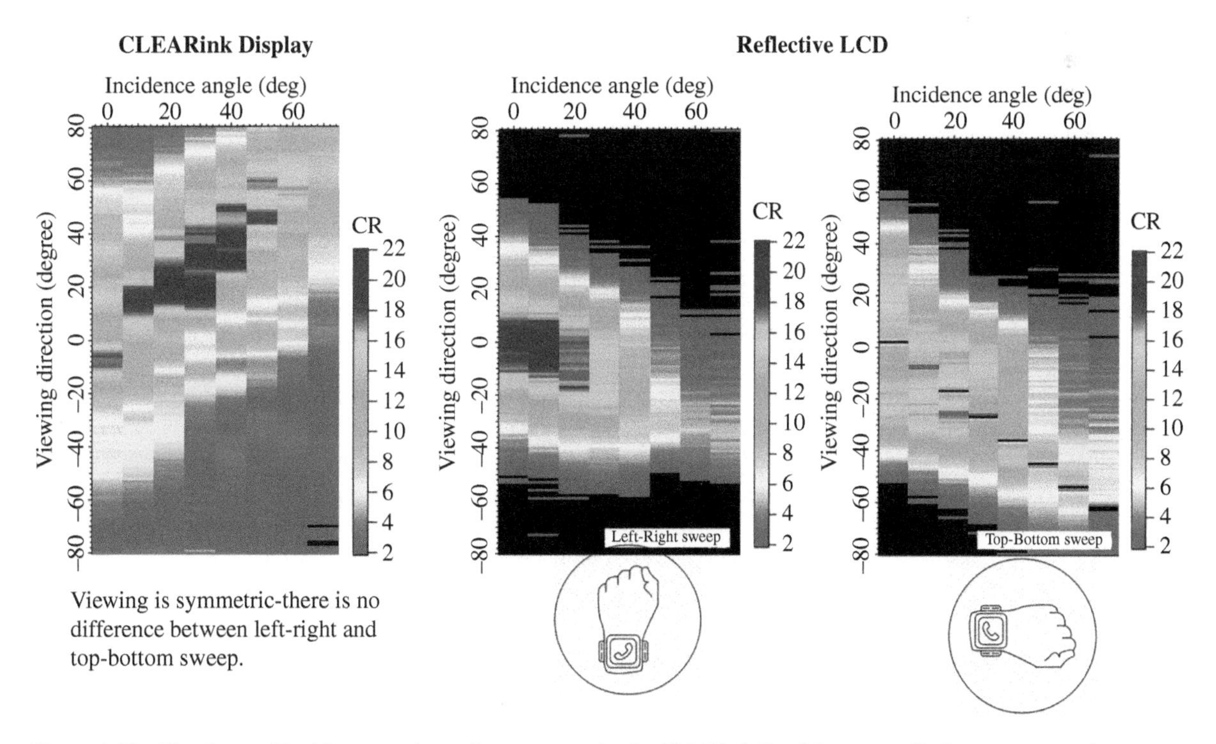

Figure 4.11 Viewing and incidence angle performance study of a CLEARink Gen 2 lens type display as compared to a reflective LCD type display. The black regions in the charts are where the contrast ratio is less than two.

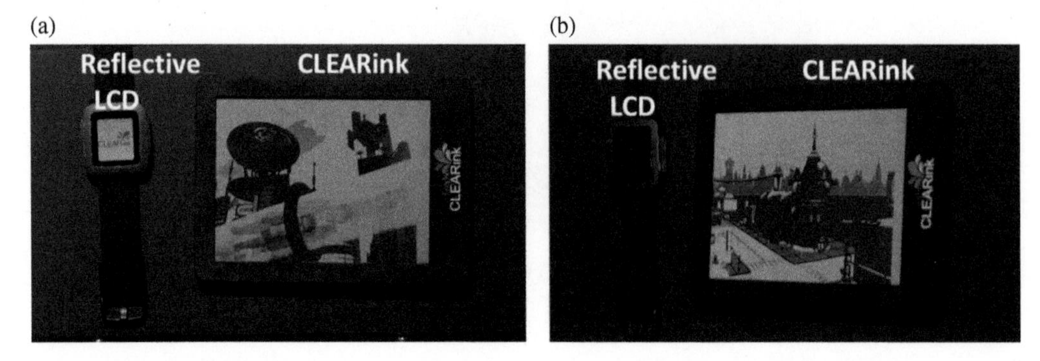

Figure 4.12 Display performance of a CLEARink semi-retroreflective display versus a reflective LCD type display. Figure (a) shows the performance with an illumination angle of (0, 20) degrees and a viewing angle of (0, 0) degrees; (b) shows the performance with an illumination angle of (40, 20) degrees and a viewing angle of (−40, 0) degrees. Note in figure (b) the reflective LCD is on but looks off, CLEARink Displays.

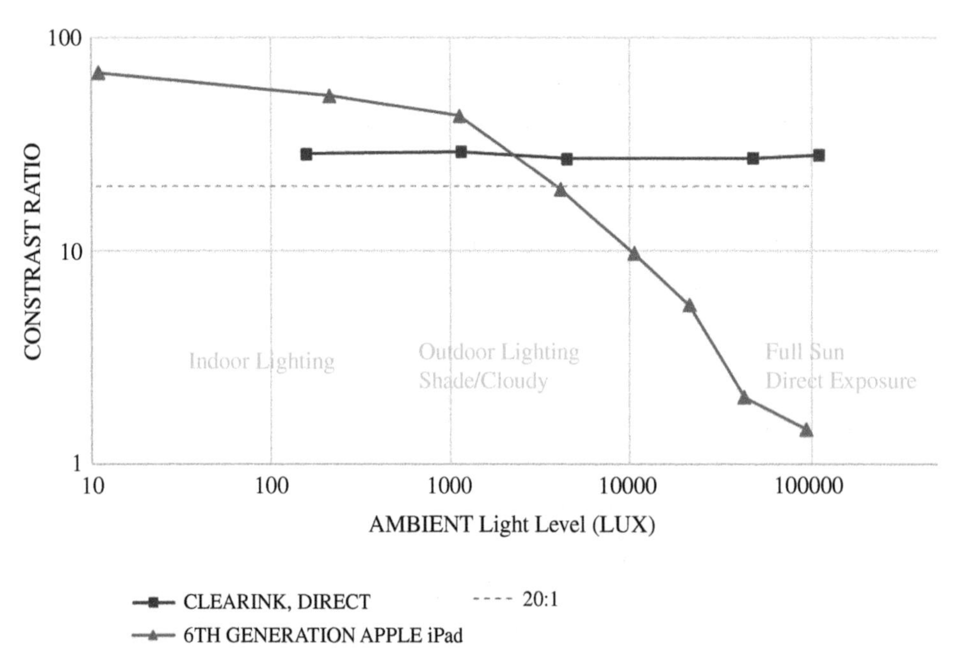

Figure 4.13 Measured contrast ratio of a CLEARink type display and an Apple iPad versus a broad range of ambient lighting conditions.

The ability to read and view content outdoors is one of the most compelling use cases for reflective displays. We conducted a study where we illuminated the surface of various displays with semi-diffuse/semi-collimated lighting and measured the white state and dark state display luminance under viewing conditions approximating the outdoor reading use case. The details of the test method and the effect on collimated versus diffuse light are reported elsewhere [8]. We tested a CLEARink Gen 3 type display and a 6th generation Apple iPad (FFS Technology "Retina" LCD) over a broad range of illuminance levels. We reported contrast ratio versus light level in Figure 4.13.

As would be expected—because we have all experienced this effect—the iPad has very good contrast in low ambient light levels, but under high ambient light, the contrast drops dramatically. This is typical for an emissive

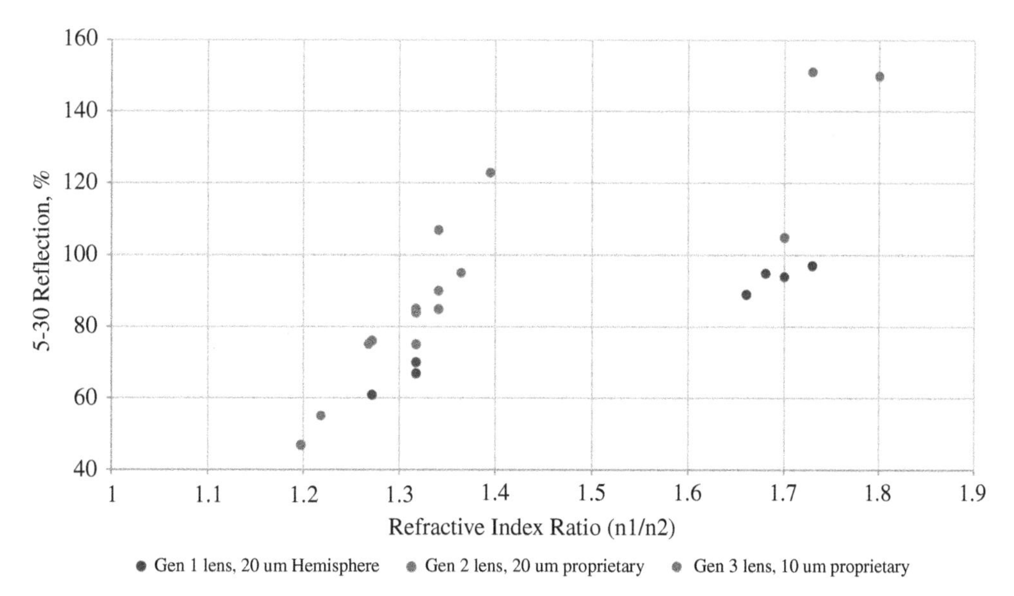

Figure 4.14 The measured reflection normalized to a Lambertian white standard as described by the 5° to 30° Ring Light Test method plotted against the ratio of the refractive index of the lens (n_1) to the electrophoretic fluid medium (n_2) for three generations of CLEARink lens designs.

type display because the dark state is very low under low lighting conditions, so modest levels of the backlight are needed to produce extremely high contrast ratios. As the ambient light increases, the dark state brightness increases proportionally, so higher and higher backlight levels are required to maintain contrast. As the ambient light increases, the backlight must also increase, which translates to an increase in power consumption [7]. In contrast, the CLEARink display maintains the same contrast ratio overall measured lighting conditions. This is because the white state reflection increases with the increase in a dark state, maintaining contrast. Device designers need to take into consideration their use-cases and the desired display performance in those uses cases as well as the resulting power demands [7].

The CLEARink semi-retroreflective display type that employs electrophoretic particles causing frustration of the TIR is a viable approach to a reflective display. CLEARink is on its third,-generation lens design, but more optical gain is achievable. As discussed above, a higher optical gain can be achieved by modifying the shape of the lens. Still, the optical gain can also be increased by decreasing the critical angle of TIR resulting in a larger number of rays entering the microstructure that is reflected. As the refractive index of the lens (Figure 4.3, n_1) increases and the refractive index of the electrophoretic particle medium decreases (Figure 4.3, n_2), the reflection will increase (based on principles outlined in Figure 4.3). In Figure 4.14 below, we show what is possible in the future as the lens's refractive index increases and the fluid's refractive index decreases (expressed as the ratio n_1/n_2). The data in Figure 4.14 covers a range of lens refractive indices and electrophoretic medium. With proper lens design, material selection, and replication quality, displays with a reflection of greater than 100% are very achievable.

4.3 CLEARink Reflective Color Displays

The CLEARink Displays' approach to color is very simple in principle. The microstructured lens array is replicated directly onto a conventional CFA used in an emissive LC-type display. To produce the desired color in an image, the particles are driven away from the lenses to allow TIR through the desired color sub-pixel, for example,

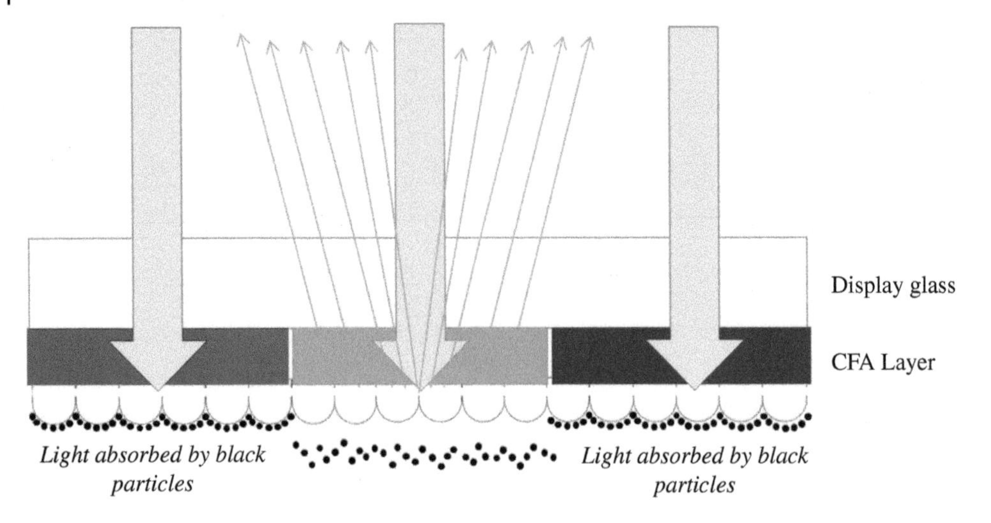

Figure 4.15 A general depiction of how color is generated using the CLEARink type display with a conventional RGB type color filter array.

the green sub-pixel in Figure 4.15. In the neighboring sub-pixels, the particles are driven toward the lenses to frustrate TIR and not allow light to reflect through the other color sub-pixels. When designing a CFA, the saturation of the colors is decreased in half to account for light traveling twice through the CFA layer, Figure 4.15. Conventional CFAs for emissive liquid crystal displays are typical of the RGB type, meaning all colors are derived from combinations of red, green, and blue subpixels. The color subpixels can take many shapes, sizes, and patterns; occasionally, a clear subpixel can be added to increase luminance. Adding a clear or "white" subpixel (not shown in Figure 4.15) can be beneficial to increase the luminance or "whiteness" of the display and give a more paper-like viewing experience.

The first CLEARink color demo was exhibited at SID Display Week in 2017 (Figure 4.16). This demo had a diagonal of 6 in., color resolution of 106 ppi, and a contrast ratio of about 8 : 1. This demo used Gen 2 type optical structures replicated directly onto an RGBW quad CFA design, as shown in Figure 4.17a,b. The quad layout was chosen because it matched the TFT available at the time. The decision was to use an RGBW design. After all, the goal was to achieve a paper-like experience and high reflected brightness. The demo is shown in Figure 4.16 below and has a brightness of 40% as referenced to a Lambertian white standard (5° to 30° Ring Light Test described previously).

Figure 4.16 Picture of CLEARink demo shown at the Society for Information Displays Display Week in 2017.

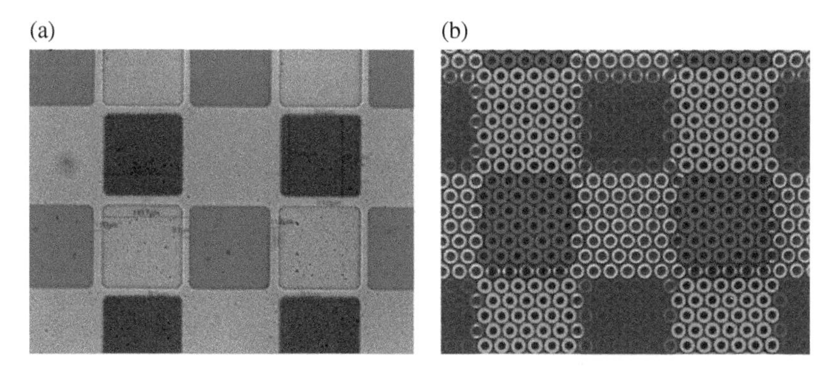

Figure 4.17 Optical micrographs of 2017 SID demo; (a) color filter array, and (b) Gen 2 type optical structures microreplicated onto the CFA.

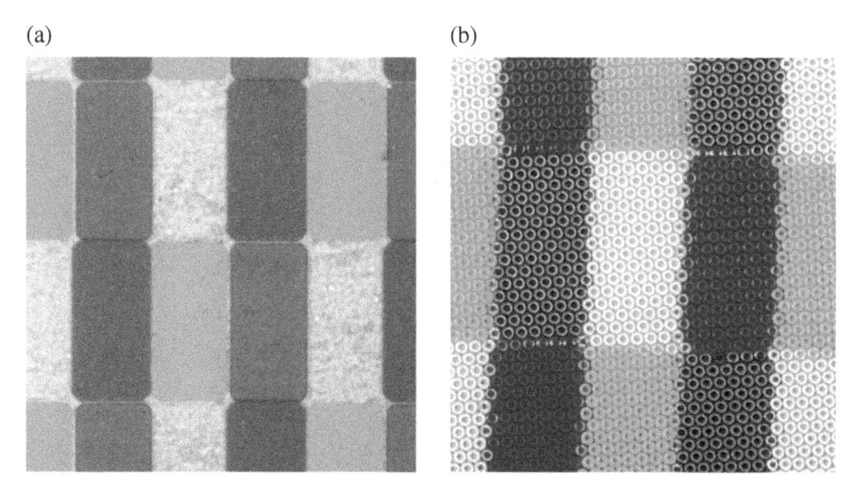

Figure 4.18 Optical micrographs of; (a) color filter array, and (b) Gen 3 type optical structures microreplicated onto the CFA.

The resolution in the 2017 demos was low, so the sub-pixel size was relatively large compared to the diameter of the lenses, as can be observed in Figure 4.17b. Some color mixing from lenses bordering the sub-pixels can be observed in Figure 4.17b, but a color gamut in the 10–15% range was achieved, as measured by %NTSC 1931 [8]. What is very important to note was that the reflected brightness was relatively high at around 40% (by the 5° to 30° Ring Light Test method), the combination of respectable color gamut, good white states, and high brightness resulted in a very nice paper-like display with full color and video capability.

Commercial products likely require higher resolution than 106 ppi, and a higher color gamut would be preferred. Going to higher resolution shrinks the pixel size. Since each microlens in the CLEARink display acts as an individual light source, it is critical to maintain the highest fill factor possible. For this reason, the decision was made to go with a PenTile™ L6W design [7, 20], and smaller microlenses were developed. The PenTile design (Figure 4.18a) has two sub-pixels per pixel and gives a 50% fill facto compared to a traditional RGB stripe with only 33%. This higher fill factor equates to higher brightness and larger sub-pixels for the equivalent effective resolution. Perhaps even more important than the improved CFA design, the Gen 3 optical structures were developed, and reduced the lens diameter. CFA and lens design improvements greatly improved the sub-pixel brightness and minimized optical cross-talk while doubling the resolution from the 2017 RGBW quad design. However, some optical cross-talk could still be observed, as shown in Figure 4.18b.

The PenTile CFA design and the Gen 3 optical structures were exhibited at the 2019 Society for Information Displays Display Week. Figure 4.19 show pictures of the demos exhibiting different content. The demos in Figure 4.19 had a 9.7″ diagonal and a resolution of 227 ppi. The color gamut was measured to be 19% NTSC xyY(1931), 17% sRGB u'v' (CIE1976), and 35% reflection (5° to 30° Ring Light Test method).

The actual CFA used in the 2019 9.7″ demo had small spaces between the color subpixels. This resulted in a whiter, more paper-like experience but compromised color gamut. The CFA shown in Figure 4.18a was built into a 1.3″ diagonal 202 ppi wearable demo and exhibited by CLEARink at SID Display Week 2018. The overall color gamut increased significantly; the color coordinates are shown in Figure 4.20. Red-Green-Blue color points were measured with an Ocean Optics Spectrophotometer, and the results are shown in sRGB and NTSC color spaces. In addition to changes in the CFA design, improvements were made in driving the subpixels, resulting in higher sub-pixel contrast. These improvements lead to a 30.7% NTSC xyY(1931) 29.5% sRGB u'v' (CIE1976) color gamut.

Improvements in color performance are certainly obtainable. Implementing higher brightness optical structures (Figure 4.14), improvements in CFA design, and improving the sub-pixel contrast ratio will enhance the

Figure 4.19 Pictures of CLEARink Display's demos exhibited at the 2019 Society for Information Displays Display Week Conference.

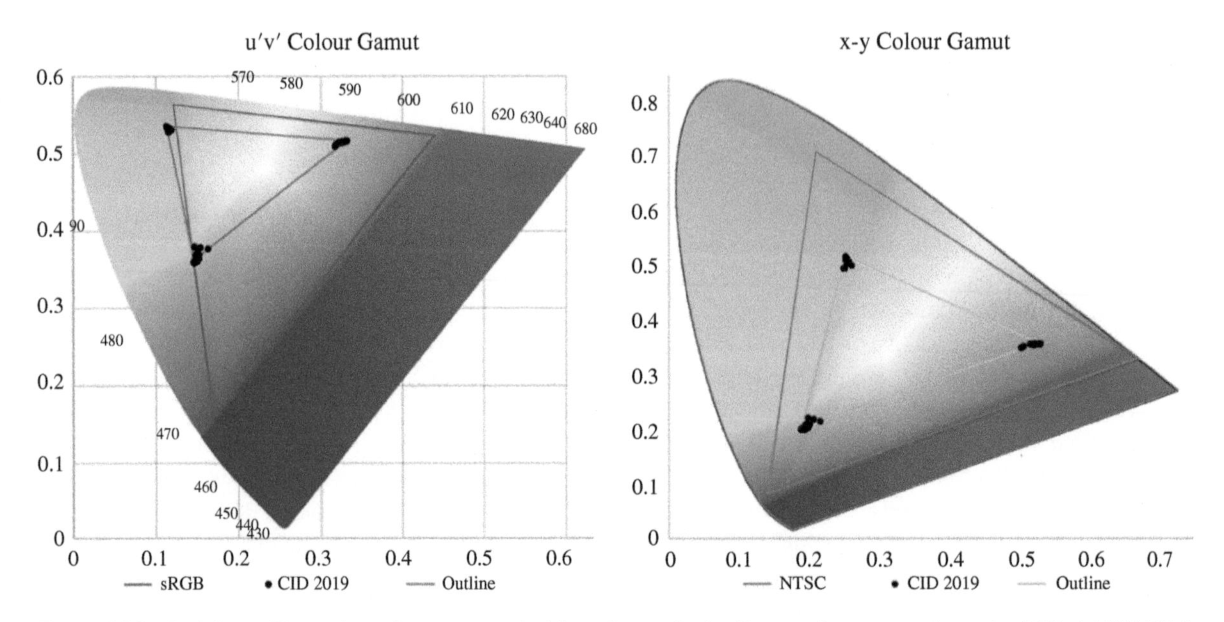

Figure 4.20 Red-Green-Blue color points measured with an Ocean Optics Spectrophotometer shown in sRGB u'v' (CIE1976) and NTSC xyY(1931) color spaces.

color gamut to 50% NTSC. The next step in CFA design is to include a black matrix between sub-pixel colors. These improvements will minimize further color "cross-talk" between sub-pixels and help with the sub-pixel contrast ratio but will reduce overall brightness depending on the resolution and width of the black matrix material. Further improvements in the sub-pixel contrast ratio can be made by improving the drive waveforms. This will be discussed in detail in the next section on device architecture and driving.

4.4 Electrophoretic Displays with CLEARink Structure

Conventional electrophoretic displays as commercialized by E Ink employ positively charged black particles and negatively charged white particles in a clear fluid [1, 21, 22]. An electric field is applied across the particles using a transparent conductor on the front plane and a conductor on the backplane. Driving either the black or white particle to the front surface will depend on the direction of the applied electric field as shown in Figure 4.21 below. The white particles are on the size scale of a micron and have a high refractive index (typically TiO_2); the small white particles being surrounded by a low refractive index fluid are highly light scattering in all directions, which gives a paper-like white appearance. When driven to the front surface. The black particles are light adsorbing and provide a dark state. Combining white and dark particles in a low index fluid leads to good contrast and a paper-like reading experience.

The concepts for an electrophoretic type display had been around for more than a decade [23] before EInk existed, but prototypes lacked stable driving and lifetimes. The key enabling advancements of the technology which allowed for commercialization were: (i) the ability to encapsulate the black and white particles, keeping them from migrating laterally; and (ii) the ability to charge the particles in such a way as to provide shelf-life stability and stable driving over long periods [1]. Encapsulation techniques typically employed to contain the particles are microcapsules [1] or microcups® [24]. CLEARink uses standard photolithography processes to build "walls" around each sub-pixel to confine the particles.

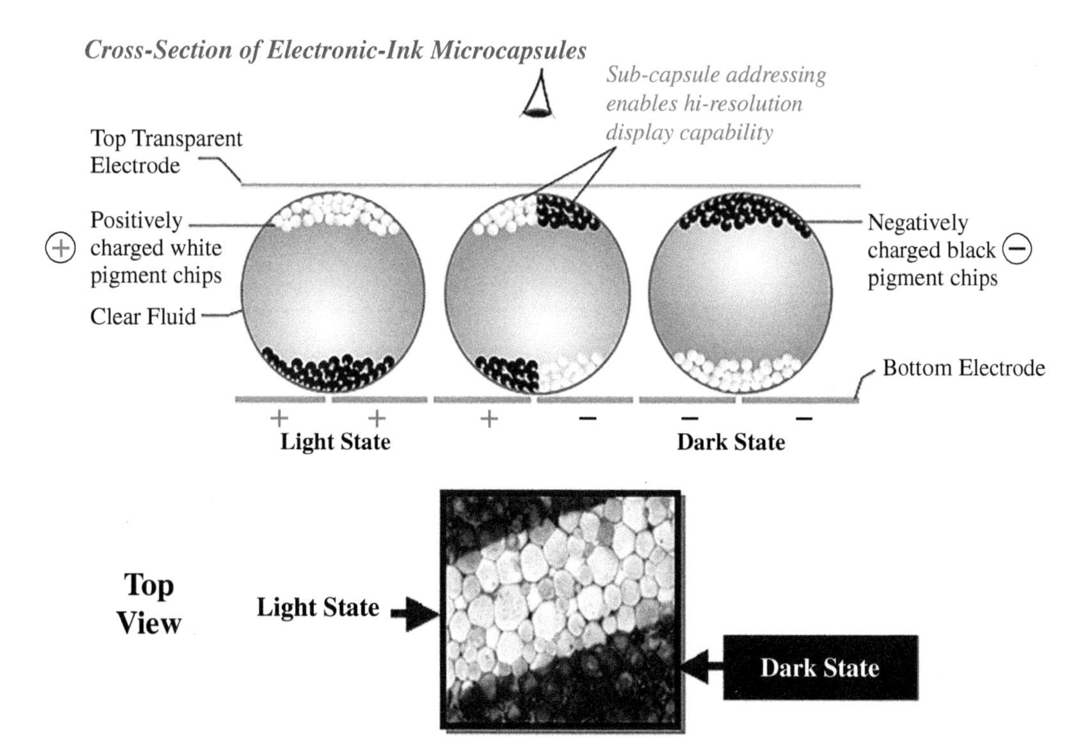

Figure 4.21 E Ink 2-particle type display. The particles are encased in microcapsules and sandwiched between a transparent top conductor and a bottom electrode; an electric field is applied between the two conductors resulting in particle motion.

Early electrophoretic displays suffered poor grayscale control, ghosting, long reset flashes, and slow switching speeds. Drive waveform strategies and improvements have been published elsewhere [25–27] and will not be reviewed here. Over time steady improvements were made in all areas, but, as of this writing, no video-capable electrophoretic display has been commercialized.

Conventional 2-particle (black and white particles) electrophoretic displays are limited because of their slow switching speeds and high drive voltages required for switching. A charged particle's velocity or response time in a fluid is proportional to its electric charge and applied electric field. The response time is inversely proportional to the fluid viscosity and size of the particles (assuming constant particle charge) [28]. Typical EPDs are driven at ±15 V and take ~500 ms (gray to gray) to achieve the desired optical state (~250 ms black to white) [25]. Driving faster without compromising contrast would require significantly higher voltages resulting in high power consumption and more expensive hardware. The high drive voltages result from the relatively large electrode gap necessary for the encapsulation of the particles and the high viscosity sufficient fluid to impart bi-stability.

The slow switching speeds are also because both the black particles and white particles have to travel 10's to a hundred microns in the distance to switch optical states to high contrast. For example, to achieve a dark state, a thick layer of black particles needs to be driven to the front surface. At the same time, the white particles need to be driven far enough away from the front surface not to influence the desired dark state. At a given velocity (or electric field), the particles' distance must travel is large, resulting in a longer optical response time. Grayscale is achieved by mixing black and white particles, and the proportion of each particle in the mix determines the gray level.

The CLEARink electrophoretic display technology operates with positively charged black particles in a low refractive index fluid [29, 30]. The white state is the natural resting state resulting from TIR when only the low refractive index fluid is in optical contact with the optical structures (Figure 4.22). The dark state is achieved by applying an electric field that moves the black particle to the surface of the optical structures, thereby frustrating TIR.

If you recall from the discussion above on TIR, surface waves propagate along the plane between the two media [14]. This surface wave is known as the *evanescent wave*. It decays rapidly in the direction of the low index medium, the rate of decay being determined by the wavelength of light, refractive indices of the two media, and the angle of incidence. The surface wave oscillates between the two media, and the electric field decreases by a factor of 1/e as it penetrates the low refractive index medium. The penetration depth for visible light can be calculated for our system to be on the order of 90–150 nm, depending on exact refractive indices and angles [9]. This method for using TIR and single-particle type to achieve white and dark states is important to distinguish between conventional multi-particle type electrophoretic displays. Using the CLEARink approach, the distance the particles need to travel to attain complete optical switching is in the range of 200–500 nm; approximately two orders of magnitude less than standard 2-particle systems. This leads to a faster optical response for switching. Very low drive voltages can also be achieved because the black particles are contained in a low viscosity fluid. The distance the particles need to travel is very short, and the cell gaps can be minimal (~5 μm). Since optical states are constantly changing when showing video content, low drive voltages are critical to managing power consumption.

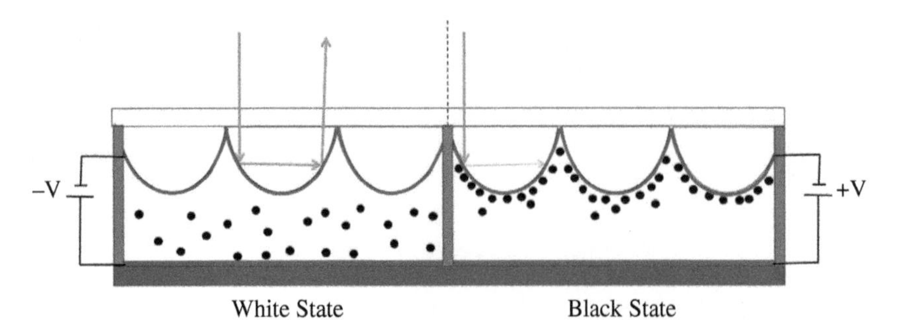

Figure 4.22 CLEARink type display device. A positive voltage is applied to the optical structures to drive the black particles to the surface of the lenses to frustrate TIR and achieve a dark state.

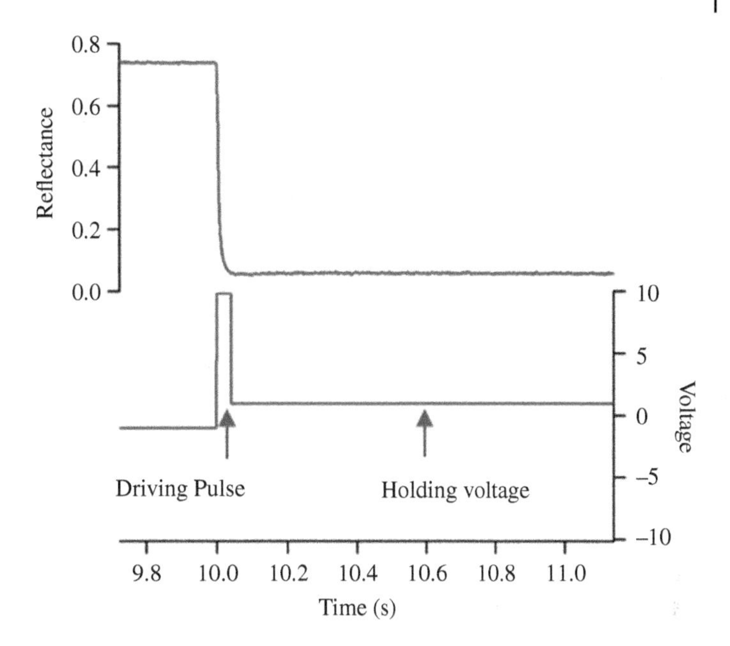

Figure 4.23 Example of a CLEARink waveform and resulting optical state.

The fTIR type display uses different voltage waveforms to drive the electrophoretic particles to the desired optical state than a standard multi-particle type electrophoretic display. The standard 2-particle type display uses only three voltages, + full voltage, − full voltage, and 0 V. A voltage waveform consisting of three stages controlling the three voltages and time, (i) erase the original image, (ii) activate the particles, and (iii) display a new image [25, 26]. The continuous variable to achieve targeted values for each stage is time (pulse width modulation),. All three stages can take hundreds of milliseconds: gray to gray level transitions take the longest [25]. Controlling voltages less than the full voltage is not useful because lower voltages would result in even slower switching speeds. The CLEARink, fTIR type display, employs waveforms that control both voltage and time. In simplest terms, to achieve the desired optical state, the basic waveform has two parts; the drive pulse and the hold pulse, as shown in Figure 4.23 below. The drive pulse is typically at full rated voltage for the physical stack and lasts for 20–50 milliseconds; the purpose of this section of the waveform is to accelerate the particles toward the desired optical state. The hold pulse is 50–90% lower in voltage than the drive pulse and lasts for milliseconds to 10 seconds, depending on how long the desired image is to be held. A reset pulse is required at the end of the hold pulse, which consists of a single "max voltage" pulse held for milliseconds to seconds depending on the previous optical state (the shorter the previous state the shorter the reset, thus allowing video rates). This approach to driving optical states provides for both high switching speeds and low power consumption. Figure 4.23 shows a drive pulse at 10 V, but more recent device stacks allow for high-speed video content using less than 5 V [7]. The CLEARink system is not genuinely bi-stable, meaning the image will not retain indefinitely without power. The only way to achieve bi-stability in an electrophoretic display is to have the charged particles in a high viscosity fluid. To achieve fast switching speeds and low drive voltages, CLEARink chose to employ a low viscosity fluid. However, when showing static images, the refresh rates can be slowed to 30 Hz, and with the low hold voltages, the power consumption will still be low. CLEARinks device architecture includes two modes of operation, a low-power ePaper mode where content and waveforms operate at lower voltages and slower speeds, and a video mode where the device runs at full voltages and speeds.

The CLEARink technology uses voltage control to achieve desired gray levels. The drive voltage and hold voltage can be tuned independently to achieve the desired gray level. Figure 4.24 below demonstrates 28 gray levels. With further waveform optimization, much higher gray levels can be achieved.

One of the major challenges in electrophoretic display-type systems is controlling charges over time. The optical states are achieved by applying DC voltages over some length of time, but when the amount of white state images

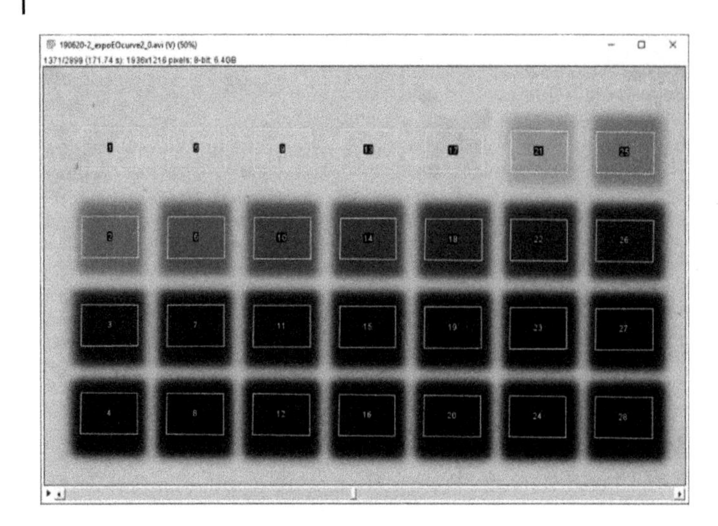

Figure 4.24 1 mm × 1 mm squares demonstrating 28 gray levels on a single active matrix device. The yellow boxes are the selected regions used for measuring reflected brightness.

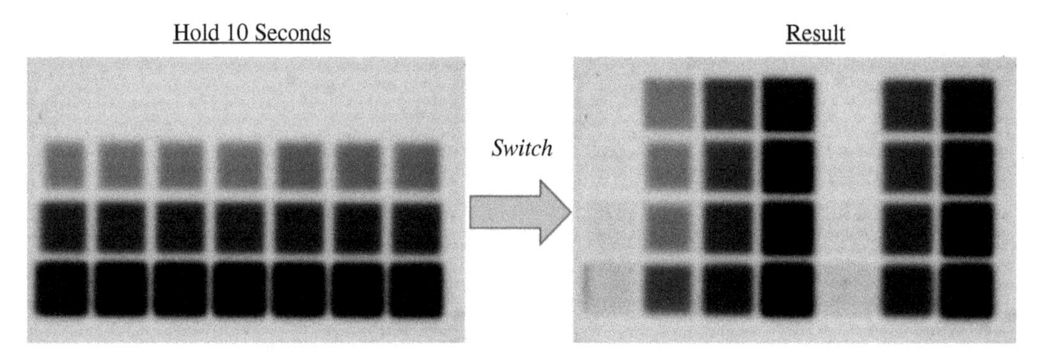

Figure 4.25 Left images shows four rows, each with a distinct gray level being held for 10 seconds. Right image shows 1 second after each row was switched to four different gray levels (three gray levels repeated).

and dark state images over time do not add up to 0, accumulation of charges can occur. This is referred to as the "DC balancing" problem. Liquid crystalline displays solved this issue by inverting the field at a certain frequency. This works because the same gray level can be achieved with a +V or a −V. This is not the case for electrophoretic systems. If the system does not stay balanced, then charges accumulate, leading to visual artifacts such as "optical drift" or "ghosting.." These visual artifacts can often be controlled by applying reset pulses to balance out the accumulation of charges. Figure 4.25 below shows a CLEARink type display with four rows of gray levels (the top row is white, so the squares are not visible) being held for 10 seconds before being switched to 7 columns of gray levels (columns 1 and 5 being white). A residual ghost image can be seen in the full black to full white switch (bottom row of columns 1 and 5), demonstrating ghosting. This can be improved with resets and optimized hold voltages.

Grayscale control is important at the millimeter scale, as shown above for monochrome displays., A color display must control the amount of light through each color sub-pixel and achieve excellent grayscale control at the sub-pixel level. CLEARink uses photolithography-defined walls around each sub-pixel (refer to Figure 4.22) to contain the particles inside said sub-pixel with minimal electrical and optical coupling. Figure 4.26 below demonstrates individual sub-pixels being driven to different gray levels. The surrounding sub-pixels are held at a constant intermediate gray level. It is essential to note the good sub-pixel fidelity, i.e., the lack of optical and electrical coupling to neighboring sub-pixels.

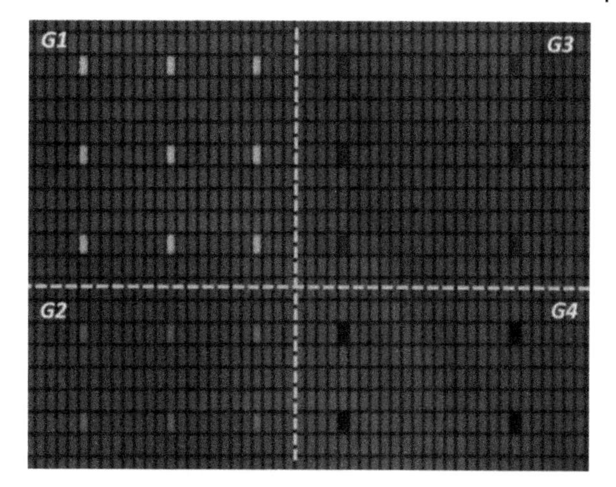

Figure 4.26 An optical micrograph showing sub-pixels driven to four different gray levels, the array is divided into quadrants by the dashed lines, each quadrant is a different gray level (G1–G4).

Driving full color is much more challenging than driving monochrome displays. High contrast has to be achieved at the sub-pixel level. This means that when the desired color sub-pixel is driven to full white, the neighboring sub-pixels must be driven to a full dark without optical or electrical coupling. Figure 4.27 shows sub-pixel level color switching. The color filter used in the device shown in Figure 4.27 does not have a black matrix between sub-pixels. Also, there is a slight misalignment between the CFA sub-pixels and the TFT sub-pixels, resulting in slight light leakage, a common problem with hand-built prototypes.

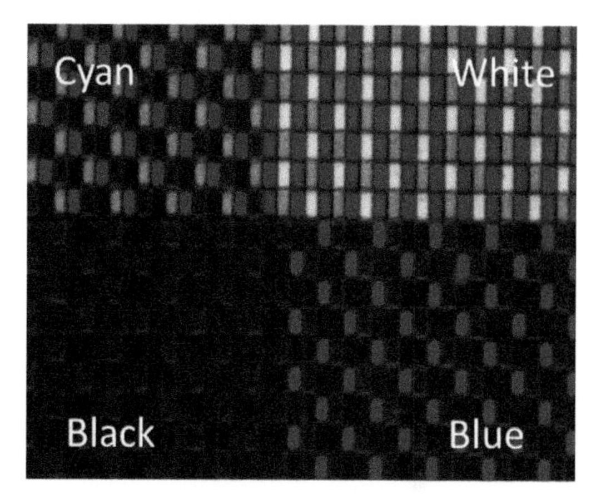

Figure 4.27 Optical micrographs of a full color device at the sub-pixel level. The upper right quadrant denotes all pixels are in the white state, in the lower left all pixels are driven dark. The upper left the green and blue subpixels are beings driven white which would give a visual cyan color. In the lower right quadrant, the blue pixel is driven white, the surrounding sub-pixels are driven dark.

4.5 CLEARink Device Architecture

The control scheme for the CLEARink active matrix display is outlined in Figure 4.28 below. The input file from an SD card or HDMI is in sRGB and must be converted to RGBW format and sub-pixel rendering. The Loo-Up Tables (LUTs) give the voltage-time sequence for the desired optical states. All of these operations are performed in an FPGA along with timing control. The output signal goes to source and gate drivers that scan the rows and columns of the thin film transistor back-plane,, which supplies the target voltages across the electrophoretic ink.

The target waveforms are shown in Figure 4.23. The rendering algorithms were developed and optimized in computation and simulation software platforms and then programmed directly into FPGA hardware for the 9.7″ tablet demonstrator.

Estimating the power consumption of a CLEARink type display is more complex than a conventional monochrome E Ink eReader display for a number of reasons. A conventional eReader is image bistable, meaning it holds an image without added power, is monochrome, and does not show video content. A CLEARink display can offer video content, color, and stable images, but occasional refresh pulses are required. For these reasons, CLEARink developed two modes of operation “video mode” and “ePaper mode. EPaper mode operates at 30 Hz and slightly lower voltage, reducing power consumption. Video mode operates at 60 Hz and higher voltages to achieve an approximately 30 fps frame rate.

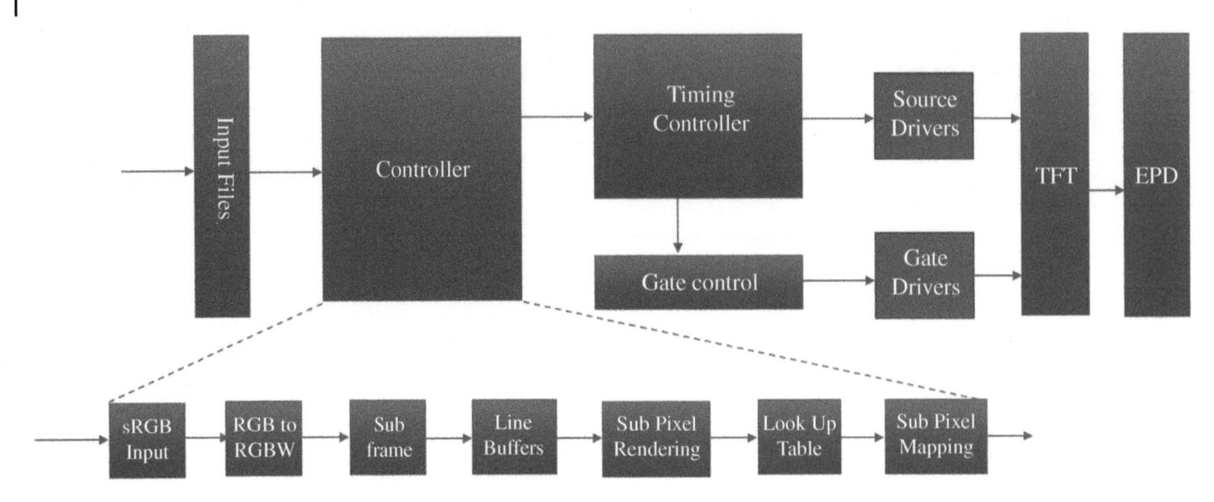

Figure 4.28 Controls scheme for CLEARink active matrix display.

The power consumption of an fTIR display, P_{total}, is estimated by

$$P_{total} = P_{driver} + P_{TFT} + P_{ink} \tag{4.1}$$

where P_{driver}, P_{TFT}, and P_{ink} are the power consumption values associated with the display driver ICs, TFT backplane, and ink, respectively. Typical power consumption ranges are summarized in Table 4.3. In our model, we estimate the static driver IC power consumption to be in the range of 70–80 mW. The power consumption is dominated by the TFT panel's dynamic switching power, which can consume between 20 and 140 mW depending on the image update rates (0–100% switching). At 100% active switching fTIR displays draw less than 1/10th the power of traditional electrophoretic displays, which typically operate at 15 V. The ink power consumption is almost negligible at less than 2 mW. This results in a total panel power consumption ranging from 91–222 mW. -For example, for a 4400 mA-hr Lithium ion battery and 80% power conversion efficiency, a battery life (display only) of 59–143 hours can be realized, enabling ePaper tablets that operate for weeks assuming a typical use case of 2 hours per day.

Display architectures are commonly optimized to their use cases. MIP reflective LCD is an example of optimizing an LCD architecture for low-power applications such as wearables. The CLEARink architecture was designed to offer two modes of driving, ePaper mode and video mode, so that customers could prioritize the modes depending on application and use-case. In ePaper mode, the CLEARink display runs lightly lower voltages at lower frame rates; in Video mode, the frame rates and voltages are higher. Table 4.4 below shows the different modes and resulting estimates on battery life. The assumed use case is two hours daily operation, no front light, and a 4400 mA-hr battery.

Table 4.3 Power consumption estimates for a 9.7″ active matrix display.

Parameter	Power [mW]
P_{driver}	70–80
P_{TFT}	20–140
P_{ink}	1–2
P_{total}	91–222

Table 4.4 Estimated battery life of an eReader using a CLEARink type display in ePaper mode and video mode.

	ePaper mode	Video mode
% Active switching	25	75
Frame rate [Hz]	30	60
Voltage [V]	1–3	2–4
Power consumption [mW]	101.9	231.7
Battery life (display-only) [days]	63.9	28.1

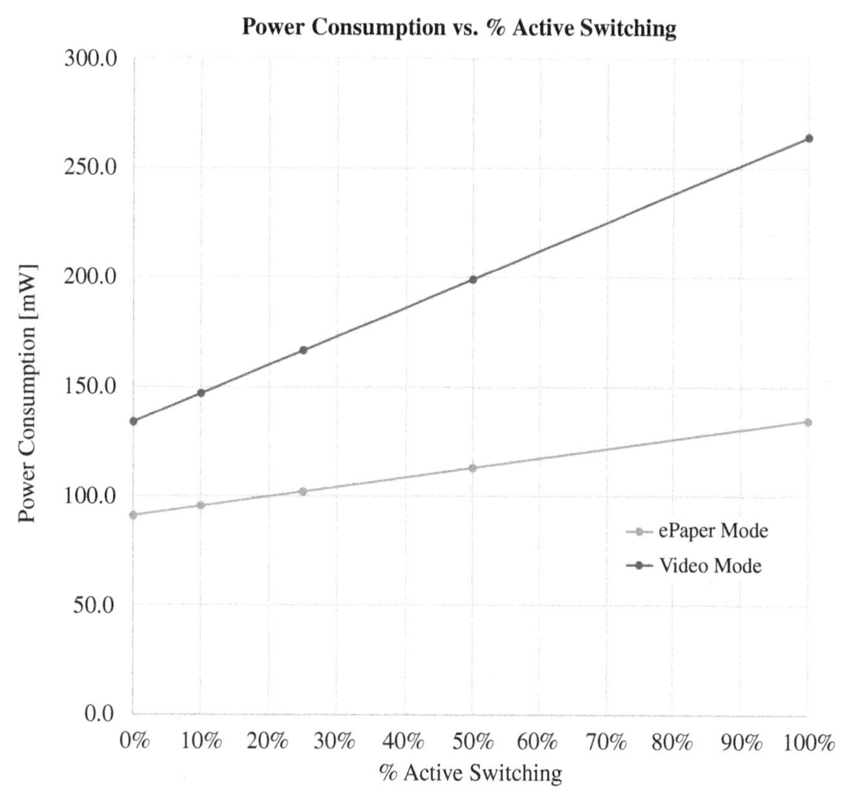

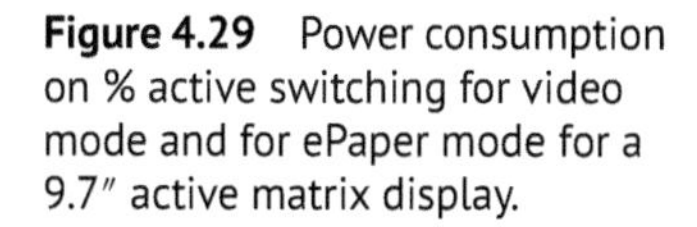
Figure 4.29 Power consumption on % active switching for video mode and for ePaper mode for a 9.7″ active matrix display.

Another important variable in overall power consumption is the total number of actively switching pixels. Given the goal of achieving a paper-like display experience, it is preferred to have a white paper-like background as a resting state, and content is actively switched from this state. In ePaper mode, to display the text, the actual percentage of the display actively switching would be quite low. In video mode, the % active switching would be high. Figure 4.29 shows the impact of power consumption on % active switching for Video mode and ePaper mode.

A front light is recommended if a brighter viewing experience is desired for indoor, low ambient light conditions. We estimate the power consumption contribution for a 9.7″ diagonal front light to range 83–109 mW for light conditions ranging 7–70 lux using typically preferred brightness levels determined by Rempel et al. [31]. CLEARink worked with a front light partner to develop a highly efficient directional frontlight leveraging the semi-retroreflective optical structures. The basic principle is shown in Figure 4.30. Figure 4.31 shows the front light in operation on a CLEARink display.

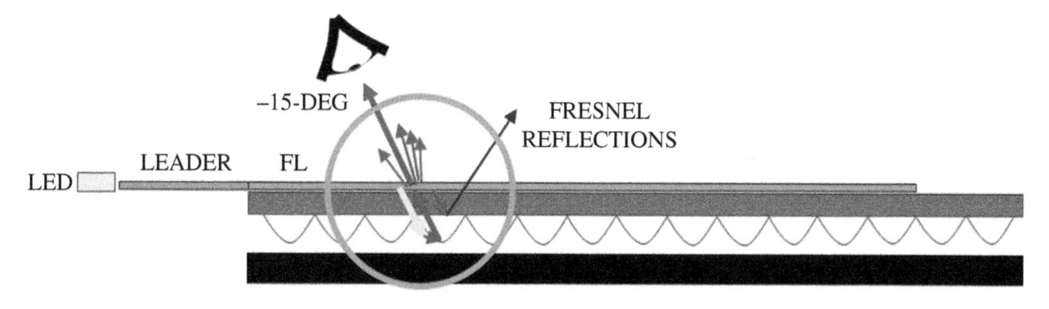

Figure 4.30 Front light guide showing light being directed out of the light guide into the CLEARink optical structures.

Figure 4.31 Picture of an image on a CLEARink display with a front light.

As ambient light conditions get brighter, emissive displays must increase the backlight power to increase the white state brightness to maintain image contrast. This increase in backlight power results in a considerable reduction in operation time between battery charges. The power savings of a CLEARink type display in outdoor, high ambient light viewing is very substantial has been reported elsewhere in detail [7].

4.6 Manufacturing and Supply Chain

The CLEARink technology was developed to "drop-in" to existing LCD fabs. The rationale was that there is an excess of capacity on Gen 5 lines or smaller due to the migration to larger Gen lines in China, Taiwan, and Japan. With this strategy in mind, the technology was developed to be compatible with standard a-Silicon backplanes, standard photolithography processes, and fill-align-seal with standard One Drop Fill (ODF) equipment. Existing LCD supply chain partners can fabricate the backplanes and CFAs, and the electrophoretic ink developed by Merck AG would be supplied to the qualified fab partners. The only significant deviation from standard LCD processes would be the micro-replicated optics. The CFA would need to be supplied to the microreplicator, then supply back to the LCD fab.

4.6.1 Status of Technology and Future Projections

CLEARink Displays demonstrated their rapidly improving technology at SID Display Week in 2017, 2018, and 2019. Each year the small team (about 1/10th the size with approximately 1/10th the funding as its competition) showed dramatic improvements in color, resolution, stability, and controls architecture. The next major milestone was to complete the buildout of an engineering developer kit to offer to future customers for evaluation. 1.3″ active matrix development samples had been built and tested for six months showing good stability during driving and also good shelf stability. The next step is to run full reliability tests on the final device stack. CLEARink Displays changed ownership in April 2020, resulting in a shift in focus. As a result of this ownership change, the author cannot comment on the current status and future direction of the technology after March 2020.

The CLEARink display technology is in its infancy from a technological maturity point of view, compared to traditional EPD, OLED, and LCD technologies. If development resources continue to be deployed, display performance would continue to improve rapidly. Improvements in lens design, quality, and materials will lead to higher optical gain, which translates directly into brightness and color performance. Further optimization of the CFA will result in an improved color gamut. Ink, waveform, and device stack development will improve video speeds and reduce ghosting. Full-stack architecture, material sets, and pixel sealing will need to be optimized to pass full reliability tests. A custom ASIC for the CLEARink architecture will need to be designed, built, and validated for the first commercial product. As of the completion of this chapter (July 2020), the market needs for low power, reflective display that is paper-like yet offers rich color and video capability is yet to be fulfilled by any technology, but the author is aware of several companies and technologies that are chasing this dream.

Acknowledgments

I would like to acknowledge the generous support in the writing of this chapter from Dr. Michele Mossman, Dr. Thomas Johansson, Dr. Bavo Robben, Bram Sadlik, Sriram Peruvemba, Samantha Phenix, Dr. Julian Bigi, Dr. Peter Kazlas, Dr. Robert Holman, and the entire CLEARink R&D and Operations teams. I would also like to thank Professor Lorne Whitehead for his invention, continued dedication to the technology, and mentorship. Thank you to the entire CLEARink team. It was a joy working with you.

References

1 Comiskey, B., Albert, J.D., Yoshizawa, H. (1998). An electrophoretic ink for all-printed reflective electronic displays. *Nature* 7 (394): 253–255.
2 Heikenfeld, J., Drzaic, P., Yeo, J.-S. et al. (2011). Review paper: a critical review of the present and future prospects for electronic paper. *J. Soc. Inf. Display* 19 (2): 129–156.
3 Uchida, T., Katagishi, T., Onodera, M. et al. (1986). Reflective multicolor liquid-crystal display. *IEEE Trans. Electron Dev.* 33 (8): 1207–1211.
4 Cummings, W. (2010). 63.1 The impact of materials and system design choices on reflective display quality for mobile device applications. *SID Symposium Digest of Technical Papers* 41(1):935–938.
5 Fiske, T. G. (2011). 64.2 MEMs-based reflective display measurements in ambient use conditions. *SID Symposium Digest of Technical Papers* 42(1):954–956.
6 Castillo, S. G., Feng, L., Bachmann, T., et al. (2019). 57.4 Solid state reflective display (SRD®) with LTPS diode backplane. *SID Symposium Digest of Technical Papers* 50(1):807–810.
7 Fleming, R., Kazlas, P., Johansson, T., et al. (2019). 36.3 Tablet-size eTIR display for low-power ePaper applications with color video capability. *SID Symposium Digest Technical Papers* 50(1):505–508.
8 Fleming, R., Peruvemba, S., Holman, R., et al. (2018). 48.2 Electronic paper 2.0: frustrated eTIR as a path to color and video. *SID Symposium Digest Technical Papers* 49(1):630–632.
9 Pedrotti, F.L., Pedrotti, L.M., and Pedrotti, L.S. (1993). Geometrical optics. In: *Introduction to Optics*, vol. 38–39 (ed. R. Henderson), 419–420. New Jersey: Prentice Hall, Englewood Cliffs.
10 Whitehead, L.A. and Mossman, M.A. (2009). Reflections on total internal reflection. *Opt. Photonics News* 2: 28–34.
11 https://www.3m.com/3M/en_US/scotchlite-reflective-material-us/industries-active-lifestyle/active-lifestyle/how-retroreflection-works/

12 https://www.3m.com/3M/en_US/road-safety-us/resources/road-transportation-safety-center-blog/full-story/~/road-signs-retroreflectivity/?storyid=328c8880-941b-4adc-a9f9-46a1cd79e637

13 Whitehead, L.A., Tiedje, J.T., and Coope, R. J. N. (1999). Method and apparatus for controllable frustration of total internal reflection United States Patent No. 5,999,307, 7 December.

14 Hecht, E. (1987). *Optics*, 2e, 95–113. San Francisco, CA, Editor Bruce Spatz: Addison Wesley.

15 Mossman, M. A., Rao, S.P., and Whitehead, L. A. (2002). P-83 Grey scale control of TIR using electrophoresis of sub-optical pigment particles. *SID Symposium Digest Technical Papers* 33(1):522–525.

16 Goulding, M., Farrand, L., Smith, A., et al. (2010). 40.1 Dyed polymeric microparticles for colour rendering in electrophoretic displays. *SID Symposium Digest Technical Papers* 41(1):564–567.

17 Goulding, M., Farrand, L., Smith, A., et al. (2012). High Performance Particle Materials and Fluids for Colour Filter Free, Colour Electrophoretic Displays. *IDW/AD* 245–248.

18 Goulding, M., Smith, N., Farrand, L., et al. (2015). 23.1 Colloidal dispersion materials for electrophoretic displays and beyond. *SID Symposium Digest Technical Papers* 46(1):326–329.

19 Fleming, R., Peruvemba, S., Sadlik, B., et al. (2019). EP2-1 ePaper 2.0 (Full Color + Video) Enable by Retro-Reflective eTIR Technoloy. Publisher: Society for Information Display, 25th International Display Workshops (IDW'18) Nagoya, Japan 3: 1252–1255.

20 Elliott, C. B. (2014). Image Reconstruction and Color Reproduction on Subpixelated Flat Panel Displays. PenTile Book 12–14 draft, May 2014.

21 Chen, Y., Au, J., Kazlas, P., et al. (2003). Flexible active-matrix electronic ink display. *Nature* 5 (423): 136.

22 Albert, J.D., Comiskey, B., Jacobson, J.M., et al. (2003). Suspended particle displays and materials for making the same. U.S. Patent No. 6,515,649 B1.2003-02-04.

23 Ota, I., Ohnishi, J., and Yoshiyama, M. (1973). Electrophoretic image display (EPID) panel. *Proc. IEEE* 61 (7): 832–836.

24 Liang, R.C., Zang, H.M., Chung, J., et al. (2003). Microcup® electronic paper by roll-to-roll manufacturing processes. *The Spect.* 16 (2): 16–21.

25 Bai, P.F., Hayes, R.A., Jin, M.L., et al. (2014). Review of paper-like display technologies. *Prog. Electromagn. Res.* 147: 96–116.

26 Kao, W.-C. (2010). Electrophoretic display controller integrated with real-time halftoning and partial region update. *J. Disp. Technol.* 6 (1): 36–44.

27 Shen, S., Gong, Y., Jin, M., et al. (2018). Improving electrophoretic particle motion control in electrophoretic displays by eliminating the fringing effect via driving waveform design. *Micromach.* 9 (143).

28 Kang, H.-L., Kim, C.A., Lee, S.-I., et al. (2016). Analysis of particle movement by dielectrophoretic force for reflective electronic display. *J. Disp. Technol.* 12 (7): 747–752.

29 Robben, B., Beunis, F., Neyts, K., et al. (2018). Electrodynamics of electronic paper based on total internal reflection. *Phys. Rev. Appl.* 10: 034041.

30 Robben, B., Strubbe, F., Beunis, F., et al. (2020). Polarity-dependent adsorption of inverse micelles. *Langmuir* 36 (23): 6521–6530.

31 Rempel, A. G., Heidrich, W., Li, H., et al. (2009). Video Viewing Preferences for HDR Displays Under Varying Ambient Illumination. Proceedings of the 6th Symposium on Applied Perception in Graphics and Visualization, New York. 45–52.

5

Bistable Cholesteric Liquid Crystal Displays – Review and Writing Tablets

Clinton Braganza and Mauricio Echeverri

Kent Displays Inc., Portage Blvd, Kent, OH, USA

5.1 Introduction

Cholesteric liquid crystals (ChLC) were the beginning of this incredible phase of matter – discovered in the late nineteenth century. It was not till the late twentieth century that researchers at the Liquid Crystal Institute at Kent State University exploit their true applications and usefulness. Today, ChLCs are well suited for ePaper and more. They have multiyear bistability and do not require any polarizers or color filters to produce images [1]. ChLCs can be encapsulated for integration into flexible substrates (Figure 5.1a), are easily electrically driven (Figure 5.1b), and can also be optically addressed to create images [2].

In the last three decades, they have been developed into formidable display architectures from direct driven signage to large area passive matrix displays for signs, instrumentation, and more. They have been developed into flexible embodiments for electronic skins, eReaders, and, most recently, writing tablets (Figure 5.1c).

More fundamentally, they can be used to create full-color images by either photo addressing, stacking layers, placing side-by-side ChLC tuned to primary colors or electrical tuning of each pixel. Displays made of ChLCs also lend themselves well to transparent modes. This permits the display to be used as a tracing paper or a form. The simplicity and ease of manufacturing and driving enable scale up in size and capacity.

ChLCs have been driven passively, actively, direct, and optically. They have been stacked, used glass substrates, plastic substrates, and even been formed on ultra-flexible fabric materials. These materials have been the subject of research groups and commercialization activities worldwide. They continue to be the focus of many such applications today.

We describe fundamental modes of operation, their effectiveness, and their unique new capability in analog writing devices created via pressure-induced flow alignment! The historical and scientific journey of ChLCs as described here is a true testament to the promise of this phase of matter that has been developed and exploited over the years!

5.2 Materials and Optical Properties

Cholesteric liquid crystals and Nematic liquid crystals administered with chiral dopants (Chiral Nematic) [3, 4] have a helical structure in which there is a slight twist along the helical axis of the liquid crystal director $\vec{n}$ [5]. The pitch, P_o is the distance for the director to rotate 2π, the period is $P_o/2$ because $\vec{n}$ and $-\vec{n}$ are equivalent.

E-Paper Displays, First Edition. Edited by Bo-Ru Yang.

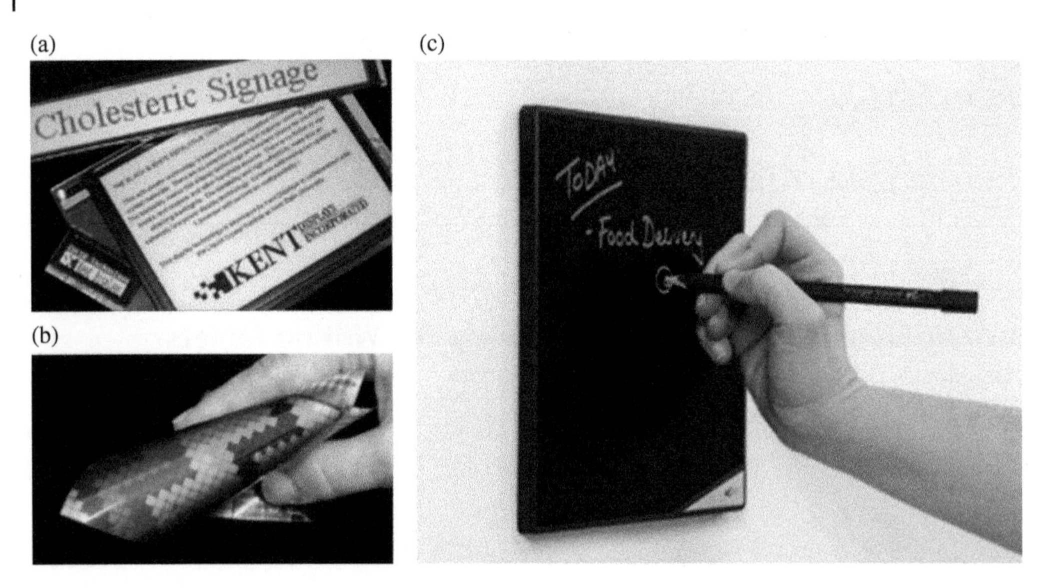

Figure 5.1 Examples of ChLC displays of different categories. (a) Glass based readers and signage, (b) flexible displays, (c) writing tablets.

Suppose one were to place these materials between two transparent substrates. In that case, the helical axis can take on the following orientations, showing different textures, depending on the treatment of the surfaces adjacent to the liquid crystal. The helical axis orients perpendicular to the substrates with strong homogeneous alignment. This texture is called the Planar texture and is transparent [6]. When the surface alignment favors the molecules to anchor perpendicular to the substrate, the helical axis either lays parallel to the substrates, showing a fingerprint texture between crossed polarizers or randomly showing a weakly scattering Focal conic texture [7]. It is possible to control the substrate surfaces to make the Planar and Focal conic textures Bistable [8].

5.2.1 Optical Properties of Cholesteric Liquid Crystals

For light incident parallel to the helical axis of the Cholesteric liquid crystals in the Planar texture, there is a Bragg reflection whose wavelength λ_o is centered at $\langle n_e - n_o \rangle P_o$, where n_e and n_o are the extraordinary and ordinary refractive index, respectively. The bandwidth $\Delta\lambda$ of this reflection is $(n_e - n_o)P_o$ [9]. The derivations of these equations are done by solving Maxwells equation for the propagation of the electric field vector through the helix [4] or those based on Bragg's reflection. The reflection is of the same handedness as the chirality of the liquid crystal, while the other handedness and wavelengths are transmitted [1]. With the Planar texture showing no defects and having typical birefringence of about 0.2, the reflectivity can be maximized by making the cell thickness at least 10 pitches. This color generation does not need polarizers, making this an ideal material for plastic substrates with birefringence. In the Focal conic texture, light is mostly forward scattered through the liquid crystal with some weak backscattering. This backscattering can be high in systems with a high polymer content. These two states, one providing color and another clear, can create stable images.

5.2.1.1 Color from Cholesteric Liquid Crystals

An absorbing layer can be placed behind these textures to further enhance the contrast of these images, with the Focal conic texture showing the color of the absorbing layer and the Planar texture showing the additive sum of the color of the liquid crystal Bragg reflection and absorber color. It is desired to place the absorbing layer as close as possible to the liquid crystal layer to limit any reduction in contrast from haze of the back substrate Figure 5.2.

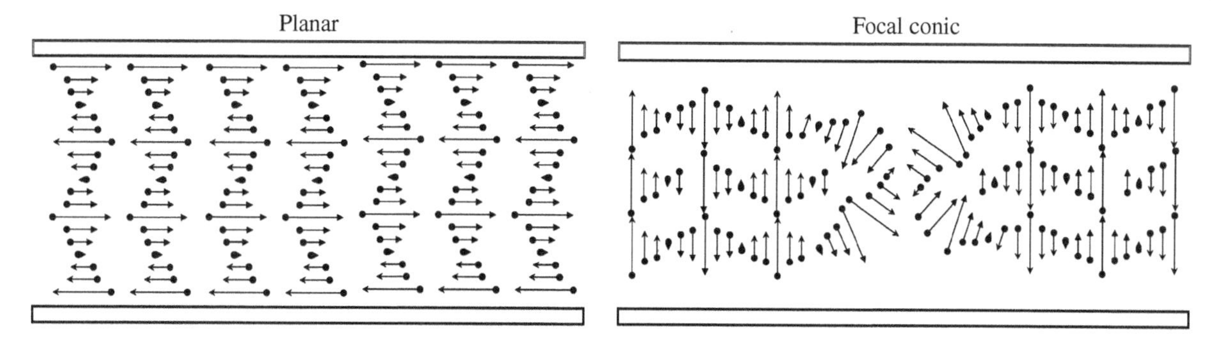

Figure 5.2 Illustrations of some structures of cholesteric liquid crystals.

The color reflected by the liquid crystal can be tuned by changing the pitch of the helix. The pitch can be modified by changing the concentration or the helical twisting power (HTP) of the chiral dopant according to the equation $P_o = c.\ HTP$. The HTP is a parameter dependent on the chiral dopant, which imparts chirality and hence handedness to the doped liquid crystal. This can be achieved either in situ or ex-situ. For example, the HTP can be temporarily changed using UV light to allow image creation, as in Figure 5.3 [2].

Recent work on color tuning application of an electrical field was shown using a new state of cholesteric liquid crystals, the Oblique heliconical structure [10]. Electrothermal color tuning has also been used for the same purpose [11]. A good review of the many methods developed to change the peak wavelength, including strain, humidity, and pH, is discussed elsewhere [12].

5.2.1.2 Optimizing Cholesteric Liquid Crystals for ePaper Applications

Although in ref. [11, 13] temperature is being used to tune the color; for most ePaper applications, the color needs to remain stable over a large temperature range (from at least 0 to 65 °C). Much work has been done to achieve this, including doping with left- and right-hand chiral dopants [14] and using dopants showing an opposite change in HTP with temperature in the same mixture [15].

The peak wavelength λ_o has a significant dependence on the incident angle, blue shifting with increasing angles θ, [16] according to $\langle n_e - n_o \rangle P_o \cos\theta$. To reduce this undesirable dependence, an imperfect Planar texture is desired. This can be done by designing the surfaces in contact with the liquid crystal to form an imperfect Planar texture [17]. In surface stabilized cholesteric displays, the cell surfaces are treated with weak homogeneous or homeotropic alignment layers to stabilize the imperfect Planar helix. Under a microscope, the perfect Planar texture is reflective with very few defects, while the imperfect Planar texture shows multiple domains of bright spots about 5–50 μm in diameter [18]. An example of this reduction in the change in peak wavelength shift as the viewing angle can be very significant, as shown in Figure 5.4. The measurement was done in an integrating sphere to simulate diffuse illumination, mimicking reflective media usage's lighting conditions.

Figure 5.3 An optically addressed cholesteric liquid crystal display.

5.2.1.3 Optical and Mechanical Effects of Polymer Networks

In Polymer Stabilized Cholesteric Textures (PSCT) displays, a small amount of polymer is used to improve the bistability or performance of the system. Still, it also disrupts the Planar

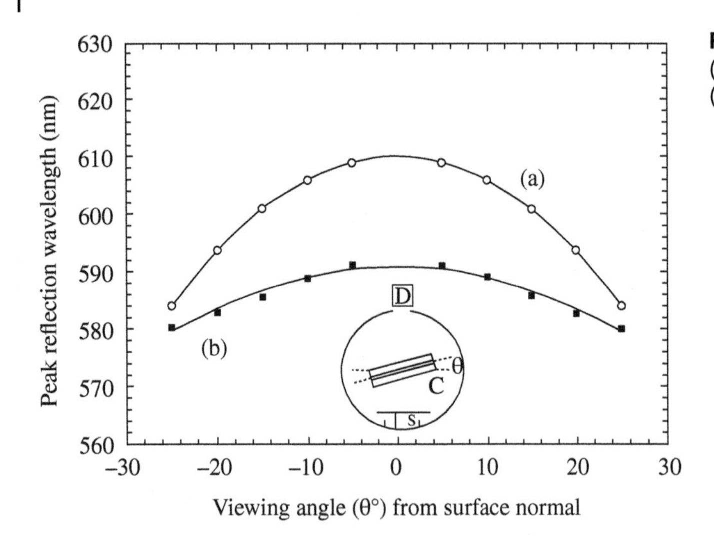

Figure 5.4 The change in peak wavelength of (a) a sample with nearly perfect Planar texture and (b) one with imperfect Planar texture.

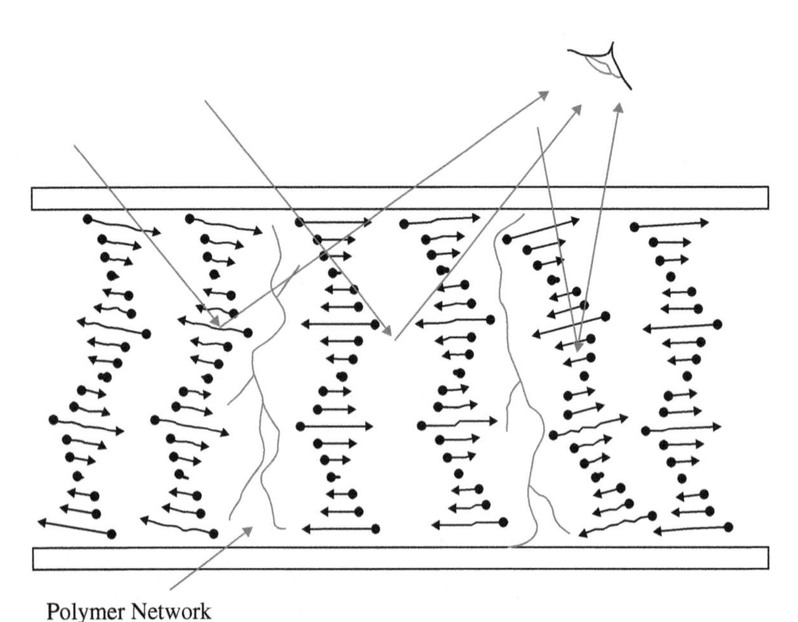

Figure 5.5 Illustration showing reflection from a cell with an imperfect Planar texture.

texture as depicted in Figure 5.5 while improving the viewing angle [8]. PSCTs are typically achieved using 0.5–10% of the polymer as long as the system remains in the liquid crystal phase. Higher polymer concentrations can generate structures where the polymer surrounds the liquid crystal-like in microemulsions and Polymer Dispersed Liquid Crystals (PDLC) [19]. A PDLC made with ChLC is shown in Figure 5.6. In these systems, the viewing angle and reflectance depend on many factors, including the droplet size, uniformity, shape, encapsulation material, polymer wall thickness [20]. For example, tiny droplets will make the Focal conic texture backscatter too much light, lowering the contrast [21]. In addition to the change in reflectivity, the peak wavelength can also be affected by the encapsulation with the peak wavelength either red or blue shifting from its value in a surface stabilized system; the system in Figure 5.7 shows a redshift. With the proper choice of materials and encapsulating conditions,,. This change in reflectivity and wavelength can be controlled, as discussed later.

The extensive work on encapsulated ChLC paved the way to one of the most exciting applications of ChLC ePapers; flexible displays have tremendous advantages such as the possibility of continuous processing and the design of flexible, rugged, and lightweight devices. Encapsulating the ChLC in polymeric networks provides mechanical robustness and prevents liquid flow, allowing for a significant number of applications. The design of the polymeric networks from a structural perspective enables the improvement of mechanical properties such as adhesion and impact resistance. Figure 5.8 shows two different polymeric structures after their flexible substrates are removed. The failure mode indicates a cohesive fracture on the left since there is a remaining polymer on the substrate surface (also present on the other substrate not shown here). The right-hand side figure indicates an adhesive failure as the polymeric network layer detaches from the substrate surface [22].

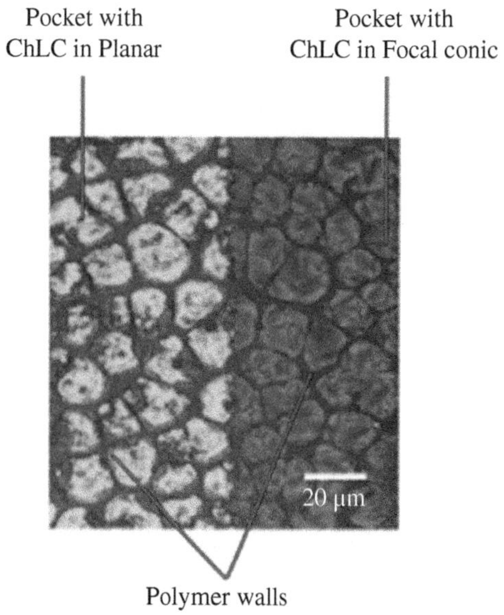

Figure 5.9 is only one example of many flexible display prototypes developed over the years [23–25]. Different levels of resolution, drapability, color gamut, and ruggedness have been achieved. However, few of them are commercially available to writing this chapter.

Sometimes it is desirable to significantly broaden the reflection bandwidth, making the reflected color more achromatic, a feature ideal for black and white ePaper. This is achieved by using polymer networks to disrupt the Planar structure to a poly-domain Planar Texture [26, 27]. One can also stack a yellow and blue cell on top of each other, attach a black

Pocket with ChLC in Planar

Pocket with ChLC in Focal conic

20 μm

Polymer walls

Figure 5.6 Microscopic picture of a PDLC where one section is electrically switched to Planar and the adjacent area is in Focal conic.

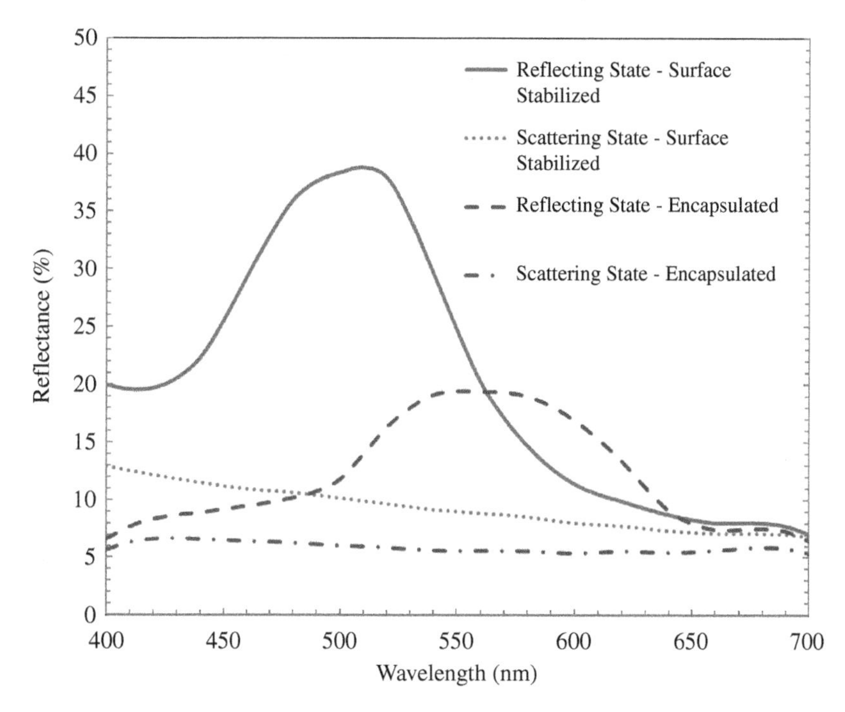

Figure 5.7 Change in reflectivity and peak reflected wavelength of a cholesteric liquid crystal after encapsulation.

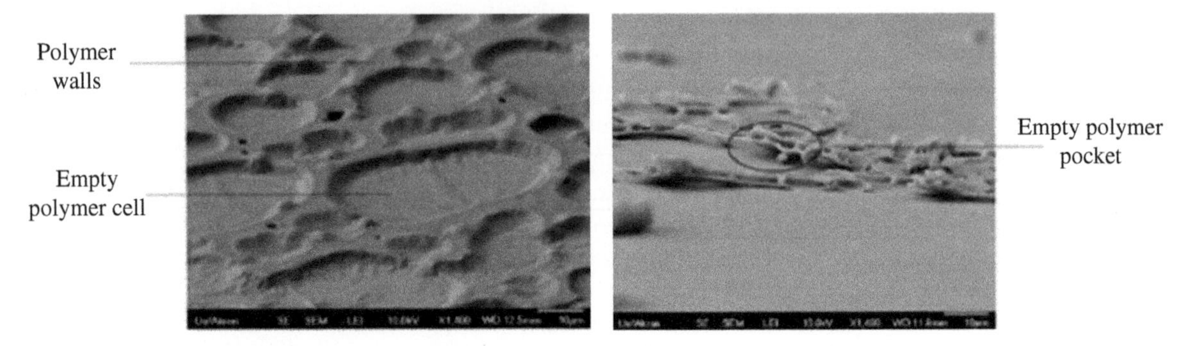

Figure 5.8 SEM micrographs of PDLC after substrates are removed and ChLC rinsed out.

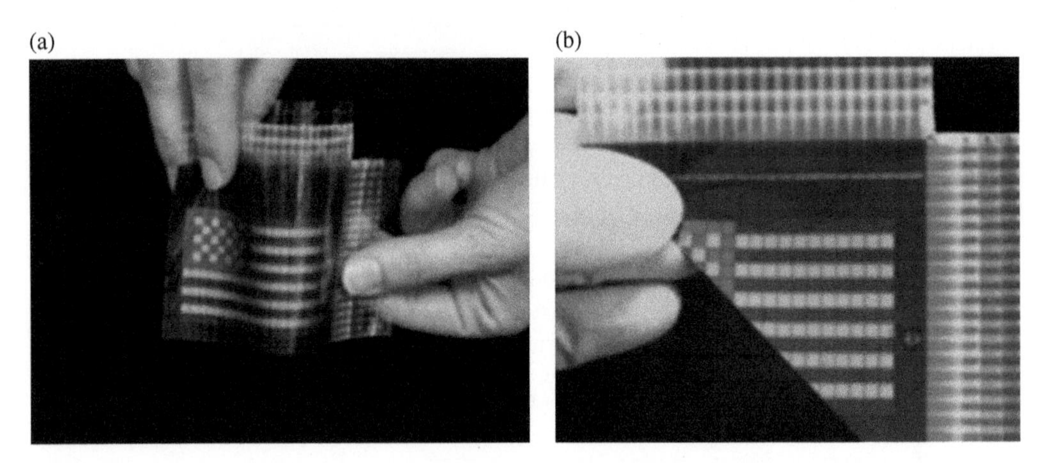

Figure 5.9 Example of flexible and rugged display showcasing (a) high flexibility. (b) ruggedness so even after cutting the display is still functional.

absorber to allow color mixing to create a white Planar texture and a black Focal conic texture. It is also possible to create pitch gradients in the Planar state by using controlled photopolymerization of mixtures monomer and cholesteric liquid crystals [28–30]. The bandwidth has also been broadened using electrical fields [31].

To achieve even higher reflectivity, one can stack cells with left- and right-handed cholesteric cells on top of each other to increase the reflectivity of the resultant cell to over 80%. It is preferred to use the larger domain size as the bottom layer in an encapsulated system.

In some cases, purer color is desired than can be achieved by controlling Δn, alignment layers, and encapsulation techniques. To compensate for the broadened bandgap and loss of color purity, reference [32] used color filters to absorb unwanted reflections at each of the color layers of a stacked RGB system to increase the gamut by 120%, as shown in Figure 5.10. Some groups use dyes in the liquid crystal layer for the same purpose.

5.3 Image Creation Using Cholesteric Liquid Crystals

5.3.1 Electrical Addressing

There are many ways to switch between cholesteric textures, including stabilizing the scattering Focal conic texture and switching it to a clear homeotropic state to make shutters and using very high pretilt, controlled cell gaps, and polarizers to make bistable displays. The more elegant and least complicated way to use them, without using

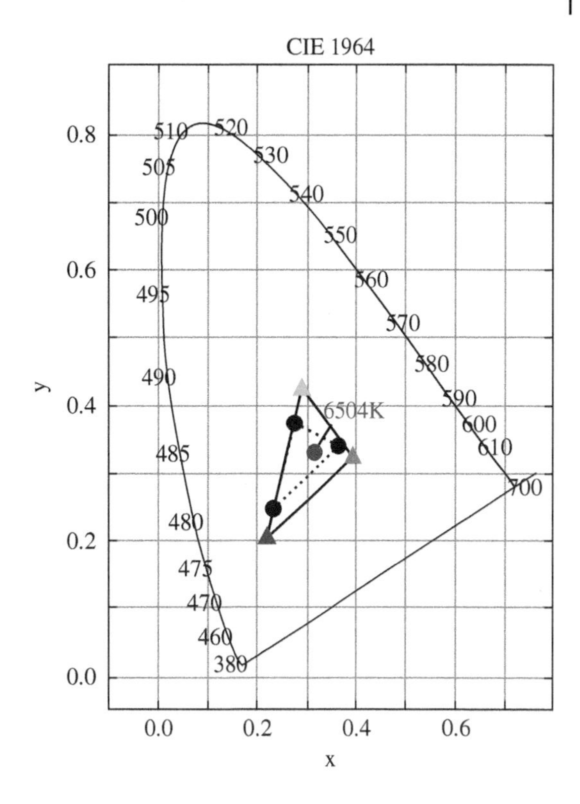

Figure 5.10 120% increase in color gamut and resultant color purity after application of cutoff filters (triangles).

polarizers, filters, complicated pretilts, or complicated manufacturing processes, is to use the cholesteric Planar texture reflecting in the visible region with both the Planar and Focal conic states being stable [24].

In the presence of an electric field, cholesteric liquid crystals minimize free energy. For a material with $\Delta\varepsilon < 0$, the helical axis will be parallel with the electric field, the opposite will be the case for $\Delta\varepsilon > 0$. For this chapter, we will consider materials with $\Delta\varepsilon > 0$.

Starting with a stabilized Planar texture, increasing fields destabilize the texture and either form oily streaks or Helfrich instability [33] to eventually form Focal conic texture after the field is turned off. The transition from Focal conic to homeotropic texture involves unwinding the helix as the liquid crystal molecules align parallel to the field. Complete unwinding occurs when the field is greater than $E_c = \pi^2 / P_0 \sqrt{K_{22} / \varepsilon_o \ \varepsilon}$, where K_{22} is the twist elastic constant of the liquid crystal and ε_0 is the permittivity of vacuum. This transition is reversible.

To return to the Planar texture, one has to increase the field above E_c to unwind the helix to homeotropic texture. From the homeotropic texture, releasing the field quickly to below $(2/\pi)\sqrt{K_{22}/K_{33}}$ Of the field required to unwind, the liquid crystal transitions from a conic helical texture to a transient Planar texture before relaxing to Planar texture due to the shorter relaxation time of this transition. Releasing the field slowly from the homeotropic texture results in the Focal conic texture [17] through the nucleation of twist fingers [34]. A schematic of this is shown in Figure 5.11.

5.3.1.1 Driving Schemes

A representative electro-optic response curve of a bistable cholesteric pixel is shown in Figure 5.12; there are two curves, one where measurements are taken with the pixel is reset to Planar texture and the other with the pixel reset the Focal conic texture. Each of the points on the curve is taken after the pixel has relaxed to its stable state, typically about 500 ms. With the pixel initially reset in the Planar texture before voltage pulse application, there is a threshold voltage, V1, before the pixel starts dimming. The region between V1 and V2 is where some domains

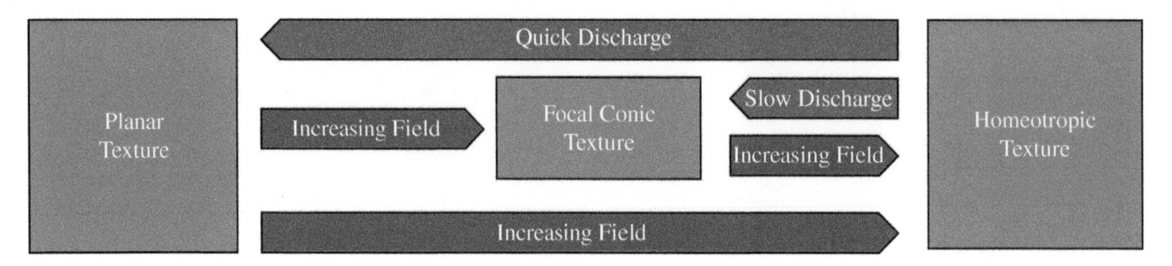

Figure 5.11 Schematic of the transitions among cholesteric textures. The intermediate textures have been omitted for simplicity.

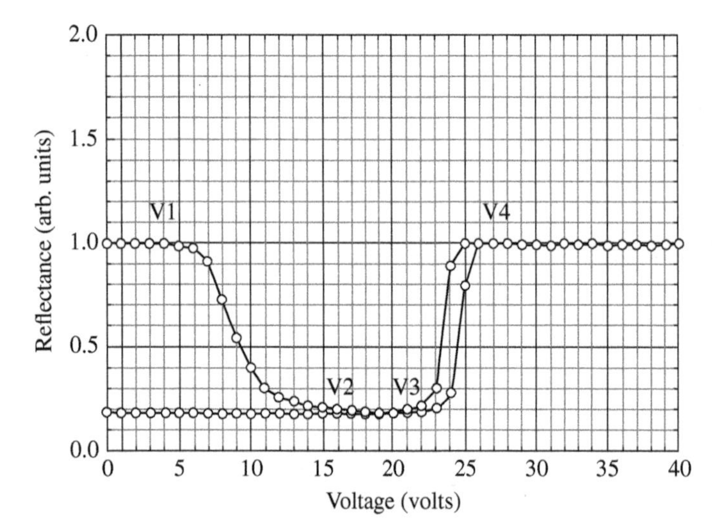

Figure 5.12 Representative response of a bistable cholesteric liquid crystal.

remain in the Planar texture and others switch to the Focal conic texture. Between V2 and V3, the pixel is in the Focal conic texture, increasing reflectance after that as some domains unwind and switch to Planar until V4, where all the domains switch to Planar. With the reset pixel initially switched to the Focal conic texture, the curve stays dark until there is sufficient voltage > V3 to get some of the domains to Planar until V4, where they all switch to Planar. There is a difference at which the initially Planar and Focal conic pixels start grayscale and all switch to Planar due to the ease at of the Planar texture switching to homeotropic texture. This curve shifts to higher values as the voltage pulse width is decreased, eventually making it difficult to switch a Planar pixel to Focal conic [17, 24].

This curve allows for direct and grayscale addressing. A pixel can be driven to either a bright state by applying a voltage > V4 or a dark state by a voltage < V3. The grayscale values between V1 and V2 are reproducible and stable.

The threshold voltage V1 allows for Bistable cholesterics to be driven very simply using simple multiplexing without active matrix backplanes. This ability and ease to multiplex cholesteric displays make it very attractive to use flexible substrate, as the flexible active-matrix backplanes are not readily available. In displays using the conventional drive scheme, Figure 5.13 shows one way that this is implemented [24, 35], where $V_{na} = 0$, $V_{a=}(V_3 + V_4)/2$, $\Delta V = V_4 - V_3$. There is no cross talk between pixels as long as $(1/2)\Delta V < V1$. Grayscale is achieved by driving the columns between $(1/2)\Delta V$ and $-(1/2)\Delta V$.

To speed up the slow addressing time of the conventional drive scheme, the dynamic drive scheme takes advantage of the fast homeotropic-transient Planar transition and the hysteresis between the Focal conic and the Planar transition [36]. This drive scheme can provide an update rate of about one page per second. The cumulative drive scheme was developed to demonstrate video for smaller resolution displays and eliminate the black line observed when updating the previous two schemes. This scheme works on the cumulative effect of

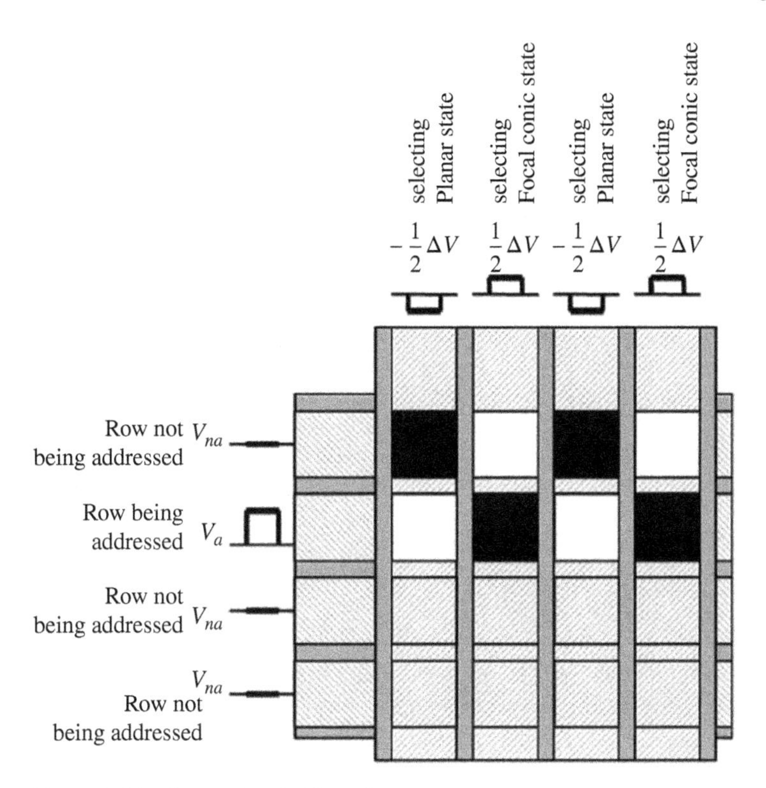

Figure 5.13 Conventional drive scheme.

repeatedly driving the same pixel multiple times, with each pulse moving the domains to the desired state [37, 38]. In addition to multiplexing, active-matrix backplanes are also used to drive the Cholesterics.

5.3.2 Photo Addressing

The bistable image created in Figure 5.3 is done without patterning the conductive coatings. The image is made using a combination of UV irradiation and a one-time electrical waveform to lock in the image. In this system, one of the chiral dopants can reduce its HTP through a trans-cis isomerization upon UV irradiation. The result is a blue shift in wavelength as the other chiral dopants in the system are of opposite handedness and do not change with UV. Upon removing UV light, the dopant reverts to its higher HTP. To maintain an image created by, say, a mask, one can use the difference in the electro-optic response shown in Figure 5.14. Applying a voltage in the shaded area will keep the unexposed area Planar and switch the exposed area to Focal conic, maintaining the image until one erases it all to the Planar texture.

5.3.3 Pressure Addressing

When a pixel of a bistable cholesteric liquid crystal is switched to the focal texture, localized pressure induces a volume change that generates flow in the film that causes it to reorient to the Planar texture. If the pitch is tuned to the visible range, the Planar texture will reflect visible light, and the Focal conic will show the color of the absorber behind it. In an unencapsulated system, this change can be uncontrolled. For example, a pencil-pushing down on a 25cm^2, $5\,\mu\text{m}$ pixels will cause about a third of the pixel to change to the bright state. The next sections will discuss controlling and harnessing this change to create images of writing surfaces and tablets that use direct pressure to the liquid crystal layer.

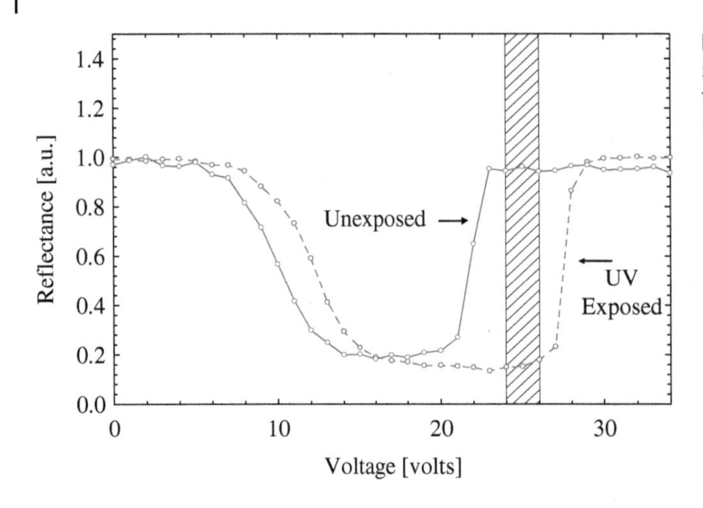

Figure 5.14 The electro optic response of the same displays that is initially reset to the Planar texture, before (solid line) and after (dashed line) UV exposure.

5.4 Applications

5.4.1 Displays for Reading

Black and white and monochrome Cholesteric displays using the technologies described above have been made on glass and plastic substrates. Full-color displays were made by stacking panels of the primary colors on top of each other and placing RGB sub-pixels on the same layer.

There are many methods of placing color into sub-pixel. One can create polymer walls in the interpixel regions to prevent the cholesterics tuned to RGB from mixing and fillinging each color row in sequence in a pixelized vacuum filing process. In another method, a cholesteric liquid crystal tuned to red is filled into a panel with polymer walls separating the future subpixels. With the appropriate UV light dosage, the HTP of an irreversible tunable chiral dopant is changed so that the appropriate pixels are tuned to green and red using the masking technique as shown in Figure 5.15 [39, 40]. In yet another method, a photo tunable chiral dopant is combined with encapsulation to create polymer walls using a photomask during polymerization [41]. As seen in references [42, 43], work in this area continues.

5.4.2 Rugged Passive Driven Displays

Using encapsulation techniques, thin, flexible patterned PET substrates, and passive driving, a reliable display can be - produced using roll to roll technique to make displays that are appropriate applications such as courtesy card and shelf labels where low power cost are key (Figure 5.16).

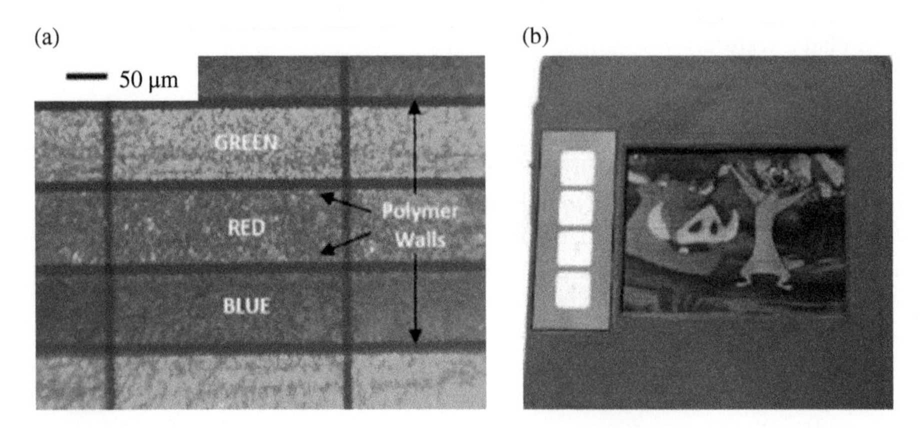

Figure 5.15 80 ppi ¼ VGA cholesteric reflective bistable display made using side by side RGB subpixels.

5.4.3 Electronic Skins

Cholesterics can be used to easily make color-changing wraps or skins. The process of manufacturing encapsulated bistable cholesteric displays is roll-to-roll compatible, including the ability to laminate layers of cholesterics reflecting the primary colors and any other cut-off filters that may be needed. A display made using this process is shown in Figure 5.17.

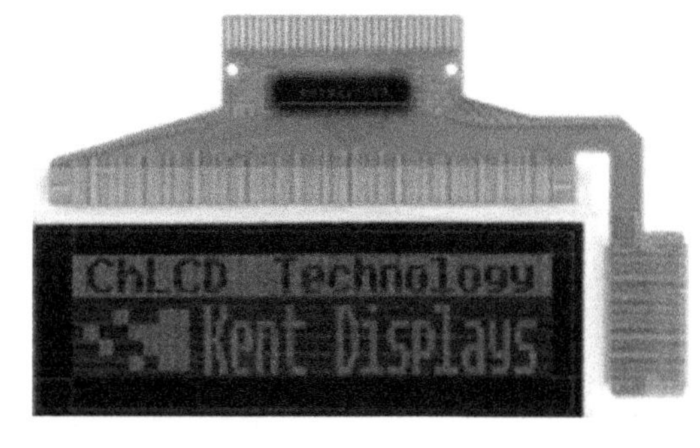

Figure 5.16 Rugged multiplexed encapsulated cholesteric displays made using plastic substrates.

5.5 Writing Tablets

ePaper technologies have transformed the user experience and interaction with electronics devices by appealing to the familiar reading experience that reflective displays offer. Cholesteric writing tablets complement this ecosystem by providing the experience of writing with pen on paper with the benefits of reusability, ruggedness, and low power associated with ChLC technology.

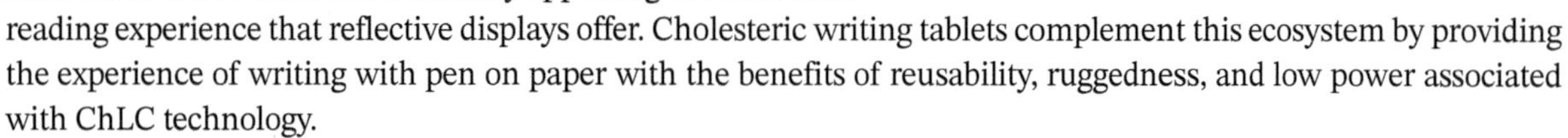

Unlike the encapsulated displays above, designed to prevent the flow of the ChLC, writing tablets take advantage of texture transitions obtained by the mechanically induced flow. They are their own category of ePaper that displays an image created by mechanical forces instead of individual pixel driving. There is no need for digitizers, force sensors, or special styli. Properties of the image created by pressure such as brightness, line width, contrast, and response to speed and pressure can be designed to target specific applications' needs. Handheld devices and medium to large size boards are just some of the examples of commercially available writing tablets.

5.5.1 Writing Tablet Structure

Figure 5.18 is a not-to-scale representation of the construction of a writing tablet panel. A polymer-dispersed ChLC layer is sandwiched between two substrates. The front substrate must be flexible to transfer the writing pressure to the polymer dispersed ChLC layer. It also needs to be transmissive so that both, Planar, and Focal conic textures, are visible. The back substrate does not need to be flexible nor transmissive, it plays a role of supporting the conductive layer, and the polymer dispersed ChLC layer. If the back substrate is transmissive, a light-absorbing

Figure 5.17 The some of the colors produced by a single pixel multilayer encapsulated cholesteric display made for applications where color change is desired. Note the hole though the display panel.

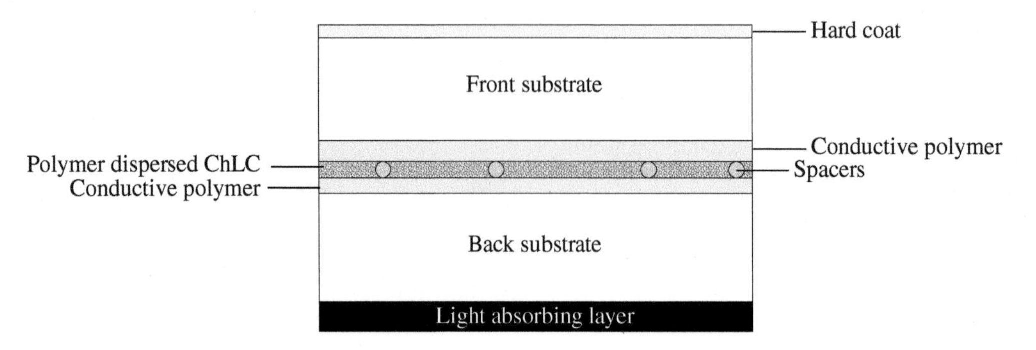

Figure 5.18 Not to scale cross section of a typical ChLC panel used in writing tablets.

layer might be needed to control the contrast between the written image and the rest of the device. The reflective and transmissive properties, and the color of the back substrate, and the light absorbing layers dictate the general appearance of the written image and the background.

Both substrates are coated with conductive polymer facing the polymer dispersed ChLC layer. There are many types of conductive polymers, but poly(3,4-ethylene dioxythiophene) polystyrene sulfonate (PEDOT: PSS) is by far the most widely used. PEDOT: PSS is easy to coat from a water-based suspension and is flexible enough to withstand tablets' writing process and manufacturing. Indium Tin Oxide (ITO) can also be used instead of conductive polymers, but its brittleness can be a disadvantage during the manufacturing process and jeopardize the final device's reliability. ITO could also require alignment or dielectric layers that prevent shorting the device.

In most cases, spacers are needed to separate the conductive layers and control the thickness of the polymer dispersed ChLC. Most typical spacers are spherical and can be made of crosslinked polymers or ceramic-like materials.

The polymer dispersed ChLC is considered the active layer of a writing tablet as it optically responds to mechanical and electrical stimuli. It consists of a ChLC. Usually, a nematic and chiral mixture is entrapped in a connected polymeric network that allows restricted fluid flow. The thickness of this layer ranges from 2 μm to around 10 μm.

Other functional layers can be added to modify the device's appearance or provide protection against environmental effects. For example, a hard coat can be coated on the front substrate to prevent premature wear and tear due to writing. The hard coat or the front substrate can also contain UV blocking agents to prevent damage by light exposure.

5.5.2 Fabrication of Writing Tablets

Unlike typical encapsulated PDLCs, the ChLC in a writing tablet is embedded in an interconnected polymer network, enabling flow without total encapsulation. It is also different from PSCT in that the presence of polymer is not only to stabilize the liquid crystal texture but also to restrict the liquid crystal preventing excessive flow. The manufacturing of writing tablets employs Polymerization Induced Phase Separation (PIPS) principles to control the morphology of the polymer network containing the ChLC. Acrylic, vinyl, or thiol-ene crosslinkers and monomers are mixed with a ChLC material in an isotropic phase with proportions typically in between PDLC's and PSCT's systems. The polymerization takes place using Ultraviolet (UV) light requiring the photoinitiator to absorb in the same frequency as the maximum emission of the light being used. The reaction is triggered following a Free Radical Polymerization (FRP) mechanism. The presence of crosslinkers, i.e. multifunctional reactive molecules, randomizes the propagation step growing a crosslinked network that starts rejecting the non-reactive ChLC components in virtue of the entropy reduction associated with the polymerization process [44, 45]. Other methods such as Temperature Induced Phase Separation (TIPS) or Solvent Induced Phase Separation (SIPS) can be used to generate the polymeric structures needed in a writing tablet. Figure 5.19 is a schematic representation illustrating that the procedure takes place after the isotropic mixture of precursors is sandwiched between the substrates and the cell gap is set with the help of spacers. In this example, the substrates are polyethylene terephthalate (P ranging from approximately 25 to 175 μm approximately.

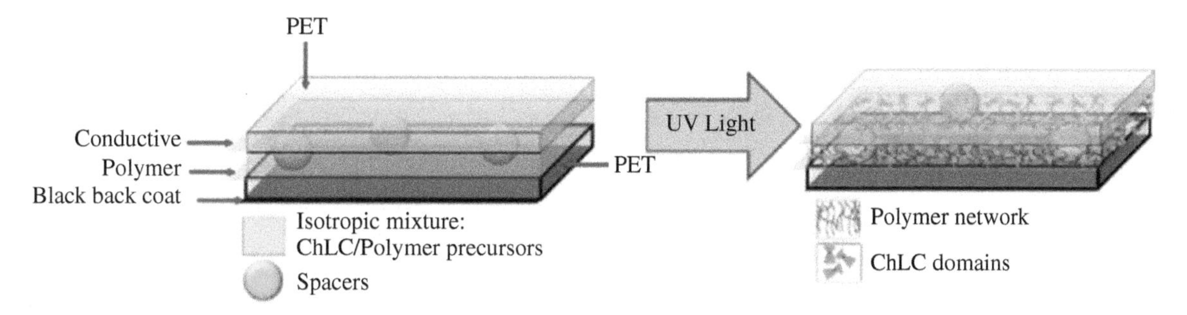

Figure 5.19 General representation of the polymerization induced phase separation process to fabricate a writing tablet.

The theoretical framework of this process proposed by different authors suggests that the kinetics of polymerization have a significant effect on the final morphology [46, 47]. This means that the crosslinker functionality and concentration, the presence of acrylates and methacrylates, the efficiency of the photoinitiation, and the curing conditions are of paramount importance to consider while formulating the initial isotropic mixture. Additionally, solubility parameters of the ChLC during the different stages of the polymerization and the viscosities and relative diffusion characteristics of the different components, are known variables of significant importance to consider that make the prediction and design of morphologies an experimental process.

As is well known, phase separation during a PIPS process can occur by nucleation and growth or spinodal decomposition. The former requires a minimum energy barrier to form nuclei that grow in time. The latter, takes place when the energy barrier is negligible, so separation proceeds through even the smallest fluctuations. Polymeric interconnected networks are often associated with phase separation by spinodal decomposition between the polymerizing structures and the liquid crystal. However, theory and experimental observations support the idea of a crossover in behavior where phase separation starts with nucleation and growth followed by emerging unstable structures corresponding to spinodal decomposition.

Furthermore, depending on the kinetics of polymerization and phenomena such as gel point, the final polymeric structure might not resemble the initial stages of phase separation. In some cases, nucleation, and growth is the dominant phase separation mechanism. Still, the morphology freezing happens later, here nucleated droplets have coalesced into semi-interconnected structures [48, 49].

The flexible properties of all the layers forming a writing tablet make it a suitable candidate for mass production using roll-to-roll processing. The most common manufacturing method in the industry starts with an isotropic mixture composed of ChLC, monomers, crosslinkers, and photoinitiators that are laminated in a cleanroom between flexible PET substrates previously coated with PEDOT: PSS. Spherical spacers control the cell gap of the laminated structure. The system is then continuously exposed to UV light that triggers the photoinitiators, subsequently initiating the polymerization. All this happens continuously at room temperature with virtually no VOC emissions or water waste. After curing, a highly precise automated laser cutting system singulates the parts without interrupting the continuity of the process. The laser cutting is engineered so high throughput can be achieved without delaminating or electrically shorting parts while, at the same time, protecting the parts from accidental delamination. The high versatility of the system is designed to perform different shapes and sizes without the need for retooling or consumables [50, 51].

5.5.3 Polymer Morphology of CHLC Writing Tablets

The ChLC response to different mechanical forces is subject to the geometry, and topological features of the resulting polymeric network after the PIPS is completed. Figure 5.20 shows SEM pictures of the morphologies of different writing tablets taken after the front and back substrate were separated and the ChLC rinsed off with isopropanol. Morphology "a" shows a tight uniform structure resembling a sandy agglomeration that restricts ChLC flow, substantially favoring the formation of thinner lines during writing. Morphology "b" combines larger

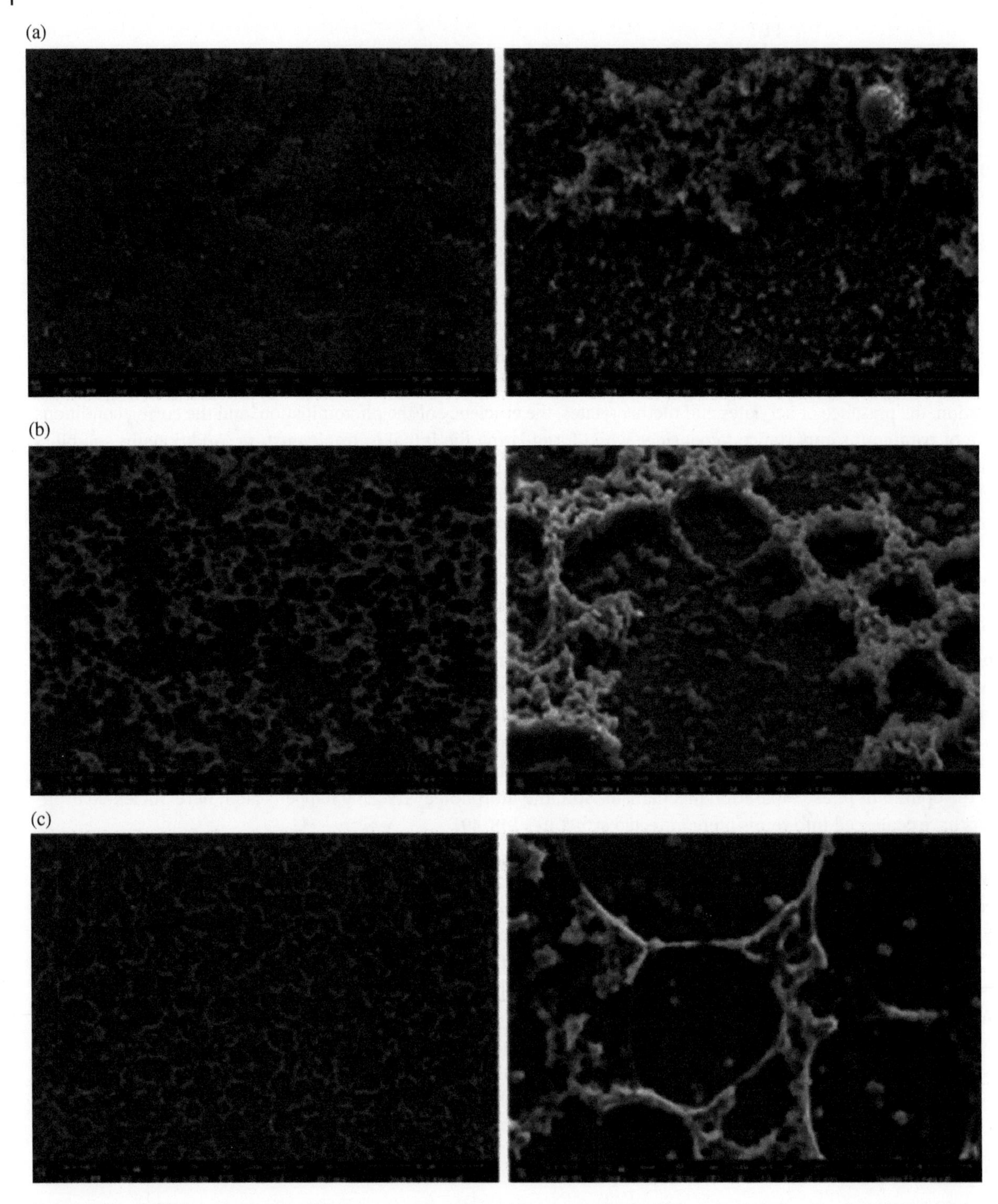

Figure 5.20 SEM pictures of three different types of writing tablet morphologies. The scale of the bar for the left-hand-side column is 50 μm and for the right-hand-side column is 5 μm.

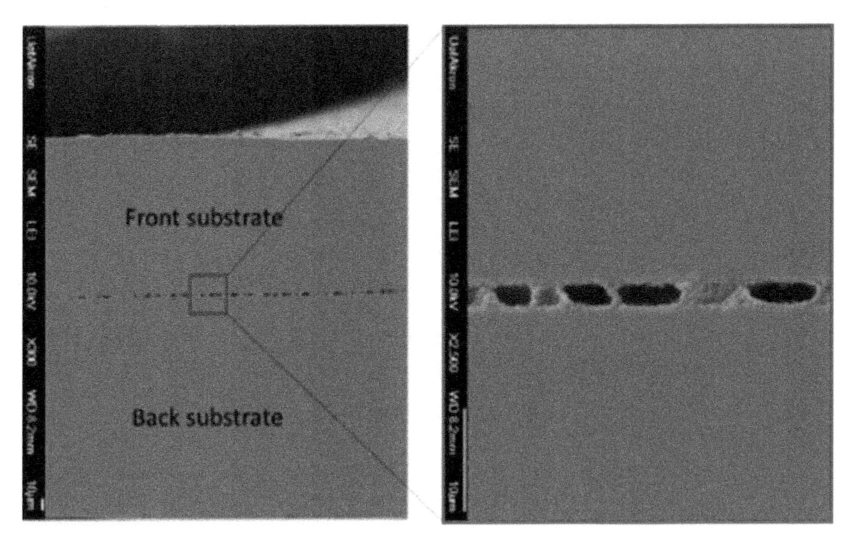

Figure 5.21 SEM micrograph of a cross section of a writing tablet. The scale bar in both figures is 10 μm.

droplets (pools) with tight tunnels connecting them, which allows for a broader line. Morphology "c" exhibits an even, more significant proportion of droplet-like reservoirs, promoting even broader and brighter written lines.

To better understand the scale of the layers in a writing tablet construction, Figure 5.21 shows a cross-section of a writing tablet obtained by SEM after preparing a sample with an ultra cryo-microtome. This morphology is similar to sample "b" in Figure 5.20. In this view, one can appreciate the polymer pockets that contained the ChLC and the tighter surrounding structure that connects them. The writing tablet of Figure 5.21 has a cell gap of 3 μm.

Figure 5.22 shows a close-up of a written mark created by applying pressure with a pointed stylus on a writing tablet. The flow created by deflecting the front substrate disturbs the formerly Focal conic textures to a bright Planar texture while the rest of the device remains in Focal conic,, allowing one to see the back substrate, which is a PET film with a black layer on the opposite side.

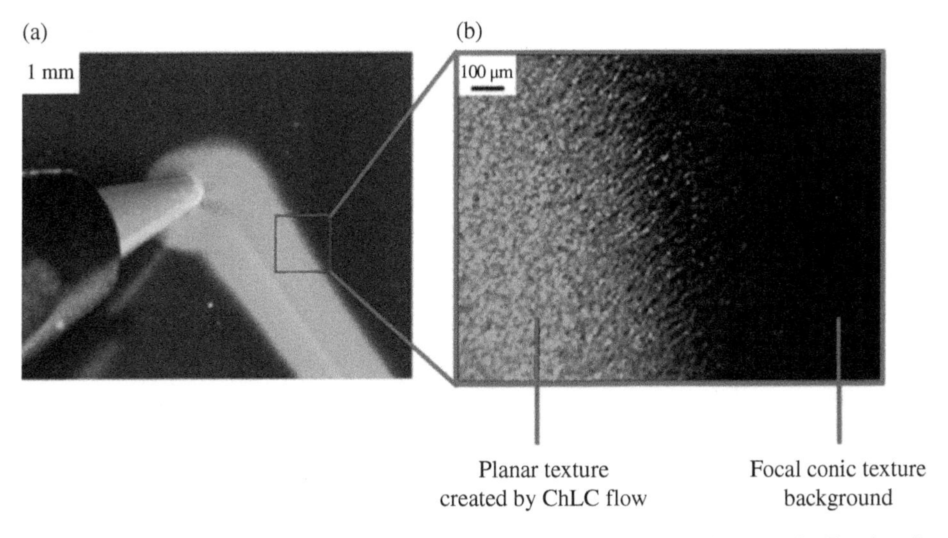

Figure 5.22 Zoom in to the border between Planar textures created by flow and the Focal conic texture of the rest of the device.

A critical aspect that separates the writing tablet technology from the more typical PDLC's and PSCT''s is that the final polymeric network must be designed to withstand high compression and shear forces involved during the writing process. For example, a polymer network with brittle properties might break irreversibly during the process. A low hardness polymer can also deform, changing the morphology of the network and, subsequently, the writing properties of the damaged area. The reliability of writing tablets will be discussed later in this chapter.

5.5.4 Characteristics of Writing Tablets

5.5.4.1 Writing Mechanism

The effect of flow on the ChLC textures has been the subject of analysis over the years. One of the first reports tracked the transmission and reflection of a layer of ChLC exposed to different shear rates [52]. The objective of the experiments was to explain the effect of shear forces on the Planar textures of the ChLC mixtures., At low shear rates, the reflectance of the ChLC layer increased and shifted toward blue, concluding that the helical structure of the ChLC behaves as a rigid unit tilting with little distortion. Intermediate shear rates seemed to create a complex stratified distribution that may include a Focal conic layer in the middle and Planar layers in contact with the shearing surfaces. Higher shear drives the system to a homeotropic state. In all cases, the system relaxed to a Planar texture after a short time.

More recently, researchers have proposed computational simulation of the microfluidic flow of cholesteric liquid crystals to study the rheology of ChLC with different anchoring conditions in the context of a Poiseuille flow. Their hydrodynamic model was solved numerically, showing other texture transitions at different pressure gradients for a ChLC with homeotropic anchoring conditions going from finger patterns at low pressures to the formation of Planar structures at larger pressure gradients [53]. To fundamentally understand the effect of flow on a polymer dispersed ChLC writing tablet, one needs to consider the complexity of the rheology of anisotropic material, such as ChLC, in confined heterogeneous geometries, such as polymeric networks (Figure 5.20). The current state of the art clarifies that it is an intricate task and an incredible opportunity for the scientific community.

As a qualitative depiction of the cycle of writing and erasing of a writing tablet, Figure 5.23 shows how an unwritten surface appears black as the weakly scattering focal conic texture allows light to go through the layers to be absorbed by the black background. If a pointed object is used to deflect the front surface of the construction, force is transferred to the polymer dispersed ChLC changing the volume and inducing localized flow on the initially Focal conic domains forcing them to reorient to a Planar texture that follows the trajectory of the writing.

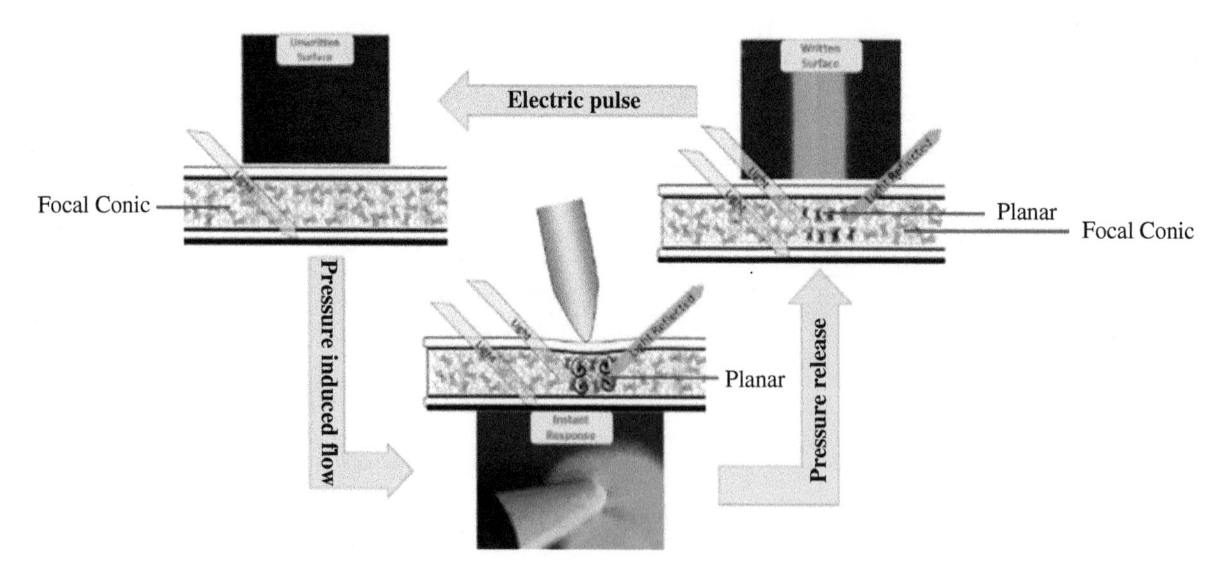

Figure 5.23 Writing and erasing cycle of a writing tablet.

The net flow of ChLC is constrained by the polymeric network providing the final shape and size of the written line that remains after the pressure is released. Notice that up to this point, there is no need for an electric field thanks to the bistable properties of the polymer-stabilized ChLC. After the writing is finished, an electric pulse can reposition all ChLC domains back to the Focal conic texture.

Human writing is a complex process that involves several neurophysiological and biomechanical variables [54]. However, for all practical purposes, one could assume that the effect of writing on a surface can be described by the writing speed and the force applied to the surface through the writing object. Previous measurements show that writing speeds and forces vary largely depending on the user, age, and writing style. However, it can be assumed that the human writing speed distribution is centered around 75 mm/s and force around 150 gF [55, 56]. These approximations will be used in the chapter to describe the general effect of human writing on the writing properties of ChLC writing tablets.

One way to characterize the writing tablet properties is to draw the line at a given weight and speed using a non-compressible stylus tip of a known radius. Its width can be measured manually or digitally. The effect of writing weight and speed on the intensity profile of written lines is shown in Figure 5.24. All measurements correspond to one writing tablet using the same instrument. Notice that the width and shape of the written mark changes

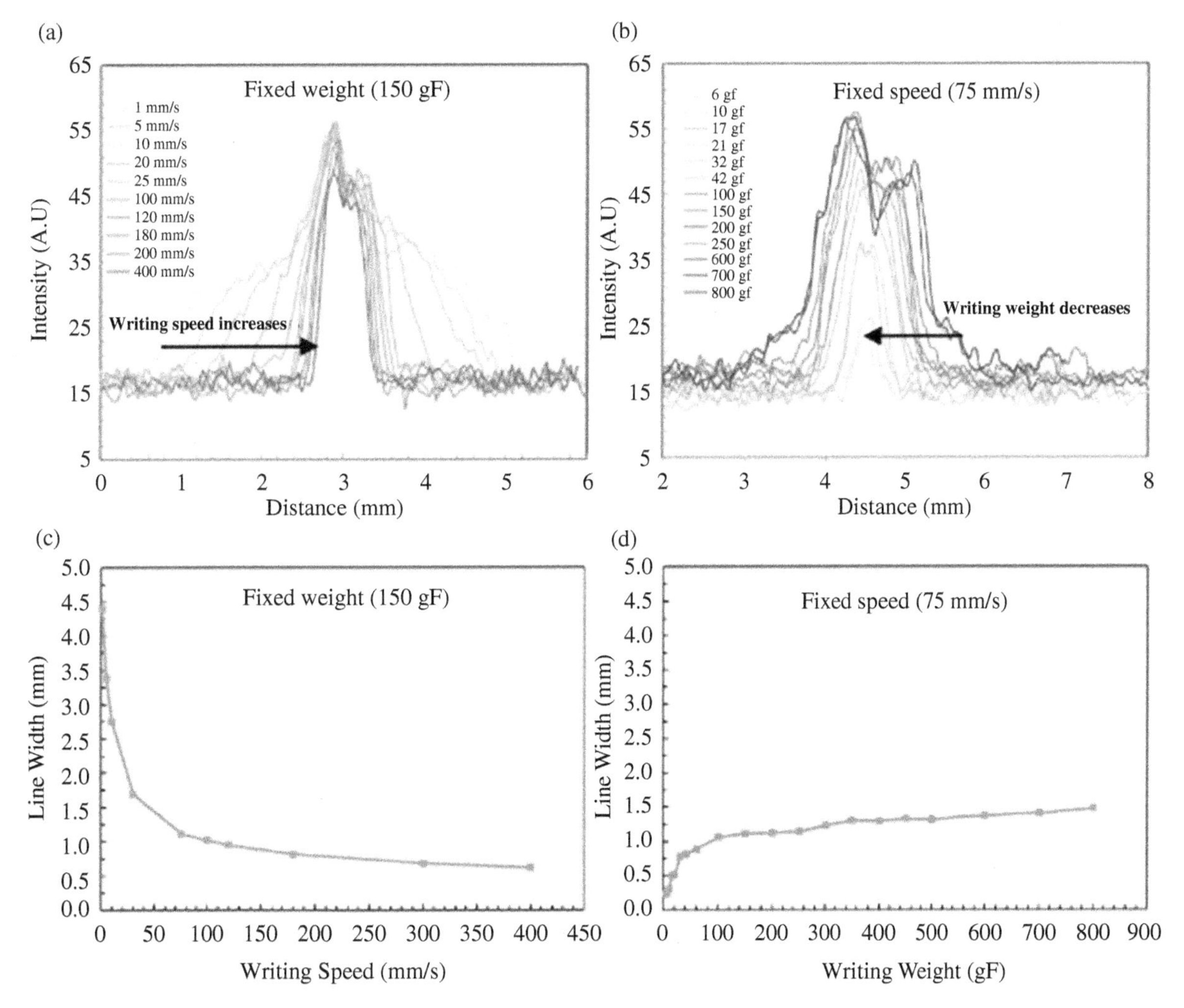

Figure 5.24 Effect on written line intensity profiles of: (a) writing speed at a fixed weight. (b) Writing weight at fixed speed. (c) and (d) are their respective plots against the line width measured at each condition.

substantially with speed at a fixed weight. In the case of fixed speed and varying weight, the shape remained similar while the intensity decreased significantly when lighter weights were applied on the writing tablet. One can infer that slower writing speeds promote larger outward flows as there is more time for liquid crystal molecules to be influenced by substrate deflection.

Interestingly, line widths cover a significant range from less than 1 mm to about 4 mm; note that the writing tip is a hard-sphere of 3 mm diameter, which means the effect is not necessarily related to the writing object size but to the flow exerted by it. The message here is that the size of the mechanically induced Planar texture is essential. The patterns and features inside the line can be of significant interest on the fundamentals of how a ChLC material responds to different forces when confined inside polymeric networks.

As hinted before, it is possible to obtain different ChLC embedded in morphologies with different features. Consequently, it is expected that the line response to weight (writing force) and speed would be different for different systems, as exhibited in Figure 5.25, where each trend represents a different polymer morphology.

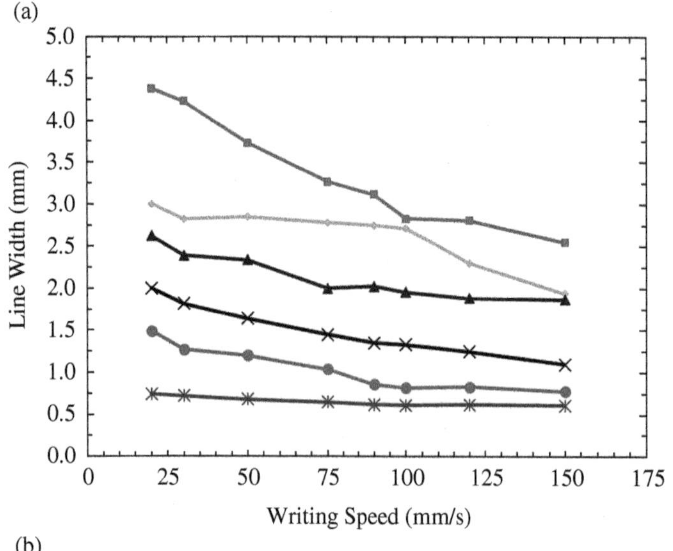

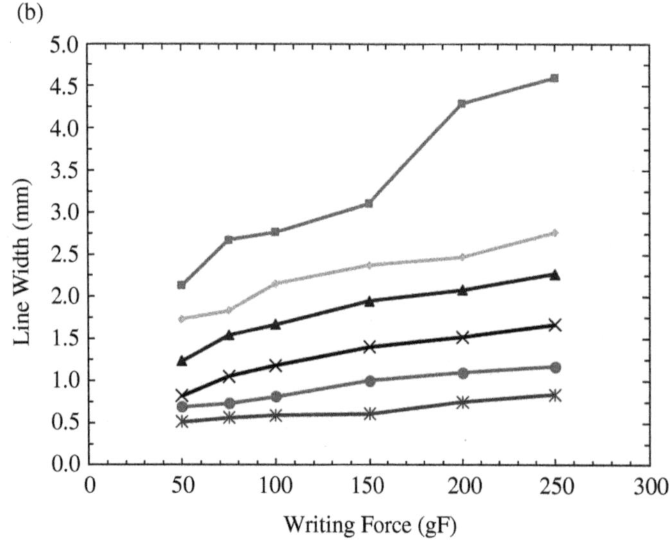

Figure 5.25 Different writing tablet morphologies exhibiting different responses to: (a) different writing speeds with writing weight fixed at 150 g. (b) different writing weights at a fixed writing speed of 75 mm/s.

It becomes evident that it is possible to obtain highly sensitive systems that respond quickly to speed or weight changes while others remain flattered over the same range of speeds and weights.

Another critical property of lines created by pressure on a ChLC writing tablet is reflectance. Generally, higher reflectance is achieved with systems that show wider line widths. However, as seen in Figure 5.26, it is also possible to obtain thinner lines with the same or higher reflectance as thicker ones. The factors affecting the reflectance of writing tablets are not entirely understood given the complexity of the morphology, alignment between the ChLC and polymer walls, interactions between ChLC and/or polymer with the surface of conductive polymer, scattering effects, etc. Writing reflectance measurement can be made through diffuse illumination at an 8° viewing cone, with

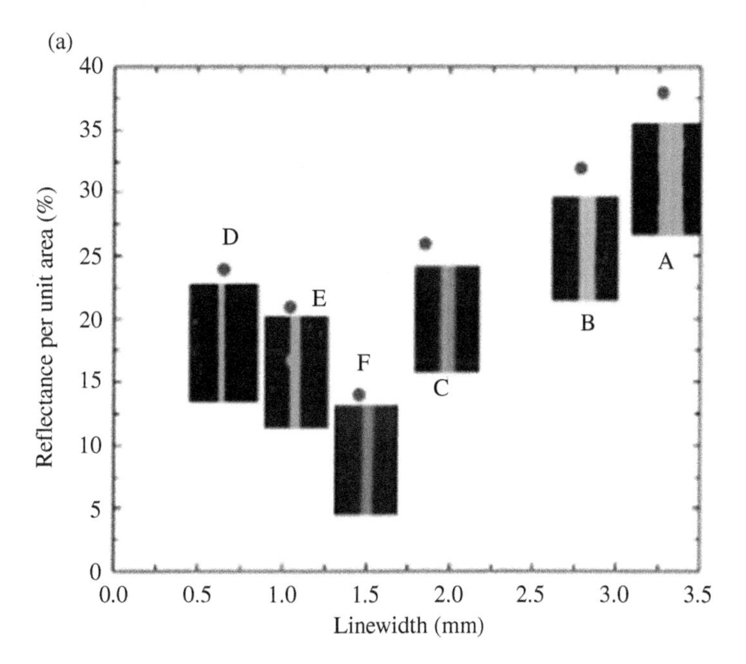

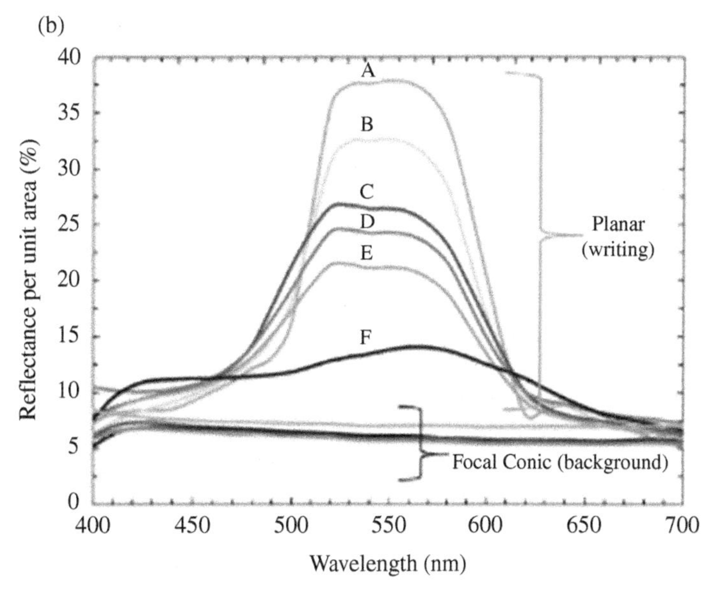

Figure 5.26 (a) Linewidth and peak reflectance per unit area plot for different writing tablets A–F. (b) Reflectance spectra of Planar and Focal conic textures of writing tablets A–F.

specular component included (SCI) under a D65 illuminant for a 10 ° (CIE 1964) observer. The measurement is made, so the aperture of the equipment covers an area that is fully written on.

Consequently, any reference to reflectance should be considered as per unit area. The comparison between the reflectance of the written line (Planar) versus the reflectance of the background (Focal conic) is a measure of the contrast of the device. The background reflectance is a combination of the optical properties of the Focal conic texture of the ChLC, polymeric network, and back substrate. As seen in Figure 5.26, contrast-enhancing is usually achieved by increasing the reflectance of the written line without increasing the reflectance of the background.

One interesting finding is that the effect of the cell gap does not follow a monotonic relationship with the reflectance of the written line. Unlike pure ChLC, writing tablets may exhibit different regimes [57]. As seen in Figure 5.27, thinner cell gaps exhibit higher reflectivity up to 4 μm, where the behavior changes perhaps now dominated by the well-known dependence on the number of pitches that can fit in the cell gap [58]. One plausible hypothesis is that thinner cell gaps maximize the alignment of ChLC molecules in the Planar texture to the substrate surface concerning the lateral polymeric walls. Similarly, one can imagine that thinner cell gaps promote the formation of ChLC pools in interconnected droplets like the structures presented in Figure 5.20; it is expected that said structures result in better and larger Planar domains.

5.5.4.2 Electro-Optic Response of Writing Tablets

Like measuring the curves with a pure electrical response, the display needs to be reset to a known initial state before measuring its voltage response. An erase and reset waveform at the drive frequency (typically around 3 Hz) that would switch the display to the weakly scattering Focal conic state are determined to obtain these curves. Typically, this waveform has two bipolar pulses, one to switch the writer to Planar and another to reset the display to the dark state. The erase pulse amplitude is set to the maximum voltage step of the measurement. The best reset amplitude is the voltage provides the darkest focal conic state. A typical result is obtained u the schematic shown in Figure 5.28 for every voltage step, a specific result is obtained as shown in Figure 5.29.

Note that, in some cases, the reflectivity of the written line is higher or lower than the reflectivity obtained by switching the display to the bright state. Looking at Figure 5.29, the reflectivity of the written line before voltage V1 is higher than the reflectivity after voltage V4. At voltages less than V1, the voltage is not high enough to switch the display to homeotropic and enable the transition to Planar. Therefore the reflectance of the written line dominates. But at voltages higher than V4, the voltage is high enough to switch the display to Planar.

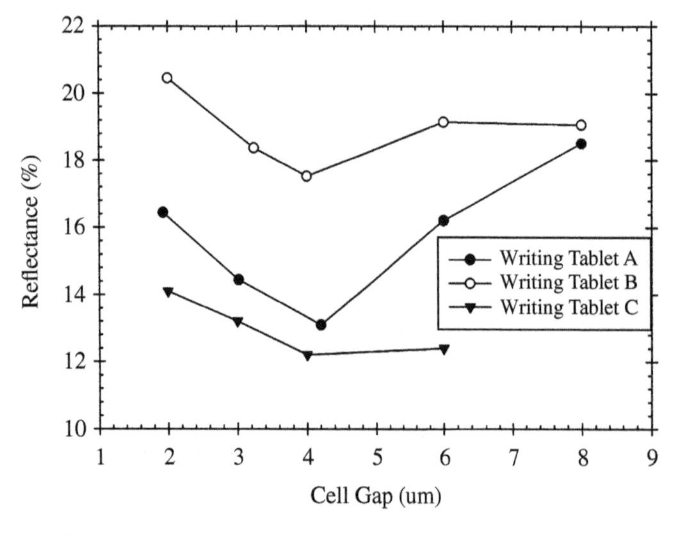

Figure 5.27 Reflectance of the written line of different writing tablets with different cell gaps With Adapted from [57].

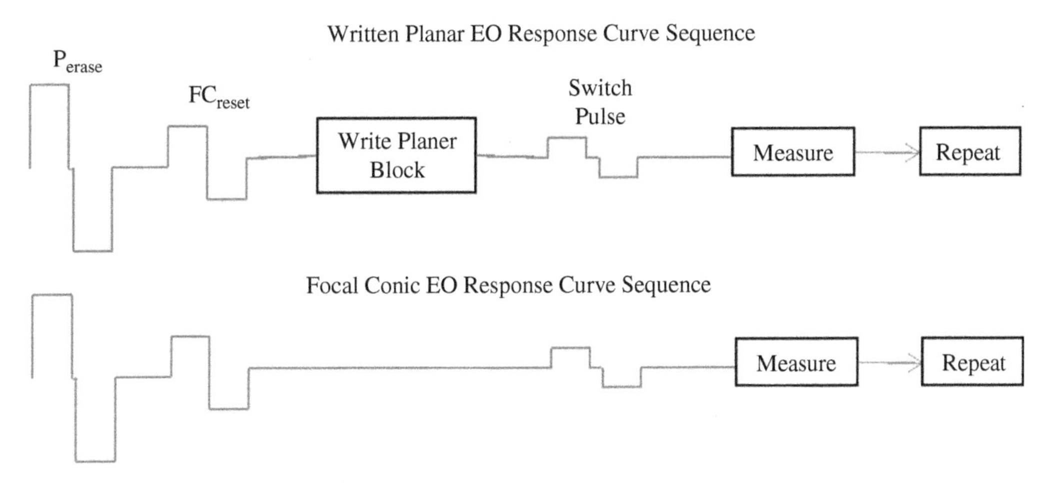

Figure 5.28 Schematic for measuring the electro-optic response of a cholesteric writing tablet.

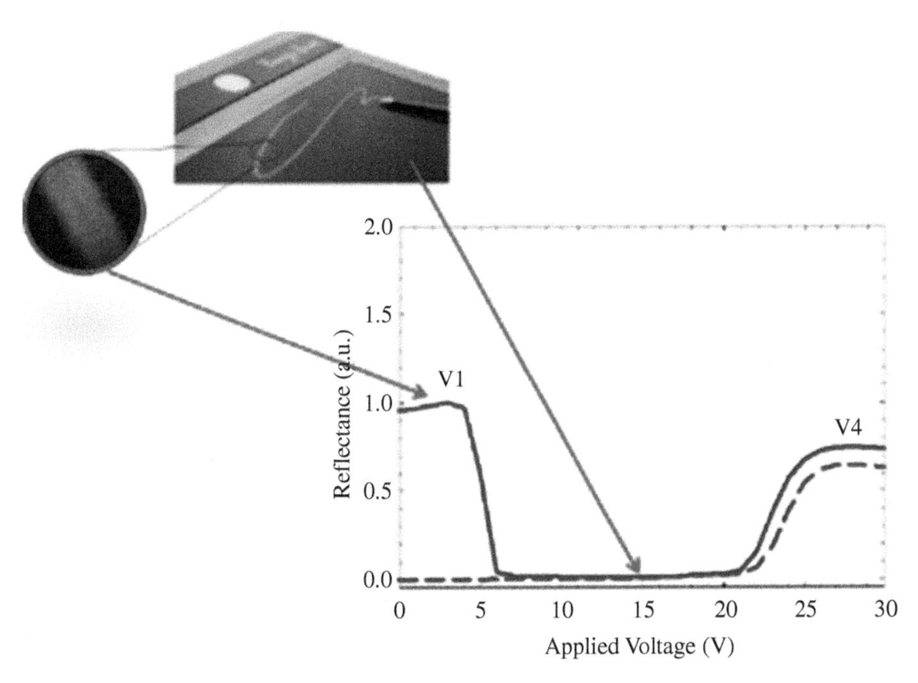

Figure 5.29 A typical electro-optic response curve of a bistable cholesteric writing tablet, the solid line is the response of the written block and the broken line is that of the background.

5.5.4.3 Exact Erase

The writing tablets described above are single-pixel devices; that is, the same electrodes are shared over the whole writing area. Therefore, the whole tablet is erased. It is possible to erase a portion of the tablet to erase minor mistakes, as shown in Figure 5.30. This type of erasing takes advantage of the pressure from the stylus that locally changes the cell gap from the nominal d_1 to a reduced d_2, as shown in Figure 5.31. This reduction in cell gap and the flow of the liquid changes the electro-optic response in that area of pressure Figure 5.32 indicates that there is a selective erase voltage, V_{EE}, where the reflectivity of the written image, R_W, is maximized and

Figure 5.30 Example of exact erase. (a) Original writing (b) Exact Erase (c) corrected writing.

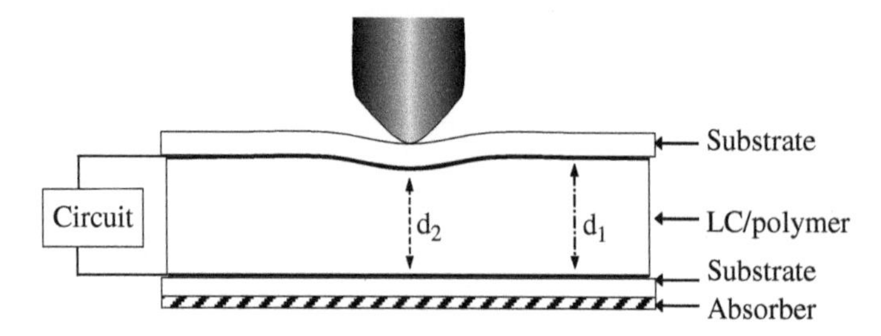

Figure 5.31 Cross section of a writing tablet showing nominal cell gap, d1, and reduced cell gap, d2, from stylus pressure.

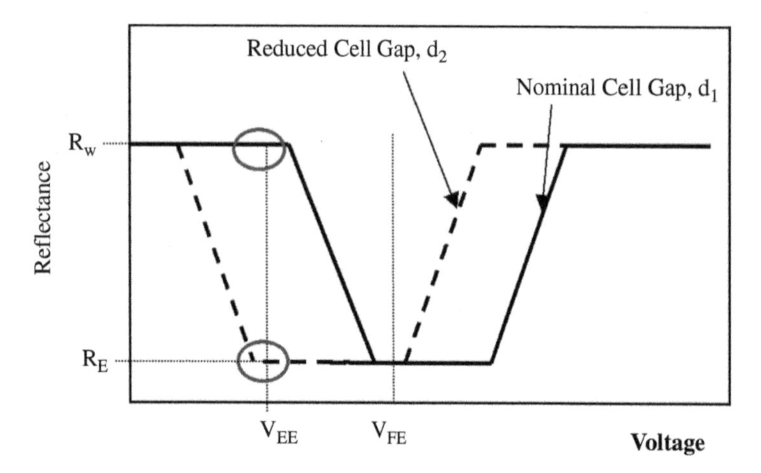

Figure 5.32 Idealized illustration of the EO curves for the regions with nominal cell gap d1 and reduced cell gap d2.

reflectivity of the erased image, R_E, is minimized. Beyond that range, it is possible to fully erase the display using, for example, voltage V_{FE} without the need for eraser pressure [59].

5.5.4.4 Writing Reliability

Unlike paper, writing tablets have to survive numerous writings during the life of the device. It would be unacceptable, for example, if the writing were illegible or altered due to previous writing. To develop a reliability test, the writing profiles of 10 individuals writing on cholesteric writing tablets were captured using a digitizer behind the tablet [60, 61].

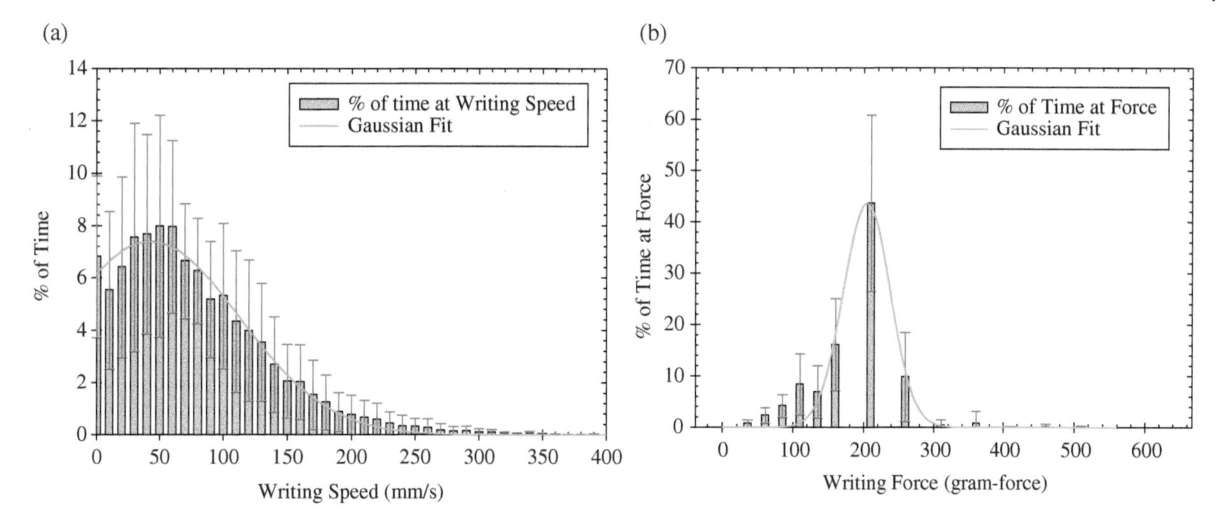

Figure 5.33 Histograms of humans writing on a writing tablets (a) writing speed and (b) writing force.

From this study (Figure 5.33), the average writing speed is 65 ± 50 mm/s, and the average writing force is 205 ± 35 g-force. Further analysis of the data shows that,, individuals only write faster than the above averages 55% of the time and only about 17% of the time at speeds greater than 115 mm/s. They also only write with forces above average about 12% of the time. The average maxima of the writing speeds and forces were also found. The average maximum writing speed is 320 ± 149 mm/s, about 5 times higher than the average writing speed above. The average maximum writing force is 390 ± 920 g-force, which is about 2 times higher than the average writing force above.

With these results, a writing test and pattern were developed (Figure 5.34a). The pattern was chosen to have some repetition to accelerate wear. The writing test is done using a robot that normally holds a stylus with a 150-gramg force attached and writing at a nominal velocity of 63 ± 33 mm/s. Using these parameters and even higher forces, writing tablets show remarkable robustness, especially at typical writing forces, as shown in Figure 5.35.

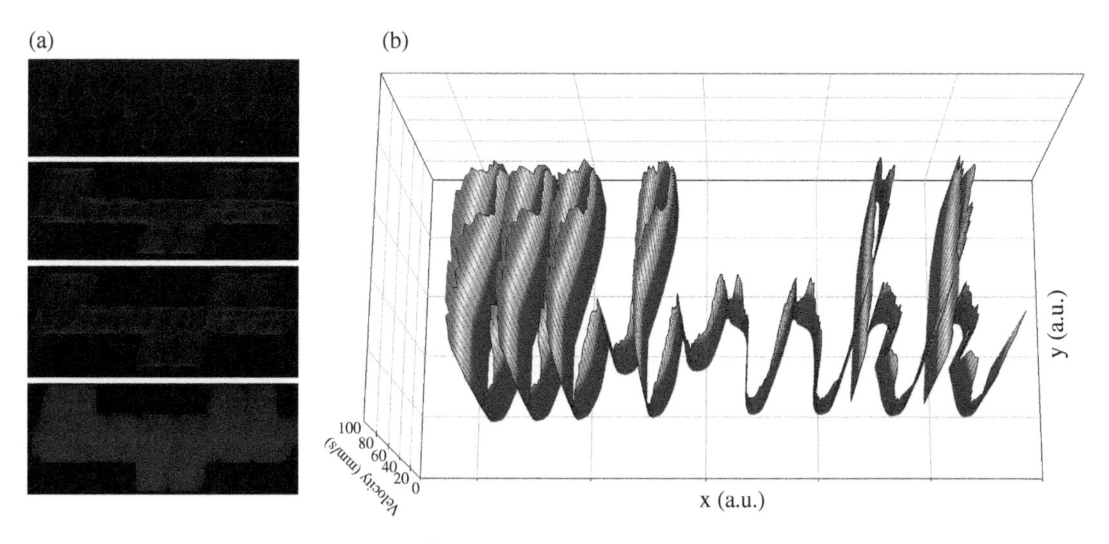

Figure 5.34 (a) Writing pattern designed for reliability testing. (b) Instantaneous velocity of the writing.

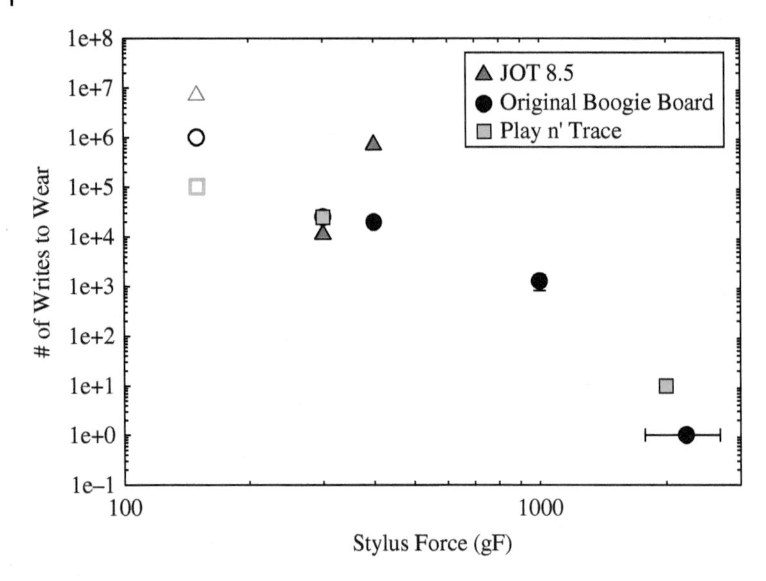

Figure 5.35 Wear points of writing tablets written on repeatedly with by a robot with increasing force. The points at the lowest force (Open Symbols) are the current iteration of the test, and not a wear point.

5.5.5 Semi-Transparent Writing Tablets

Reducing the absorption of the light absorbing layer behind the writing tablet allowed for a new use model for displays. Figure 5.36 shows writing tablets made with absorbing layers of varying degrees of transparency. Note that even with a bright reflector behind the display, the contrast of writing in the portion of the display with 50% absorption is surprisingly high. To construct the optimal semitransparent eWriter, there are multiple

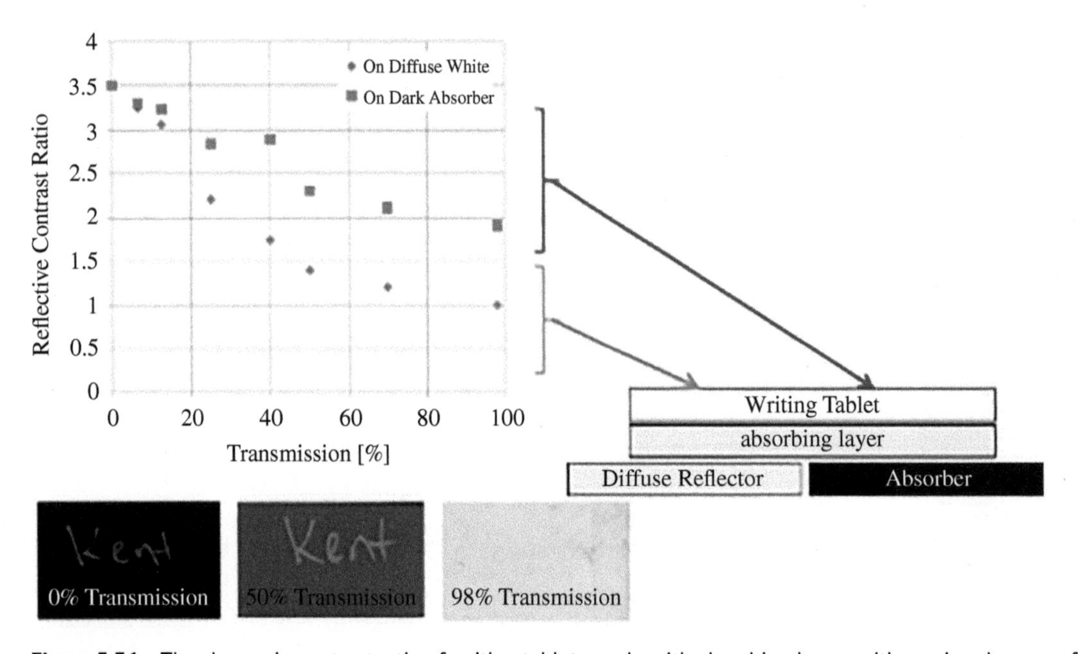

Figure 5.36 The change in contrast ratio of writing tablets made with absorbing layers with varying degrees of transparency.

parameters that need to be optimized. These parameters include the color and transmission of the absorbing layer, the haze and clarity of the liquid crystal layer in the Focal conic state, and the transmission, haze, and clarity of the substrates used to make the display [62].

The color of the absorbing layer is usually a matter of preference as almost any color can be selected as long as there is sufficient color difference between the color reflected by the written Planar texture of cholesteric liquid crystal and the color of the absorber, as shown in Figure 5.37.

An adequately designed semi-transparent writing tablet has high contrast in both reflection and transmission mode and has high clarity in transmission mode. This allows one to view objects behind the eWriter, while in use. With these properties, a semi-transparent writing tablet can be used as a stand-alone tablet and an electronic tracing paper, as shown in Figure 5.38. When coupled to another display, say an eReader, the writing tablet can be used to fill in forms harnessing its excellent writing experience, as shown in Figure 5.39.

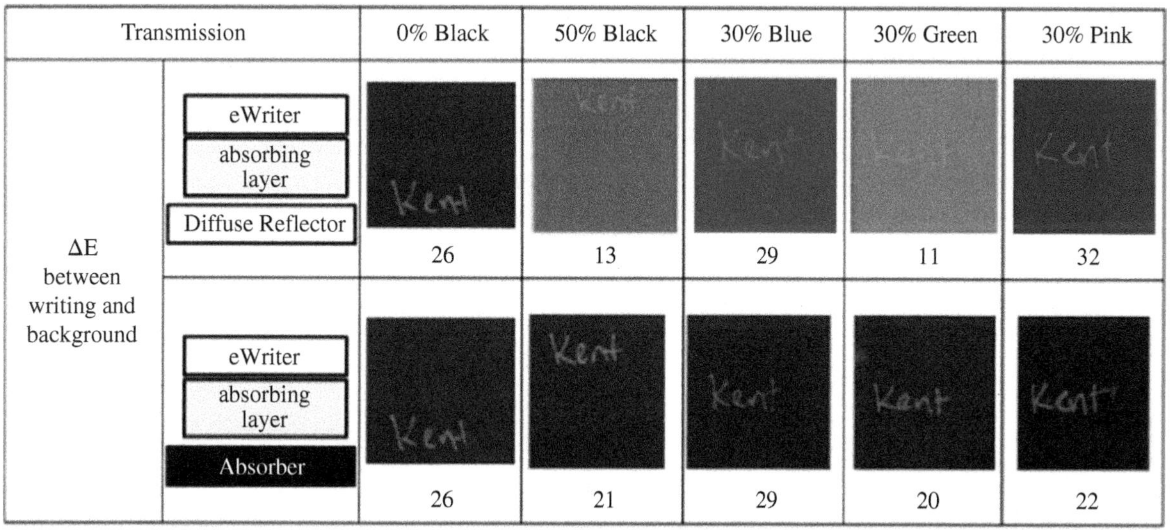

Transmission		0% Black	50% Black	30% Blue	30% Green	30% Pink
ΔE between writing and background	eWriter absorbing layer Diffuse Reflector	26	13	29	11	32
	eWriter absorbing layer Absorber	26	21	29	20	22

Figure 5.37 Color contrast of semitransparent writing tablets as quantified by color difference ΔE from the 1976 CIE L*a*b* Space.

Figure 5.38 Semi-transparent writing tablet used in tracing mode (left) and writing mode (right).

Figure 5.39 A semi-transparent display coupled to an eReader display.

5.5.6 Multi-Colored Writing

Just as the semi-transparent writing tablet above replaces tracing paper, it is possible to make a writing tablet that mimics better scratch art paper. Scratch-art paper reveals hidden patterns and colors after a messy black wax coating is scratched off. The writing tablet has a uniform background that shows bright iridescent colors that change across the writing surface. The writing surface is the color of the absorbing background until written on, at which time the color of the writing will depend on the color reflected by the cholesteric liquid crystal layer and the absorbing background at that particular writing location, as shown in Figure 5.40.

To fabricate this writing tablet, the same encapsulation technique described earlier is used to place two or more dispersions of cholesteric liquid crystals that are tuned to reflect different colors next to each other in the same layer, as shown in Figure 5.41. Note that the electrodes are shared among the colors Therefore, to erase the written image, the electro-optic response curves need to be measured for each color region. With this data, a voltage is chosen so that when applied, no matter what the original state, written, or unwritten, the field will erase all the regions of tablet to the Focal conic state [63].

5.5.7 Large Format Writing Surfaces

Bistable cholesteric liquid crystal writing tablets can be scaled up to large size formats. Some approaches involved tiling two or more panels together to allow for ease and economy of manufacturing while providing erase of the individual panels as detailed in reference [64]. Other approaches shown in Figure 5.42 are made from a single

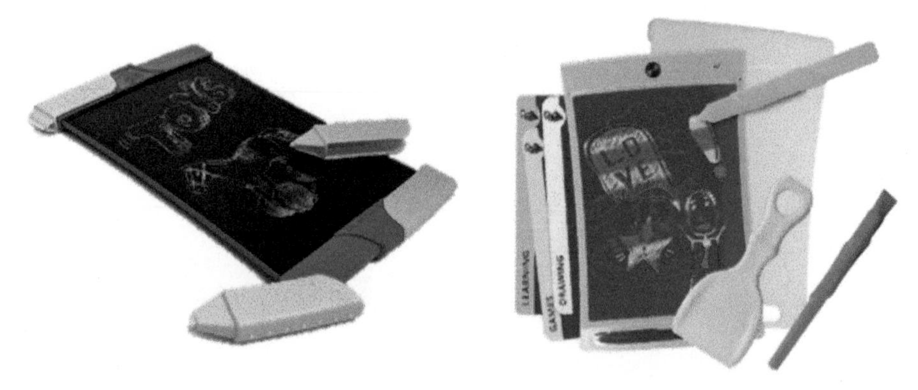

Figure 5.40 A writing tablets revealing multiple colors. The one on the left has an opaque absorber while the right is semi-transparent.

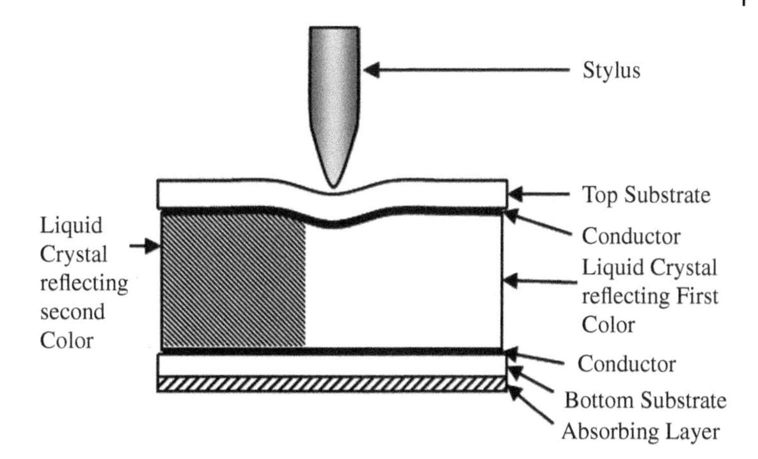

Figure 5.41 A cross section of the writing tablet revealing multiple colors.

Figure 5.42 55-in. diagonal bistable writing tablet.

panel with one button to erase the whole single 55-in. diagonal writing surface. As it is a low-power device, it only needs 4 batteries to last over 50000 erases, allowing for ease of installation and mobility. The materials and cure conditions of the panel are chosen so that the response to stylus pressure creates marks bright and wide that can be read across a room, making it an ideal collaboration tool.

5.5.8 Saving the Written Image

There are many methods of capturing the writing on the writing tablet. The simplest is by simply taking a photograph of the written image; the surface of the device usually has an antiglare coating that allows for very good camera capture. A resistive touch panel can also be placed behind the writing tablet to capture the writing. This method works well with smaller devices, but larger panels' palm rejection is an issue. Using EMR digitizers, one can eliminate this issue as the digitizer can distinguish between touch and pen inputs.

The most elegant and effective way of capturing the writing is placing a print with location information encoded into it behind a semi-transparent writing tablet and using a pen with a camera to read the location code. The writing tablet being transparent to IR allows the camera to read the code with ease with no loss in positional information. The camera can be connected to a mobile device to share the captured image as shown in Figure 5.43.

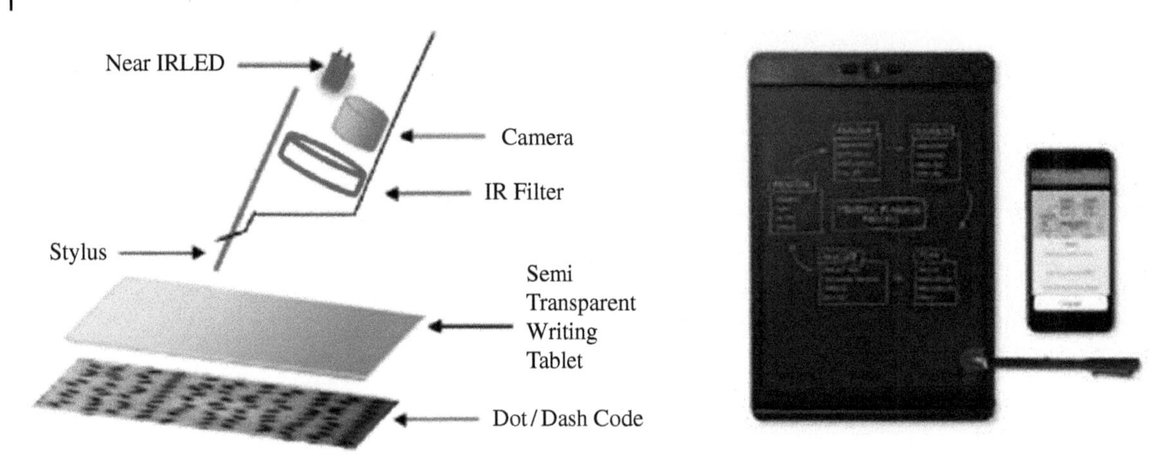

Figure 5.43 Capturing the written image using dot/dash encoded print behind a semitransparent writing tablet.

5.6 Conclusions

Cholesteric liquid crystals are very well suited for various ePaper applications as they have all the properties needed, including low power consumption, very small bill of materials, ability to use birefringent substrates, bistability, ruggedness, flexibility, and scalability. The displays made from ChLC's can be passively driven, withstand harsh outdoor conditions, made into readers and billboards. They can be put into wallets or used to change the color of a phone case.

Writing tablets are a very successful example of the use of ChLC displays that cater to writing as an undeniably essential human activity complementary to reading and still very much dependent on the use of paper. The simple but elegant use of ChLCs results from years of evolutionary and revolutionary developments that resulted in the practical, manufacturable, and low-cost variety of products currently in the market that has been sold by the millions.

The study of the phenomena involved in the writing process, understanding the effect of ChLC flow in confined complex geometries, is still an intriguing topic ready to welcome the academic and industrial interest to keep propelling innovations around these fascinating materials.

References

1 de Gennes, P.G. and Prost, J. (1993). *The Physics of Liquid Crystals*, 2e. Oxford: Clarendon.

2 Venkataraman, N., Magyar, G., Montbach, E. et al. (2009). Thin flexible photosensitive cholesteric displays. *J. Soc. Inf. Disp.* 17 (10): 869–873. https://doi.org/10.1889/JSID17.10.869.

3 Bahadur, B. (ed.) (1992). *Liquid Crystals: Applications and Uses*, 3e. World Scientific p. 424.

4 Booth, C.J. (1998). Chiral nematic liquid crystals: the synthesis of chiral nematic liquid crystals. In: *Handbook of Liquid Crystals*, 303–334. https://doi.org/10.1002/9783527620555.ch4a.

5 Wu, S.-T. and Yang, D. (2001). *Reflective Liquid Crystal Displays*. Wiley.

6 Lehman (1889). Uber Fliessende Krystalle. *Z. Phys. Chem.*. Stoechiom, Verwandtschafsl, 4 (62).

7 Dierking, I. (2003). The nematic and cholesteric phases. *Textures of Liquid Crystals.* 51–74. https://doi.org/10.1002/3527602054.ch5.

8 Yang, D.-K., West, J.L., Chien, L.-C., and Doane, J.W. (1994). Control of reflectivity and bistability in displays using cholesteric liquid crystals. *J. Appl. Phys.* 76 (2): 1331. https://doi.org/10.1063/1.358518.

9 Blinov, L.M. and Chigrinov, V.G. (1994). *Electrooptic Effects in Liquid Crystals*. Springer-Verlag.

10 Lavrentovich, O.D. (2020). Electromagnetically tunable cholesterics with oblique helicoidal structure [invited]. *Opt. Mater. Express* 10 (10): 2415. https://doi.org/10.1364/ome.403810.

11 Froyen, A.A.F., Wübbenhorst, M., Liu, D., and Schenning, A.P.H.J. (2021). Electrothermal color tuning of cholesteric liquid crystals using interdigitated electrode patterns. *Adv. Electron. Mater.* 7 (2): doi: 10.1002/aelm.202000958.

12 Mulder, D.J., Schenning, A.P.H.J., and Bastiaansen, C.W.M. (2014). Chiral-nematic liquid crystals as one dimensional photonic materials in optical sensors. *J. Mater. Chem. C* 2 (33): 6695–6705. https://doi.org/10.1039/c4tc00785a.

13 Huang, Y., Zhou, Y., Doyle, C., and Wu, S.-T. (2006). Tuning the photonic band gap in cholesteric liquid crystals by temperature-dependent dopant solubility. *Opt. Express* 14 (3): 1236. https://doi.org/10.1364/oe.14.001236.

14 Göbl-Wunsch, A. and Heppke, G. (1979). Temperaturfunktionen der durch Mehrfachdotierung mit chiralen Verbindungen in einer nematischen Phase induzierten Helixganghöhe. *Zeitschrift fur Naturforsch. Sect. A J. Phys. Sci.* 34 (5): 594–599. https://doi.org/10.1515/zna-1979-0512.

15 Schadt, M. (1997). Liquid crystal materials and liquid crystal displays. *Annu. Rev. Mater. Sci.* 27 (1): 305–379. https://doi.org/10.1146/annurev.matsci.27.1.305.

16 Chandrasekhar, S. (1992). *Liquid Crystals*. New York: Cambridge University Press.

17 Yang, D.K., Huang, X.Y., and Zhu, Y.M. (1997). Bistable cholesteric reflective displays: materials and drive schemes. *Annu. Rev. Mater. Sci.* 27 (1): 117–146. https://doi.org/10.1146/annurev.matsci.27.1.117.

18 Doane, J.W. and Khan, A. (2005). Cholesteric liquid crystals for flexible displays. In: *Flexible Flat Panel Displays*, 331–354. https://doi.org/10.1002/0470870508.ch17.

19 Doane, J.W. (1990). Polymer dispersed liquid crystal displays. In: *Liquid Crystals – Applications and Uses* (ed. B. Bahadur), 361–395. World Scientific.

20 Schneider, T., Montbach, E., Davis, D. et al. (2008). UV Cured Flexible Cholesteric Liquid Crystal Displays, vol. LCD, pp. 2–12.

21 Kitzerow, H.-S. and Crooker, P.P. (1993). UV-cured cholesteric polymer-dispersed liquid crystal display. *J. Phys. II Fr.* 3 (5): 719–726.

22 Echeverri, M. (2012). Phase Diagram Approach to Fabricating Electro-Active Flexible Films: Highly Conductive, Stretchable Polymeric Solid Electrolytes and Cholesteric Liquid Crystal Flexible Displays.

23 Crawford, G.P. (2005). Encapsulated liquid crystal materials for flexible display applications. In: *Flexible Flat Panel Displays* (ed. G.P. Crawford), 313–330. Wiley.

24 Yang, D.K. (2006). Flexible bistable cholesteric reflective displays. *IEEE/OSA J. Disp. Technol.* 2 (1): 32–37. https://doi.org/10.1109/JDT.2005.861595.

25 Schneider, T., Nicholson, F., Khan, A. et al. (2005). Flexible encapsulated cholesteric LCDs by polymerization induced phase separation. *SID Symp. Dig. Tech. Pap.* 36 (1): 1568. https://doi.org/10.1889/1.2036311.

26 Yang, D.-K. and Ma, R. (1999). Polymer stabilized black-white cholesteric reflective display. no. 19.

27 Miller, N. et al. (2003). 54.1: Ultra low power black and white cholesteric display with COG and solar array. *SID Symp. Dig. Tech. Pap.* 34: doi: 10.1889/1.1832557.

28 Broer, D.J., Lub, J., and Mol, G.N. (1995). Wide-band reflective polarizers from cholesteric polymer networks with a pitch gradient. *Nature* 378 (6556): 467–469. https://doi.org/10.1038/378467a0.

29 Li, Y., Zhan, T., and Wu, S.-T. (2020). Flat cholesteric liquid crystal polymeric lens with low f-number. *Opt. Express* 28 (4): 5875. https://doi.org/10.1364/oe.387942.

30 Li, Y., Zhan, T., Yang, Z. et al. (2021). Broadband cholesteric liquid crystal lens for chromatic aberration correction in catadioptric virtual reality optics. *Opt. Express* 29 (4): 6011. https://doi.org/10.1364/oe.419595.

31 Tondiglia, V.T., Natarajan, L., Bailey, C.A. et al. (2011). Electrically induced bandwidth broadening in polymer stabilized cholesteric liquid crystals. *J. Appl. Phys.* 110 (5): 53109. https://doi.org/10.1063/1.3632068.

32 Montbach, E., Pishnyak, O., Lightfoot, M. et al. (2009). 4.2: Flexible electronic skin display. *SID Symp. Dig. Tech. Pap.* 40 (1): 16. https://doi.org/10.1889/1.3256657.

33 Helfrich, W. (1971). Electrohydrodynamic and dielectric instabilities of cholesteric liquid crystals. *J. Chem. Phys.* 55 (2): 839–842. https://doi.org/10.1063/1.1676151.

34 Clinton Braganza, D.-K. Y. (2005). Dynamics of the Growth of Twist Fingers in Cholesteric Liquid Crystals. *APS March Meeting*. http://meetings.aps.org/link/BAPS.2005.MAR.B37.9.

35 Yang, D.K., Doane, J.W., Yaniv, Z., and Glasser, J. (1994). Cholesteric reflective display: drive scheme and contrast. *Appl. Phys. Lett.* 64 (15): 1905. https://doi.org/10.1063/1.111738.

36 Huang, X., Yang, D., and Doane, J.W. (1995). Transient dielectric study of bistable reflective cholesteric displays and design of rapid drive scheme. *Appl. Phys. Lett.* 67 (9): 1211–1213. https://doi.org/10.1063/1.115010.

37 Zhu, Y.M. and Yang, D.-K. (1998). P-82: Cumulative drive schemes for bistable reflective cholesteric LCDs. *SID Symp. Dig. Tech. Pap.* 29 (1): 798–801. https://doi.org/10.1889/1.1833882.

38 Khan, A., Miller, N., Ernst, T. et al. (2004). 22.4: Novel drive techniques and temperature compensation mechanisms in reflective cholesteric displays. *SID Symp. Dig. Tech. Pap.* 35 (1): 886–889. https://doi.org/10.1889/1.1821411.

39 Braganza, C., Bowser, M., Krinock, J. et al. (2011). 30.2: Single layer full color cholesteric display. *SID Symp. Dig. Tech. Pap.* 42 (1): 396–399. https://doi.org/10.1889/1.3621334.

40 Liang, C.-C., Liao, Y.C., Chen, C.J. et al. (2011). 30.4: A low cost, full color cholesteric LCD with low voltage and crosstalk-free drive scheme. *SID Symp. Dig. Tech. Pap.* 42 (1): 404–407. https://doi.org/10.1889/1.3621336.

41 Chien, L. -C. and Doane, J. W. (1997). Pixelized Liquid Crystal Display Materials Including Chiral Materal Adopted To Change Its Chrality Upon Photo-Irradation. US Patent 5,668,614.

42 Jin, Y., Hong, Z., Jeon, C.W. et al. (2014). A novel flexible color display using two-step UV exposure method. *Mol. Cryst. Liq. Cryst.* 595 (1): 83–91. https://doi.org/10.1080/15421406.2014.917792.

43 Chin, C.-L., Cheng, K.-L., Wang, Y.-R. et al. (2012). 40.2: Novel Phototunable chiral materials for single layer color cholesteric display. *SID Symp. Dig. Tech. Pap.* 43 (1): 548–550. https://doi.org/10.1002/j.2168-0159.2012.tb05838.x.

44 Kim, J.Y., Cho, C., Palffy-Muhoray, P. et al. (1993). Polymerization-induced phase separation in a liquid crystal-polymer mixture. *Phys. Rev. Lett.* 71 (14): 2232–2236.

45 Nwabunma, D., Chiu, H., Kyu, T., and Introduction, I. (2000). Theoretical investigation on dynamics of photopolymerization-induced phase separation and morphology development in nematic liquid crystal/polymer mixtures. *Chem. Phys.* 113 (15).

46 Nwabunma, D., Kim, K.J., Lin, Y. et al. (1998). Phase diagram and photopolymerization behavior of mixtures of UV-curable multifunctional monomer and low molar mass nematic liquid crystal. *Macromolecules* 9297 (98): 6806–6812.

47 Duran, H., Meng, S., Kim, N. et al. (2008). Kinetics of photopolymerization-induced phase separation and morphology development in mixtures of a nematic liquid crystal and multifunctional acrylate. *Polymer (Guildf).* 49 (2): 534–545. https://doi.org/10.1016/j.polymer.2007.11.039.

48 Maschke, U. and Coqueret, X. (2000). Phase diagrams and morphology of polymer dispersed liquid crystals based on nematic-liquid-crystal – monofunctional-acrylate mixtures. *Phys. Rev. E* 62 (2): 2310–2316.

49 Soule, E.R. and Rey, A.D. (2019). Chapter 2: Phase diagrams, phase separation mechanisms and morphologies in liquid crystalline materials: principles and theoretical foundations. In: *RSC Soft Matter*, vol. 2019, no. 8,, 19–36. Royal Society of Chemistry.

50 Gregg, A., York, L., and Mark, S. (2005). Roll-to-roll manufacturing of flexible displays. In: *Flexible Flat Panel Isplays* (ed. G.P. Crawford), 409–445. New York: Wiley.

51 Montbach, E. and Davis, D. (2018). Roll-to-roll manufacturing of flexible displays. In: *Roll-to-Roll Manufacturing: Process Elements and Recent Advances* (ed. J. Greener, G. Pearson and M. Cakmak), 285–323. Wiley.

52 Pochan, J.M. and Marsh, D.G. (1972). Mechanism of shear-induced structural changes in liquid crystals-cholesteric mixtures. *J. Chem. Phys.* 57 (3): 1193–1200. https://doi.org/10.1063/1.1678376.

53 Wiese, O., Marenduzzo, D., and Henrich, O. (2016). Microfluidic flow of cholesteric liquid crystals. *Soft Matter* 12 (45): 9223–9237. https://doi.org/10.1039/C6SM01290F.

54 Chau, T., Ji, J., Tam, C., and Schwellnus, H. (2006). A novel instrument for quantifying grip activity during handwriting. *Arch. Phys. Med. Rehabil.* 87 (11): 1542–1547. https://doi.org/10.1016/j.apmr.2006.08.328.

55 Hsu, H.M., Lin, Y.C., Lin, W.J. et al. (2013). Quantification of handwriting performance: development of a force acquisition pen for measuring hand-grip and pen tip forces. *Meas. J. Int. Meas. Confed.* 46 (1): 506–513. https://doi.org/10.1016/j.measurement.2012.08.008.

56 Caligiuri, M.P., Teulings, H.L., Dean, C.E. et al. (2010). Handwriting movement kinematics for quantifying extrapyramidal side effects in patients treated with atypical antipsychotics. *Psych. Res.* 177 (1–2): 77–83. https://doi.org/10.1016/j.psychres.2009.07.005.

57 Montbach, E., Khan, A. A., Echeverri, M., and Braganza, C. (2016). Enhanced brightness ewriter device, US10287503B2, November.

58 St, W.D., John, W.J., Fritz, Z.J.L., and Yang, D.K. (1995). Bragg reflection from cholesteric liquid crystals. *Phys. Rev. E.* 51 (2): 1191–1198. https://doi.org/10.1103/PhysRevE.51.1191.

59 Braganza, C., Lightfoot, M., Echeverri, M. et al. (2017). 38-2: eWriter with select erase functionality. *SID Symp. Dig. Tech. Pap.* 48 (1): 539–541. https://doi.org/10.1002/sdtp.11697.

60 Braganza, C., Schlemmer, M., DeMiglio, A. et al. (2013). P.70: Durability and reliability of eWriters. *SID Symp. Dig. Tech. Pap.* 44 (1): 1254–1256. https://doi.org/10.1002/j.2168-0159.2013.tb06460.x.

61 Braganza, C., Lightfoot, M., Schlemmer, M. et al. (2014). 54.3: Human and mechanical writing performance of eWriters. *SID Symp. Dig. Tech. Pap.* 45 (1): 789–792. https://doi.org/10.1002/j.2168-0159.2014.tb00207.x.

62 Braganza, C., Montbach, E., Khan, A. et al. (2015). 23.3: Flexible Semitransparent eWriter Displays. *SID Symp. Dig. Tech. Pap.* 46 (1): 334–337. https://doi.org/10.1002/sdtp.10239.

63 Braganza, C., Echeverri, M., Kammer, K. et al. (2016). P-110: eWriter featuring color burst technology. *SID Symp. Dig. Tech. Pap.* 47 (1): 1539–1542. https://doi.org/10.1002/sdtp.10994.

64 Montbach, E., Krinock, J., Liu, L. et al. (2013). Large area, seamlessly tiled, flexible eBoard. *Dig. Tech. Pap. SID Int. Symp.* 44 (1): 1250–1253. https://doi.org/10.1002/j.2168-0159.2013.tb06459.x.

6

The Zenithal Bistable Display

A Grating Aligned Bistable Nematic Liquid Crystal Device

Guy P. Bryan-Brown[1] *and J. Cliff Jones*[2]

[1] *New Vision Displays Inc., Malvern, Worcestershire, UK*
[2] *School of Physics and Astronomy, University of Leeds, Leeds, UK*

6.1 Introduction

Paper has several attributes that any replacement technology must target. Its properties include having high reflectance (>70%), sufficient contrast (>8 : 1), light weight, and flexibility. By its nature, paper is inherently high resolution, well above the limit of human vision. Although most often used for black and white text images, it can display full color images of photographic quality. Once written on, the image is retained for years. Paper is usually considered to be low cost, though its economic and environmental costs are not as clear when compared to an electronic display because of its write-once-read-many nature.

It is the ability to update the image that drives electronic paper requirements. The update will be occasional for many applications, although regular updates are often required. Electronic paper technology is concerned with developing displays that add updateability to as many of these attributes of paper as possible. Of these paper-like qualities, it is essential to display the desired image without continuous electrical power. This necessitates electronic paper to be a bistable, reflective display. Being a reflective display limits the image quality possible with any electronic paper, and image quality is often compromised when compared with paper. Of bistable reflective displays, image quality and economical price are often the differentiators between the technological decisions, and the various other features of conventional paper are considered secondary and application specific. Many of these other paper-like attributes are yet to be realized commercially. Some compromise of image quality over paper is usually tolerated for update ability. No bistable reflective display technology has successfully reproduced full color images in anything other than high-cost prototypes. Some applications demand limited color, perhaps two or three additional colors to black and white. The flexibility, durability, and lightweight offered by plastic displays have not been accomplished commercially, despite years of research.

Various technological approaches have been taken to create electronic paper [1], including encapsulated electrophoretic inks [2], electrophoretic powders [3], electrofluidic [4], electrowetting [5], and electrochromic [6]. Each of these methods aims to produce absorption areas on a background with Lambertian reflection to mimic

E-Paper Displays, First Edition. Edited by Bo-Ru Yang.

the attractive appearance of conventional paper. As novel display systems, they require significant outlay to become commercially successful, a barrier only yet transcended by electrophoretic, with the success of *E-Ink Corp* and the *Kindle*. The great majority of electronic displays sold worldwide use liquid crystals to modulate incident light [7], being 50 years since the first commercially successful products [8]. The first liquid crystal displays (LCD) were reflective but had limited multiplexibility and hence were restricted to segmented information and unable to display complex images. Inducing bistability in liquid crystals was a topic of much early research effort since it promised high image content through unlimited multiplexing without the need for additional non-linear elements (usually thin-film transistors) at each pixel [9]. Bistability was achieved firstly in cholesteric liquid crystals [10], Smectic A [11], and ferroelectric chiral Smectic C liquid crystals [12], but efforts to induce bistability in the nematic liquid crystal phase usually deployed in displays [13] did not lead to commercial application. These included bistable twisted [14] and splay-bend [15] nematic devices. However, significant commercial progress of a bistable nematic LCD did not occur until the invention of the Zenithal Bistable Display (ZBD) in 1995 [16–19]. This uses a surface relief grating to induce bistable pretilts of the nematic director. Other than using a grating surface to align the liquid crystal, the device is similar to the standard reflective twisted-nematic device used in watches and calculators. This simplicity, and the low cost it facilitates, has led to electronic shelf-edge label displays for the retail sector [20, 21]. The operation and optimization of this device concern this chapter.

6.2 Operating Principles and Geometries

Liquid crystals represent phases of matter that have intermediate long-range order between the crystalline solid and isotropic liquid. Thermotropic liquid crystal behavior occurs in rigid, anisotropic organic molecules, with the degree of order being related to the temperature of the material. Nematics are the simplest liquid crystal phases, with only long-range orientational order of the long molecular axes, which tend to point parallel to each other, defining a unit vector called the director, **n**. This order leads to the anisotropy of macroscopic properties such as the dielectric permittivity and refractive index, which depends on the degree of order, quantified by the parameter S:

$$S = \langle 3\cos^2\theta - 1 \rangle \tag{6.1}$$

where θ is the angle between a molecular axis of symmetry and the director, and the brackets < > represent the mean over the thermodynamic ensemble. For calamitic liquid crystals, θ is defined with respect to the long axis of the rod-like molecules. Although most LCD use calamitic nematics as the active element of the display, discotic liquid crystals are also often used in a polymer form as passive optical compensation films; therein, θ is defined with respect to the short, symmetry axis of the disk-like molecules. A typical nematic LCD works by creating an aligned director profile using some surface treatment layer, overcoming the orientational elastic constants for splay k_{11}, twist k_{22}, and bend k_{33}, to disrupt that profile by coupling the applied electric field to the dielectric anisotropy $\Delta\varepsilon$ and observing the effect on polarized light through a change in the optical birefringence Δn. Each of these devices' relevant, anisotropic physical properties, $\Delta\varepsilon$, Δn, and k_{ii} (i = 1, 2, 3), are related to S.

Critical to LCD behavior is achieving uniform alignment of the quiescent director profile through the judicious arrangement of appropriate surface layers. Commercial LCDs use bespoke alignment polymers to achieve either homeotropic alignment (**n** || **s**, where **s** is the surface normal) or tilted-planar homogenous alignment ($\mathbf{n.s} = |\sin\theta_p|$, where θ_p is the pretilt angle of the director to the plane of the surface). For example, the common twisted nematic (TN) liquid crystal device still used in watches and calculators uses two low pretilt planar alignment layers formed from unidirectionally rubbed polyimide arranged with the rubbing directions at 90° to each other. The contacting nematic director then has a 90° twist from one surface to the other, leading to modulation of incident polarized light due to the effect of the liquid crystal birefringence. For example, light transmitted between crossed polarisers T is given by [7, 22]:

$$T = \frac{\sin^2\left(\frac{\pi}{2}\sqrt{1+\left(\frac{2\Delta n.\mathrm{d}}{\lambda}\right)^2}\right)}{1+\left(\frac{2\Delta n.\mathrm{d}}{\lambda}\right)^2} \tag{6.2}$$

Choosing the cell gap d to meet the condition [7, 22]:

$$\frac{\Delta n.d}{\lambda} = \sqrt{m^2 - \frac{1}{4}}, m = 1,2,3,\ldots, \tag{6.3}$$

where the wavelength λ is assumed to be 550 nm for white light operation, then causes the polarization direction of the light to be rotated through 90°, so that the device transmits ambient light when sandwiched between crossed polarizers (or appears black for parallel polarizers). The TN display works by ensuring that the field is applied parallel to the surface normal **s** (i.e. from transparent ITO electrodes immediately below the alignment layer on the opposing surfaces) and the material has a positive dielectric anisotropy Δε. The orientational free energy F_n associated with the director **n** is the sum of both elastic and electric terms in the direction of the applied field (taken as z here) and given by:

$$F_n = \frac{1}{2}\int_0^d \left[k_{11}(\nabla.\boldsymbol{n})^2 + k_{22}(\boldsymbol{n}.\nabla\times\boldsymbol{n})^2 + k_{33}(\boldsymbol{n}\times\nabla\times\boldsymbol{n})^2 - \varepsilon_0\Delta\varepsilon(\boldsymbol{n}.\boldsymbol{E})^2 \right] dz$$

$$= \frac{1}{2}\int_0^d \left[k_{11}\left(\frac{\partial n_x}{\partial x} + \frac{\partial n_y}{\partial y}\right)^2 + k_{22}\left(\frac{\partial n_y}{\partial x} - \frac{\partial n_x}{\partial y}\right)^2 + k_{33}\left(\left(\frac{\partial n_x}{\partial z}\right)^2 + \left(\frac{\partial n_y}{\partial z}\right)^2\right) - \varepsilon_0\left(\varepsilon_\perp + \Delta\varepsilon\sin^2\theta\right).E^2 \right] dz \tag{6.4}$$

where θ is the tilt angle from the xy plane. What Eq. (6.4) shows is that, if the material has a positive Δε, the electric term is minimized if θ increases toward θ = 90° and the director lies parallel to the field. Because the director is bound at the alignment surfaces, field-induced reorientation of the director only occurs away from the surfaces and induces an elastic deformation across the device. Eq. (6.4) predicts the balance between reorientation to minimize the electric free energy, which is opposed by an elastic cost given by the terms in k_{ii} that oppose that distortion.

ZBDs are examples of surface bistability. The surface itself has two stable director configurations rather than a fixed surface orientation unaffected by the applied field and induced elastic distortion. For zenithal bistability, these states have two different surface pretilt angles: in one state, the director is near planar, as for the TN device, but in the second state, the surface alignment has a much higher pretilt, close to the homeotropic condition and parallel to the surface normal. Unlike conventional nematic LCD, such as the TN described above, the surface director can be switched between states and retain that state after the switching event, such as an electrical pulse, has occurred, when the device is said to be "latched." The director in the device's bulk may be affected by electric fields in the same fashion as the conventional TN, modulating the light but without latching to the opposing state.

Surface bistability in ZBD is realised using a surface grating as one of the liquid crystal alignment layers. The grating is formed from, or coated with, a material that induces the director's homeotropic (i.e. normal) alignment to the local surface [16]. This induces a splay/bend elastic deformation as the director field follows the curvature of the surface, Figure 6.1a. This deformation decays away from the surface, and the director field becomes uniformly vertical aligned at a distance h, which is similar in magnitude to the grating pitch P. If the grating is designed with a suitably high ratio of grating amplitude A to pitch P, the elastic energy can be relieved by forming topological defects of strength +½ at the points of highest negative curvature (i.e. the grating

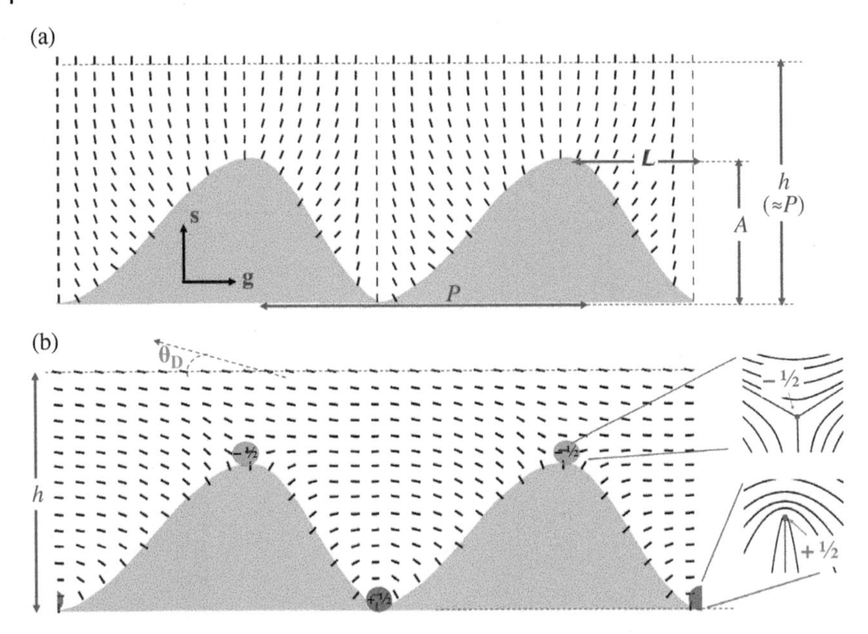

Figure 6.1 Zenithal bistable states of the nematic director close to a deep homeotropic surface relief grating of pitch P, amplitude A and blaze asymmetry L (if L < P/2). (a) The Continuous or C-state, where the director has uniform vertical alignment (i.e. **n** || **s**) at the distance h from the lower surface. (b) The Defect or D-state, wherein the director is again uniform at h from the lower surface but with a near-planar alignment, with pretilt θ_D related to the degree of asymmetry L/P. Inset are sketches of the director profile around −½ and +½ topological defects in the director **n** field.

groove bottoms) and −½ strength defects at points of highest positive surface curvature (i.e. the grating ridge tops), Figure 6.1b). In 3D, these defects are lines parallel to the grating ridges and troughs (i.e. perpendicular to the grating vector, **g**). Again, at the distance *h* from the surface, the director profile is uniform but now almost planar aligned. For a shallow surface (low A/P), the Continuous (or C-) state of Figure 6.1a is favored, since the elastic deformation is low. For a surface with deep grooves (high *A/P*), the elastic deformation is relieved by introducing the defects, and the Defect (or D-) state of Figure 6.1b is favored. At some intermediate range of *A/P*, these two states can be arranged to have similar energies separated by an energy barrier, and the surface is bistable. The term "zenithal" refers to the angle between the director and the surface **s** in the plane containing **g** and **s**. At the plane of uniform director profile *h*, the C-state has a vertical tilt (or nearly vertical if the surface is blazed), whereas the D-state has a lower tilt. The director in the plane acts as if it were a flat alignment surface, with pretilt θ_p, zenithal $W\theta$ and azimuthal $W\phi$ anchoring energies. Thus, the deep, homeotropic grating gives zenithally bistable states. It should be observed that zenithal bistability may also result from a planar surface if the director can be forced to lie in the plane perpendicular to the grooves and undergo the elastic deformation. Of course, the director would usually lie parallel to the grooves (i.e. $\mathbf{n} \perp \mathbf{g}$), as first described by Berreman [23]. However, if a deep planar bigrating is used, such as an array of micro-posts, the elastic deformation cannot be avoided, and zenithal bistability is achieved [24]. In such instances, the C-state has planar homogeneous alignment ($\theta_p = 0°$), and the D-state gives pretilts typically around $\theta_p = 30°$.

As a bistable surface, the grating can be used opposite a variety of surfaces to form the LCD, providing a number of different alignment geometries, each of which is bistable. The two most common geometries are shown in Figure 6.2. The first ZBD results were published using a switchable half-wave plate [17], with the opposing surface being monostable (i.e. flat) homeotropic alignment, Figure 6.2a).

The bistable C- and D-states correspond to vertically aligned nematic (VAN) and hybrid aligned nematic profiles. The VAN state should appear dark between crossed polarisers, whereas the HAN state transmits light when the grating, hence the D-state director, is oriented at 45° to the polariser. Ignoring the director variation close to

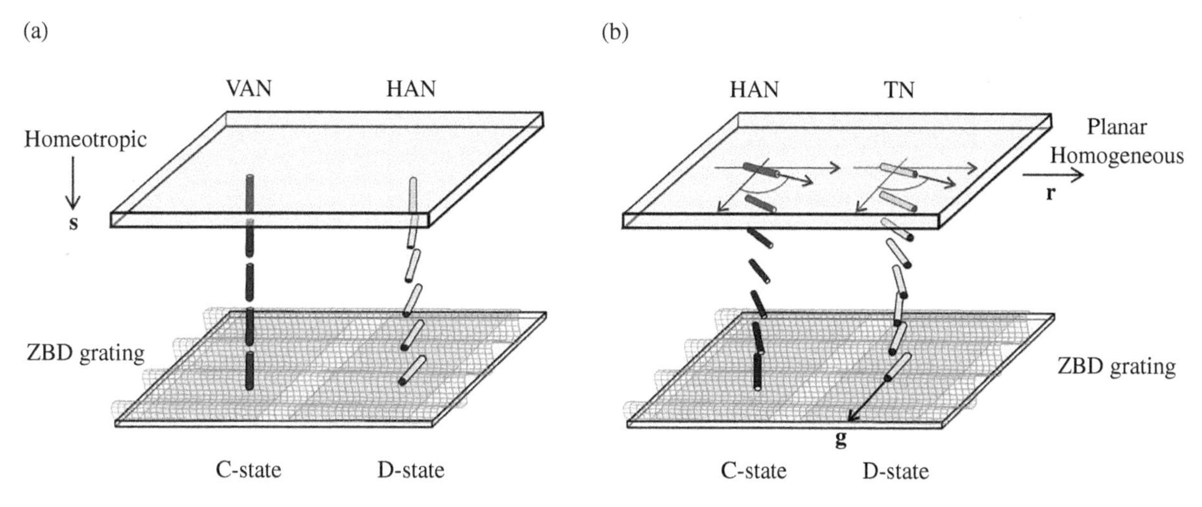

Figure 6.2 Example modes of a ZBD, based on choice of the opposing monostable surface. (a) Opposite a monostable homeotropic surface, the surface bistable C- and D-states lead to vertically aligned nematic (VAN) and hybrid aligned nematic (HAN), respectively. (b) Opposite a monostable planar homogeneous surface with the preferred alignment direction **r** angled with respect to the grating vector **g**, gives the HAN in the C-state, and a twisted nematic in the D-state.

the grating surface and treating the director profile as linear (i.e. $k_{11} = k_{33}$), the optimum cell gap is given by the half-wave condition:

$$d_{HAN} = \frac{\lambda}{\Delta n_D}(2m+1); m = 0,1,2,\ldots, \tag{6.5}$$

where the average tilt is taken as half way between the homeotropic condition and the D state (accurate if the splay and bend elastic constants are similar), with a birefringence related to the pretilt θ_D through:

$$\Delta n_D = \frac{n_e n_o}{\sqrt{n_e^2 \sin^2\theta_D + n_o^2 \cos^2\theta_D}} - n_o. \tag{6.6}$$

Commercial devices use the bistable grating opposite a Planar Homogeneous [19], leading to a HAN configuration for the C-state and a 90° TN for the D-state. Here, the rubbing direction of the planar homogeneous surface **r** is arranged perpendicular to **g**, Figure 6.2b), and the nematic is weakly doped with chirality to ensure a single twist direction is maintained. For convenience, these homeotropic and planar homogeneous geometries are termed VAN and TN ZBD geometries, respectively.

In addition to the bistable surface, a second difference between ZBD and conventional nematic LCD is the switching mechanism. Devices such as the TN respond to the root-mean-square of the applied voltage taken over a time sufficient for the director to reorient (e.g. the frame time). ZBD, on the other hand, responds to the electric field: it latches if an electrical pulse has sufficient magnitude, duration, and the correct polarity. Rather than coupling to the dielectric anisotropy, $\Delta\varepsilon$, latching is realized using the flexoelectric polarization, the polarity associated with splay and bend deformations. That is, the free energy of Eq. (6.4) contains an additional term that is linear in the applied field given by:

$$F_{flexo} = -\frac{1}{2}\int_0^d \left[e_1(\mathbf{n}.\nabla.\boldsymbol{n})\boldsymbol{E} + e_3(\boldsymbol{n}\times\nabla\times\boldsymbol{n})\boldsymbol{E}\right]dz$$

$$= -\frac{1}{2}\int_0^d \left[(e_1+e_3)\sin(2\theta)\frac{\partial\theta}{\partial z}E\right]dz. \tag{6.7}$$

Although the sum of splay e_1 and bend e_3 flexoelectric coefficients is small, around $100\,\mathrm{pC\,m^{-1}}$ for the strongly dipolar mesogens used for a ZBD [25], the elastic deformation and its associated electric polarity is concentrated close to the grating surface. This is further aided by the effect of a strongly positive $\Delta\varepsilon$, which presses the elastic distortion still closer to the surface with applied field. Thus, typical latching times are around 1 ms for pulses of amplitude $2\,\mathrm{V\,\mu m^{-1}}$ [26]. Of course, the nematic liquid crystal used to form the cell will also switch due to the coupling between the dielectric anisotropy $\Delta\varepsilon$ and the applied RMS voltage, causing changes to the director orientation in the bulk of the device and a corresponding optical response.

The nematic continuum theory of Oseen, Frank, Ericksen, and Leslie [27] is a triumph of theoretical physics used by device scientists and engineers to understand and design ever more complex display devices. Perhaps, without such an accurate and useful theoretical basis, the modern LCD display industry would not have advanced to its current dominance. However, these theories usually treat the director as a simple unit vector, such as in the static-free energy expression in Eq. (6.4). Such theories do not allow changes to the nematic order parameter S that high elastic strains may induce. Nematic defects are a good example of such strains: these are ≈50 nm regions where the director is undefined, and the order is effectively $S = 0$. Modeling of such systems then requires a Q-tensor approach [28], where both orientation and order parameter magnitude may vary. This method has been applied to the ZBD device [29, 30], and most accurately with a direct comparison with experimental results in [31]. This work showed that the defects nucleated at the steepest parts of the grating and then moved across the surface toward the points of highest surface curvature under the influence of a positive electric field coupling to the positive flexoelectric coefficients. The defects' trajectory could either follow the surface closely, as in [30–32], or separate from the surface into the bulk liquid crystal between the grating ridges, as shown in the example given in Figure 6.3. The path taken depends on the properties of the liquid crystal material, the grating surface,

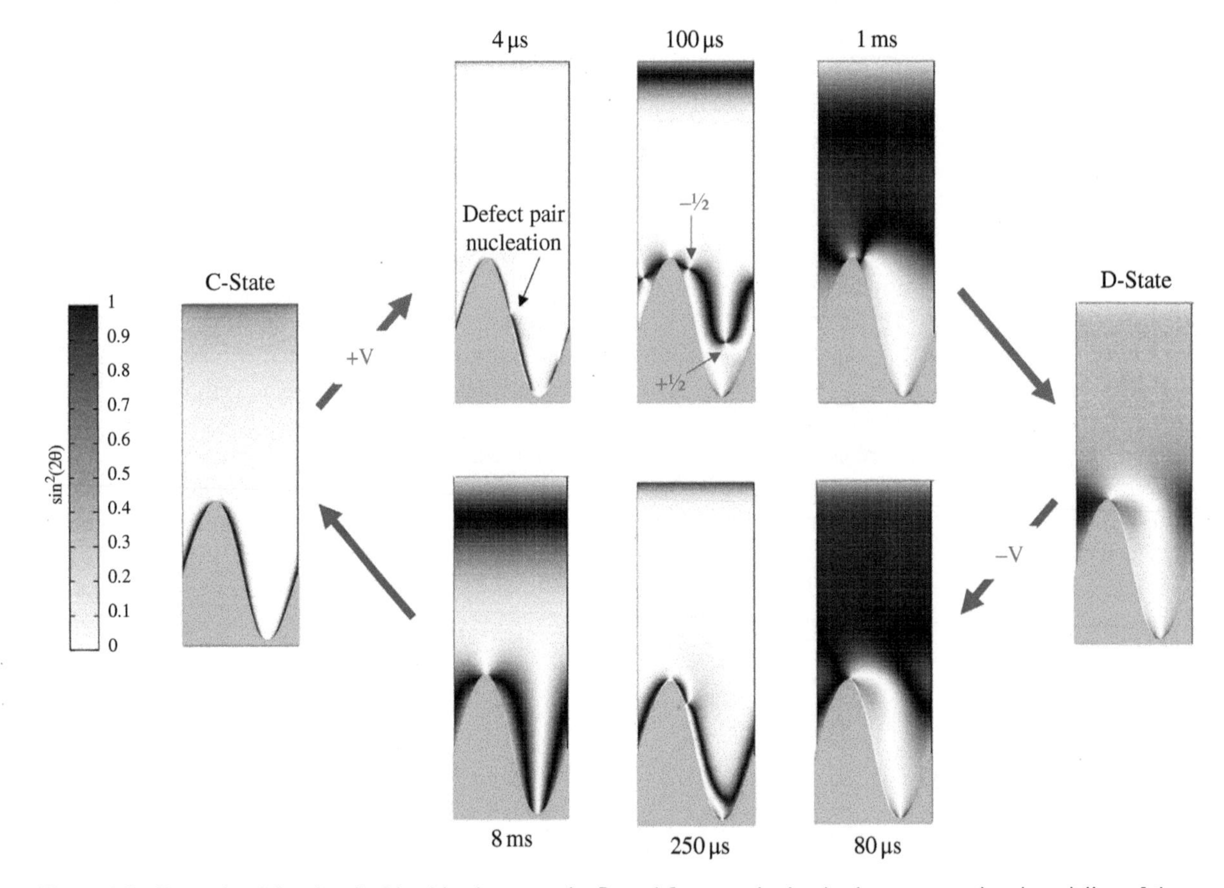

Figure 6.3 Example of the electrical latching between the D- and C-states obtained using computational modeling of the Q-tensor [31].

and the electrical pulses' timing [30]. Latching occurs at some point where the defects are sufficiently close to the regions of high surface curvature that the defects continue their path due to the influence of that curvature. A negative electrical pulse causes the defects to move toward each other until they annihilate on the surface. The negative defect is far more mobile than the positive [33], and hence the point of annihilation is close to the grating ridge.

Figure 6.4 shows the optical response to a series of pulses for typical ZBD devices, wherein a positive pulse represents the electrical polarity applied to the grating electrode (a convention retained throughout this chapter). The transmission of ZBD in the VAN geometry ZBD to monopolar pulses that change sign after every two pulses is given in Figure 6.4a). The commercial liquid crystal E7 was used on a grating that produced a pretilt of $\theta_D = 20°$ so that the half-wave-plate condition calculated using Eqs. (6.5) and (6.6) gives $d_{HAN} = 2.7\,\mu m$. There is a momentary drop in transmission for each pulse due to the conventional RMS switching response of the positive $\Delta\varepsilon$ nematic material. However, positive pulses cause a change to the transmitting HAN state that does not decay after the pulse has been removed. The device has been latched from the non-transmissive C-state (VAN) to the transmissive D-state (HAN in the optical half-wave plate condition). This pulse sequence of alternating pairs of opposite polarity shows that the latching is strictly polarity dependent: a negative pulse applied to the device already in the C-state causes a transient response, as the director close to the grating aligns parallel to the field, but the C-state is retained due to the negative polarity. Only when a positive pulse of sufficient time and magnitude is applied does the device latches to the D-state. Ensuring that the pulse duration and amplitude required to latch

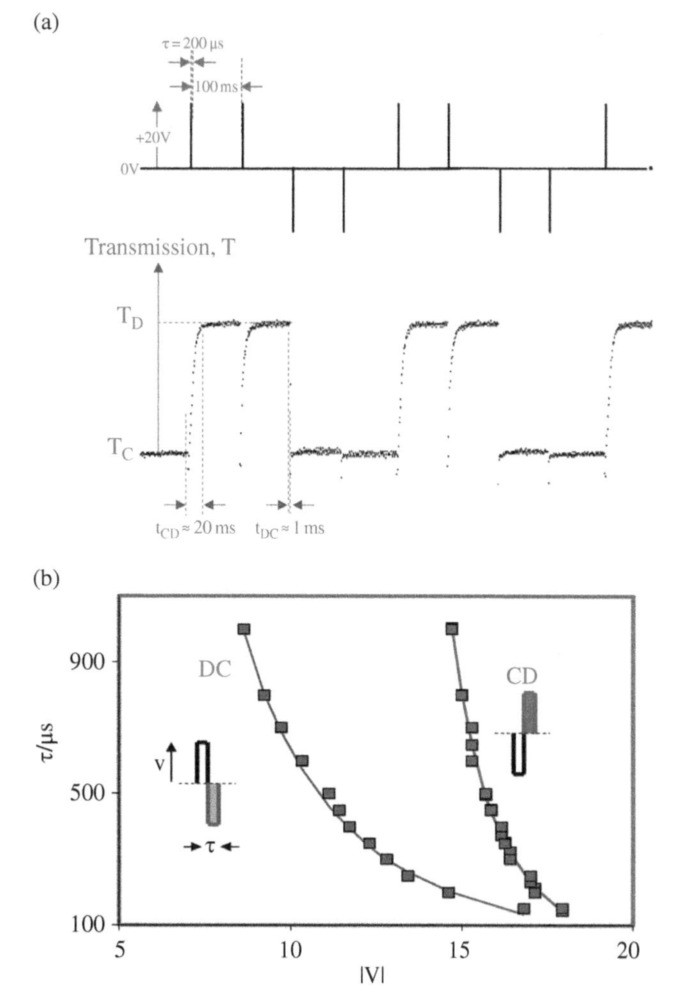

Figure 6.4 The electro-optic response of typical Zenithal Bistable Devices. (a) Transmission T versus time response to a series of |20V| 100 μs pulse pairs of opposite polarity for a 2.7 μm spaced ZBD with the commercial liquid crystal E7 in the VAN mode geometry [16]. (b) Latching response to bipolar pulses of Mixture B [26] measured in a 5.7 μm spaced ZBD in the TN geometry at 23 °C. Latching from D to C occurs when the trailing pulse is negative, and from C- to D-state when the trailing pulse is positive.

between either state is important for display optimisation. Early results for commercial STN mixtures produced typical latching with 40 V and 1 ms pulses [18]. However, low viscosity materials with higher dipole moments, and therefore flexoelectric polarisations, were developed [26], providing latching pulses of 16 V and 500 μs, such as those shown in Figure 6.4b).

In practice, the grating is designed with asymmetry and blaze, which is done for two reasons. Firstly, defects may be nucleated on the steep side of the surface either side of the grating ridge. Both pairs then interact with each other and prevent latching into the D-state. With a blazed grating, the defect pair on the steeper side forms first and reaches the lines of highest surface curvature to latch before the second pair has moved far from its point of nucleation [31, 32]. The second reason for including blaze is to induce some pre-tilt into the director of the D-state. At h, the pretilt from the surface plane is approximately given by [34]:

$$\theta_D = 90°\left(1 - \frac{2L}{P}\right). \tag{6.8}$$

For a symmetric sinusoidal grating $L = P/2$, the +½ and −½ defects are uniformly spaced and Eq. (6.8) shows that $\theta_D = 0°$. In practice, a small amount of tilt is required for the domain-free operation of the TN [8, 35], and a near-sinusoidal grating is used. Note, it is important that the pretilt is not too high, since this will affect deleteriously the viewing angle associated with the D-state, and gratings are usually designed to be nearly symmetric, with $P = 0.80\,\mu m$ and $L \approx 0.38\,\mu m$ to give $\theta_D \approx 6°$.

Diagrams such as those of Figures 6.1 and 6.3 show the topological defects as points in a two-dimensional space in a plane containing the director. However, the ±½ strength defects are lines running orthogonal to the plane of the director. The defects run along the grating groove ridges and troughs and meet and annihilate at two points to form a defect loop. Although held by the surface curvature by the grating, the defects must either run along the flat sides of the grating at the ends of the loop or into the bulk of the liquid crystal. One would expect the loops to be pinned at microscopic inhomogeneities in the cell that would occur on a practical grating surface. Without such pinning sites, the loops would only be stable if the deformation energies for the D- and C-states are equal; otherwise, the loop would grow for a D-state stable device (e.g. for deep gratings or high temperatures, where S is low, and defects are more stable) or shrink if the device is C-state stable (shallow gratings or lower temperatures). Even if both states are of equal energy, disruption of the director profile through flow, such as that induced by pressing the front glass surface of the display, will cause the desired image to be erased. High levels of resistance to mechanical shock and a wide range of temperature operations are obtained in practiceby including regular π-phase slips in the grating design [36], as shown in Figure 6.5. These provide concave and convex surfaces perpendicular to the grating, which stabilize the structures at regular intervals (usually spaced by several micron, as shown in Figure 6.5a). When considering the latching in 3D [37], the defects will nucleate and annihilate at the inflection points within each slip, and the defect moves along the surface of appropriate curvature, rather than as shown in the over-simplified model of Figure 6.3. This type of defect motion is observed experimentally with high-resolution polarized light microscopy.

6.3 Grating Fabrication and Supply Chain

Compared with other LCD alignment surfaces, either rubbed or photoaligned polymers, the ZBD surface requires considerable engineering to achieve consistent device operation. In a typical LCD manufacturing process, the time taken for each processing step is typically around 20–30 seconds. Photolithographic patterning of a sub-micron grating directly onto LCD glass is not possible within this limit, so the mass production of ZBD LCD required the establishment of a high-speed replication process. This is done by embossing a photopolymer using a structured film with the required shape.

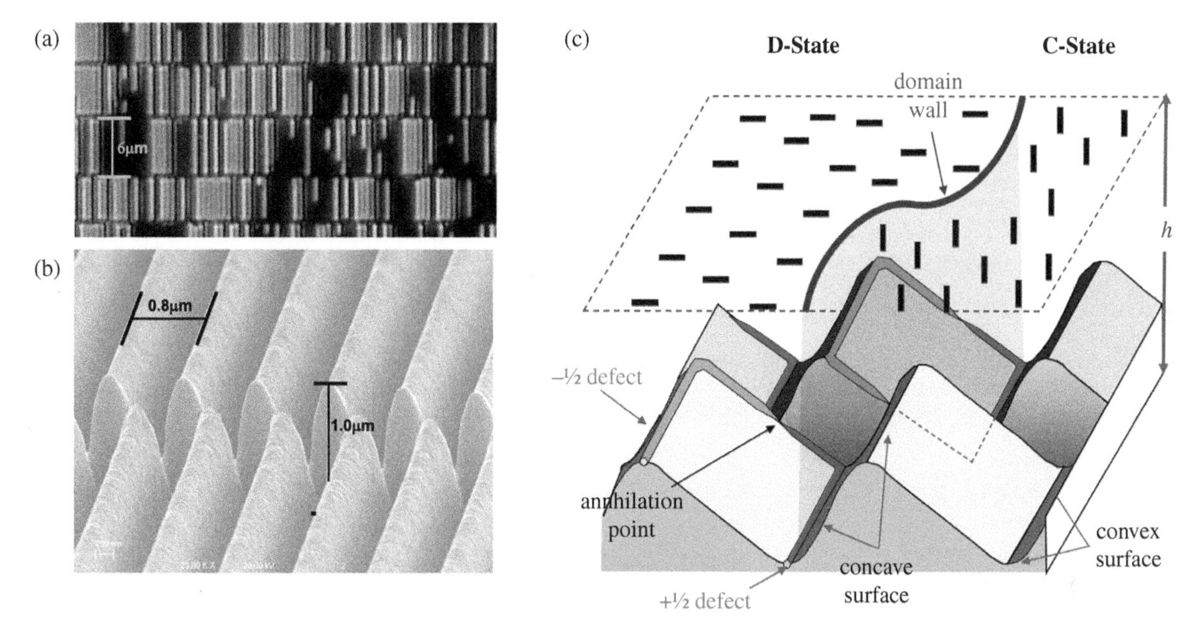

Figure 6.5 Three-dimensional structure of the ZBD grating. (a) Polarizing optical microscope image of a partially latched pixel. White lines are gratings in the D-state and dark regions are C-state. (b) Scanning Electron Microscope image of the ZBD grating structure showing the phase-slips in the grating structure. (c) Principal of operation of the grating phase slip. Defect loops are stabilized by the concave and convex surface corners that result in the direction parallel to **g**, in addition to the concave and convex curvature of the groove bottoms and ridge tops.

This replication process has several steps that are summarized in Figure 6.6. These are:

1) Formation of photoresist master.
2) Transfer of master pattern into nickel.
3) Mass production of grating film from nickel.
4) Formation of grating profile on LCD glass using grating film.

To make the photoresist master, a large area of glass substrate is thoroughly cleaned and coated with a perfect (pinhole-free) photoresist layer. Chrome on glass mask with the required grating pattern pre-etched is then placed in hard contact with this photoresist layer, and exposure is carried out using off-axis UV radiation. This mask is typical of the type used in the photolithographic industry for deep UV patterning. It is typically 140 mm × 140 mm, suitable for producing small displays. However, for larger electronic paper applications, accurate steps and repeat are required to produce larger grating areas free from visible seams. After this exposure, the mask is moved by a precisely controlled distance across the photoresist before re-establishing hard contact and carrying out a second exposure, Figure 6.7. This process is repeated until the target overall grating dimensions are exposed. Following subsequent thermal processing, the photoresist layer is developed to reveal a surface profile grating in which the seams between each exposure are invisible to the naked eye.

Figure 6.8 shows a typical grating profile. Off-axis UV exposure generates a small amount of asymmetry in the grating groove profile (grooves tilt to the right), which is sufficient to control ZBD latching such that the C to D transition only leads to a single low tilt state. A typical seam is also shown.

After completing the mastering step, the following process involves transferring this profile into a nickel shim. This technology is well established in the CD/DVD mastering industry, which relies on creating nickel stampers to mass-produce the disc media [38, 39]. This step ensures that a single master can be replicated into many km of film without damage.

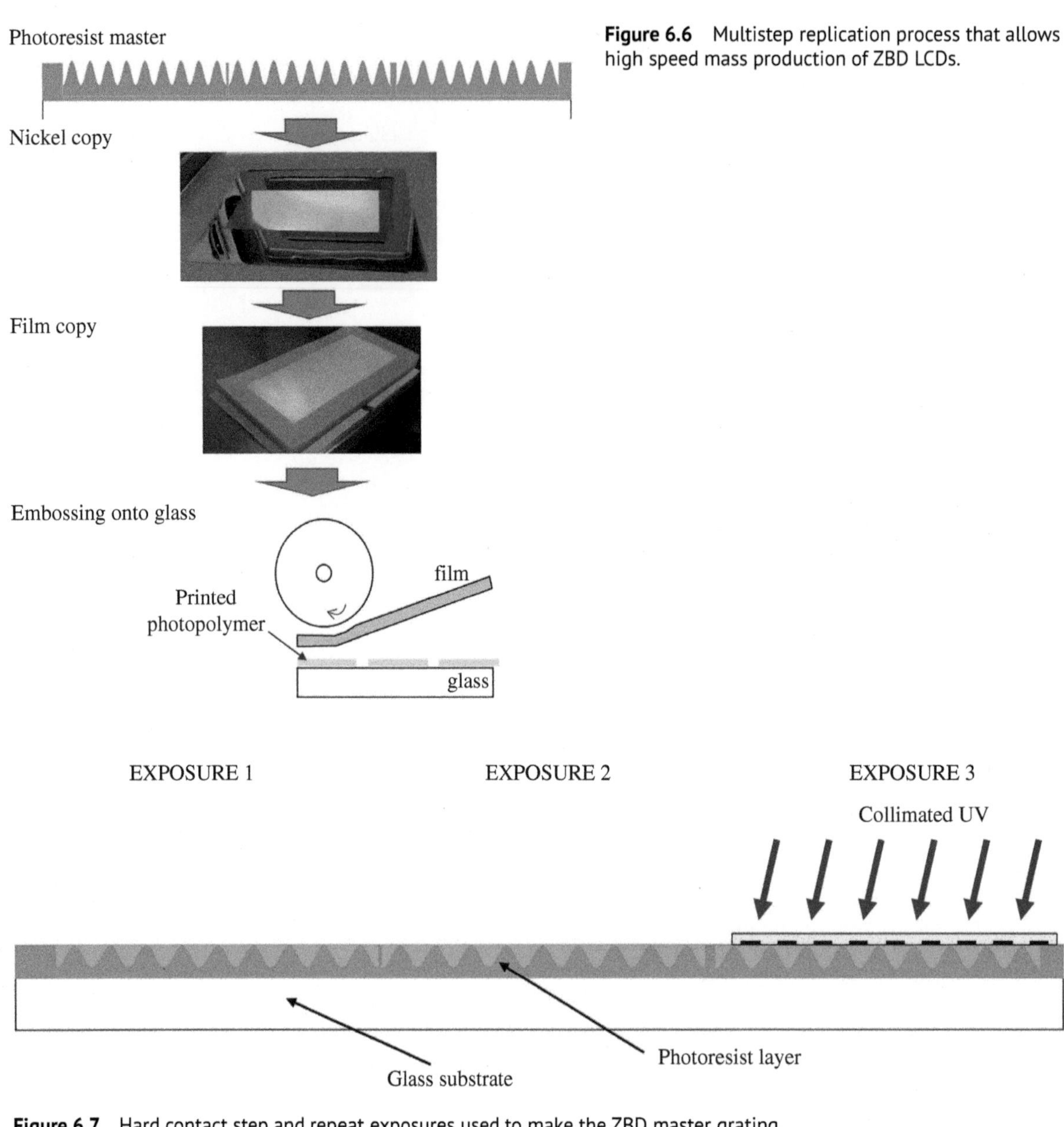

Figure 6.6 Multistep replication process that allows high speed mass production of ZBD LCDs.

Figure 6.7 Hard contact step and repeat exposures used to make the ZBD master grating.

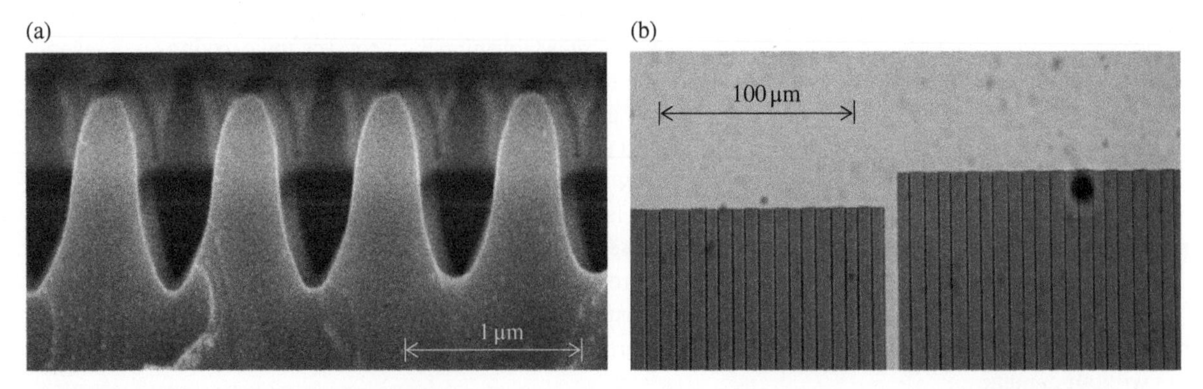

Figure 6.8 (a) Scanning Electron Micrograph of a typical groove profile of ZBD master grating showing slight asymmetry. (b) Photograph of a typical 5 μm seam between two grating regions. The vertical slip lines can be seen in this optical image but not the horizontal grating grooves.

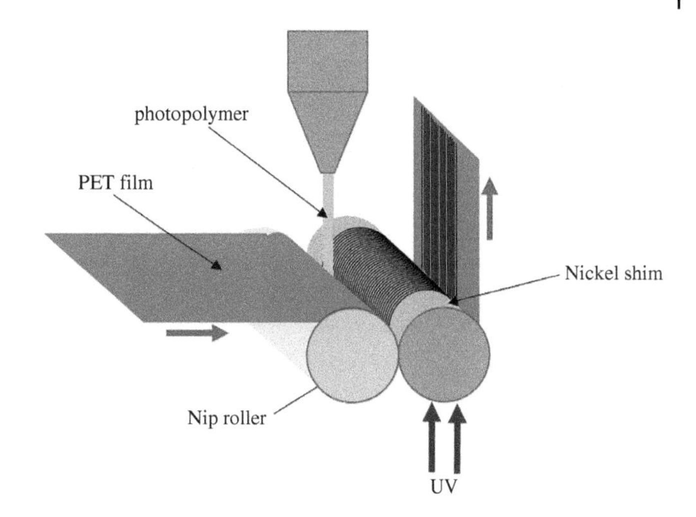

Figure 6.9 Schematic of process for transferring grating profile from nickel to film.

The photoresist master is first coated with a thin seed nickel layer via sputtering. This surface becomes one electrode in a large galvanic growing tank containing an optimum mix of electrolytes and other additives. Electrochemical deposition is done until the target nickel thickness is achieved, typically 150 to 450 μm. This 1st generation copy is then peeled from the glass substrate and is cleaned in a plasma chamber to remove photoresist material retained within the nickel copy's grooves. Thus, only one 1st generation copy can be made from each photoresist master. However, it is possible to then grow many 2nd generation copies from a single 1st generation nickel shim and many 3rd generation copies from each 2nd generation shim, and so on. In each case nickel-nickel release is facilitated by the growth of a precisely controllable nickel oxide layer, which can be formed by wet processing before each galvanic growth process. This unique nickel property has made it the primary choice for multigenerational micropatterned tooling creation.

Direct embossing onto LCD glass using nickel is not well suited to the particular requirements of ZBD LCD. Therefore, another step is required in which the high-cost nickel is converted into low-cost plastic film. Figure 6.9 shows a schematic of this process. The nickel shim is mounted on a roller which is compressed against a second nip roller. A polymer film material (typically PET) is inserted between these rollers, and a UV-sensitive photopolymer is poured into the gap between the PET film and the nickel shim. Roller rotation leads to casting the grating profile into a photopolymer layer formed between the nickel and the PET, which is then cured into a solid by high-intensity UV. The photopolymer is designed to adhere firmly to the PET film while also readily releasing from the nickel surface. A protective release liner is added over the grating surface before wind-up and subsequent sheeting. The resulting film is low cost, easy to transport, and can be stored for many years without performance degradation.

6.4 ZBD LCD Manufacturing Processes

The ZBD LCD is a passive matrix display technology, and much of the manufacturing process follows the same process steps as TN and STN displays. In fact, one substrate (the "P" plate) is processed with a rubbed polymer layer identically to TN/STN while the other "G" plate has some modified steps summarized in Figure 6.10. These steps are now described in more detail.

The basic process is to roller emboss the grating into a photopolymer layer by pressing the grating film into the uncured photopolymer and then hardening the grating through UV curing. The film is designed so that the grating area covers the glass so that the same film can be used regardless of the dimensions of the display being produced. Therefore, a first challenge is patterning the grating onto LCD glass so that the grating regions cover only

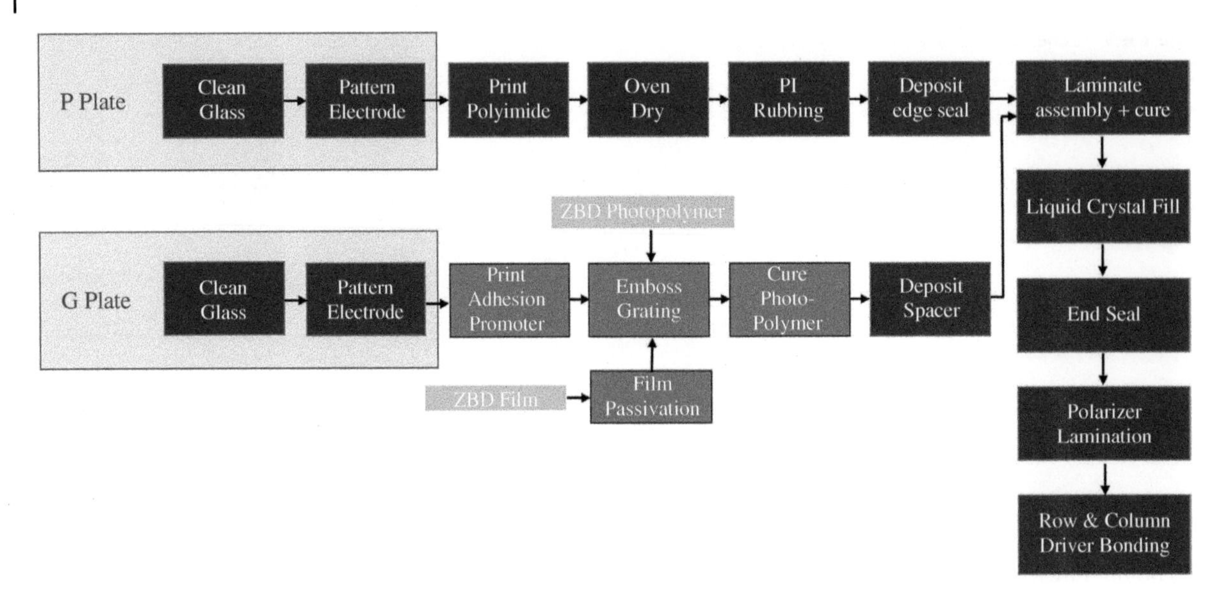

Figure 6.10 A simplified process flow for the manufacture of ZBD LCDs.

the viewing area of each of the module locations on the mother-glass. The grating must not cover the edge seal regions since seal materials are optimized for glass–glass bonding, and obviously, the grating should not cover bonding ledges since this would compromise the connection of driver ICs to row/column electrodes. Several methods have been proposed for controlling the grating extent [40], but the most successful in production has been selective adhesion. This method relies on controlling the relative adhesion energies at the interfaces between the relevant layers in the embossing process. A surfactant adhesion promoter is patterned onto the substrate using either ink-jet or flexo-printing to define the areas where the grating is retained on the mother-glass. Figure 6.11 shows a schematic of these layers and the desired retention of the final LC alignment grating (ZBD photopolymer) after cure and peel. For the process to operate correctly, the relative adhesion between the photopolymer and glass E_{PS}, film E_{PF} and adhesion promoter E_{PA} must each be considered, and the following relationship holds:

$$E_{PA} > E_{PF} > E_{PS} \tag{6.9}$$

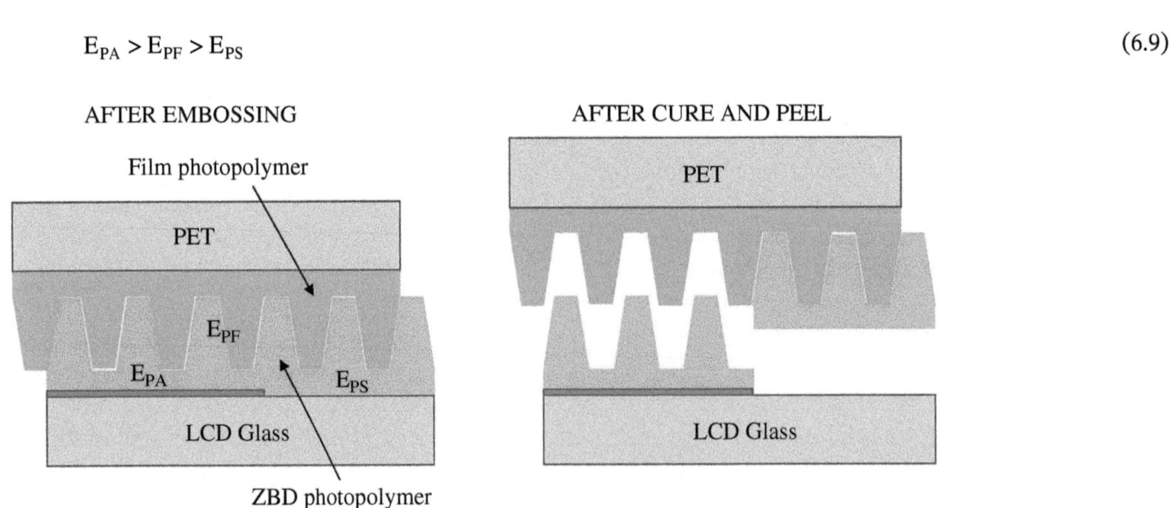

Figure 6.11 Definition of adhesion energies and subsequent layer retention after peeling. Subscripts denote uncoated substrate S, grating film F and adhesion promotor A.

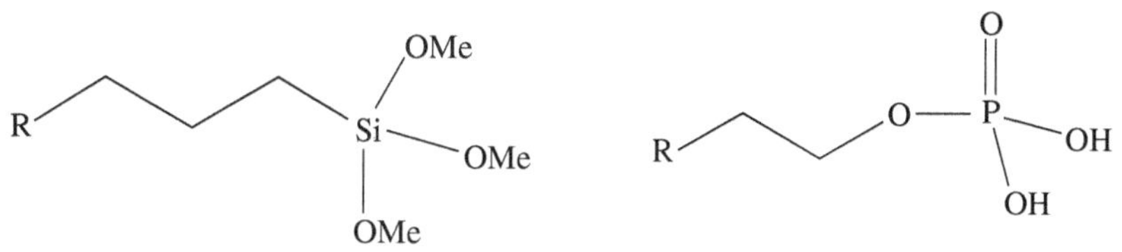

Figure 6.12 Examples of typical active components to promote adhesion of ZBD photopolymer to SiOx (left) and ITO (right) regions of the substrate. Group R is compatible with photoactive groups in the ZBD photopolymer material.

E_{PS} is naturally low ($<1.0\,J/m^2$) since the ZBD photopolymer has no intrinsic adhesion to the LCD substrate surface. The adhesion promoter contains two active components, Figure 6.12, in glycol solvents selected for optimum adhesion to both SiOx and ITO portions of the LCD glass surface. Such mixes are also effective on glass-like hard coat layers, often applied over ITO electrodes in passive matrix LCDs. Once applied, the adhesion promotor solution is dried and cured at 130 °C to ensure full crosslinking of the silane material with itself as well as with the phosphate material.

The final challenge is to reduce the size of E_{PF}. Since both the film photopolymer and the ZBD photopolymer use similar photoactive groups, the former is able to crosslink into the latter leading to an inseparable bilayer. However, it is possible to passivate the surface groups in the film photopolymer and reduce E_{PF} to the range of 5–10 J/m^2, which is well below E_{PA} (typically $>50.0\,J/m^2$) [40].

The grating is formed on the mother-glass employing a roller embossing process, as shown in Figure 6.13. Following the application and cure of the adhesion promotor, the LCD substrate is coated in a thin layer of ZBD photopolymer using inkjet printing. A proprietary photopolymer is used in production, which gives homeotropic alignment with the appropriate anchoring energy. Alternatively, a conventional photopolymer can be used and coated with a silane surfactant using a vapor deposition process [41]. Control of the anchoring energy then results by varying the surface density of the pendant groups. Following the photo-polymer printing step, the mother-glass is then laminated against film using a pair of rollers of optimum shore hardness that operate with an optimum applied pressure (P) and speed (V). Shortly after embossing, the photopolymer layer is cured via UV radiation, and the film is peeled to reveal a perfect copy of the grating structure on the LCD glass. If the embossing speed is too high or pressure too low, then the cured grating structure will include a significant "DC offset" of polymer below the modulation, and this leads to an undesirable voltage drop across the grating that is made worse by its relatively low electrical permittivity to the LC material [42]. With optimum range of conditions, the grating shown in Figure 6.14 is obtained.

The adhesion conditions of Eq. (6.9) ensure that unwanted photopolymer in areas outside the grating/display viewing areas is removed from the mother-glass and remains attached to the film. Thus, the glass is spontaneously patterned in the peeling process. Once the grating has been formed, the substrate is ready to receive spacer beads and proceed through the remaining steps in the same fashion as a conventional LCD plate, Figure 6.10. Other than the adhesion promoter, grating film, and ZBD photopolymer, other display materials are common with those used for TN/STN displays. These include ITO glass, seal material, spacer beads, LC material, polarisers, and driver ICs.

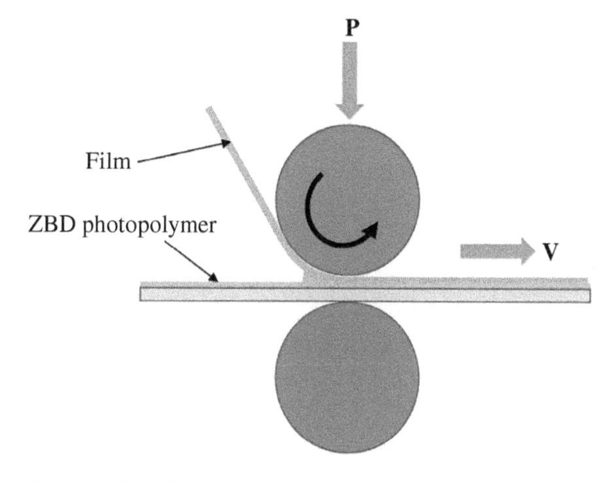

Figure 6.13 Schematic of production embossing process.

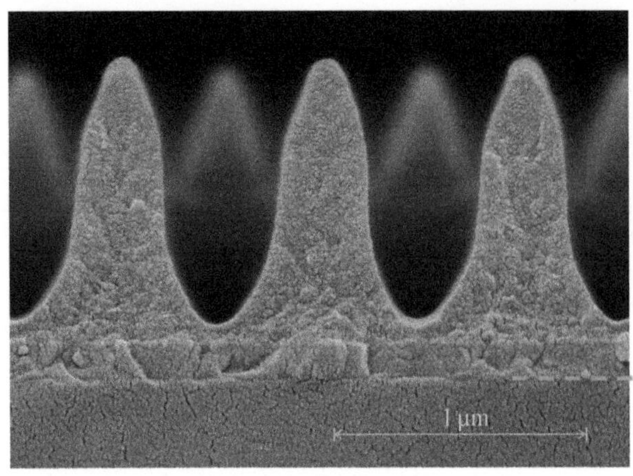

Figure 6.14 SEM section showing an optimum ZBD grating. The pitch of the grating is 0.80 μm and the amplitude 1.05 μm.

6.5 Electrical Addressing

For any matrix display, the electric signals are applied from silicon driver ICs on the edges of the display. Conventional passive-matrix addressed TN and STN LCD use the principles of RMS addressing to share signals across a matrix of row and column electrodes on the opposing display substrates to give the desired pattern of appropriate voltages when averaged over a frame time [7]. Usually, one of N rows has a strobe voltage $+V_S$ applied for an Nth fraction of the frame time, and this is synchronized with data voltage either $+V_D$ to not switch (due to the resultant $V_S - V_D$ appearing across the pixel) or $-V_D$ to switch (due to the resultant $V_S + V_D$). For the remainder of the N-1 rows of the frame, the row experiences the non-discriminatory data for the other rows. Pixels in the switched state begin to decay toward the unswitched state unless refreshed in the next frame. This method means that discrimination results from only one N^{th} of each frame, thereby limiting the number of rows that can be addressed according to the steepness of the display's electro-optic characteristic. For conventional LCD, overcoming this multiplexing limit relies on thin-film-transistors (TFT) located at each pixel to amplify the voltage signals received from the drivers. The thousands of rows can be addressed to form complex, high-resolution images. Indeed, some candidates for electronic paper, such as electrophoretic displays, are not truly bi-stable but are metastable. This means that they retain the image (and hence can operate with very low power) even after the signals are removed. However, a pixel will begin to latch into the opposing state with any small voltage, and hence a TFT is still required to isolate the pixel from the data voltages being applied to other rows of the panel.

The basics of driving a truly bi-stable display, such as ZBD, are similar to conventional LCD RMS, except the strobe voltage V_S is designed to lie close to the latching threshold. The data signal $\pm V_D$ is applied synchronously with the polarity designed to give the appropriate resultant to the potential difference of row – column voltage. If $V_S + V_D$ is sufficient, the pixels on that row will latch into the new state. Discrimination is achieved when $V_S - V_D$ is not sufficient to cause that same change, and the pixels in the row do not latch. Thus, the image is formed line-by-line. The data voltage $\pm V_D$ being applied to the other rows is far lower than the threshold and so, once latched, the row of pixels remains in their stored conditions, and the addressing signal can be applied to other rows without changing the latched state. Each addressing signal is arranged to give DC balance to prevent electrolytic damage to the liquid crystal, and avoid unwanted latching caused by long sequences of the same data signal.

Of course, the signal is applied to either change the state or leave it unchanged; a second signal is required to address each row for the other state that responds to the opposite polarity. There are several ways that this may be implemented [32]:

a) *Two-field Addressing*. A frame is divided into a positive field (selecting the D-state) and a negative field (selecting the C-state).

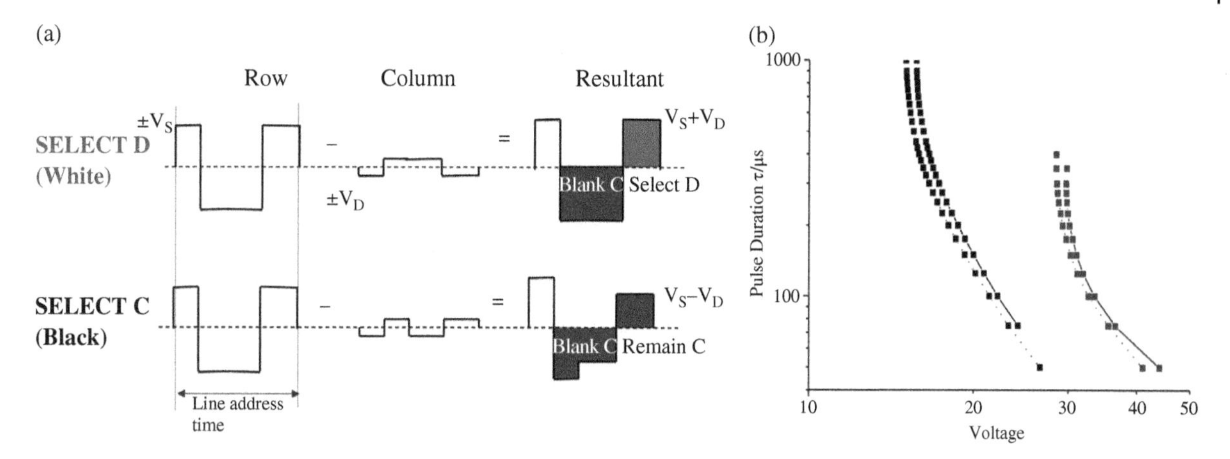

Figure 6.15 (a) Progressive line blanking (PLB) addressing scheme for ZBD. (b) Comparison of a typical ZBD latching characteristic for simple bipolar and PLB pulse sequences. PLP has a faster, lower voltage response since the blanking portion also couples to the positive $\Delta\varepsilon$ of the material, distorts the director profile closer to the grating and hence increases the flexoelectric polarization during the select portion.

b) *Blanking*. The frame is divided into two fields, but the first field indiscriminately blanks the screen to one state (say black) so that the white pixels can be selected discriminately in the second field. The blanking is usually done line-at-a-time, similarly to two-field addressing, or a page /section blank may be used.
c) *Progressive Line Blanking*. Rather than dividing the frame into two fields, each line is addressed in sequence with a blanking portion, followed by a selection portion. An example of such a scheme used for ZBD is shown in Figure 6.15.
d) *Line-ahead blanking*. The blanking pulse can have sufficient electrical impulse to blank into one state (usually C) regardless of the applied data. Thus, a row can be blanked during the addressing signal for the previous row or rows. Although potentially faster, this type of scheme isn't usually done using conventional STN type drivers since they are restricted to four voltage levels only.
e) *Strobe*-extension. As with Line-ahead blanking, the selection strobe may be extended into the following lines to help improve response time further. Again, this is not used with STN drivers due to the four-level restriction.

For many simple applications, such as shelf-edge label displays, the slow speed of the basic addressing schemes such as Two-field or Line-at-a-time Blanking is adequate and readily applied using standard STN drivers. Where addressing speed is important, faster schemes such as line-ahead blanking allow line addressing times significantly faster than 1 ms per line to be achieved.

6.6 Optical Configurations

Although it is possible to create a scattering mode ZBD LCD [43], this technology is usually configured to operate by modulating polarized light [44]. Before considering the bulk optical properties of the LCD alignment states, it is important to realize that the grating itself has a strong optical effect due to both diffraction and form birefringence. In a practical LCD, the design is chosen to minimize these effects since they inevitably add coloration to the white state and/or reduce the contrast ratio. This is achieved by ensuring that the bistable grating surface is adjacent to the front polariser in the LCD cell (Figure 6.16) and further that the polarizer is aligned so that its transmission axis is perpendicular to the LC slow axis (so-called "o-mode"). Since the LC remains in the plane of the page for both bistable configurations, the incident light only samples n_o of the LC material. By engineering the ZBD

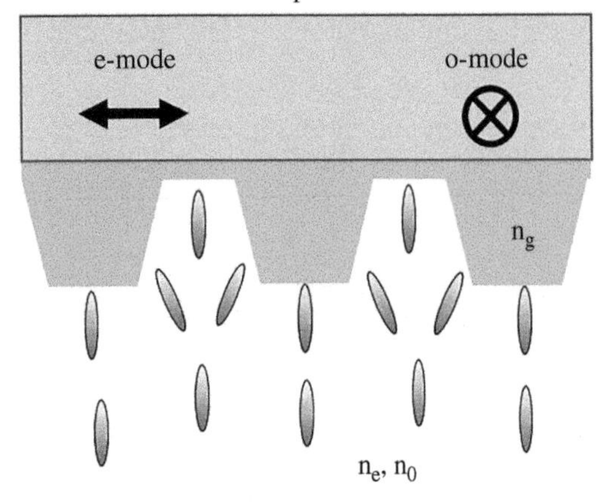

Figure 6.16 Two possible arrangements (e-mode or o-mode) of the front polariser with respect to the LC and grating.

photopolymer such that its refractive index (n_g) is equal to n_o, it is thus possible to make the grating-LC interface optically invisible. With the grating effects removed, it is possible to readily model the optical viewing of the ZBD LCD.

Regarding Figure 6.16, the front (transmissive) and rear (reflective) polarisers are typically placed in the crossed configuration such that the TN state is white and the HAN state is black. The viewing properties of these two states are shown in Figure 6.17. For this example, the bistable grating tilt angles are 5° and 90°, the rubbed polymer tilt is 5°, $\Delta n = 0.12$, and the cell gap is set to the first Gooch and Tarry minimum [22] given by Eq. (6.3). The wide viewing of the bright TN state is perhaps surprising since TN LCD is generally regarded as a narrow viewing angle mode. However, it should be recognized that multiplexed TN usually operates between two non-zero voltages leading to higher LC director tilt in the mid-cell region, which significantly compromises viewing. In contrast, the configurations in Figure 6.17 are both maintained at zero applied voltage, and so the TN state supports a low director tilt (averaging about 3°) throughout the bulk of the device.

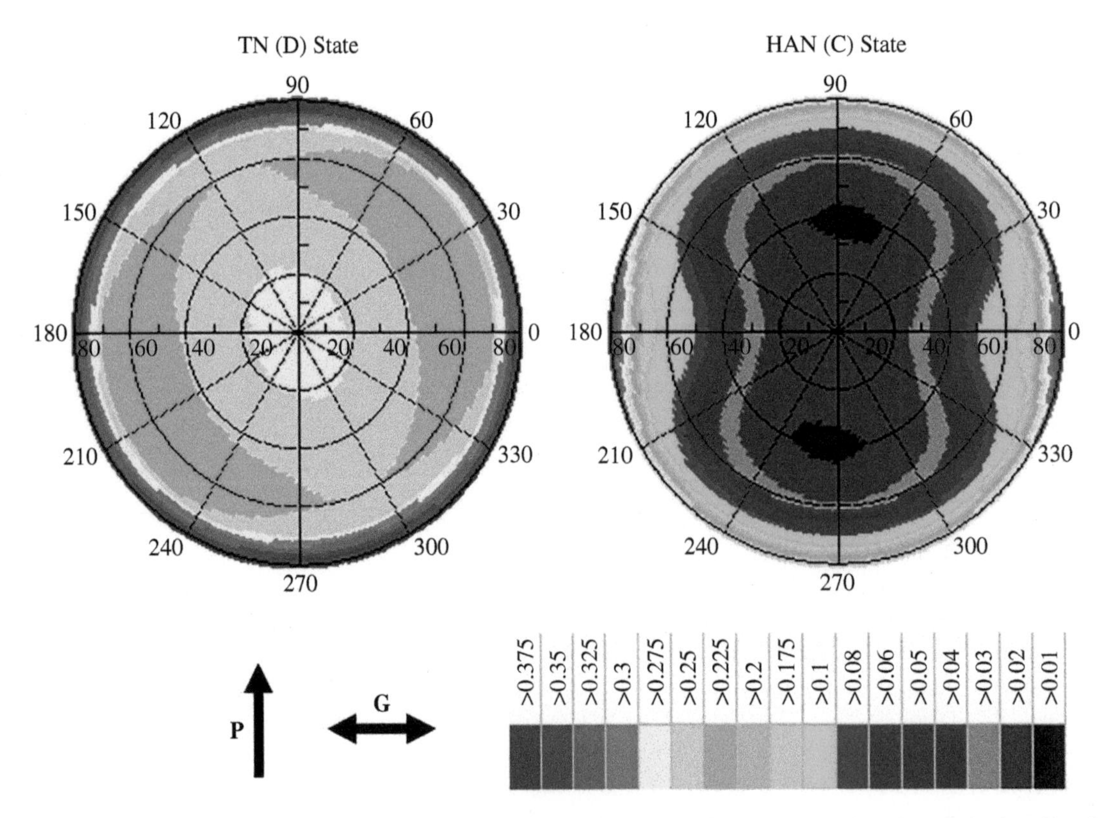

Figure 6.17 Modeled reflectivity of the bright (TN) state and the dark (HAN) state as a function of viewing direction. The orientation of P plate rubbing direction and G plate alignment is also shown. The model assumes specular reflection.

The dark HAN state, has the significant tilt in the mid-cell region, which leads to narrowing the viewing properties in the horizontal plane. Thus, the display's contrast ratio is worse in one orientation compared to the other. Usually, this direction can be chosen to best match the intended use of the display. It is also possible to increase this viewing with compensation layers if required. One other advantage of using the TN (D) state as the white state of the display is that this is the configuration that the ZBD LCD cools into after LC filling and annealing (to a temperature above the liquid crystal to isotropic transition, T_{NI}). Therefore, the unlatched interpixel gaps across the active area of the display will remain white, thus adding to the overall reflectivity of the display.

As well as viewing properties, it is important to consider the optimization of white state color balance and viewing properties. For a given LC and polarisers, it is possible to adjust the color hue of the display by changing the optical retardation, Δnd. In particular, to raise it to a level in-between the 1st and 2nd Gooch-Tarry minimum values of Eq. (6.2). Figure 6.18 shows the modeled and measured color coordinates (CIE1931) for the TN state in response to changes in the LCD cell gap ($\Delta n = 0.12$). The modeled color passes very close to the D65 white point for an adequate LC layer thickness of 6.3 μm.

Figure 6.17 shows the reflective efficiency of the ZBD mode (assuming specular reflection). However, in practice, there must be consideration of how incident light is reflected. In particular, the level and form of light diffusion are key to avoiding the "metallic" specular appearance observed in some reflective display modes. Conventional diffusers are included in the reflective polarizer materials and typically consist of high index beads in a low index polymer matrix. These provide a Gaussian scattering profile (Figure 6.19) that leads to high reflectivity only in the vicinity of the specular reflection angle; this is undesirable for wide viewing. However, recently many reflective LCDs have utilized structural diffusers [45], which produce a "flat-top" diffusion profile that can also be asymmetric around the specular angle to take advantage of particular display illumination conditions.

The effect of structural diffusers is demonstrated in Figure 6.20, which displays both diffuser types. These were photographed with a fixed camera and lighting conditions with the displays at four different tilt angles. The structural diffuser enables much higher reflective intensity, especially at angles well away from the specular condition. Indeed, the introduction of structural diffusers can be considered one of the leading performance breakthroughs for reflective LCDs (including ZBD) in the last 10 years. They allow LCDs to exhibit reflective properties closer to a Lambertian reflector and thus provide a more "paper-like" appearance to the display.

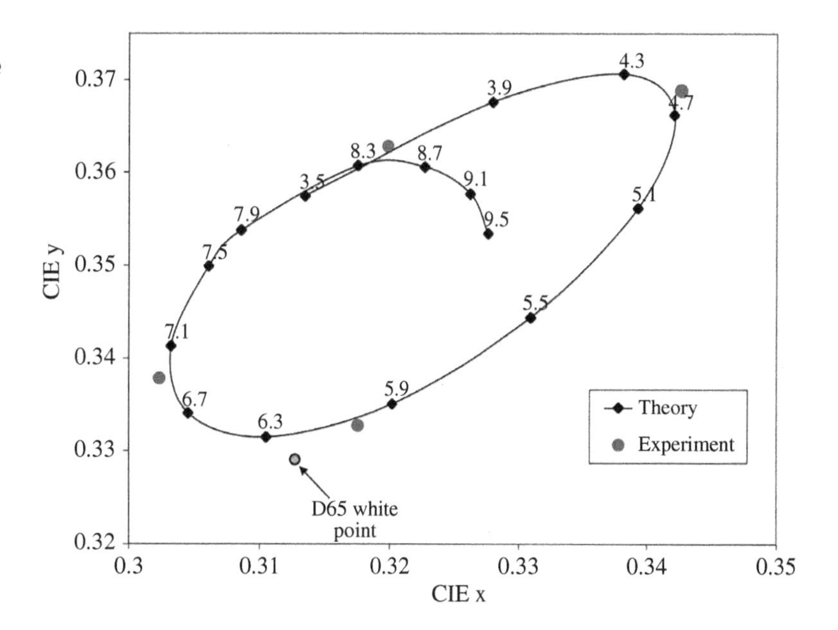

Figure 6.18 Modeled and measured color coordinates of the ZBD LCD white state as a function of cell gap (labels shows gap in microns).

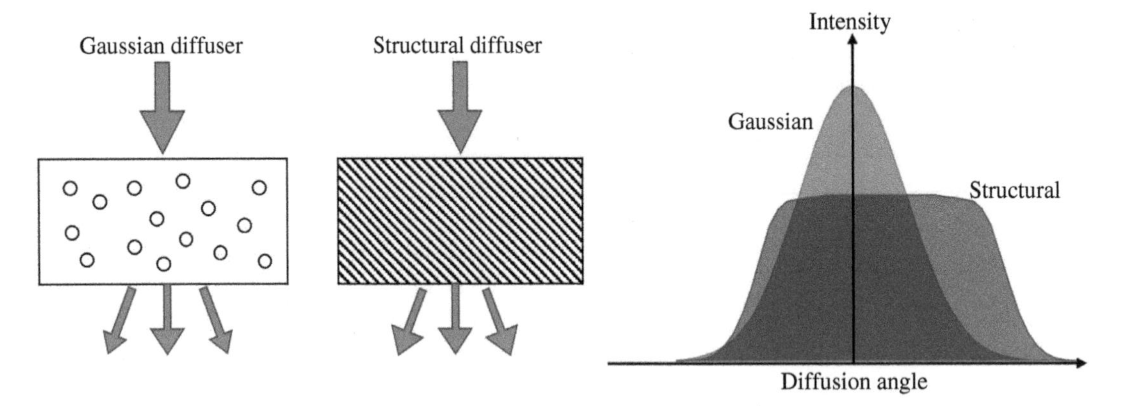

Figure 6.19 Scattering from conventional (Gaussian) diffusers and structural diffusers. The latter provide a "flat-top" diffusion profile which can also be asymmetric with respect to the incident light direction.

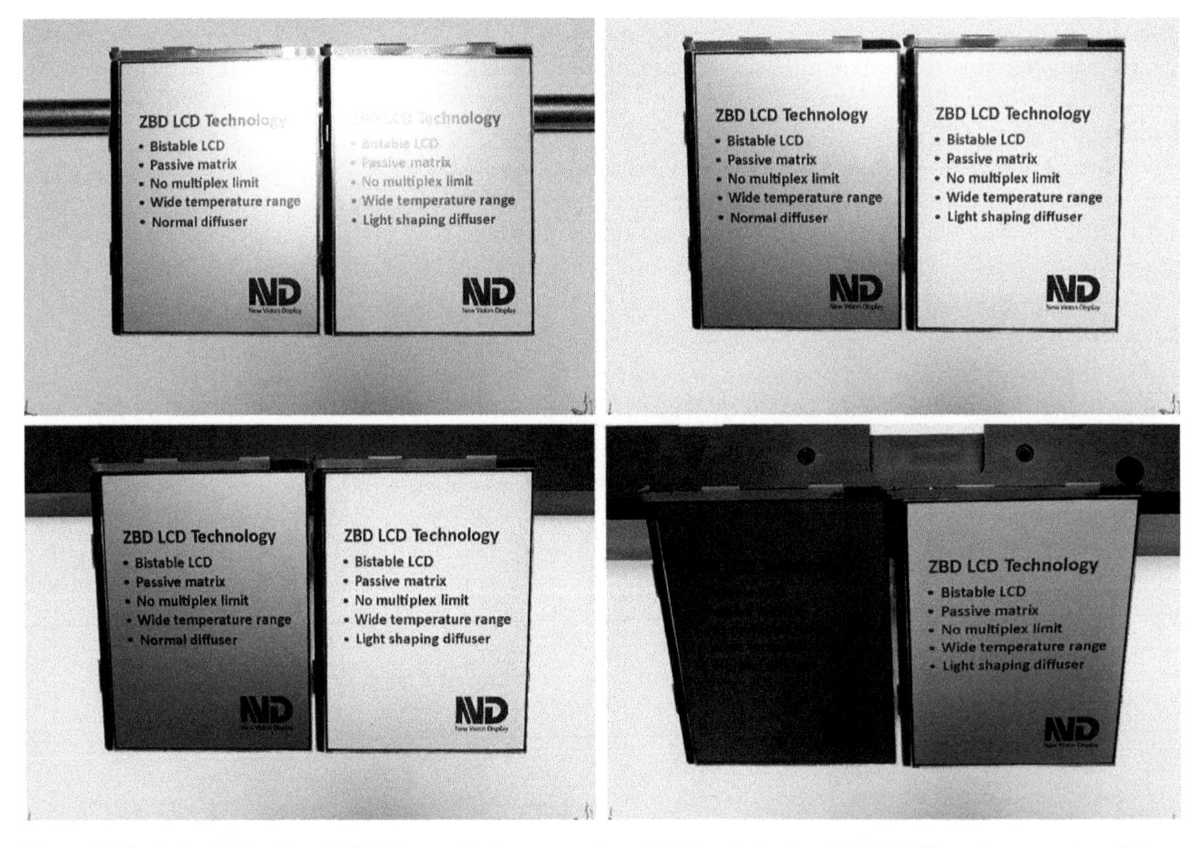

Figure 6.20 Reflectivity of two ZBD LCDs employing conventional (left) and structural (right) diffuser layers on top of the rear reflective polariser. Viewing conditions are varied from specular (top left photo) to around 45° from specular (bottom right photo).

6.7 Novel Arrangements

As well as a two-polariser fully reflective LCD mode, the ZBD technology can be deployed in other configurations, which can have merits for particular applications. A useful example of this is the dual-layer display, the configuration for which is shown in Figure 6.21. This display mode [46] combines a low power/zero power reflective display on top of a full color emissive OLED display (or a transmissive LCD). This allows a-mode display with a low-power black/white mode or a higher power color video mode. Usually, the OLED display incorporates a wide-band quarter-wave plate at the front as an anti-reflection coating. The polariser/waveplate pair is required to suppress the very high reflectivity from the metal tracks on the OLED backplane. The idea [46] is to replace the static quarter-wave plate with a bistable retarder to allow the display to operate in an electronic-paper mode with the OLED switched off. ZBD is well suited to such a configuration since it can be tuned to act as a bistable waveplate and a has a very high aperture ratio (passive matrix). A HAN/VAN ZBD mode is then inserted between the optical layers and the OLED backplane. In the VAN state, the ZBD layer does not disturb the antireflection properties of the polariser/waveplate layer, thus maintaining a low reflectivity. However, a suitably optimized HAN state can cancel out the fixed waveplate and lead to a bright reflectivity that can be suitably diffused by the insertion of an appropriate structural diffuser. Figure 6.21 shows the operation of such a prototype that can will enable a display product that can display a zero-power fixed image while also allowing full color multimedia operation when required.

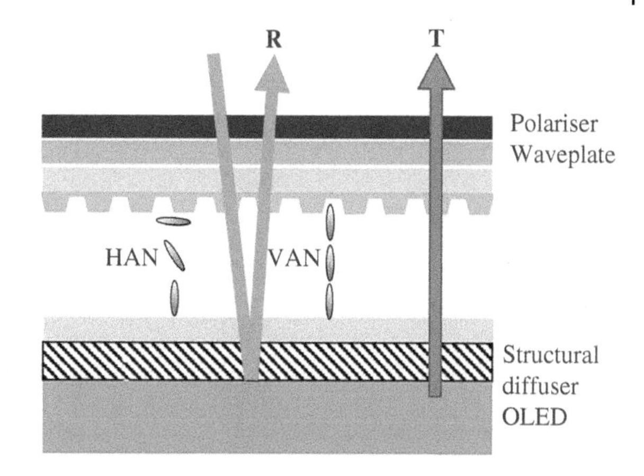

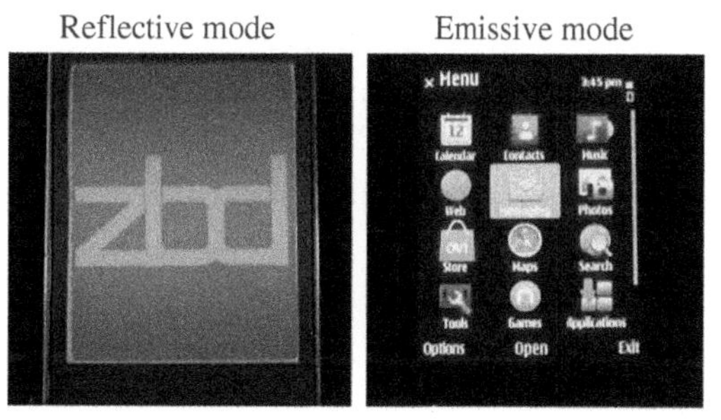

Figure 6.21 Configuration of ZBD/OLED dual layer display and resulting reflective and emissive performance.

A different dual-mode device can be considered in which a bistable mode such as ZBD is combined with a TFT backplane [47]. A typical configuration is shown in Figure 6.22, in which a HAN/VAN mode ZBD is constructed onto a TFT backplane that carries transflective electrodes. Operation of the TFT at high ($\approx$ 15 V) voltage allows the ZBD surface to be latched into the HAN (white) or VAN (black) states. In addition, video-rate greyscale operation is available by first latching the whole display into the white state and then using standard low voltage TFT addressing to modulate the HAN state brightness. A typical prototype is shown in Figure 6.22, and a close-up of the pixels latched into both the bistable states. Suitable diffuser and color filter layers should also be added to operate fully. This mode can also be considered using a fully reflective electrode and a frontlight.

Both modes shown in this section have practical limitations which are yet to be overcome. Firstly, both use single-polariser modes in which the polarizer is at 45° to the grating grooves. This leads to diffraction and form birefringence effects. While the latter can be compensated with optical films, the latter be reduced by reducing the grating periodicity to sub-optical values. This requires utilization of emerging grating mastering techniques. Secondly, to achieve practical cell gaps (3–4 μm or greater), single polariser modes should use LC materials with low Δn (0.08–0.10). Such materials are challenging to utilize in a ZBD mode while maintaining a usefully wide operating temperature range.

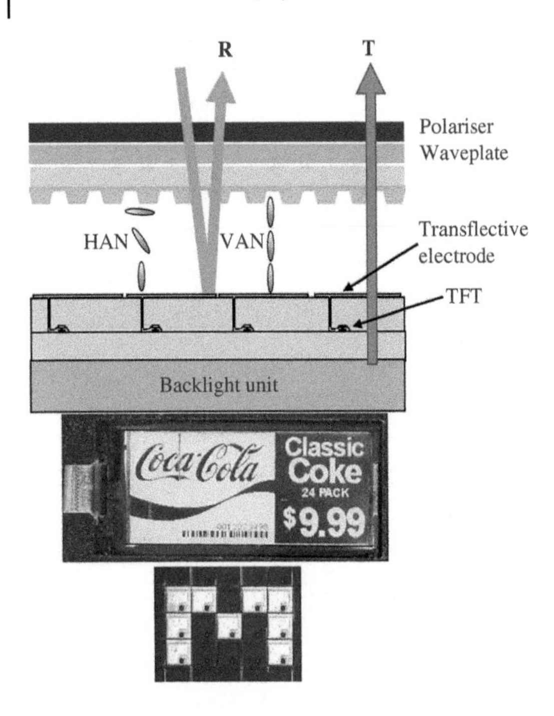

Figure 6.22 Dual mode display in which a ZBD LCD is fabricated on an active-matrix backplane to allow a fixed image mode or a video mode. A working prototype is shown along with a close-up view of pixels in two bistable states.

6.8 Conclusions

One of the most important paper features, which all electronic -paper must aspire to, is its ultra-low-cost. Among the lowest cost displays available are simple reflective TN LCD. They are passive matrices and do not require TFTs to drive them. This low cost is delivered because the liquid crystal materials, glass, alignment layers, polarisers, and electronic drivers are commoditized components. The ZBD is a bistable equivalent to the TN option, with only the additional step of the grating fabrication required. The large-scale replication of the ZBD grating described here provides a means of maintaining that low cost while offering unlimited image complexity through its line-at-a-time passive matrix addressing. Of course, electronic paper must also offer high reflectivity across a wide range of viewing angles. The contrast of ZBD is world-beating for a bistable technology. Near the specular angle, the light reflected is also high and typically higher than a Lambertian reflector. However, many electronic paper applications require a wide range of viewing angles, for which the properties of the paper-like Lambertian reflector are advantageous. For such applications, improved viewing has been achieved using additional optical films, albeit at some small additional cost to the panel. Table 6.1 summarizes the current state of commercially available ZBD devices.

Table 6.1 Operating parameters of a typical commercial ZBD display.

Parameter	Value
Temperature range	−10 to +70 °C
Operating voltage (typical)	20 V (20 °C), 40 V (−10 °C)
Optical response time	t_{on} = 20 ms, t_{off} = 0.1 ms (5.0 μm cell gap)
Cell gap	4.0–8.0 μm (2 polariser mode)
Reflectivity	34% (fully diffuse illumination)
Display update speed	4.0 ms per line (@20 °C).
Display complexity limit	Infinite number of lines possible.
Bistable image lifetime	>10 years (between −40 and +85 °C)

Acknowledgments

JCJ wishes to thank RCUK (EPSRC) for an Advanced Manufacturing Fellowship (EP/S029214/1).

References

1 Heikenfeld, J., Drzaic, P., Yeo, J.-S., and Koch, T. (2011). A critical review of the present and future prospects for electronic paper. *J. Soc. Inf. Disp.* 19 (2): 129–156.
2 Comiskey, B., Albert, J.D., Yoshizawa, H., and Jacobson, J. (1998). An electrophoretic ink for all-printed reflective electronic displays. *Nature* 394: 253–255.
3 Hattori, R., Yamada, S., Masuda, Y., and Hihei, N. (2003). Novel type of bistable reflective display using quick response liquid powder. *Proc. SID Int. Symp. Dig. Tech. Pap.* 34 (1): 846–849.
4 Heikenfeld, J., Zhou, K., Kreit, E. et al. (2009). Electrofluidic displays using Young–Laplace transposition of brilliant pigment dispersions. *Nat. Photonics* 3: 292–296.
5 Hayes, R.A. and Feenstra, B.J. (2003). Video-speed electronic paper based on electrowetting. *Nature* 425: 383–385.
6 Back, U., Corr, D., Lupo, D. et al. (2002). Nanomaterials-based electrochromics for paper-quality display. *Adv. Mater.* 14 (11): 845–848.
7 Jones, J.C. (2017). Liquid crystal displays. In: *The Handbook of Optoelectronics. Second Edition. Volume 2: Enabling Technologies. Chapter 6* (ed. J.P. Dakin and R.G.W. Brown), 137–224. Boca Raton, FL: CRC Press.
8 Jones, J.C. (2018). Fifty years of the liquid crystal display. *Liq. Cryst. Today* 27 (3): 44–70.
9 Jones, J.C. (2012). Bistable LCDs. In: *Handbook of Visual Display Technology* (ed. J.C. Janglin, W. Cranton and M. Fihn), 1507–1543. Berlin Heidelberg: Springer Verlag.
10 Greubel, W. (1974). Bistability behavior of texture in cholesteric liquid crystals in an electric field. *Appl. Phys. Lett.* 25 (1): 5–7.
11 Coates, D., Crossland, W.A., Morrissy, J.H., and Needham, B. (1978). Electrically induced scattering textures in smectic A phases and their electrical reversal. *J. Phys. D. Appl. Phys.* 11: 2025–2034.
12 Clark, N.A. and Lagerwall, S.T. (1980). Submicrosecond bistable electrooptic switching in liquid crystals. *Appl. Phys. Lett.* 36: 899–901.
13 Berreman, D.W. and Heffner, W.R. (1981). New bistable liquid crystal twist cell. *J. Appl. Phys.* 52 (4): 3032–3039.
14 Dozov, I., Nobili, M., and Durand, G. (1997). Fast bistable nematic display using monostable surface switching. *Appl. Phys. Lett.* 70 (9): 1179–1181.
15 Thurston, R.N., Cheng, J., and Boyd, G.D. (1980). Mechanically bistable liquid-crystal display structures. *IEEE Trans. Electron Dev.* ED-27 (11): 2069–2080.
16 Bryan-Brown, G.P., Brown, C.V., and Jones, J.C. (2001). *US Patent 6,456,348*. Priority date 16th October 1995.
17 Bryan-Brown, G.P., Brown, C.V., Jones, J.C. et al. (1997). Grating aligned bistable nematic device. *Proc. SID Int. Symp. Dig. Tech. Pap.* 28: 37–40.
18 Jones, J.C., Bryan-Brown, G.P., Wood, E.L. et al. (2000). Novel bistable liquid crystal displays based on grating alignment. *Proc. SPIE 3955, Liq. Cryst. Mater. Dev. Flat Panel Displays* 3955: 84–93.
19 Wood, E.L., Brett, P.J., Bryan-Brown, G.P. et al. (2002). Large area, high resolution portable ZBD display. *Proc. SID Int. Symp. Dig. Tech. Pap.* 33 (1): 22–25.
20 Jones, J.C. (2008). The zenithal bistable display: from concept to consumer. *J. Soc. Inf. Disp.* 16 (1): 143–154.
21 Jones, J.C. (2017). Defects, flexoelectricity and RF communications: the ZBD story. *Liq. Cryst.* 44 (12): 2133–2160.
22 Gooch, C.H. and Tarry, H.A. (1975). The optical properties of twisted liquid crystal structures with twist angles less than or equal to 90 degrees. *Appl. Phys.* 8: 1575–1584.
23 Berreman, D.W. (1971). Solid surface shape and the alignment of an adjacent nematic liquid crystal. *Phys. Rev. Lett.* 28 (26): 1683–1685.
24 Kitson, S. and Geisow, A. (2002). Controllable alignment of nematic liquid crystals around microscopic posts: stabilization of multiple states. *Appl. Phys. Lett.* 80 (19): 3635–3637.

25 Kischka, C., Elston, S.J., and Raynes, E.P. (2008). Measurement of the sum (e1+e3) of the flexoelectric coefficients e1 and e3 of nematic liquid crystals using a hybrid aligned nematic (HAN) cell. *Mol. Cryst. Liq. Cryst.* 494: 93–100.

26 Jones, J.C., Beldon, S., Brett, P. et al. (2003). Low voltage zenithal bistable devices with wide operating windows. *Proc. SID Int. Symp. Dig. Tech. Pap.* 34: 954–957.

27 Stewart, I.W. (2004). *The Static and Dynamic Continuum Theory of Liquid Crystals*. CRC Press.

28 de Gennes, P.G. and Prost, J. (1993). *The Physics of Liquid Crystals*. Oxford: Clarendon Press.

29 Parry-Jones, L.A. and Elston, S. (2005). Flexoelectric switching in a zenithally bistable nematic device. *J. Appl. Phys.* 97: 093515.

30 Spencer, T.J. and Care, C. (2006). Lattice Boltzmann scheme for modeling liquid-crystal dynamics: zenithal bistable device in the presence of defect motion. *Phys. Rev. E* 74 (061708): 1–14.

31 Spencer, T.J., Care, C., Amos, R.M., and Jones, J.C. (2010). A zenithal bistable device: comparison of modelling and experiment. *Phys. Rev. E* 82 (021702): 1–13.

32 Jones, J.C. (2014). Bistable nematic liquid crystals. In: *The Handbook of Liquid Crystals. 2nd Edition. Volume 8, Chapter 4* (ed. J.W. Goodby, P.J. Collins, T. Kato, et al.), 87–145. New York: Wiley.

33 Tang, X. and Selinger, J.V. (2019). Theory of defect motion in 2D passive and active nematic liquid crystals. *Soft Matter* 15: 587.

34 Wood, E.L., Hui, V.C., Bryan-Brown, G.P., et al. (2006) *US patent 7,053,975*; priority date 20th July 2000.

35 Raynes, E.P. (1974). Improved contrast uniformity in twisted nematic liquid-crystal electro-optic display devices. *Electron. Lett.* 10 (9): 141–142.

36 Jones, J.C. (2011). *US patent* 8,199,295; priority date 2nd July 2003.

37 Day, S.E., Willman, E., James, R., and Fernandez, A. (2008). Defect loops in the zenithal bistable display. *Proc. SID Int. Symp. Dig. Tech. Pap.* 39 (1): 1034–1037.

38 Orlic, S. (2017). Optical information storage and recovery. In: *The Handbook of Optoelectronics. Second Edition. Volume 2: Enabling Technologies. Chapter 13* (ed. J.P. Dakin and R.G.W. Brown), 505–540. Boca Raton, FL: CRC Press.

39 https://en.wikipedia.org/wiki/Compact_Disc_manufacturing.

40 Bryan-Brown, G.P., Walker, D.R.E., and Jones, J.C. (2009). Controlled grating replication for the ZBD technology. *Proc. SID Int. Symp. Dig. Techn. Pap.* 40: 1334–1337.

41 Jones, S.A., Bailey, J., Walker, D.R.E. et al. (2018). Method for tuneable homeotropic anchoring at microstructures in liquid crystal devices. *Langmuir* 34 (37): 10865–10873.

42 Amos, R.M., Bryan-Brown, G.P., Wood, E.L., et al. (2010). *US patent 7,824,516*, priority date 13th July 2003.

43 Jones, J.C. (2013). *US patent 8,384,872*, priority date 30th Nov 1999.

44 Jones, J.C. and Bryan-Brown (2010). Low cost zenithal bistable device with improved white state. *Proc. SID Int. Symp. Dig. Tech. Pap.* 41: 207–211.

45 Ishinabe, T., Kusama, K., Fujikake, H., and Shoshi, S. (2013). Wide-color-gamut and wide-viewing-angle color reflective LCD with novel anisotropic diffusion layer. *Proc. SID Int. Symp. Dig. Tech. Pap.* 44: 350–353.

46 Bergquist, J. (2011). Patent application WO/2011/107826, priority date 1st March 2010.

47 Jones, J.C. (2006). *US patent 7,019,795*, priority date 20th June 2002.

7

Reflective LCD with Memory in Pixel Structure

Yoko Fukunaga

Senior Engineer, Panel Design Department, Japan Display Inc.

7.1 Introduction

Low-temperature polycrystalline silicon (LTPS) is a well-established backplane thin film transistor (TFT) technology used for flat panel displays. LTPS TFT technology allows us to integrate complementary MOS (CMOS) circuits onto the glass. Development of the system-on-glass (SOG) display using LTPS TFT liquid crystal displays (LCDs) has been hot technology topics in the late 1990s and early 2000s. SOG technology allows us to integrate various functional circuits onto display panels, such as timing controllers, display drivers, and decoders.

Memory in pixel (MIP) LCD is one of the unique applications of SOG technologies. It is one of the best ways to reduce LCD driving power because this approach needs no image data refreshment when displaying still images [1–7].

Reflective LCD technology is also one of the best ways to reduce display power consumption. The backlight consumes a large amount of transmissive LCD's power. A reflective LCD utilizes environmental light and has no need to turn on an auxiliary light when the environment is bright enough.

MIP technology and reflective LCD technology are good combinations because MIP circuits can be integrated behind the reflective pixel electrode without sacrificing the reflective aperture ratio.

LTPS SOG technology makes it possible to integrate display interface circuits onto a glass border area. An MIP LCD with display interface circuits is a good example of a fully integrated digital display system. It works as a digital memory device that can display memory data on the screen. A fully digital interface display requires fewer control and data signals to drive it. This reduces the number of flexible-on-glass (FOG) pads required, and allows the system to be driven directly by a microcontroller. In other words, it does not require an external display driver IC, which consumes large amounts of power for display operations. Reflective displays are suitable for maximizing outdoor visibility with low power consumption. MIP reflective LCD technology is advantageous for products that require a small FOG pad design and super low power operation. Therefore, they have been widely used for sports watches over the last decade.

This chapter traces the development of MIP reflective display technologies and their applications. It includes a review of the basic technology behind an MIP circuit and its pixel design and LCD optical design and the design characteristics of available market products and their super low power operations. The chapter also reviews selected applications of MIP reflective LCDs and previews future possibilities for this technology and its applications.

The following section provides an overview of the MIP technology, how it works, and how to drive it.

E-Paper Displays, First Edition. Edited by Bo-Ru Yang.

7.2 Memory in Pixel Technology and Its Super Low Power Operation

MIP technology is one of the best ways to reduce LCD driving power because this approach needs no image data refreshment when displaying still images [1–7].

Figure 7.1 shows the pixel design concept of a MIP reflective LCD. A static random-access memory (SRAM) circuit is embedded behind the reflective electrode in each sub-pixel.

Figure 7.2 compares a conventional active-matrix LCD with a MIP LCD.

Figure 7.2a shows a system block diagram and an equivalent pixel circuit of a conventional active-matrix LCD. Each pixel datum is written through the signal line when the gate line is on. The pixel datum is stored as a charge of the capacitor between the pixel electrode and the common electrode (COM). When the gate is off (voltage holding period), the pixel voltage decays depending on the leak current of the pixel TFT and the voltage holding ratio of the liquid crystal (LC) cell with the storage capacitor in parallel. The pixel data need to be refreshed at an appropriate cycle, such as 60 Hz, to avoid flickering issues caused by voltage decay. The Horizontal Driver (H-Driver) and the Vertical-Driver (V-Driver) have to continuously refresh all the pixel data.

Figure 7.2b shows a system block diagram and an equivalent pixel circuit of a MIP LCD, while Figure 7.3 shows an equivalent MIP circuit with applied signals to drive the LC.

The SRAM memory datum of each sub-pixel is written through the signal line when the gate is on. Each sub-pixel datum is stored in the SRAM. The OUT or the XOUT switch is selected according to the memory datum. As a result, the pixel electrode is connected to a line that conveys an FRP or an XFRP signal. An FRP signal is equivalent to a signal that drives a common electrode (VCOM signal), while an XFRP signal is the inverse of a VCOM signal. In a normally-black mode LCD, when an FRP signal is selected, there is no voltage applied to a LC pixel, which results in a black image being shown. Conversely, the selection of an XFRP signal results in a white image being shown. The voltage polarity applied to LC is toggled at an appropriate cycle by changing the VCOM/FRP and XFRP signal polarity to avoid flickering and image sticking issues caused by the offset voltage of the LC capacitor.

As shown in Figure 7.2, all the display image data are stored in SRAMs placed in the active display area (A.A.). There is no need to operate the H-Driver and the V-Driver when displaying a still image.

Generally, power consumption (P) is estimated by

$$P = CV^2f\big(C: \text{Capacity}, V: \text{Voltage}, f: \text{frequency}\big) + \text{static current} .$$

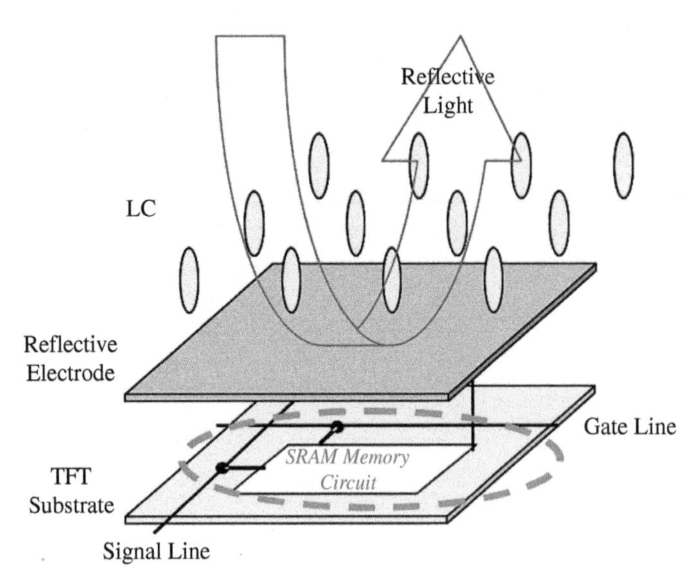

Figure 7.1 Design concept of memory in pixel (MIP) reflective LCD.

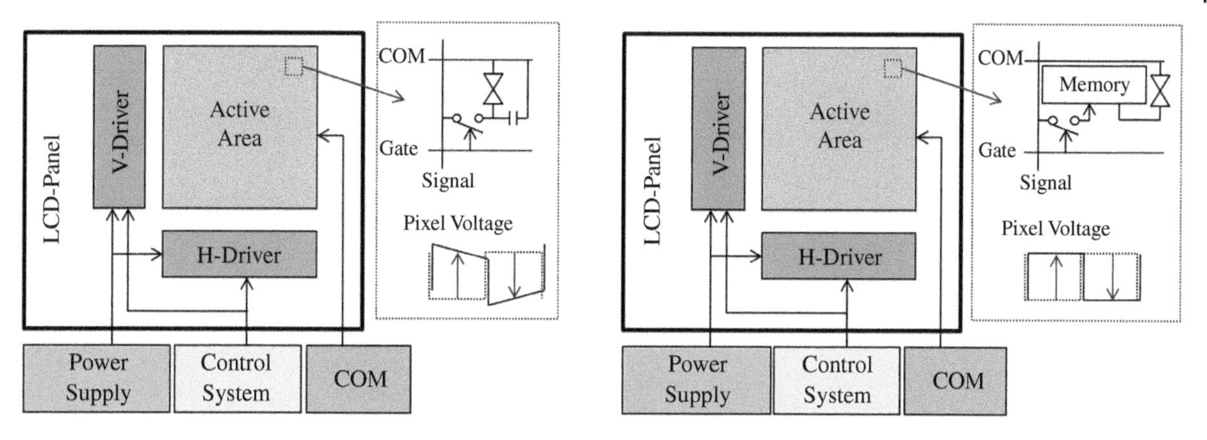

Figure 7.2 System block diagram and equivalent pixel circuit of a conventional active-matrix LCD and a memory in pixel LCD.

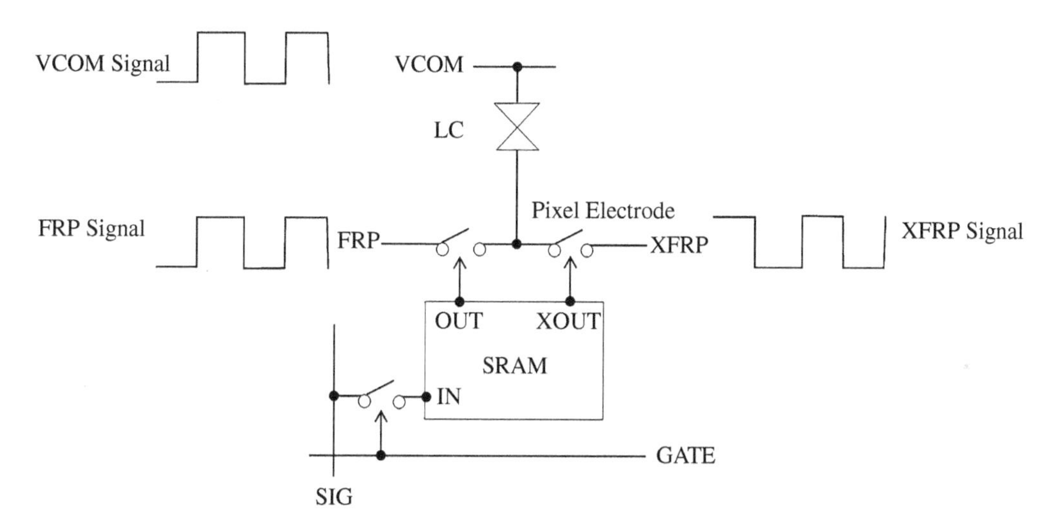

Figure 7.3 Equivalent MIP circuit with applied signals to drive a liquid crystal.

The majority of power in a conventional active-matrix LCD is consumed to drive signal lines. The charging and discharging current to signal lines, which act as a large capacitor, is required at a high-frequency level.

On the other hand, when displaying a still image, a MIP LCD consumes only a small amount of charging and discharging current to the LC capacitor at a low enough frequency to avoid flickering and image sticking. This results in an ultra-low-power operation.

The main difference between a MIP reflective LCD and other low power memory display modes, such as an electrophoretic display and a cholesteric LCD, is that the display mode itself does not have memory characteristics. Generally, memory display modes have larger capacitance per unit area. They consume larger amounts of power to refresh an image, and they cannot show moving images.

A MIP LCD utilizes an SRAM circuit as digital memory, and an LC on the memory operates as a digital optical shutter. The combination of the separated functions allows the system to achieve super-low power operation to display moving images.

Even when displaying moving images, a MIP LCD consumes lower power than a conventional LCD. The following section describes the reason why.

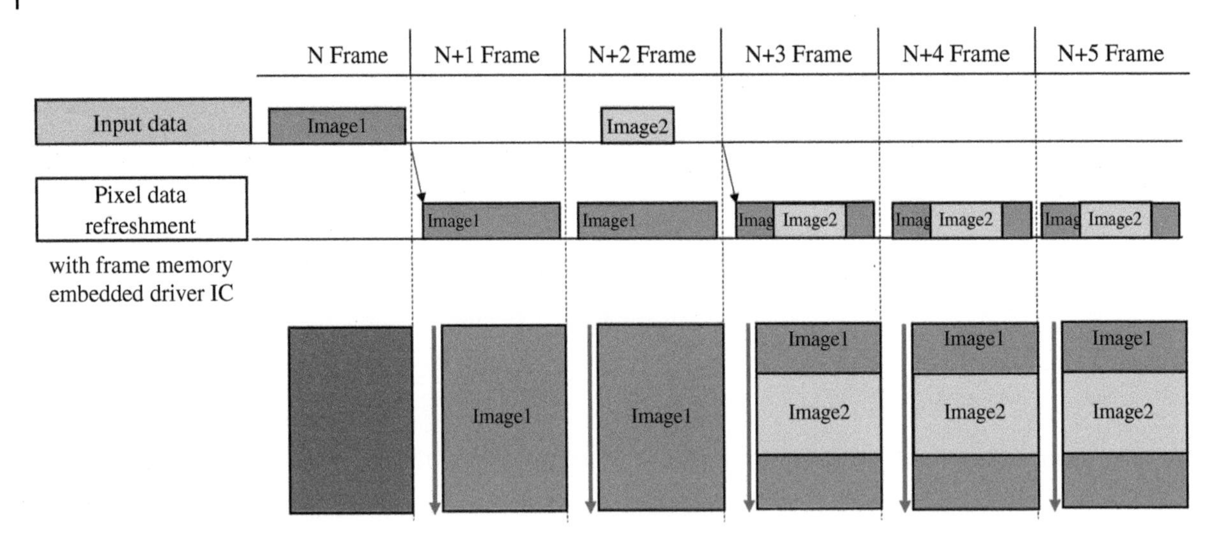

Figure 7.4 Display image data transfer of a conventional active-matrix LCD.

Figure 7.4 illustrates how a conventional active-matrix LCD transfers input image data display via a display driver IC with frame memory. For example, full-screen image data ("Image 1") are sent to the display driver IC, and the data ("Image 1") are stored in the frame memory. The frame data ("Image1") are converted to analog signals and transferred to all the pixels via the signal lines in the next frame period. When some portion of the image data ("Image2") needs to be changed, only the data, "Image 2," are sent to the display driver IC, and the portion is overwritten on the "Image 1" data. The driver IC autonomously refreshes the full-screen data at a frame rate using the data stored in the frame memory.

Figure 7.5 illustrates how a MIP LCD transfers input image data to a display area. Full-screen image data ("Image 1") are sent to the active area through the display interface circuits embedded on the border area. The data are directly stored in the SRAMs behind the pixel electrodes. When some portion of the image data ("Image2") needs to be changed, the gate lines corresponding to the area ("Image 2") are selected and overwritten. The power consumption to overwrite the display is proportional to the gate lines to be written. This operation reduces power consumption when images need to be partially updated.

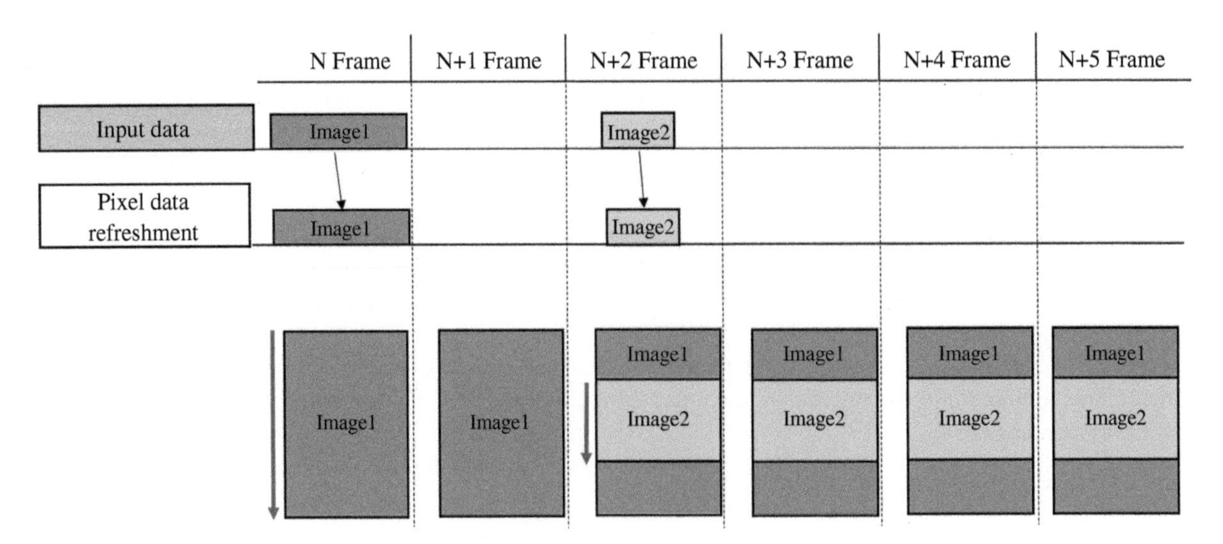

Figure 7.5 Display image data transfer of an MIP LCD.

With a conventional active-matrix LCD, the gradation of each subpixel is controlled by analog voltage. The driver IC needs to equip digital/analog converters and analog amplifiers to drive the signal lines, continuously consuming bias current.

In contrast, a MIP LCD is a fully digital display. It is driven by digital signals.

To return to the equation to estimate power consumption (P),

$$P = CV^2 f\left(C : \text{Capacity}, V : \text{Voltage}, f : \text{frequency}\right) + \text{static current} .$$

A MIP LCD system can eliminate static current to generate analog signals and reduce the frequency to drive signal lines (f), and reduce the display area (C) to be rewritten. This contributes to super-low levels of power operation.

However, because of this full digital display feature, each subpixel of MIP LCD cannot express grayscale. The following section addresses the challenge of showing grayscale using MIP LCDs.

7.3 Sub-Pixel Pattern to Show Gray Scale

A MIP full digital LCD design is advantageous to realize super-low power operation, but expressing grayscale is a major challenge. The so-called "area coverage modulation" method expresses grayscale by combining dots, as has been the case with printing devices. However, increasing pixel bits and resolution is disadvantageous, and there is a risk of pattern noise.

To apply this method to a MIP reflective LCD, each pixel has to be divided into the number of grayscale bits, and the bit number of SRAM has to be laid out within the pixel area. As shown in Figures 7.2 and 7.3, MIP circuits are more complicated than conventional active-matrix pixel circuits. For the reasons described above, MIP LCDs have severe limitations when trying to achieve high levels of resolution. Another risk is pattern noise. If the point of gravity of each grayscale expression does not coincide with that of the pixel itself, pseudo patterns with the adjacent pixels are often recognized as pattern noise [8]. Reflective LCDs utilize environmental light, so it is impossible to compensate for any decrease in the effective pixel electrode area by increasing backlight luminance. It is important to divide pixels to maintain the center of gravity while minimizing the loss of the effective pixel electrode area.

Figure 7.6a shows a schematic diagram of the "3 divided patterns" for 2-bit gradation. Each pixel is comprised of R, G, and B sub-pixels. Each subpixel electrode is divided spatially into 3 equal parts and electrically into two parts [7]. The most significant bit (MSB) area2 times bigger than the lowest significant bit (LSB). The pixel electrode of each bit is connected to an SRAM. Figure 7.6b shows a schematic diagram of 2-bit gradation expressed

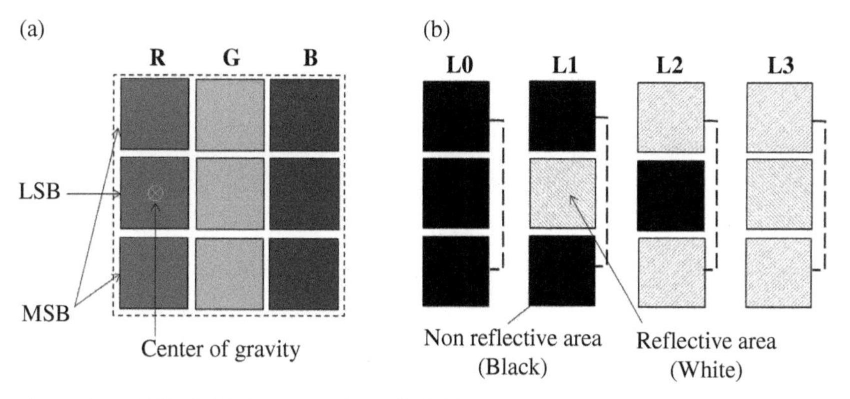

Figure 7.6 (a) "3 divided patterns," and (b) 2-bit gradation expressed by "3 divided patterns."

by the "3 divided patterns" by area coverage modulation. The "3 divided patterns" make it possible to coincide the centers of gravity for LSB and MSB while minimizing the loss of the effective pixel electrode area.

The following section describes the optical design features of reflective LCDs suitable for MIP LCDs.

7.4 Reflective LCD Optical Design

Development of active-matrix color-reflective LCDs occurred in the 1990s [9], and portable game consoles and personal digital assistants (PDAs) with this technology were commercialized in the late 1990s. However, smartphones have replaced these applications with high image quality transmissive LCDs or organic light-emitting Diodes (OLEDs) during the 2010s. MIP reflective LCDs utilize the technology and manufacturing resources that were established in the 2000s.

A reflective LCD needs to have light scattering functionality because, without this property, it looks like a mirror. The light-scattering properties have to be optimized depending on how customers use the products, such as the ambient light condition and the viewing angle range. The voltage-reflectance (V-R) curve needs to be optimized for MIP digital driving to achieve super low power operations. Auxiliary lighting is required in dark environments. The following section describes the technology to address these issues.

7.4.1 Light Scattering Functionality

There are 2 methods used to add light scattering functionality in the market available MIP reflective LCDs. One is by scattering reflective electrodes [9–11], and the other uses a light control film on the front of the LCD cell [6, 7, 12]. Achieving enough reflectance at the appropriate viewing angle range is the most important design issue for reflective LCDs.

Figure 7.7 illustrates the structure of a color reflective LCD with scattering reflective electrodes. The microstructures under the reflective electrodes are fabricated using a conventional photolithography process. The pattern pitch and shape of the microstructures are controlled by the mask design and process conditions.

Figure 7.8 illustrates the structure of a color reflective LCD with a light control film. A light control film, which has an asymmetric scattering feature controlled by random internal structures with different refractive indices [13], is put on the front side of the LC. This design makes it possible to get desirable anisotropic scattering properties [6, 7, 12].

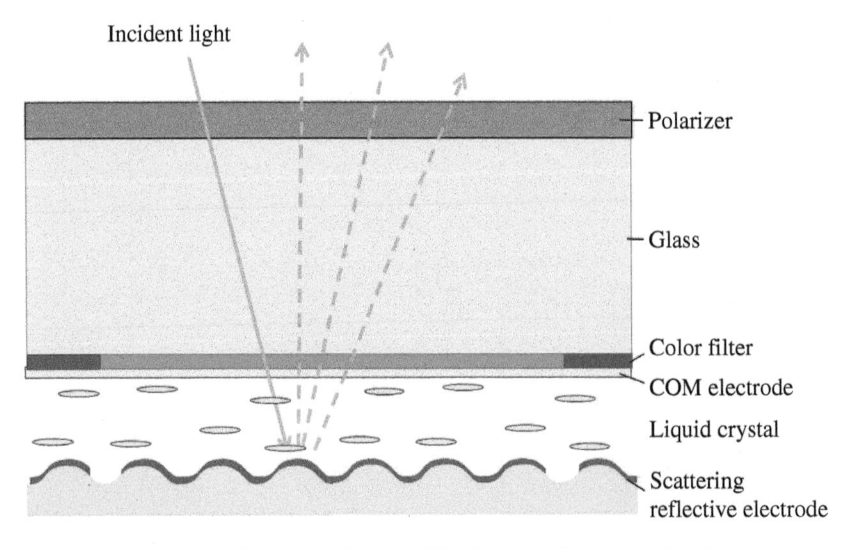

Figure 7.7 Structure of a color reflective LCD with scattering reflective electrodes.

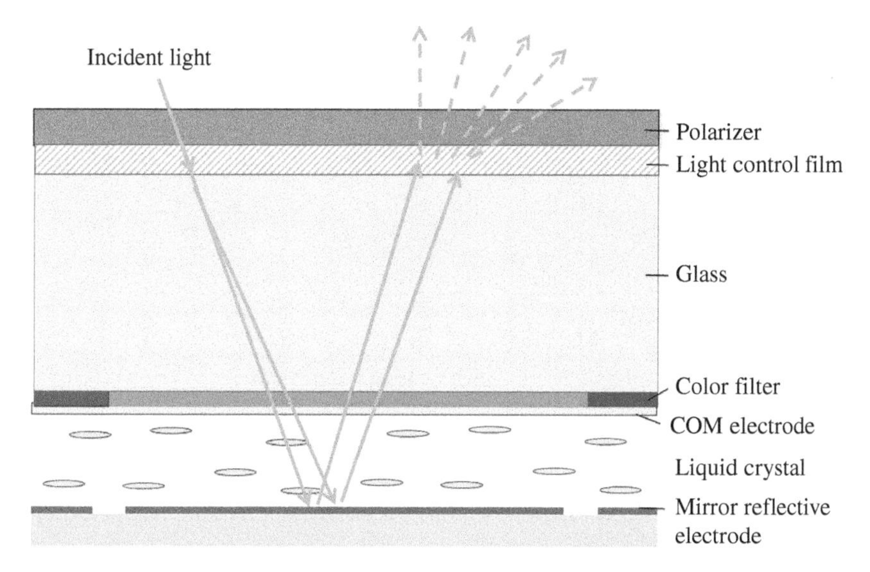

Figure 7.8 Structure of a color reflective LCD with a light control film.

The scattering reflective electrodes method is advantageous for high-resolution LCDs to avoid image blurring issues because its scattering function is in the LC. However, it is challenging to obtain anisotropic scattering properties. Moreover, it also has a tendency to show rainbow diffraction patterns under sunlight caused by the periodic pattern of the microstructure.

A light control film method can achieve high reflectivity at a specified viewing angle range without causing rainbow diffraction patterns. On the other hand, this method is unsuitable for larger LCDs because this specified viewing angle range is recognized as gradation of reflectance. Image blurring, which is dependent on the pixel pitch versus the color filter glass thickness, is another issue for high-resolution devices. Practically, it is not an issue for MIP reflective LCDs because MIP circuits have more stringent limitations to achieve high levels of resolution.

The following section describes a LC and an optical design suitable for MIP reflective LCDs.

7.4.2 LCD Cell Design

Electrically controlled birefringence (ECB) and vertical alignment (VA) modes are widely used for reflective and transflective LCDs [5–7, 9–12]. Both modes have been used in currently market-available MIP reflective LCDs. Figure 7.9 shows a basic example of an optical switching mechanism corresponding to both ECB normally white and VA normally black reflective modes. Unpolarized ambient light changes to linearly polarized light after passing through the polarizer. A wide-band quarter-wave plate film ($\lambda/4$) is added between the polarizer and the LC cell to avoid light leakage when showing a black image. The LC cell is designed to work as a quarter-wave ($\lambda/4$) optical switch. Since the light passes through the LC layer two times in reflective mode, the reflective LC cell works as a half-wave optical switch ($\lambda/2$). This results in rotating linearly polarized light 90° or not. Thus, it is possible to show black and white images.

VA and ECB normally white modes are advantageous to achieve a high contrast target ratio. The black level of these modes is insensitive to cell gap variation because LC molecules are vertically aligned when showing black (Figure 7.9, left). The ECB mode is advantageous for higher reflectivity and low voltage driving because of its V-R characteristics.

Since reflective LCDs are not visible in a dark environment, the necessity of auxiliary lighting remains unchanged. It is not ideal to sacrifice reflective electrodes and MIP circuit areas to minimize the operating time of auxiliary light. The following section describes the challenge of adding an auxiliary lighting function.

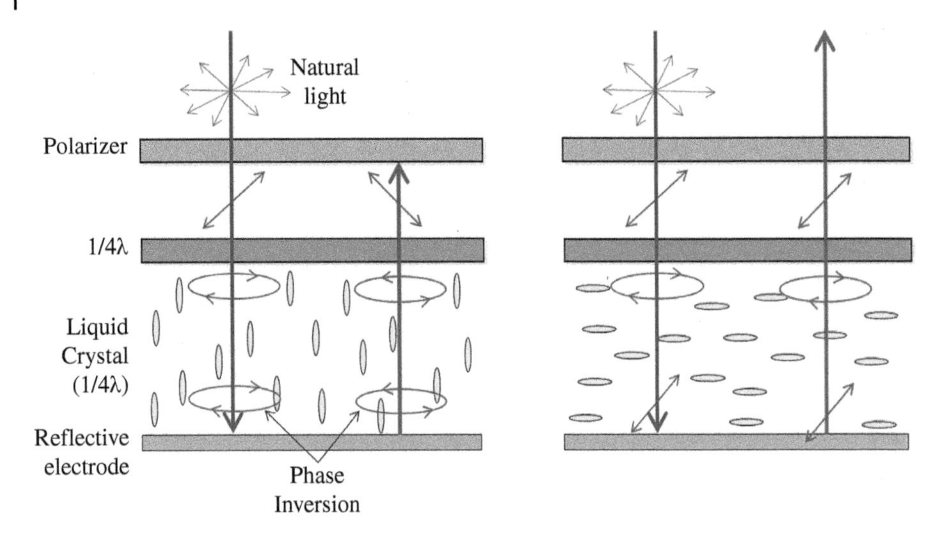

Figure 7.9 Reflective optical switching mechanism of ECB and VA mode LCDs.

7.4.3 Pixel Design for Auxiliary Lighting

There are two ways to add auxiliary lighting. One is by adding front-light, and the other is backlight. Front lighting is considered an excellent way to light reflective displays because it does not sacrifice reflectivity. Reflective LCDs with front-light were introduced in the 2000s, but they were soon replaced by transflective LCDs with backlight because of the poor reflective image quality and the manufacturing and module design issues associated with front-lights. Backlighting is better because backlight technologies are now well-established and are readily available. However, securing the transmissive area without sacrificing reflective areas is the biggest challenge.

Figure 7.10 shows a cross-sectional diagram of a conventional transflective LCD pixel, which is composed of a dual cell gap normally black ECB mode [7]. The retardation of the transmissive area is set to 2/λ, while the reflective area is set to 4/λ to coincide the voltage dependence on transmittance (V-T) with the reflectance (V-R). Since the transmissive area is put inside the pixel electrode, there is an inevitable trade-off between the reflectance and the transmittance. This design utilizes signal and gate lines for light-shielding layers in transmissive mode to achieve a higher contrast ratio target.

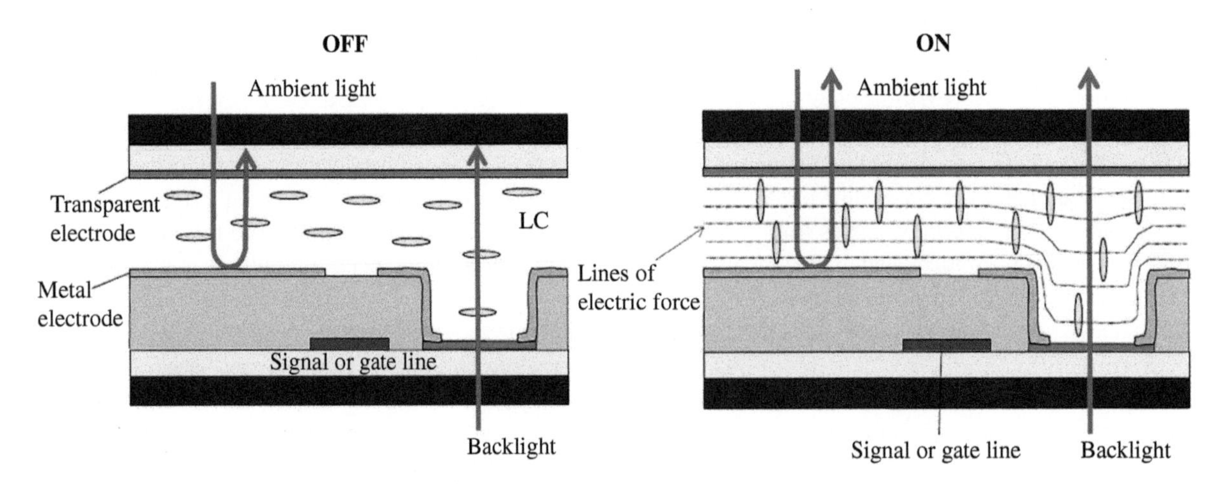

Figure 7.10 Cross sectional diagram of a conventional transflective LCD pixel.

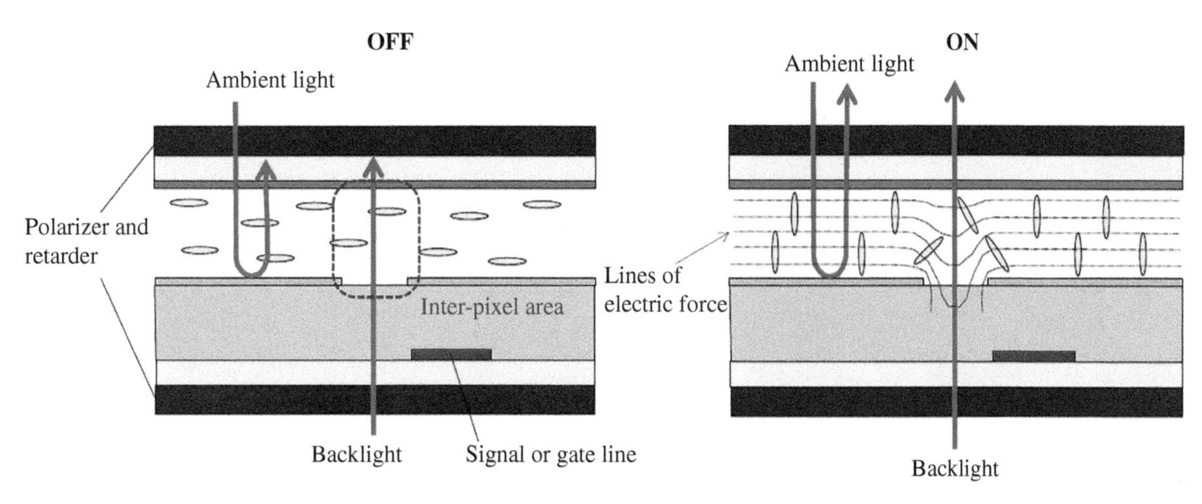

Figure 7.11 Cross sectional diagram of a MIP reflective LCD pixel.

Figure 7.11 shows a cross-sectional diagram of a MIP reflective LCD pixel, composed of a single cell gap normally black ECB mode [7]. The gate and signal lines to drive MIP circuits are set under the pixel electrode as much as possible to avoid LC responses caused by the leakage of their electric fields. Hence, the inter-pixel areas can be utilized as transmissive areas. This approach makes it possible to keep transmissive areas without sacrificing any reflective areas. The directions of LC molecules on the inter-pixel area are controlled by the fringe field from the adjacent pixel electrodes. A normally-black mode is inevitable for this approach to avoid uncontrolled LC molecules causing light leakage when displaying a black image.

As shown in Figure 7.11, a single-cell gap LC cell is employed for this approach. Figure 7.12 shows the typical V-R and V-T curves of a single gap ECB normally black LCD, which is applied to commercially available MIP reflective LCDs. The V-T curve is not saturated in this voltage range because the cell gap is not optimized for the transmissive mode. The compatibility of V-R and V-T curves is a big issue for conventional active-matrix transflective LCDs. However, the area coverage modulation method utilizes only black and white states, which is not critical for MIP reflective LCDs.

Figure 7.13 illustrates a schematic diagram of 2-bit gradation in transmissive mode expressed by the fringe of the 3 divided sub-pixel electrodes, where k denotes the length of one side of a square of a divided pixel electrode. Since the

Figure 7.12 Typical V-R and V-T curves of a single gap ECB normally-black LCD.

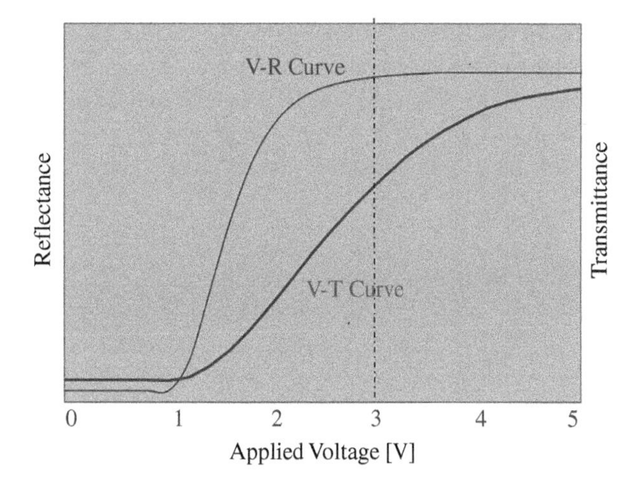

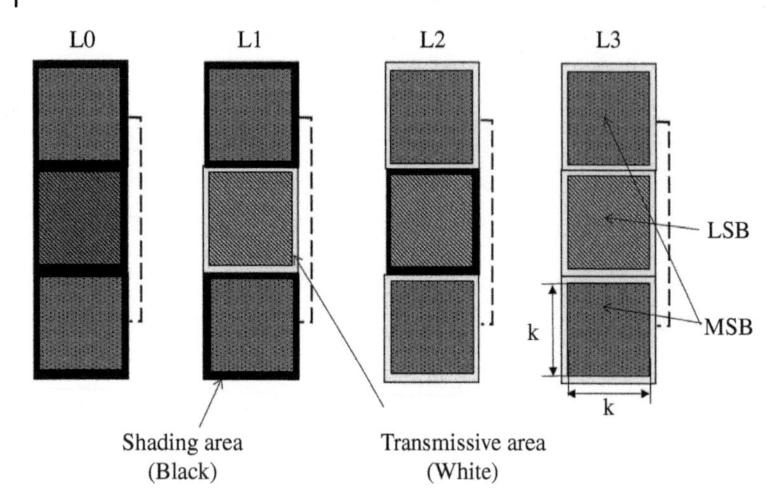

Figure 7.13 2-bit gradation expressed by the fringe of the pixel electrode.

directions of the LC molecules of the transmissive area are controlled by the fringe field of the divided pixel electrodes, transmittance of each sub-pixel is proportional to the perimeter of the white pixel electrodes [7].

Table 7.1 lists the area, the perimeter, and the ratios of LSB and MSB for the "3 divided patterns." LSB/MSB ratios for both the area and the perimeter are 0.5. The "3 divided patterns" design is suitable for reflective mode and transmissive mode to express 2-bit gradation [7].

As described in the previous section, MIP LCD consumes only a small amount of charging and discharging current to the LC capacitor at low frequency to avoid flickering and image sticking. Power consumption (P) is estimated by

$$\text{“}P = CV^2f\left(C:\text{Capacity},\ V:\text{Voltage},\ f:\text{frequency}\right) + \text{static current”}.$$

It is desirable to drive LC at low voltage and low frequency as much as possible. From the viewpoint of low voltage driving, ECB mode is more advantageous when compared with VA mode. If the power consumption is the number one priority, then ECB mode is the best choice.

The MIP reflective LCD pixel design has major advantages in preventing flickering and image sticking issues compared with a conventional active-matrix LCD pixel design. Since LC pixels of MIP LCDs are driven by static voltage, we do not need to consider a TFT leak current and voltage holding ratio of the LC. Furthermore, each LC subpixel is shown in only a black or a white state.

Figure 7.12 shows the V-R curve of an ECB normally black LCD. Since the V-R curve is fully saturated at 3 V, COM frequency can be reduced to 0.5 Hz without any issues. On the other hand, flickering is observed in VA mode at the same VCOM frequency because the V-R curve is not saturated at 3 V.

The coincidence of V-R and V-T curves is not an issue for the area coverage modulation method, but avoiding flickering issues in transmissive mode is problematic. The driving method to solve this issue is described below.

Table 7.1 The area and the perimeter of LSB and MSB for "3 divided patterns."

	LSB	MSB	LSB/MSB
Area	k^2	$2k^2$	1/2
Perimeter	4k	8k	1/2

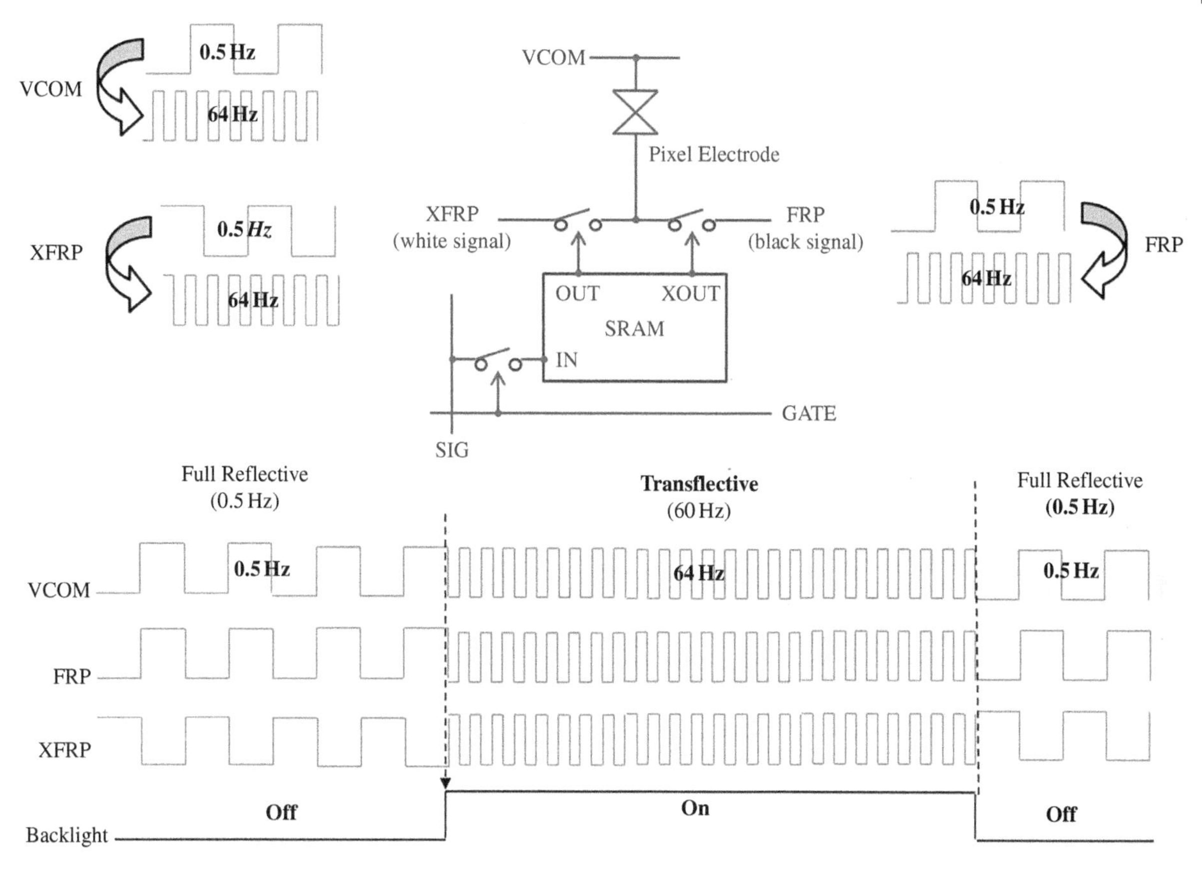

Figure 7.14 VCOM driving method to avoid flickering issue with backlight.

7.4.4 Super Low Power Driving for Auxiliary Lighting

Flickering and image sticking issues are caused by an offset voltage of the LC cell. The main causes are the difference between the LC interface potential on the COM electrode and on the reflective electrode, including the charge-up of the LC. The V-T curve in Figure 7.12 shows that the transmittance is sensitive to the voltage shift at the driving voltage (3 V). So, flickering is easily observable in transmissive mode.

Figure 7.14 shows the driving method to avoid flickering issues in transmissive mode while keeping ultra-super low power operation in reflective mode. Human eyes recognize flickering light as continuous light when the frequency is higher than the "flicker fusion threshold," around 50–60 Hz. Flickering issues can be avoided by changing the VCOM frequency to synchronize with the lighting period. This driving method has been applied to wearable devices, which can be powered by a button battery with the battery lasting for up to one year without charge.

As described in the previous section, MIP LCD is limited in the number of colors it can display. The following section shows how to display natural images with limited colors.

7.5 How to Show a Natural Image

There is an intrinsic problem for the MIP LCDs' image quality deteriorates when converting full-color data to lower bit color data. This section describes the method to show natural images with limited color bits.

Figure 7.15 Comparison of how color depth reduction and error diffusion processes affect image quality.

The error diffusion and dithering methods are widely used when reducing color depth [14, 15]. Figure 7.15 shows how color depth reduction and error diffusion processes affect an image. When the original image with 24-bit (8-bit/color) is converted to a 6-bit (2-bit/color) by discarding the lower 6-bit/color, large flat areas and the loss of detail are noticeable. To minimize the deterioration in image quality, error diffusion or dithering data processing is required before sending data to the MIP. Figure 7.16c depicts the 6-bit (2-bit/color) image after applying the Floyd-Steinberg dithering method for error diffusion. As seen in the figure, the image quality is enhanced with this method. However, fluttering is an issue when showing moving images. Ordered dithering is necessary for moving images. While the image quality processed by ordered dithering is slightly inferior to images created by error diffusion, the difference is less perceptible, especially when images are changed successively.

The following section introduces the specifications of selected market available MIP reflective LCD devices and the driving method to achieve super low power operations.

7.6 Design Characteristics of Current Market-Available Products and Their Super Low Power Operations

Figure 7.16 shows a picture and key specifications of a 1.2 in. 240 × 240 2-bit/color MIP reflective LCD. The difference between the circumradius and the active area, which is the most critical parameter to achieve narrow border watch design, is only 2.4 mm. Thanks to the small number of input pins and lack of driver IC, the FOG pad size is small enough to achieve a small border size. The LCD module consists of LCD glass, front and rear polarizers, and an FPC without a component. The display interface is a parallel interface with 6 data bus lines customized for MIP LCD design. When all gate lines are refreshed, the maximum refresh rate is 30 fps. The power consumption needed to show a static image is only 4 μW. This design architecture is widely used for GPS sports watches because of the narrow border design and adequate maximum refresh rate.

Figure 7.17 shows a picture and key specifications of a 1.04″ 208 × 208 1-bit/color MIP reflective LCD. The LCD module consists of LCD glass, front and rear polarizers, and an FPC without a component. The display interface

	Specifications
LCD type	ECB, normally black with light control film
Active area size	1.2 inch (ϕ30.24 mm) round
Number of pixel	240 (H) × RGB × 240 (V)
Pixel Pitch	42.0 μm (H) × 126.0 μm (V) (201.6 ppi)
Display interface	6 Bus parallel, direct interface
Power supply voltage	3.2 V, 4.5 V
Interface voltage	3.2 V / 0 V
Number of signal pins	Total 14 — Control signal : 6 Data signal : 6 VCOM control : 2
Maximum refresh rate	30 fps (All gate lines refresh)
Power Consumption @RT White image fVCOM=1 Hz	4 μW @ Static Image 9 μW @ 1 fps Image Refresh 160 μW @ 30 fps Image Refresh
Color gamut	Typ. 23%
Contrast ratio	Typ. 20:1
Number of colors	64 color (2bit/color)
Reflectance	Typ. 23% Point light source 30°–0°

Figure 7.16 Picture and key specifications of a 1.2 in. 240 × 240 2bit/color MIP reflective LCD.

	Specifications
LCD type	ECB, normally black with light control film
Active area size	1.04 inch (ϕ26.29 mm) round
Number of pixel	208 (H) × RGB × 208 (V)
Pixel Pitch	42.1 μm (H) × 126.4 μm (V) (200.9 ppi)
Display interface	SPI
Power supply voltage	3.0 V
Interface voltage	3.0 V / 0 V
Number of signal pins	Total 6 — Control signal : 4 Data signal : 1 VCOM control : 1
Maximum refresh rate	10 fps (All gate lines refresh)
Power Consumption @RT White image fVCOM=1 Hz	3 μW @ Static Image 18 μW @ 1 fps Image Refresh 150 μW @ 10 fps Image Refresh
Color gamut	Typ. 23%
Contrast ratio	Typ. 20:1
Number of colors	8 color (1bit/color)
Reflectance	Typ. 25% Point light source 30°–0°

Figure 7.17 Picture and key specifications of a 1.04″ 208 × 208 1-bit/color MIP reflective LCD.

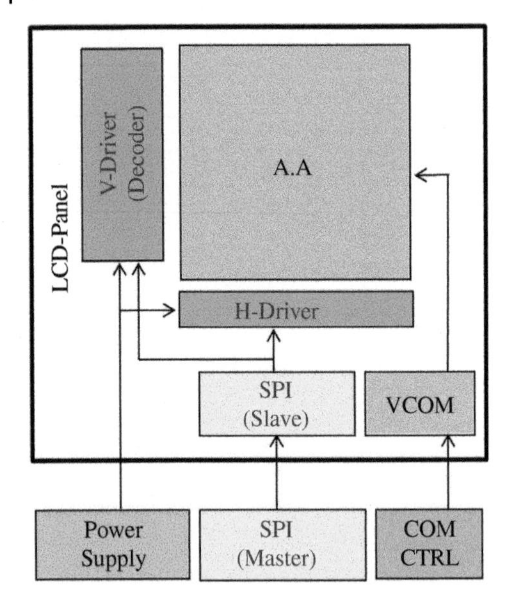

Figure 7.18 System block diagram of a 1.04″ 208 × 208 1-bit/color MIP LCD.

is a Serial Peripheral Interface (SPI), one of the simplest display interfaces, which almost all super low power microcontrollers (MCUs) support. The interface includes four control signals (SCS, SCK, SI, DISP), one data signal (SI), and one VCOM timing control signal (COMIN), and it can be driven by a single power source (3 V). When all gate lines are refreshed, the maximum refresh rate is 10 fps. The power consumption needed to show a static image is only 3 μW.

Figure 7.18 shows a system block diagram of a 1.04″ 208 × 208 1-bit/color MIP LCD. It has a SPI(Slave) and a VCOM driving signal generator on the glass border. All the commands and data are sent directly from an MCU that supports SPI. Received SPI commands and data are processed in the SPI circuit and the decoder, located on the border area. Because of the SPI and single 3 V power supply features, this system architecture is used for activity trackers powered by a coin battery without charging.

Figure 7.19 shows how VCOM/FRP and XFRP signals are generate from the COMIN input signal in LCD. A timing signal is needed to change the polarity of the VCOM/FRP and XFRP (COMIN).

Since a MIP LCD can be driven directly from MCU, reducing the power to generate control signals in an MCU is important to achieve super low power as a system. Since a MIP LCD stores display image data in the active area, only the voltage supply to keep SRAMs working and COMIN signal to toggle LC polarity is required (Figures 7.2, 7.3, and 7.19) when displaying a still image. This means that the MCU can go into sleep mode when displaying a still image. It is important to reduce the power consumption to generate the COMIN signal. Recently, every electronic device has included a super low-power real time clock (RTC) circuit. By utilizing the RTC output to generate a COMIN signal, MIP reflective LCDs can be lit at super low power as a system.

Figure 7.20 shows how to generate a COMIN signal from RTC output. Every RTC has a 32.768 kHz oscillator and frequency dividers to count seconds. By utilizing these circuits, COMIN timing can be generated at ultra-super low power consumption.

Super low power consumption is a key feature of MIP LCDs. This feature is summarized in the following section.

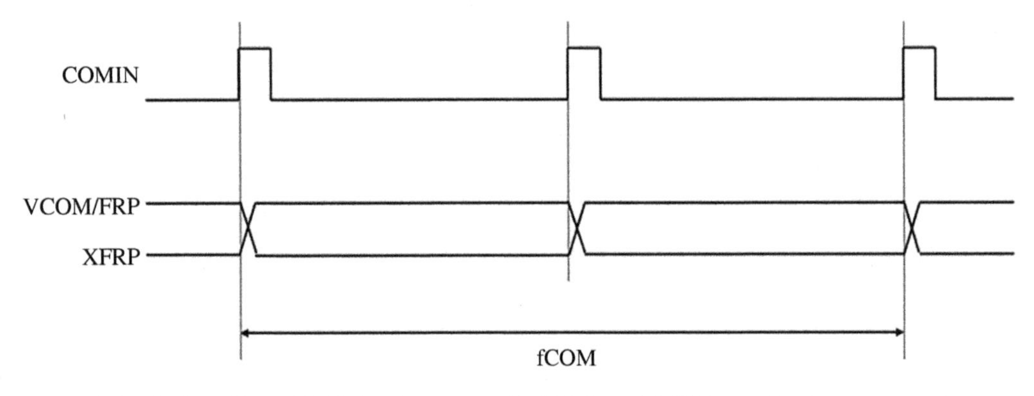

Figure 7.19 VCOM/FRP and XFRP signals generated in LCD from COMIN input signal.

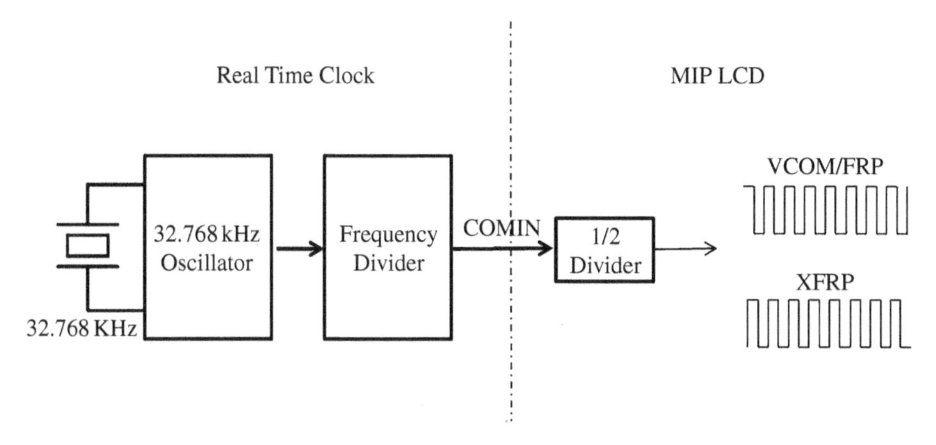

Figure 7.20 VCOM control signal generated by real time clock circuits to drive MIP reflective LCD.

7.7 Summary of Power Consumption

Figure 7.21 shows power consumption levels of both transmissive LCD and reflective LCDs, with and without MIP. Large amounts of a transmissive LCD's power are consumed by the backlight. Reflective LCDs utilize environmental light and have no need for a lit backlight when the environment is bright enough. Under the same conditions as described in Figure 7.16, 1.2 in. 240 × 240, the power consumption of a conventional active-matrix transmissive LCD fabricated by the LTPS process is estimated to be around 50 mW, of which 20% (10 mW) is consumed for data refreshing through the driver IC, and 80% (40 mW) is consumed for backlight. A conventional active-matrix LCD equips a driver IC to generate voltages and timing signals to operate the H-driver, V-driver, and COM driver, as well as analog signals to drive signal lines. The majority of the power consumed in the driver IC is used to generate these signals and voltages. The MIP reflective LCD described in Figure 7.16 consumes only 160 μW for a 30 fps moving picture and 4 μW for a still image. A MIP reflective LCD is controlled directly from an MCU. The MCU engine can go into sleep mode when displaying a still image. However, a VCOM toggling timing signal is needed even when displaying a still image. To generate this VCOM toggling timing, power consumption can be reduced by utilizing low-power RTC signals. MIP reflective LCD's power consumption is less than 1% of conventional transmissive LCDs when displaying still images.

The following section describes how MIP reflective LCD technology has been applied to devices in the market.

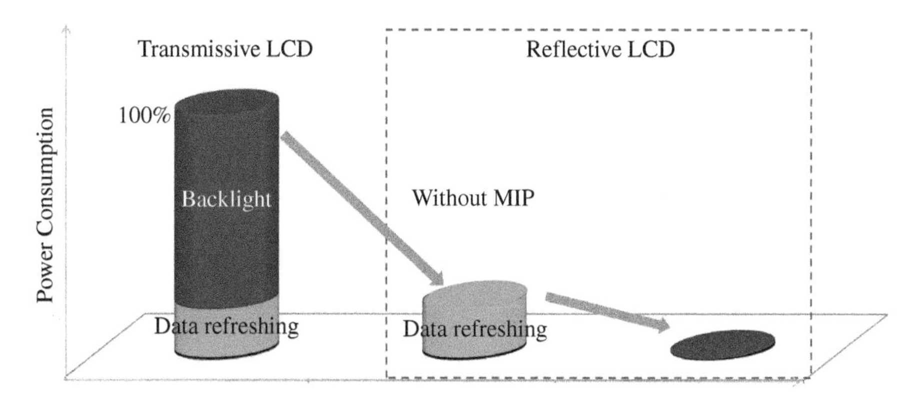

Figure 7.21 Power consumption comparison of transmissive LCD, and reflective LCDs with and without MIP.

7.8 Applications

The main features of the MIP reflective LCDs include:

- Super-low power operation
- Outdoor visibility
- Capability of showing moving images
- Simple structure: LCD glass + Front and rear polarizer + FPC

By utilizing these properties, color MIP reflective LCDs have been used for GPS-equipped sports watches, wristbands, and smartwatches, from 2013 onwards. Other technology applications that include larger displays are cycle computers and golf carts.

However, smart watches that use -OLED technology are now challenging MIP reflective LCDs for market share due to their excellent optical properties. “Always-on display” is the “new normal” for smartwatches. The power consumption of an OLED wearable display when displaying a watch face is several milliwatts. This is lower than conventional transmissive LCDs, but far bigger than MIP reflective LCDs. Moreover, smartwatches equipped with always-on OLED displays need to be charged daily.

Recently, due to increased awareness of the need for health management monitoring, watches are being modified to include health care applications that monitor step count, walking distance, heart rate, blood oxygen levels, stress levels, and sleep patterns. From a design viewpoint, current smartwatches tend to be too bulky to be worn 24/7. The largest component inside the smartwatch case is the battery. With super low power technology, it is possible to reduce the watch size by reducing the battery size dramatically, keeping the frequency of charging to over one week.

Due to their super-low power characteristics, MIP reflective LCDs are suitable for 24/7 health monitoring devices. The following section previews future possibilities for this technology.

7.9 Future Expectations

Display technology has evolved to improve image quality, offering high definition, high color gamut, high contrast ratio, and wide viewing angle. Mobile communication technology and application processor of handsets have also been developed to improve data transfer and processing rates with the growth of the smartphone market. While low power consumption is important, mobile devices have not been the first priority. People are accustomed to having to charge their mobile devices every day. As a result, transmissive LCD and OLED are now the main players in the display market.

With the expectation of Internet of Things (IoT) technologies, infrastructure for low power communication is becoming a reality. Bluetooth low energy (BLE) has already been widely used in many applications of IoT. Low power wide area network (LPWAN) services are now availabl enabling devices to communicate over large geographical areas where power supply cables are unavailable. Ultra-super low-power MCUs incorporating BLE functionality have been available for the last decade. Recently, software development kits (SDK) with LPWAN connectivity have been released.

As more devices connect to the internet at low power communication, the sensor data, including environmental data, human activity (e.g., position, motion, and health data), and pictorial images and information interpreted from the images are all reported the internet. The data are analyzed by artificial intelligence (AI) utilized in human behavior analysis and abnormality detection. Accumulating and utilizing these data by AI makes it possible to efficiently allocate human and social resources. For example, individual health data can now be automatically analyzed and sent to medical health specialists directly, which improves the quality and timeliness of health advice.

Super low power displays are necessary to interface humans with IoT edge devices that report sensor data. With LTPS SOG technologies, it is possible to incorporate various sensor features on the glass, such as ambient light sensors, in-cell touch sensors, and fingerprint or finger vein recognition functions. A MIP reflective LCD for a wearable device consumes only several microwatts of power to show an image, and it can be lit by the power generated by a solar cell incorporated in the watch.

We believe that MIP reflective LCD technology will play an important role in the development of the IoT market in the future.

References

1 Kimura, H., Maeda, T., Tsunashima, T. et al. (2001). A 2.15-in. QCIF reflective color TFT-LCD with digital memory on glass (DMOG). *SID Symp. Digest.* 32: 268–271.

2 Senda, M., Tsutsui, Y., Sasaki, A. et al. (2003). Ultra-low-power polysilicon AMLCD with full integration. *J. Soc. Inf. Disp.* 11: 121–125.

3 Nakajima, Y., Teranishi, Y., Kida, Y., and Maki, Y. (2006). Ultra-low power LTPS TFT-LCD technology using a multi-bit pixel memory circuit. *J. Soc. Inf. Disp.* 14: 1071–1075.

4 Harada, K., Kimura, H., Miyatake, M. et al. (2009). A novel low-power-consumption all-digital system-on-glass display with serial interface. *SID Symp. Digest.* 40: 383–386.

5 Fukunaga, Y., Shima, T., Nakao, T., Teranishi, Y., Nakajima, Y. (2015). Ultra-low power reflective LCD technology and its application for wearable devices. AM-FPD. Digest. 209–212.

6 Fukunaga, Y., Tamaki, M., Mitsui, M. et al. (2013). Low power, high image quality color reflective LCDs realized by memory-in-pixel technology and optical optimization using newly-developed scattering layer. *SID Symp. Digest.* 44: 701–704.

7 Tamaki, M., Fukunaga, Y., Mitsui, M. et al. (2015). A memory-in-pixel reflective-type LCD using newly designed system and pixel structure. *J. Soc. Inf. Disp.* 22: 251–259.

8 Nanno, Y., Senda, K., Tsutsu, H. (2001). Image analysis of area ratio gray scale method for active-matrix LCDs. Proc. IDW '01. 419–422.

9 Uchida, T. (1995). Trend of reflective color liquid crystal displays. AM-LCD'95. 23–26.

10 Kubo, M., Ochi, T., Narutaki, Y. et al. (2000). Development of "Advanced TFT-LCD" with good legibility under any ambient light intensity. *J. Soc. Inf. Disp.* 8 (4): 299–304.

11 Kimura, N. (2001). High-performance reflective TFT-LCD. Proc. AM-LCD'01. 55–58.

12 T. Ishinabe and T. Uchida. (2008). A bright and wide color gamut reflective full-color LCD using diffused light control technology. Proceedings of International Meeting on Information Display. 1377–1380.

13 Honda, M., Hozumi, S., and Kitayama, S. (1993). A novel polymer film that controls light transmission. *Prog. Polym. Sci.* 3: 159–169.

14 Floyd, R. W. and Steinberg, L. (1976). An Adaptive Algorithm for Spatial Grayscale. Proceedings of the Society of Information Display. 17 (2): 75–77.

15 Bayer, B. (1973). An optimum method for two-level rendition of continuous-tone pictures. IEEE Int. Conf. on Communications. 11–15.

8

Optically Rewritable Liquid Crystal Display

Wanlong Zhang[1,2], Abhishek Srivastava[2], Vladimir Chigrinov[2], and Hoi-Sing Kwok[2]

[1] Nanophotonics Research Centre, Shenzhen Key Laboratory of Micro-Scale Optical Information Technology, Shenzhen University, Shenzhen 518060, China
[2] State Key Laboratory of Advanced Displays and Optoelectronics Technologies, Hong Kong University of Science and Technology, Hong Kong S. A. R., China

8.1 Introduction

The idea behind electronic paper (e-paper) is to store and display information on a paper-like carrier with ultra-low or zero power consumption during the non-updating period. Flexibility, readability, and multi-functionality are the main expectations for e-paper [1–3]. In particular, electrophoretic displays (EPDs), in which charged particles are electrically controlled, are the most widely used technology for e-paper, with applications in e-readers, shelf labels, and so forth [4–7]. However, with liquid crystal displays (LCDs) dominating the flat panel display market for decades, several liquid crystal (LC)-based candidates have also been developed for e-paper, the most mature being cholesteric LCDs [8–11], zenithal bistable devices (ZBDs) [12–14] and bistable nematic (BiNem) displays [15–18]. Generally, the power consumption of traditional LCDs is related to the driving frequency, and the applied voltage on individual pixels must be maintained even for a steady image. However, similar to EPDs' inherent behavior of bistability, LC e-paper only consumes power if the image is updated, and the refreshed image can be maintained for years with zero power [1, 2]. LC molecules between cross polarizers work as the shutter with an applied voltage to realize bright or dark states. Still, LC e-paper suffers from the high-level complexity of the driving electronics, which often fail for flexible displays due to the insufficient durability of the flexible conductor and contact bonding [19].

Recent developments of optically rewritable (ORW) LCDs and progress in LC photoalignment have made it possible to separate the e-paper display unit and the driving optoelectronics, significantly reducing ORW LCD structure's complexity as well as making device properties and cost both paper-like [20–23]. An ORW LCD panel consists of two substrates with different alignment materials. One alignment material is photostable and keeps the alignment direction on one substrate, while the other is photosensitive, the photoalignment layer, and can change its alignment direction with exposure to polarized light through the substrate. Instead of applying an external voltage on individual pixels as in traditional LCDs, the switching between the bright and dark states can be easily achieved by changing the polarization plane of the exposure light to form electrically controlled birefringence (ECB) mode or twisted nematic (TN) mode. The different modes come from changing the alignment direction on the photoalignment layer, which further induces the orientation of the nearby LC molecules. Thus, the inherent bistability allows information to be displayed for long periods with zero power consumption, making ORW LCDs suitable for e-paper applications. Moreover, separating the electronic elements from the ORW display panel makes the ORW LCD very durable and ready to meet the flexibility challenge of e-paper [19].

E-Paper Displays, First Edition. Edited by Bo-Ru Yang.

This chapter concentrates on the progress in developing bistable ORW LCDs based on photoalignment technology over the past few years. Firstly, a brief review of photoalignment technology is introduced, which is essential for understanding ORW technology. Subsequently, the features of ORW LCDs are presented, including rewriting speed, gray levels and full-color generation, and flexible demonstration. In particular, as it does not have a highly complex structure, the 3D ORW LCD exhibits a unique advantage among the various kinds of e-paper. With the characteristics mentioned above, and based on a well-developed LCD industry, the ORW LCD has huge potential for applications in various modern display, security, and photonic devices.

8.2 Photoalignment Technology

LC photoalignment, first introduced in 1988, is considered an excellent alternative to the conventional LC alignment method [24, 25]. To realize the effect of LC photoalignment, the substrate must contain a layer of photosensitive material whose molecules and/or photo products are capable of orientational ordering under light irradiation. Due to anisotropic interfacial interaction, this ordering determines the direction of the easy orientation axis of the adjacent LC, which means that LC alignment can be controlled by light irradiation on the alignment layer [26–29]. Compared with the conventional rubbing polyimide method, the non-contact photoalignment technique avoids mechanical damage, electrostatic charge, or dust contamination. Thus, photoalignment is more promising for high-quality, high-uniformity, large-area, and high-resolution LC alignment, even for small-area multi-domain pattern-alignment. It can also provide efficient LC alignment on curved, flexible substrates or even inside of fibers [30–32].

LC photoalignment has made tremendous improvements over the past 30 years, but most new materials, techniques, and LCD prototypes based on photoalignment technology have appeared recently [26, 27]. Since LC photoalignment originates from the anisotropic intermolecular interactions between the alignment agent with a photo-induced order and adjacent LC molecules, the photoalignment mechanisms are of four main types: (i) photochemical reversible cis-trans isomerization in azo-dye containing polymers [33–37], monolayers [38] and pure dye films [24, 39]; (ii) pure reorientation of the azo-dye chromophore molecules or azo-dye molecular solvates due to diffusion under the action of polarized light [40–42]; (iii) topochemical crosslinking in cinnamoyl side-chain polymers [43, 44]; (iv) photodegradation in polyimide materials [45–48]. Processes (i) and (ii) involving azo-dyes present reversible transformations, while the other two processes require irreversible photo-chemical changes, which are not beneficial for rewritability. For (i), the cis-trans isomerization process, change of the absorption spectra is observed after illumination, which is not the case for the photochemical stable azo-dye molecules, described in process (ii), which are involved in a reorientation and solvate formation process under the action of polarized light. Additionally, the wide absorption band of azo-dye molecules enables the usage of low-cost 450 nm light-emitting diodes (LEDs) instead of an expensive ultraviolet (UV) lamp, which makes the photo-sensitive azo-dye material practical for e-paper application [20].

This chapter focus on a typical photoalignment material, namely, sulfonic azo-dye SD1 (produced by Dainippon Ink and Chemicals Ltd). Its chemical structure is shown in Figure 8.1. The reorientation of the azo-dye molecules is perpendicular to the polarization plane of the incident light and further characterized as a rotational diffusion mechanism [40, 41].

When SD1 molecules are optically pumped by a polarized light beam, the probability of absorption is proportional to $cos^2\theta$, where θ is the angle between the absorption oscillator of the SD1 molecules and the polarization plane of the light, as illustrated in Figure 8.1. Therefore, the SD1 molecules, which have their absorption oscillators (chromophores) parallel to the polarization plane of the irradiation light, will most probably get an increase in energy, resulting in their reorientation from the initial position. This produces an excess of chromophores in a direction at which the absorption oscillator is perpendicular to the polarization plane of the irradiation light. Since the chromophore of SD1 molecules is parallel to the long molecular axis, the molecules tend to align their long axes perpendicular to the polarization plane of the irradiation light, which can be a 450 nm blue LED or UV light [25].

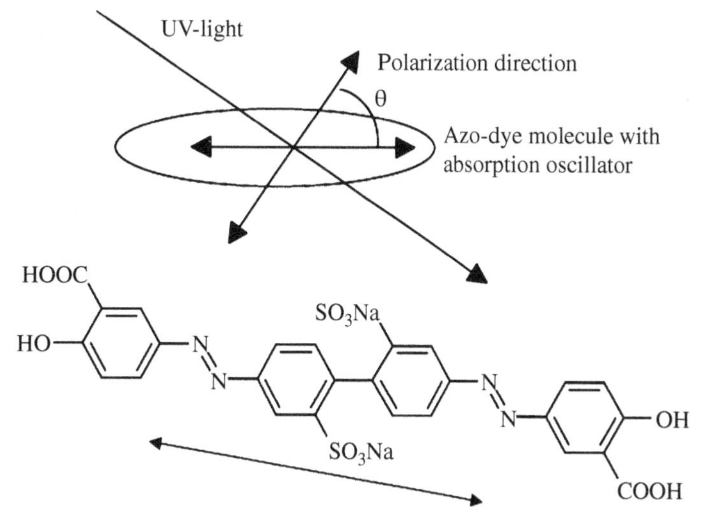

Figure 8.1 Qualitative interpretation of the photo-induced order in photochemically stable azo-dye films. Above: the geometry of the effect. Below: azo-dye molecule SD1, having the absorption oscillator (chromophore) parallel to the long molecular axis. *Source:* Chigrinov et al. 2004 [40]. Copyright 2004. Reproduced with permission of the American Physical Society.

Here, the order parameter of the SD1 molecules in the photoalignment film is determined to be the thermodynamic average $S = \langle 3cos^2\theta - 1\rangle$ and is proportional to several key parameters: (i) irradiation light power (W/cm^2) in the direction of the absorption oscillator, (ii) the absorption coefficient ($1/cm$), (iii) the molecular volume [40], and (iv) relative humidity [49, 50]. Since the absorption coefficient is one of the material intrinsic properties, to obtain a high order parameter of the SD1 molecules in the photoalignment film, larger exposure energy, and thinner film are beneficial.

The aligned SD1 molecules induce anisotropic phase retardation $\delta = 2\pi\Delta nd/\lambda$, where Δn and d are the birefringence and SD1 film thickness, respectively. λ is the measurement laser wavelength. Thus, the phase retardation can be measured by experiment to be directly proportional to the relative order parameter S of the photoaligned SD1 layer. The dependence of the phase retardation versus exposure time for different powers of an irradiating Ar^+ laser is shown in Figure 8.2 [40].

In Figure 8.2, the experimental results and numerical solution illustrate the monotonic increase of the photo-induced phase retardation from the isotropic state up to the saturation value. Meanwhile, the saturation photo-induced phase retardation level is found to be proportional to the power of the UV-irradiation power. Since SD1 molecules are photochemically stable, the reproducible rewriting process also follows the same diffusion mechanism [51–54].

The anchoring energy is one of the most important parameters. It characterizes the LC alignment quality and usually consists of two parts, azimuthal and polar, which describe how easy it is to change the alignment of the LC director at the surface [55]. The thickness of the photoaligned SD1 layer strongly affects the LC alignment quality [56]. Another factor determining the anchoring energy of the photoaligned SD1 layer is the order parameter of the SD1 molecules, which is proportional to the exposure time. In particular, the azimuthal LC anchoring energy of the photoaligned SD1 layer can be substantially varied by changing the exposure energy, as shown in Figure 8.3 [57].

In Figure 8.3, the exposure energy of a photoaligned SD1 layer larger than 1 J/cm^2 is shown to provide a high anchoring torch to the adjacent LC molecules and also makes the photoaligned SD1 comparable to the conventional rubbed polyimide layer. The controllable anchoring energy also enables the photoaligned SD1 to act as an alignment layer for ferroelectric LC [58–63] and dual-frequency LC [64, 65]. Moreover, sub-micron domain size photoalignment is achieved [66, 67]. With the advantage of its reorientation ability, the photoalignment SD1 material facilitates flexible and multi-domain LC alignment for photonic applications, such as tunable 1D and 2D gratings [68–71], fork gratings [61, 72–74], Dammann gratings [75, 76], Fresnel lenses [58, 64, 77–80], etc.

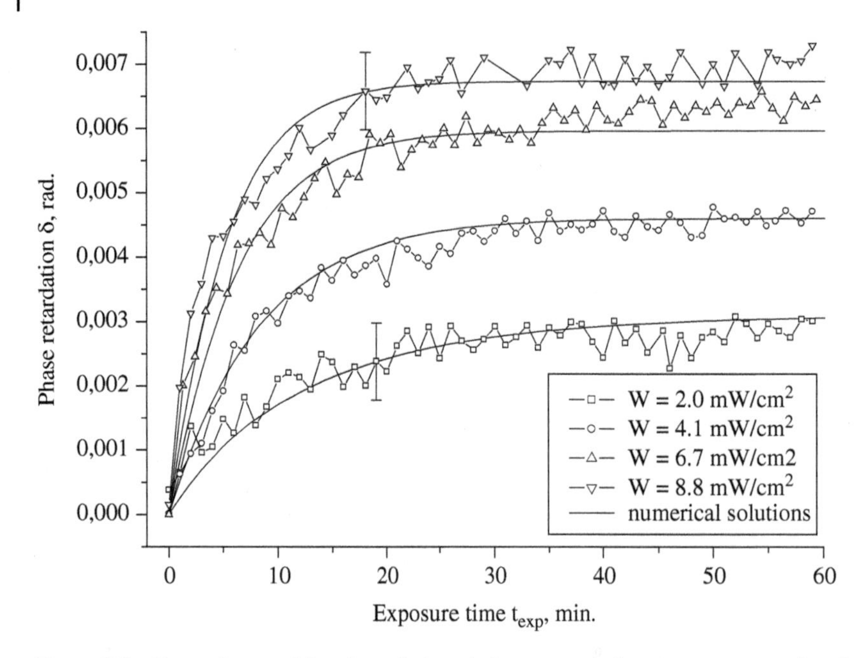

Figure 8.2 Dependence of the photo-induced phase retardation δ on exposure time for various powers W of the irradiation UV-light. Solid lines indicate the numerical solution of the diffusion model. *Source:* Chigrinov et al. 2004 [40]. Copyright 2004. Reproduced with permission of the American Physical Society.

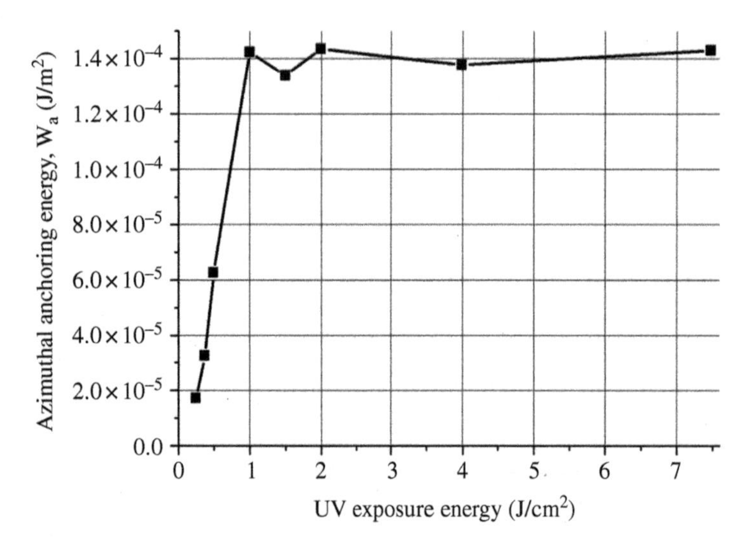

Figure 8.3 Azimuthal LC anchoring energy of the photoaligned SD1 layer as a function of the exposure energy. *Source:* Chigrinov et al. 2008 [25]. Copyright 2008. Reproduced with permission of John Wiley and Sons.

8.2.1 Optically Rewritable Liquid Crystal Display

Unlike other LC-based e-paper, ORW technology is a modified method of azo-dye photoalignment that possesses a traditionally high azimuthal anchoring energy and has the unique feature of reversible in-plane alignment direction reorientation. Indium tin oxide (ITO) electrodes are also unnecessary for ORW LCDs in the absence of applied voltage. The typical ORW LCD panel operation principle is shown in Figure 8.4.

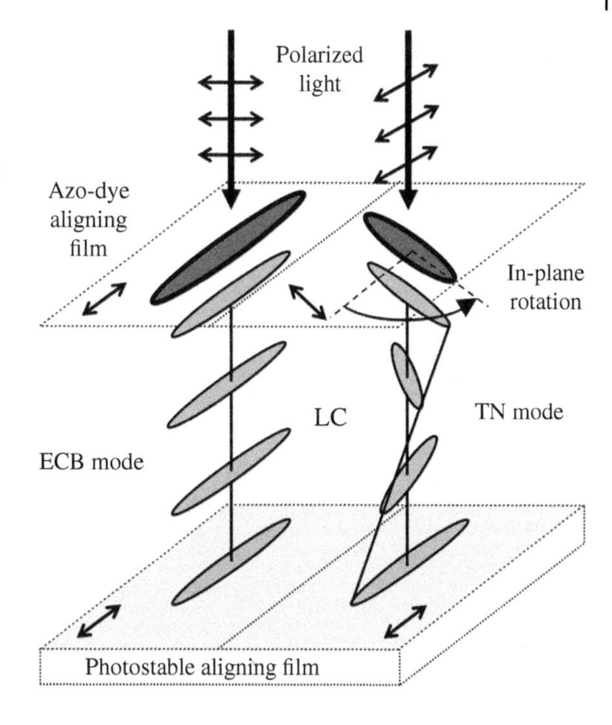

Figure 8.4 Operational principle of an ORW LCD: Azo-dye alignment film rotates its alignment direction in-plane, staying perpendicular to the polarization plane of the incident light. The LC molecules follow the top alignment direction, switching between the ECB and TN modes. *Source:* Muravsky et al. 2007 [20]. Copyright 2007. Reproduced with permission of John Wiley and Sons.

An ORW LCD panel consists of two substrates with different alignment materials. One alignment material is photo-stable and keeps the alignment direction on one substrate, while the other is photo-sensitive SD1 and can change its alignment direction when irradiated with polarized light through the substrate. Therefore, we can obtain the specified bistable twist angle in the ORW LCD panel and keep it at zero power consumption for a long time as the LC structure is always stabilized by the surface. Thus, an ORW LCD operates in the ECB and TN configurations by in-plane switching of the alignment direction of SD1, caused by the controlled irradiation of the incident polarized light [20].

The fabrication process flows of a photoaligned SD1 substrate and polyimide substrate are illustrated in Figure 8.5. As shown in Figure 8.5a, the azo-dye SD1 material solution (1% wt/wt in N, N–Dimethylformamide) is deposited on a clean glass substrate by a spin-coater. The coated substrate is baked on a hot plate at 100 °C for 10 minutes to evaporate the solvent. A 450 nm linearly polarized blue light (denoted as $\theta = 0°$) is applied to align the SD1 molecules unidirectionally with ~1 J/cm^2 exposure doses initially.

The photostable alignment film is usually made by conventional polyimide, as shown in Figure 8.5b. The polyimide solution is also deposited on a clean glass substrate by a spin-coater. After 1.5 hours' hard baking, the polyimide substrate is mechanically rubbed to provide the desired alignment direction for LC molecules with a strong polar anchoring.

As illustrated in Figure 8.6, the polyimide substrate (lower one) and the photoaligned SD1 substrate (upper one) are anti-parallelly assembled to face each other with 10 μm spacers dispersed therebetween. The LC material is filled into the sealed LC cell and forms ECB mode immediately. So that information can be displayed on the ORW LCD panel, a photomask is applied on the top for subsequent irradiation with a 90° rotation of the polarization plane of the light (here, $\theta = 90°$), which reorientates the easy axis of the SD1, in-plane. By restricting the exposed area through a photomask, separate domains with different alignment directions are fabricated. Note that the difference in θ between the first and second irradiation defines the angular difference in the easy axis within different alignment domains of the pattern, which is demonstrated as the display information on the ORW LCD.

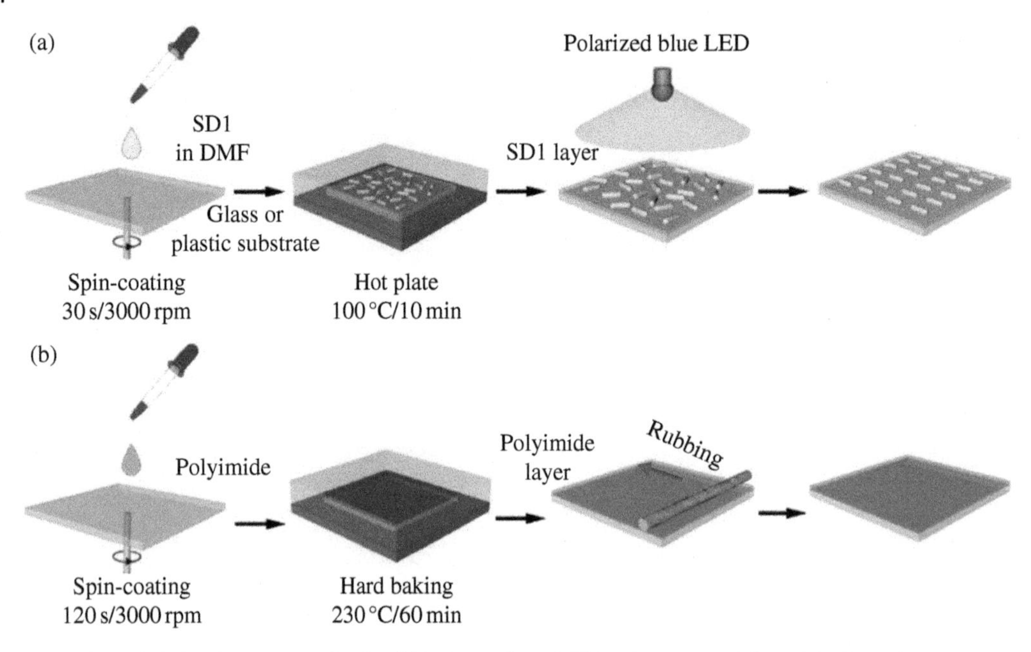

Figure 8.5 Fabrication process flows of (a) photoaligned SD1 substrate and (b) polyimide substrate. *Source:* Chigrinov et al. 2020 [81]. Copyright 2020. Reproduced with permission of MDPI.

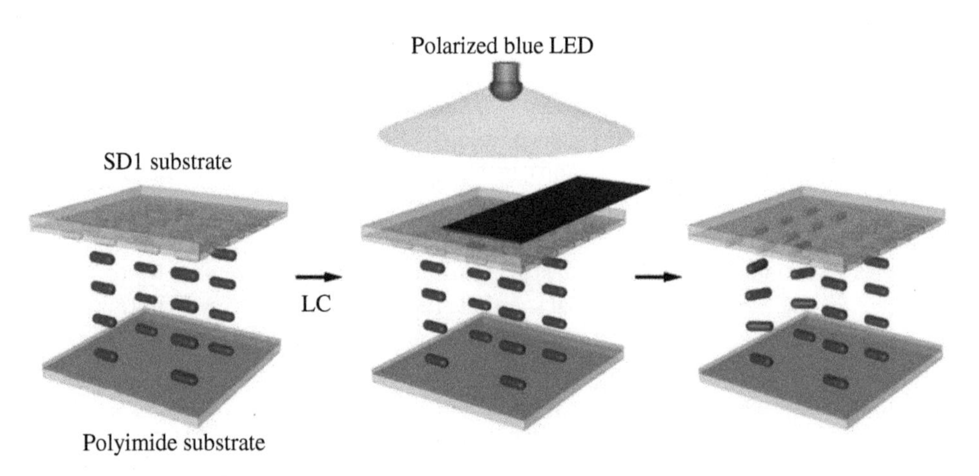

Figure 8.6 Assembly of ORW LCD panel and rewriting process. *Source:* Chigrinov et al. 2020 [81]. Copyright 2020. Reproduced with permission of MDPI.

The overall schematic structure of the ORW LCD is shown in Figure 8.7a, with the analyzer applied on the top and the reflective polarizer at the bottom of the panel. The displayed information is demonstrated in Figure 8.7b. Since both ECB and TN modes are formed in the ORW LCD panel, the active and passive display images are demonstrated with parallel and cross polarizers, respectively [83].

Photo-sensitive SD1 material can also be stabilized to replace the conventional polyimide alignment layer and form an all-photoaligned LC cell. There are two popular approaches for achieving SD1 stabilization without affecting the alignment property. The first approach is associated with mixing SD1 and reactive mesogen (RM) in optimized proportions and using this composition for photoalignment. It has been shown that such alignment

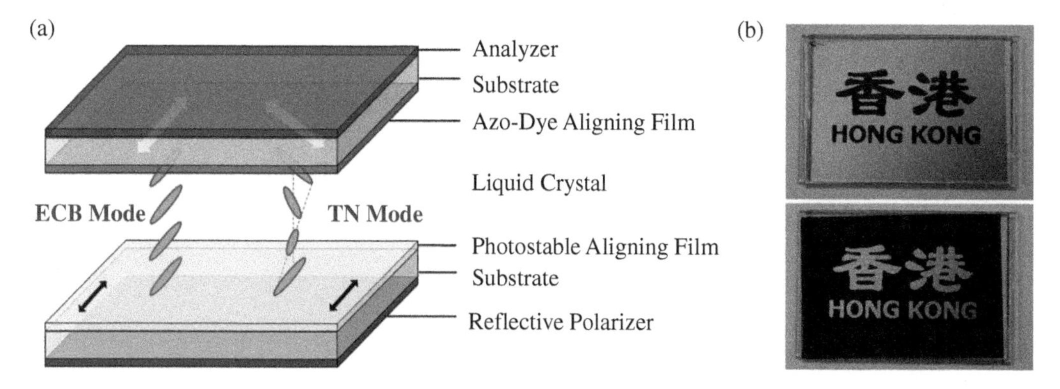

Figure 8.7 (a) The schematic ORW LCD panel in which two alignment domains with ECB and TN modes are created after irradiation by polarized light. (b) The ORW LCD displaying binary "Hong Kong" between parallel and crossed polarizers, respectively. *Source:* Sun et al. 2014 [82]. Copyright 2014. Reproduced with permission of the Optical Society.

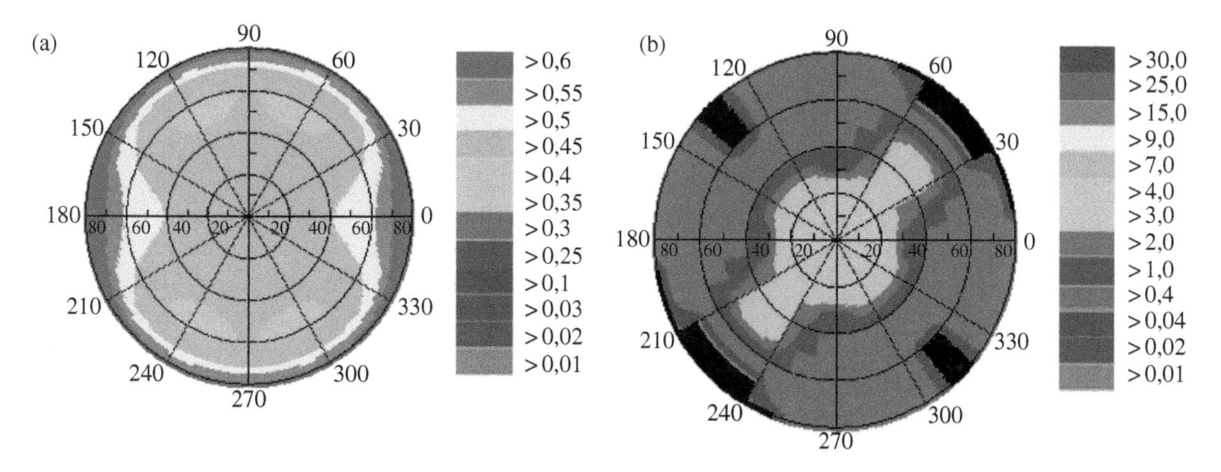

Figure 8.8 Angular dependence of the ORW LCD: (a) reflectance coefficient, (b) contrast ratio. *Source:* Chigrinov et al. 2020 [87]. Copyright 2020. Reproduced with permission of Taylor & Francis.

layers provide good alignment stability and electro-optical performance of LCDs [84]. The other approach is to protect the SD1 photoalignment layer with RM-coating. In this case, the RM layer is aligned by the SD1 photoalignment layer and subjected to photo-crosslinking to fix the alignment direction in the film, which further passes the anchoring force to align the adjacent LC bulk [85, 86]. Both approaches present high potential, not only for ORW LCD application but also to replace polyimide in the conventional LCD industry.

As a display, the contrast ratio and viewing angle are key factors to evaluate ORW LCD performance. The achieved values of the normalized reflectance coefficients range from ~0.03 (dark state) to ~0.43 (bright state) according to the simulation results, indicating a contrast ratio of 10 : 1. Meanwhile, the simulated angular dependence of the reflectance coefficient and contrast ratio are shown in Figure 8.8 [87].

8.2.2 Optical Rewriting Speed Optimization

The operational speed of an ORW LCD is represented by the rewriting time, which is defined as the time it takes for the normalized dose-dependent transmittance to reach 90% through the ORW LCD panel. For a TN cell operated in the Mauguin regime [55], the light transmittance can be written as $T = T_0 sin^2\varphi$, where φ is the twist angle

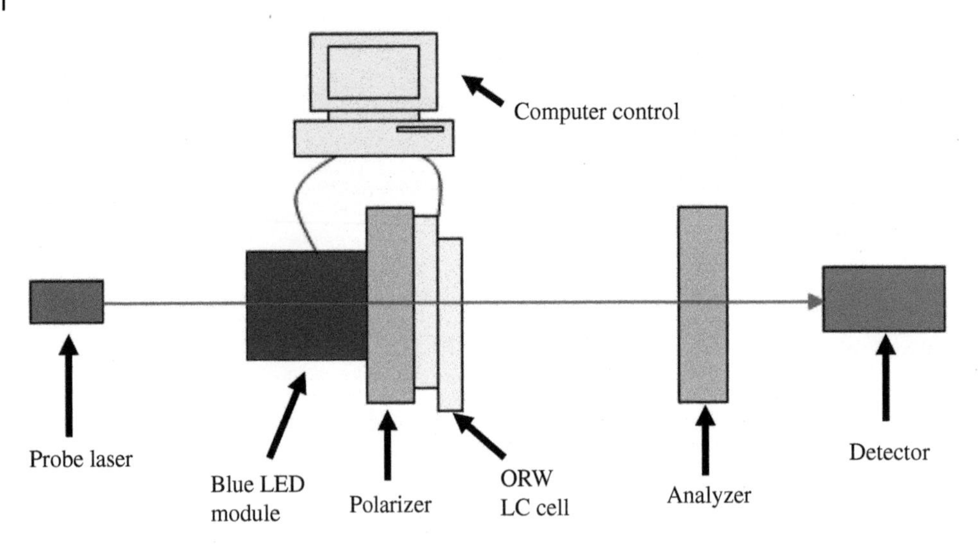

Figure 8.9 The experimental arrangement for the optical rewriting time measurement of the ORW LCD panel. *Source:* Geng et al. 2012 [89]. Copyright 2012. Reproduced with permission of Chinese Physical Society.

of LC [83]. Meanwhile, the twist angle φ is a key parameter determined by the azimuthal anchoring energy of the alignment layers. Since this energy is mainly dependent on the order parameter of the photoaligned SD1 molecules, it is further proportional to the irradiated light energy. Thus, the optical rewriting time of the ORW LCD panel is highly dependent on the speed of obtaining enough exposure energy on the rewriting photoaligned SD1 layer. Since the exposure energy is a product of the irradiation power and time, a larger irradiation power results in a faster-rewriting speed. However, a larger irradiation power will dramatically increase the cost of the light source and may also induce tilt in the SD1 molecules due to multiple reflections from several surfaces and the divergence of the light source. Thus, to increase the rewriting speed of the ORW LCD, it is necessary to study the effect of the azimuthal anchoring energy $A\varphi$ of the photoaligned SD1 substrate under fixed irradiation power [40, 81, 83, 88–94].

The general experimental measurement setup is shown in Figure 8.9. The blue light (450 nm, 80 mW/cm^2) from the blue LED module is polarized by a wire-grid polarizer and then illuminates the ORW LCD panel to photo align the SD1 molecules, which thus aligns the LC molecules in the cell. A He-Ne laser (632 nm), which propagates through a hole at the center of the LED module and is polarized by the same wire-grid polarizer, acts as a probe to inspect the transmittance of the ORW LCD panel in-situ of the rewriting process. The analyzer is placed in front of the detector with a polarization plane perpendicular to the previous polarizer [89].

The measurement process is controlled and synchronized by a computer. The ORW LCD panel is irradiated by the polarized blue LED for 30 seconds to obtain an initial photoalignment. The LC cell is rotated 70° by a motor, and then the rewriting process with the LED starts again. The red laser propagates through the LC cell, and the detector recorded the transmittance change. In this case, the transmittance of the probe laser changes from dark to bright or the reverse, which indicates the twist angle of the LC bulk changes from 0° to 70° or the reverse, respectively. The processing time in seconds is noted as the rewriting speed of the ORW LCD panel.

8.2.3 Optical Rewriting Time with Different Liquid Crystals

Theoretically, for a given ORW LCD panel, the azimuthal anchoring energy $A\varphi$ can be denoted as a function of the twist angle $A\varphi = 2K_{22}\varphi/(dsin(\Phi - \varphi))$, where Φ, φ, K_{22} and d is the angle between defined directions of the LC orientation on the top and bottom substrates, twist angle, twist elastic constant of LC, and the cell gap, respectively [25].

Table 8.1 Parameters of six liquid crystals.

Parameters	Δn	K_{11} (pN)	K_{22} (pN)	K_{33} (pN)
5CB	0.1918	6.4	3	10
ZLI-5700-100	0.1581	11.3	5	15.3
E7	0.2250	11.6	7	7.5
MCL-6080	0.2024	14.4	7.1	19.1
ZCM-5063xx	Not available	6.1	8.4	20.2
TN-0403	0.2580	12.6	20.2	20.3

Source: Geng et al. [89]. Copyright 2012. Reproduced with permission of Chinese Physical Society.

The measured value of the anchoring energy is obtained with the knowledge of two parameters: φ and K_{22}. φ is related to the overall transmittance of the ORW LCD panel and is proportional to the exposure energy, while K_{22} is an LC-related property that is different for each material. Thus, as shown in Table 8.1, a larger twist angle can be realized with larger anchoring energy for the given LC materials, as shown in Figure 8.10a. Since the K_{22} LC E7 and MCL-6080 are quite similar. The plots are somewhat overlapping as well [90].

Figure 8.10b presents the normalized transmittance of the ORW LCD panel versus rewriting time. As introduced previously, the rewriting time is defined as the time it takes for the normalized dose-dependent transmittance to reach 90% through the ORW LCD panel. Since the power of the LED is fixed, a longer rewriting time induces larger exposure energy on the photoaligned SD1 layer. The rewriting time of 5CB is about 6 seconds, which is the shortest among all the LCs because of its smallest K_{22} value. The similar K_{22} values of E7 and MLC-6080 obtain the same rewriting time of 11 seconds, even though they have quite different values of K_{11} and K_{33}. The above analyses show that K_{22} is the key parameter to reduce optical rewriting time. With a smaller K_{22} value, the rewriting speed of an ORW LCD can be increased [89–92].

Moreover, adding chiral dopant inside LC material E7 induces the average twist elastic energy value to become about 1.5 times higher than the original E7 without the chiral dopant. The experimental results also present the rewriting speed to be about 1.5 times faster than the original E7, decreasing the rewriting time dramatically [94].

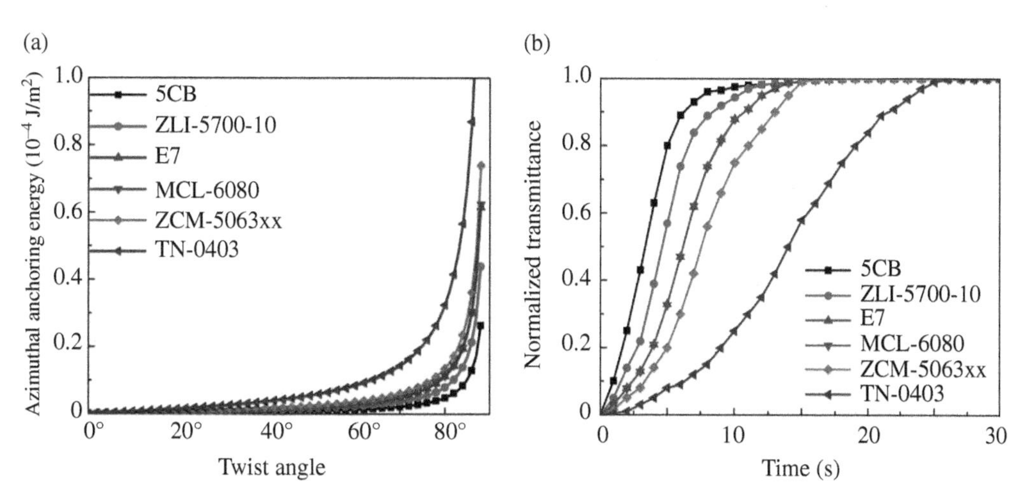

Figure 8.10 (a) Plot of azimuthal anchoring energy in 10 μm cell gap ORW LCD panel angle. (b) Half rewriting cycle for different LCs. *Source:* Geng et al. 2012 [89]. Copyright 2012. Reproduced with permission of Chinese Physical Society.

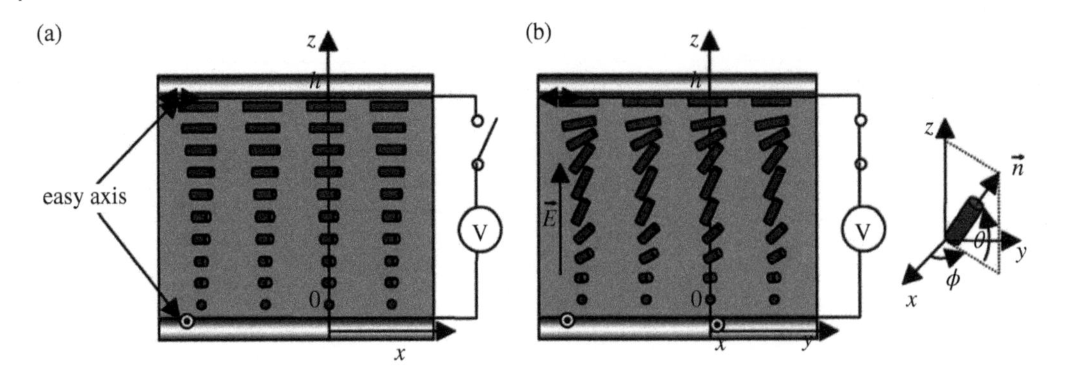

Figure 8.11 Schematic diagram of twisted nematic LC cell (a) without and (b) with applied external electrical field. *Source:* Wang et al. 2018 [93]. Copyright 2018. Reproduced with permission of Taylor & Francis.

8.2.4 Optical Rewriting Time with Application of External Electrical Field

Other than choosing LCs with a smaller twist elastic constant K_{22} value, applying an external electrical field [93] or changing operating temperature [81] can also decrease ORW LCD panels' rewriting time.

For the particular case of the ORW LCD panel in Figure 8.11, the twist angle is 90°, and the twisting is counter-clockwise when looking down from the top. In the absence of an external electrical field, the LC molecules are in the planar twist state, where the LC directors twist at a constant rate from the bottom to the top of the cell, as shown in Figure 8.11a. When a sufficiently high electrical field is applied across the cell, as shown in Figure 8.11b, the LC molecules are tilted along the electrical field direction, and the LC deformation can be considered to have a pure twist only. In this case, with the exposure energy E_{exp}, the rewriting time can be denoted as $\tau = \gamma_1^* / (\chi_E E_{\text{exp}} - 2A_\phi / d)$, where γ_1^* is the effective rotational viscosity, and χ_E is the absorption coefficient of the photoaligned SD1 layer [93]. Thus, with a certain ORW LCD panel of fixed exposure energy, it can be easily found that the rewriting time is strongly proportional to the effective rotational viscosity γ_1^*.

For the given ORW LCD panel filled with 5CB, when the applied voltage is larger than the threshold voltage V_{th}, the LC molecules will start to rotate along the electrical field from 0 V to 45 V, and the effective rotational viscosity γ_1^* will drop from 80.8 cP to 20.8 cP based on the theoretical calculation, indicating the rewriting time will decrease about four times with an external electrical voltage up to 45 V, as shown in Figure 8.12 [93].

The rewriting time is quite long due to the power of the irradiating LED being 20 mW/cm^2, which is smaller than in previous results (125 mW/cm^2) [91], but considering the factor of irradiation time, the overall exposure energy is comparable.

8.2.5 Grayscale Optically Rewritable LCD

Grayscale images are essential for display applications. Several methods to display grayscale images on an ORW LCD have been studied, including (i) controlling the twist angle of the LC bulk and (ii) generating spatial gray levels on each pixel [52, 95–97].

As introduced earlier, the light transmittance through an ORW LCD is proportional to the twist angle φ of the LC bulk. Thus, a straightforward method to generate grayscale images on an ORW LCD is to control the twist angle of each pixel, and the twist angle is determined by the azimuthal anchoring energy of the alignment layers. Meanwhile, the azimuthal anchoring energy on a photoaligned SD1 layer varies with different exposure energies, which further controls the individual pixels' twist angles an ORW LCD [98]. However, the azimuthal anchoring energy is also affected by other factors, such as the photo-aligned SD1 layer thickness [40], operating temperature [81], LC cell

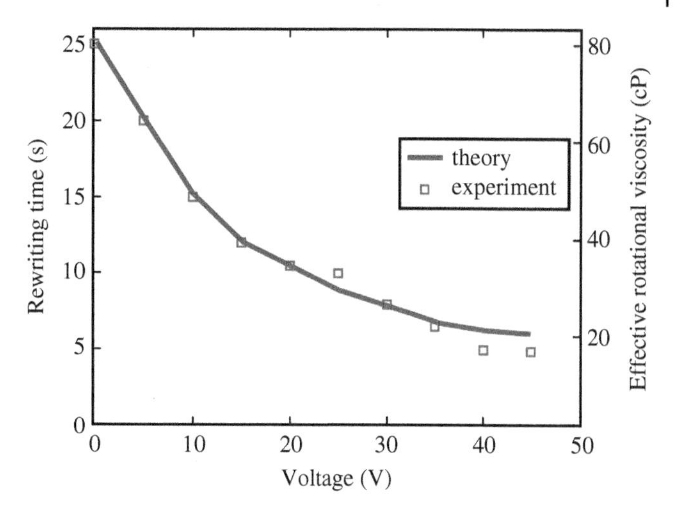

Figure 8.12 Average effective rotational viscosity $\dot{\gamma}_1$ (right axis) and rewriting time (left axis) versus different applied voltages on the ORW LCD panel. *Source:* Wang et al. 2018 [93]. Copyright 2018. Reproduced with permission of Taylor & Francis.

fabrication process [83], etc. The uniformity of the irradiating light source is also crucial for this approach, as non-uniformity would increase the cost dramatically.

Another method is to precisely control the photoaligned SD1 direction on individual pixels, which can be easily realized by a uniform LED source together with an LC-based polarization rotator [52, 95, 97] or digital micromirror device (DMD) with a rotational polarizer [99, 100]. The costly DMD with a rotational polarizer is utilized as a precise micro-patterning system for the photoaligned SD1 substrate, which will be introduced later, while the LC-based polarization rotator is an easy approach for use grayscale light printer for ORW LCDs.

Figure 8.13 presents the LC-based light printer, which can rotate the polarization azimuth of the linearly polarized irradiation light (450 nm). As shown in Figure 8.13a, the polarization plane of the irradiation light is set at 45° to the azimuthal orientation of the LC alignment. A vertically aligned nematic (VAN) LCD panel with thin-film transistors as the driving electrodes to control the phase retarder can control the ellipticity of the output light of

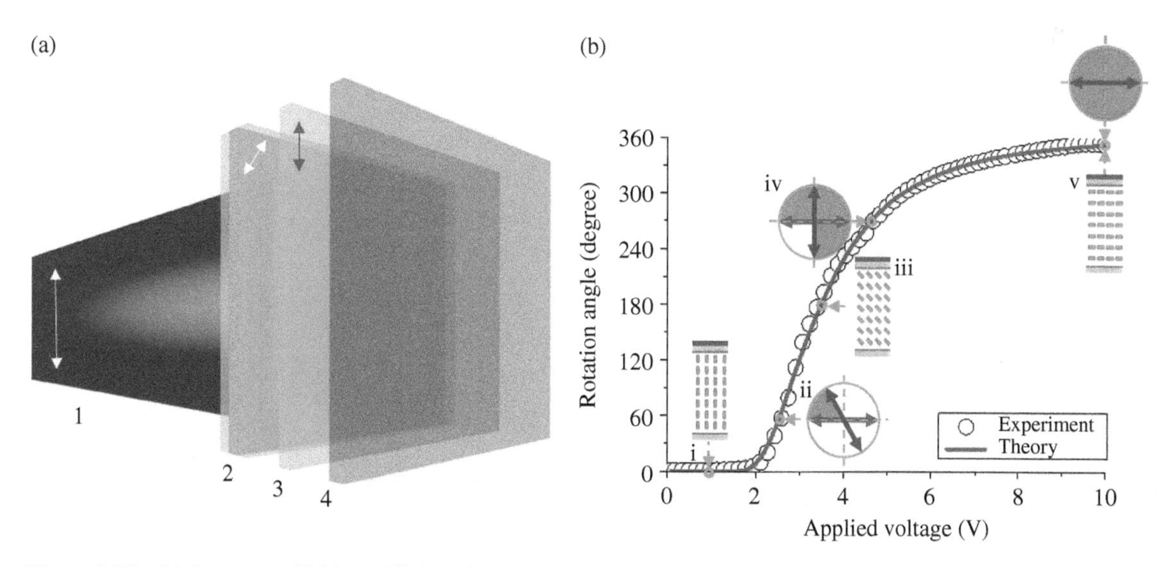

Figure 8.13 (a) Structure of LC-based light printer, which consists of 1) polarizer, providing linearly polarized irradiation light (450 nm), 2) vertically aligned nematic LCD, 3) 450 nm quarter wave plate, 4) ORW LCD. (b) Plot of experimentally determined and calculated polarization azimuth rotation angles (β) as a function of applied voltages (V). *Source:* Zhang et al. 2018 [101]. Copyright 2018. Reproduced with permission of John Wiley and Sons.

each pixel independently. With a 450 nm quarte-wave plate (Q attached to the LC cell in a parallel configuration to the polarization plane of the laser, the elliptically polarized light is converted to linearly polarized light with the desired polarization plane.

Figure 8.13b plots the experimental data for the polarization azimuthal rotation angle as a function of the applied voltage (for the VAN LC cell with an 8 μm cell gap). The three inserted graphics labeled *i*, *iii*, and *v* in the figure illustrate the corresponding orientation of the LC molecules at different applied voltages, showing three different polarization azimuthal angles β of 0°, 180°, and 360°, respectively. In the circular schematics labeled *ii* and *iv*, the red arrow represents the azimuthal polarization angle for the irradiation light, and the blue arrow presents the same after passing through the LC-based light printer at selected applied voltages. The pink shadows highlight the rotation routes [101].

Figure 8.14 exhibits grayscale images on an ORW LCD printed by the LC-based light printer. In Figure 8.14a, the approximate twist angle of each character is marked in red and controlled by selecting the polarization plane of the light printer. The characters "H" of twist angles below 40° present too low contrast to be seen. Figure 8.14b presents a 4-bit grayscale image on a 2-in. ORW LCD as a high resolution of 127 ppi (pixels per inch) [95].

A further approach to generate gray levels on an ORW LCD is pixel division, which is similar to the functioning of a commercial black/white inkjet printer. The gray levels are generated by controlling the binary dots sizes of each pixel. Thus, a simple way to display a grayscale image on an ORW LCD is to first print an 8-bit grayscale image on a transparent film with an inkjet printer. The transparent film is utilized as a photomask and written to the ORW LCD by the polarized blue LED.

Figure 8.15 presents the grayscale images on an ORW LCD based on the pixel division mechanism. A 4-in. 8-bit grayscale image on an ORW LCD is shown in Figure 8.15a, while the inset shows different dot sizes, which control different gray levels of individual pixels. Figure 8.15b exhibits a 2-in. reflective ORW LCD under sunlight. High reflectance and a good contrast ratio are obtained on the ORW LCD, indicating the potential for outdoor application [96].

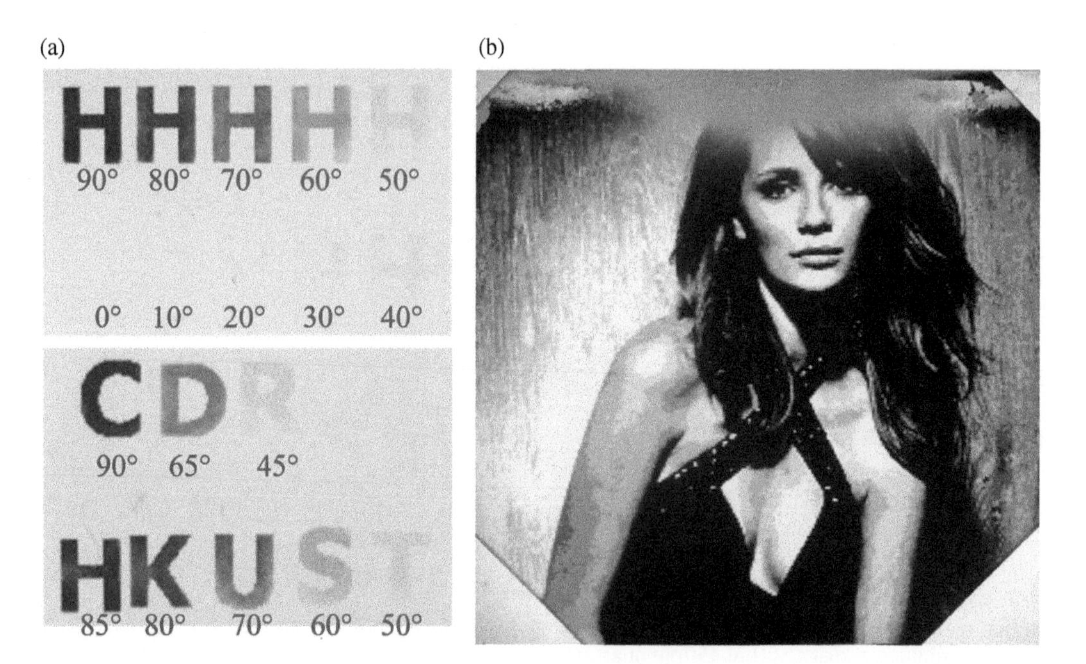

Figure 8.14 Grayscale images on ORW LCD by LC-based light printer: (a) gray levels at different twist angles (in red); (b) a 2-in. ORW LCD. *Source:* [95] Zhang et al. 2016 / Reproduced with permission of John Wiley and Sons.

(a)

(b)

Figure 8.15 Grayscale images on ORW LCD by pixel division mechanism. (a) 4-in. 8-bit grayscale image on ORW LCD; inset illustrates the pixel division. *Source:* [95] Zhang et al. 2016 / Reproduced with permission of John Wiley and Sons. (b) 2-in. 8-bit grayscale image on reflective ORW LCD under sunlight. *Source:* [96] Sun et al. 2019 / Reproduced with permission of Taylor & Francis.

8.2.6 Three-dimensional Optically Rewritable LCD

A Three-dimensional (3D) display is capable of conveying depth perception to the viewer through stereopsis for binocular vision, which is now becoming a trend in electronic products. Several attempts to generate 3D information on an ORW LCD have been developed [52, 82, 97, 102, 103].

Stereoscopic 3D effects can be achieved by displaying two different but related images with different colors [104] or orthogonal polarization planes [105] of light into the different eyes of a viewer wearing goggles. The 3D effect is generated by the human brain integrating the two images. To generate 3D effects on an ORW LCD, panels can be stacked one over another to overlap the left and right images. The final image quality may be sufficient for some applications. However, dual cells are needed to define the 3D images, and crosstalk is also induced [52]. Another method to generate stereoscopic 3D effects is to divide the ORW LCD panel into three parts with different appearances of the image, one for the left eye, a second for the right eye, and a third for the background and front the image. The complete 3D image can be written on the ORW LCD panel in three steps with photomasks and after that can be permanently stored without consuming any power [82].

Figure 8.16 shows the schematic diagram of fabricating a 3D image on an ORW LCD panel. The 3D image on the ORW LCD panel has to be divided into multiple domains with three different twist angles by three-step irradiation. First, the panel is set to provide a 0° twist angle regarding the bottom alignment layer. Next, in the second step, the ORW LCD panel is irradiated through the right-hand image mask by linearly polarized light with the polarization plane azimuth at +45° concerning the first generated exposure the image for the right eye. Afterward, the same ORW LCD panel is irradiated through a periodic amplitude mask (with a period of 100 μm), stacked with the left-hand image mask, by the linearly polarized light with the polarization plane azimuth orthogonal to that in the second step. Consequently, the easy axis of the area underneath the open window reorients orthogonally to the easy axis with the second exposure. Thus, multiple domains with three different twist angles are obtained on the ORW LCD panel.

Figure 8.17a illustrates the director profile in each domain. The dotted arrow shows the easy axis on the bottom substrate, while arrows in regions 1, 2, 3, and 4 represent the relative twist angle in different domains. A QWP is placed on the ORW LCD panel with its optical axis aligned parallel to the easy axis of the bottom substrate. The polarization states of the light in different regions after propagating through the QWP are shown in Figure 8.17b.

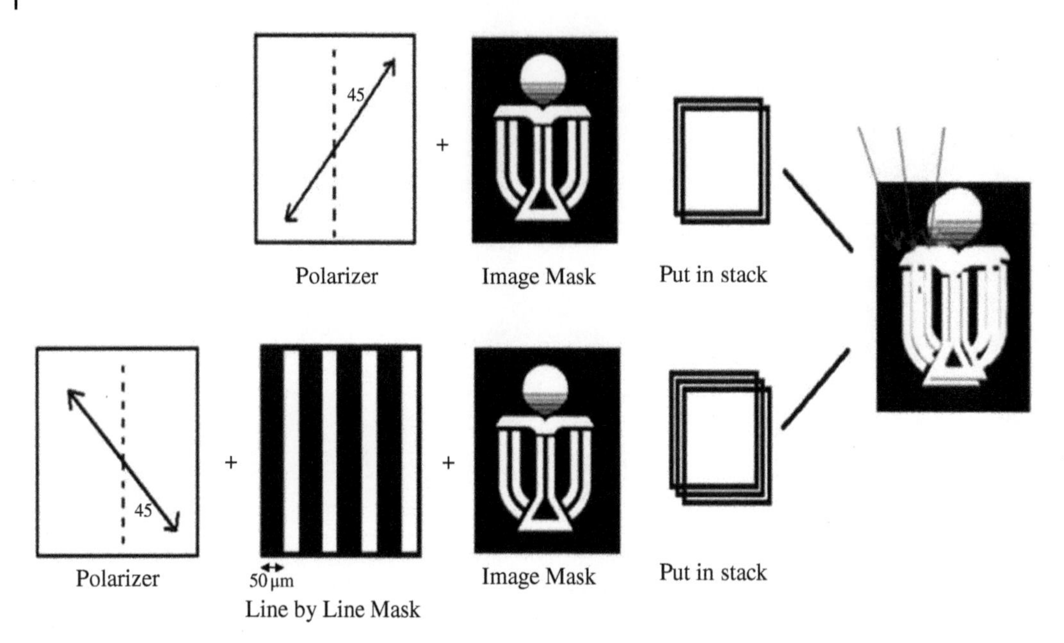

Figure 8.16 Schematic diagram to illustrate the fabrication of three alignment domains in simple steps by irradiation. *Source:* Sun et al. 2014 [82]. Copyright 2014. Reproduced with permission of the Optical Society.

Figure 8.17c and d show the binary and grayscale images, respectively, for the left eye and right eye with orthogonal circularly polarized goggles. Thus, the two images for the different eyes offer the 3D image scenes.

An autostereoscopic 3D ORW LCD is also possible, presenting a spatial image to a viewer without any goggles. This characteristic is very important for advertising, gaming devices, and other applications. Figure 8.18a illustrates the autostereoscopic 3D ORW LCD panel structure. A lenticular lens is inserted and well aligned between the ORW LCD panel and the front polarizer. The lenticular lens is an array of 70 line/in. magnifying lenses that magnify images viewed from slightly different angles, as shown in Figure 8.18b. The left and right images are divided into many strips and integrated into one image that matches the period of the lenticular lens. The image is binary printed on a transparent film to be transferred onto the ORW LCD panel as described before. Thus, with the help of the lenticular lens array, the strips are separated by a small angle and, when observed by the two eyes, appear to have 3D depth from a certain distance [102].

Autostereoscopic 3D images are shown in Figure 8.18c–f. Figure 8.18c, d are images of a dragon for the left and right eyes, respectively. The dragon's claw encircled by a solid red line shows an obvious difference between the images. Similarly, (e) and (f) are images of a tiger for the left and right eyes, respectively. The autostereoscopic 3D effect can be clearly observed from a certain distance.

8.2.7 Full-color Optically Rewritable LCD

The full-color display is a big challenge for e-paper applications. Conventional LCDs embed a red, green, and blue (RGB) color filter to generate a full color. Though the absorption of color filters is up to about 70% percent, dramatically decreasing the overall transmittance, conventional commercial LCD products still rely on them [106, 107]. Thus, an attempt has also been made to embed a color filter into an ORW LCD panel to realize a full-color display, as shown in Figure 8.19.

Figure 8.19a illustrates the structure of a reflective full-color ORW LCD. A reflective paint color array is inserted between the ORW LCD panel and analyzer. Like a color filter; the reflective paint background serves as a

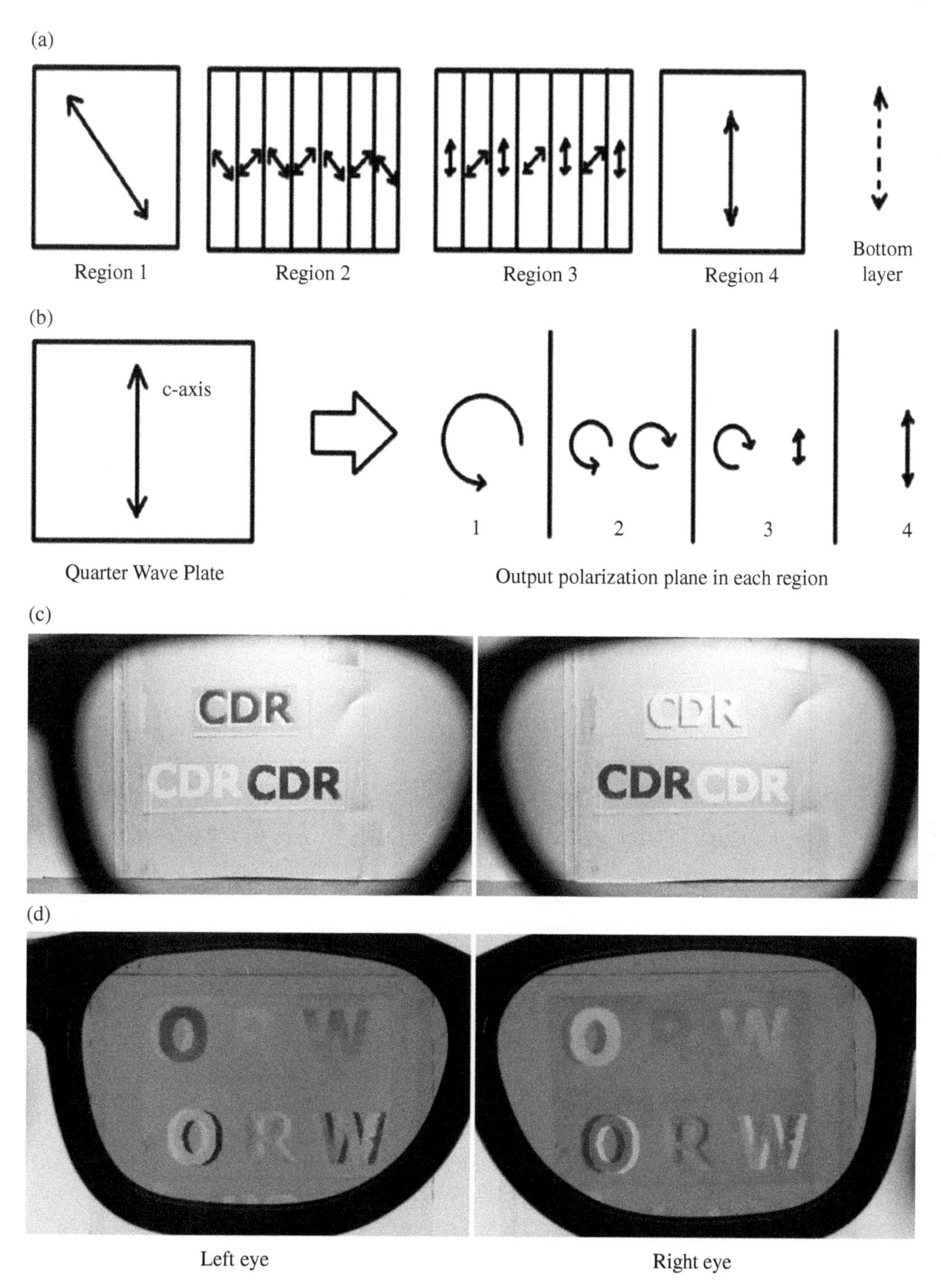

Figure 8.17 (a) The output polarization plane azimuths of the light from the three regions. (b) The polarization plane azimuths of the light from the three regions after placing a QWP on top of the ORW LCD panel. *Source:* Sun et al. 2014 [82]. Copyright 2014. Reproduced with permission of the Optical Society. Two (c) binary images and (d) grayscale images for different eyes with orthogonal circularly polarized goggles. *Source:* [95] Zhang et al. 2016 / Reproduced with permission of John Wiley and Sons.

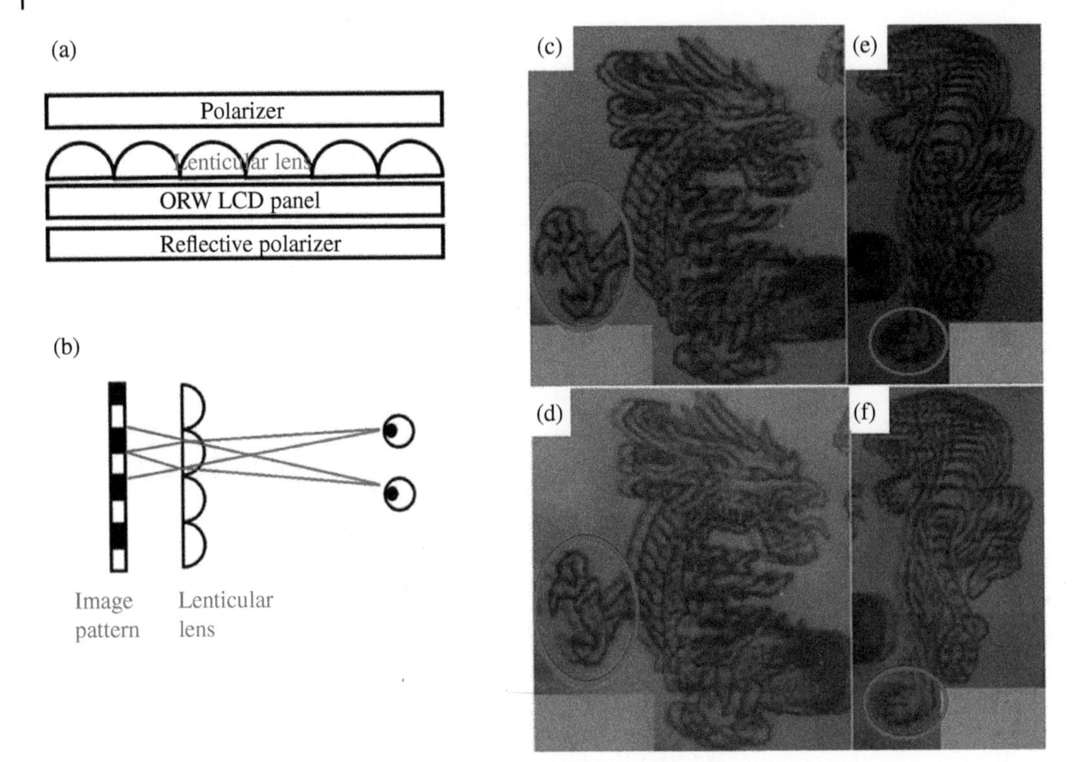

Figure 8.18 (a) The structure of the autostereoscopic 3D ORW LCD panel. (b) The principle of the 3D ORW LCD panel using the lenticular lens. (c)–(f) Autostereoscopic 3D image on an ORW LCD panel. (c) Image of a dragon for the left eye. (d) Image of a dragon for the right eye. (e) Image of a tiger for the left eye. (f) Image of a tiger for the right eye. *Source:* [102] Wang et al. 2013 / Reproduced with permission of John Wiley and Sons.

patterned color array and a reflector to create a full-color display panel. Figure 8.19b demonstrates a 2-in. reflective full-color ORW LCD with an RGB flower, and Figure 8.19c exhibits a 2-in. transmittance full-color ORW LCD with an absorption RGB color filter inside. Since the size of each color line is around 1 mm in width, the resolution of the full-color ORW LCD is limited. In principle, a high-resolution ORW LCD can be implemented by a high-resolution color filter film and reflective paint film. Moreover, the color performance of the full-color ORW LCD is determined by the CIE 1931 Chromaticity Diagram, and a color triangle of 21% NTSC is obtained, which is a good start full-color e-paper application [108].

8.3 Flexible Optically Rewritable LCD

Flexible flat panel display technologies offer many potential advantages, such as very thin profiles, lightweight, the ability to flex, curve, conform, and fold, etc. [109], and a flexible ORW LCD has also been demonstrated with many of these advantages [110]. As introduced earlier, ORW technology enables bistable switching of the LCD using an external light printer instead of complex electrodes on the display panel. Since a high-temperature process for driving electrodes like thin-film transistors is not needed for ORW LCDs, low-cost plastic substrates like polyethersulfone (PES) films and even polarizer films are applied for flexible ORW LCDs [51, 52, 110, 111]. Apart from flexible substrates, the spacers, which determine the thickness of the LCD panel, also play an important role in flexible displays. However, realizing flexible ORW LCDs is still challenging since the LC molecules flow between the two substrates on bending, and the cell gaps change appreciably.

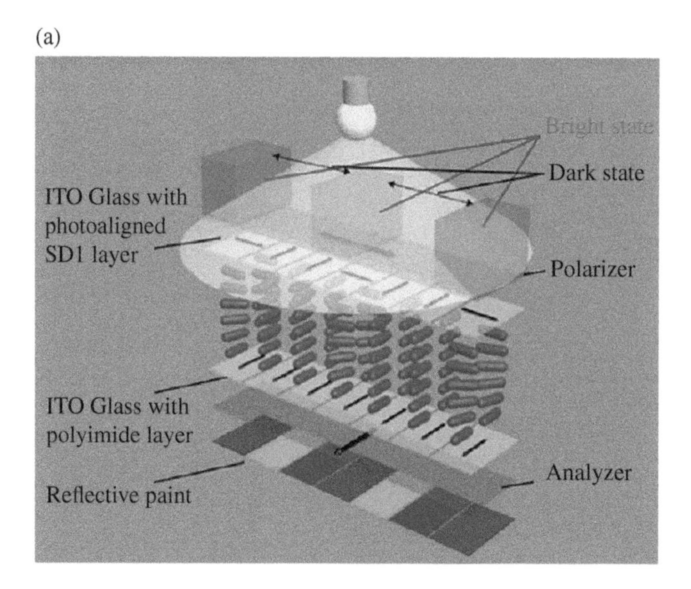

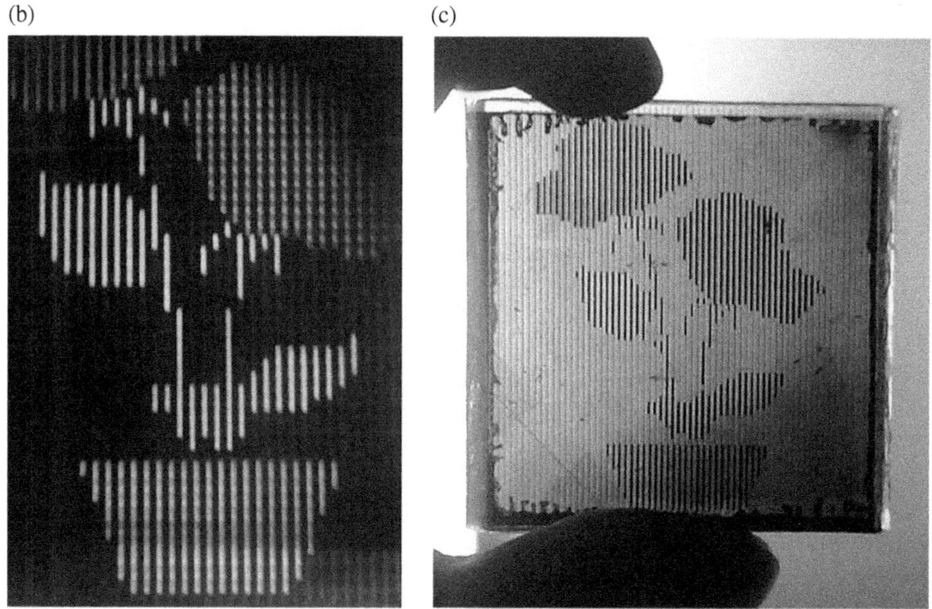

Figure 8.19 (a) Schematics of ORW LCD for realizing red with patterned reflective paint. (b) 2-in. full color reflective ORW LCD. (c) 2-in. full color transmissive ORW LCD. *Source:* [108] Ma et al. 2020 / Reproduced with permission of Taylor & Francis.

The stamp printing process has been developed to regularly pattern spacers inside an ORW LCD panel, shown in Figure 8.20a. A silicon stamp with the same structure as the spacers is fabricated, with each raised square supporting a drop of epoxy. The spacer height is controlled by pressing the silicon stamp onto the PES substrate. UV light is used to cure the epoxy to form solid spacers, shown in Figure 8.20b. The PES substrate with the spacer pattern is then deposited on the photoaligned SD1 layer with the normal process, as introduced before, and then assembled with another plastic substrate with a photo-stable alignment layer. After filling with LC, the fabricated flexible ORW LCD panel is observed under an optical polarization microscope, and the dark (left) and bright (right) states

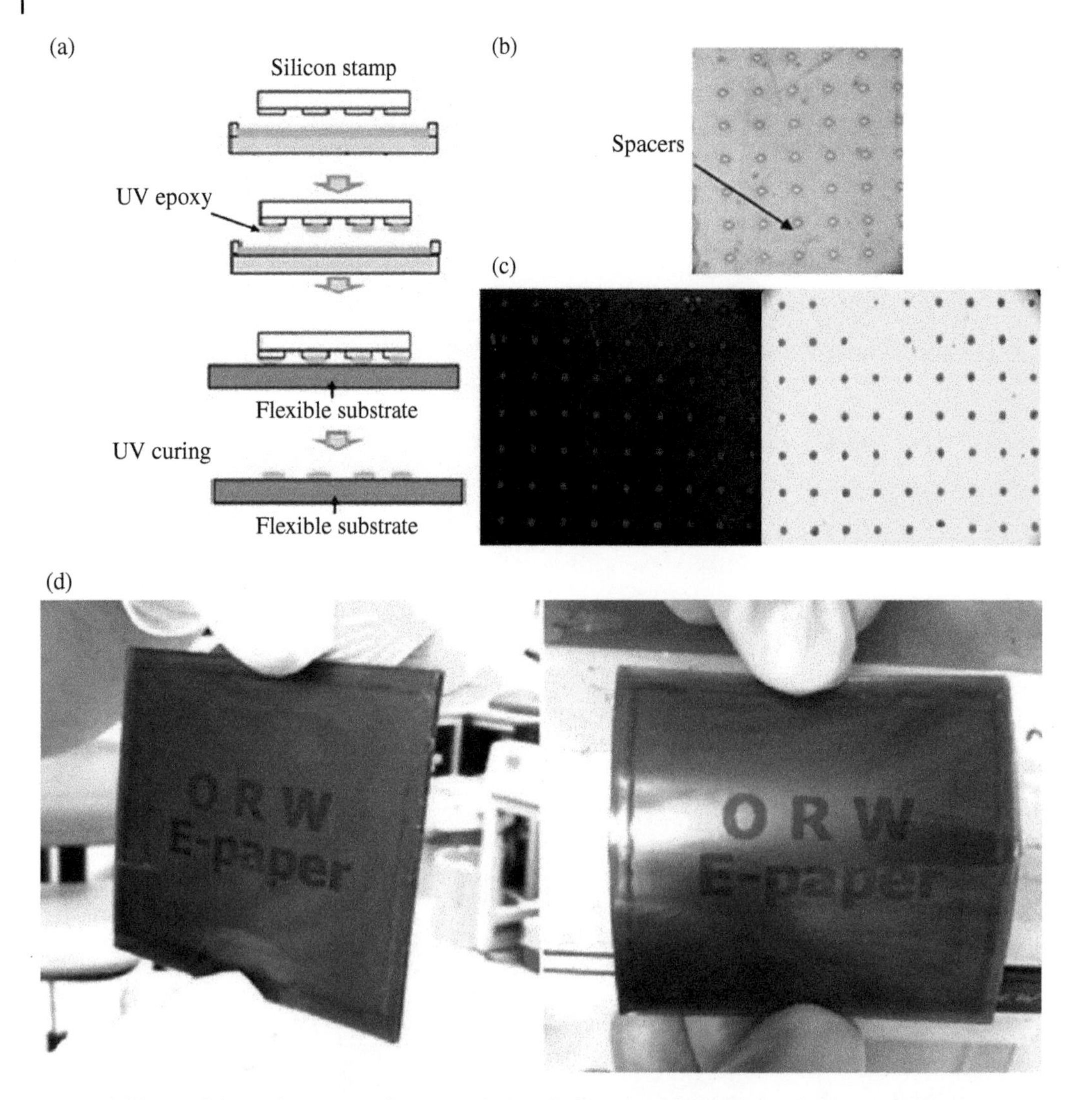

Figure 8.20 (a) Schematic process of spacer printing. (b) Top view of spacer distribution on PES substrate. (c) Dark state (left) and bright state (right) on ORW LCD panel under optical polarization microscope. (d) The fabricated 2-in. flexible ORW LCD prototype. *Source:* [110] Zhang et al. 2018 / Reproduced with permission of AIP Publishing.

are shown in Figure 8.20c. The spacer size is 20 μm, and the distance between two spacers is 200 μm. Figure 8.20d demonstrates a fabricated 2-in. flexible ORW LCD with good quality. The content on the panel is not deformed during the bending process, and the uniformity of the cell gap and LC molecule distribution is maintained. Moreover, due to the low-temperature process, it is possible to directly use commercial transmissive and reflective polarizers as the substrates. This unique advantage will further reduce the cost and increase the potential for real application.

8.4 Dye-Doped Optically Rewritable LCD

The front polarizer is still the main disadvantage for ORW LCDs since it has to be removed for each erasing and rewriting, which is inconvenient or impractical in most applications. To eliminate the front polarizer, the dichroic dye has been doped into the LC bulk to serve as a polarizer in the guest-host system [112–115]. Due to the dichroism of the dye molecules, they strongly absorb the incident light with the polarization plane parallel to their absorption axis and transmit the rest. By doping dichroic dyes into the LC bulk, the dye molecules can follow the alignment direction of the LC molecules via the guest-host effect.

The schematic structure of the dye-doped ORW LCD panel is shown in Figure 8.21a, with the mixture of dichroic dye and LC molecules filled into the ORW LCD panel forming the guest-host system. Figure 8.21b shows the plots of the dichroic ratio and contrast ratio as a function of the ORW LCD cell gap. The expression of the dichroic ratio is given by $DR = A_{\parallel}/A_{\perp}$, where $A_{\parallel}$ and $A_{\perp}$ are defined as the effective absorbance parallel and perpendicular to the rear polarizer axis, respectively [112]. The contrast ratio is denoted as a function of the dichroic dye's anisotropic absorption: $CR = 10^{-A_{\perp}}/10^{-A_{\parallel}}$. The dichroic ratio remains constant at a value around 5.0 when the cell gap is less than 10 μm, whereas the contrast ratio increases with a larger cell gap. Images on the dye-doped ORW LCD panel with different cell gaps are shown in Figure 8.21c. Figure 8.21d demonstrates an image on a 1-in. dye-doped ORW LCD with an optimized cell gap of 10 μm. The grayscale picture is transferred from the photomask, as introduced in Section 8.2.5, and the "25" embedded with the logo of The Hong Kong University of Science and Technology is demonstrated with several gray levels [112]. Although the contrast ratio is lower than that of the conventional ORW LCD, the advantage of being front polarizer-free benefits its ultimate application for e-paper.

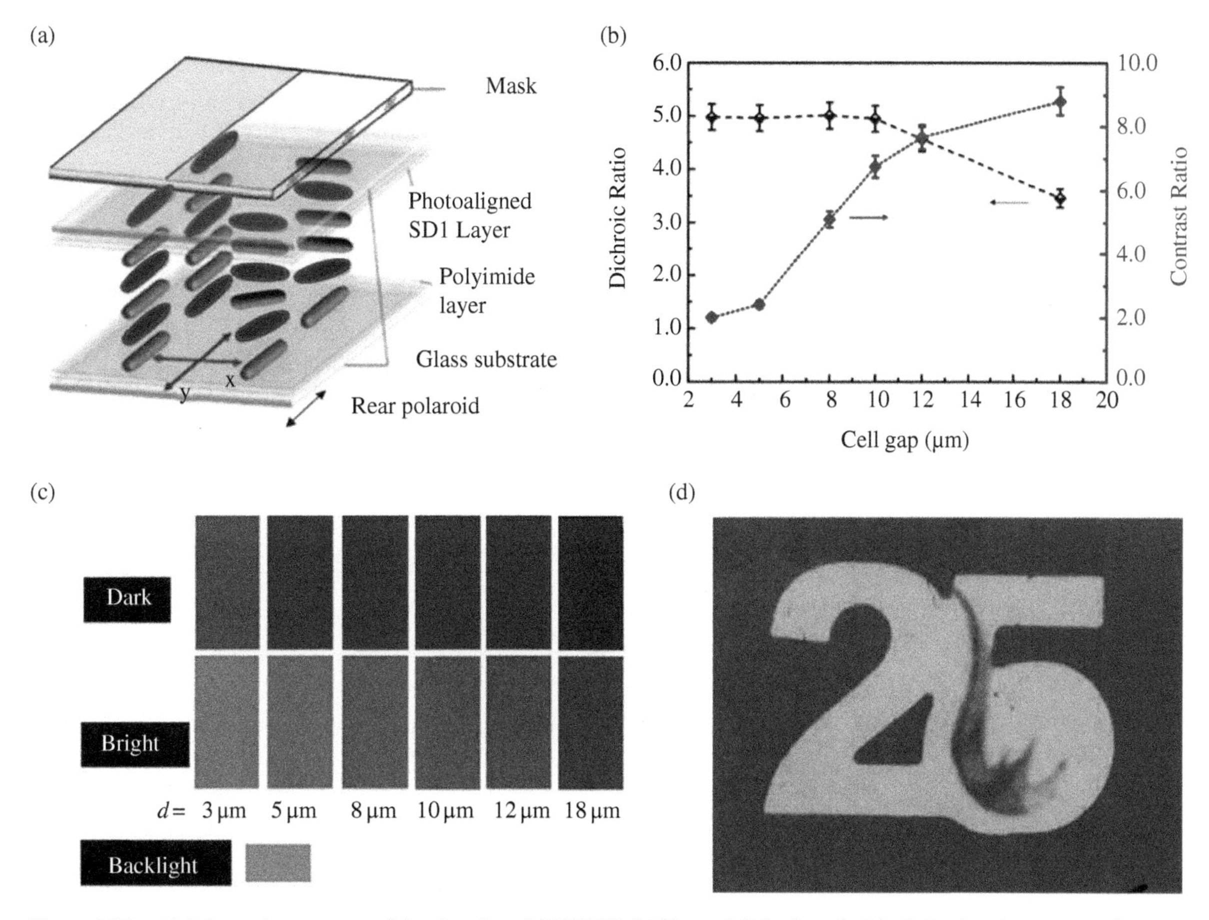

Figure 8.21 (a) Schematic structure of the dye-doped ORW LCD. (b) Plots of dichroic ratio (black dots) and contrast ratio (blue dots) as a function of the cell gap of the dye-doped ORW LCD panel. (c) Dark and bright states of the dye-doped ORW LCD panel with different cell gaps. (d) Image on 1-in. dye-doped ORW LCD panel. *Source:* [112] Meng et al. 2018 / Reproduced with permission of the Optical Society.

8.5 Conclusion

ORW technology is a modified method of azo-dye photoalignment that possesses a traditionally high azimuthal anchoring energy. With the photochemical stability, the photoalignment material SD1 has the unique feature of reversible in-plane alignment direction reorientation perpendicular to the polarization plane of the incident light. By introducing photoaligned SD1 as one ORW alignment layer and another photo-stable alignment layer, namely, rubbed polyimide, the assembled ORW LCD panel exhibits the unique feature of changing between the bright and dark states with linearly polarized light. Different from conventional LCDs, the bistability characteristic enables ORW LCDs to display information for a long time with zero power consumption. Moreover, the simple structure of an ORW LCD panel without any electronic components provides a large-scale, low-cost solution for electronic paper (e-paper) applications.

This chapter briefly introduces recent research about ORW LCDs for e-paper applications. As the foundation of ORW technology, the diffusion model of the photoalignment material SD1 explains the in-plane reorientation mechanism, indicating several methods to speed up the rewriting process, including choosing LC materials with a small twist elastic constant, increasing the average twist elastic energy by doping chiral dopants, and reducing rotational elastic energy by applying an external electric field. The faster rewriting speed decreases the display content refreshing time down to several seconds by using a linearly polarized LED light printer with a power density of $125\,mW/cm^2$. The specially designed light printer embedded with an LC-based polarization rotator panel can directly write a 4-bit grayscale image on the ORW LCD panel in one step with a resolution of 127 ppi (pixels per inch). The spatial grayscale is another solution, providing 8-bit gray levels on the ORW LCD. The displayed images, with high resolution and 256 gray levels, are comparable with those on real printed paperstable under bright sunshine. Attempts to realize a full-color and flexible ORW LCD panel have also been presented, although the performance is not practical for commercial applications, and research is ongoing for improvement.

The 3D ORW LCD is unique among all other e-paper devices. Two main approaches are introduced in this chapter, based on stereoscopic and autostereoscopic 3D effects. The stereoscopic 3D effect can be clearly seen on a single ORW LCD panel with circularly polarized goggles by dividing the alignment domains into three twist angles. An autostereoscopic 3D ORW LCD has also been demonstrated. By inserting a lenticular lens array, images can be observed from different viewing angles.

Beyond display applications, ORW technology, which enables the reconfiguration of LC molecules with space-varying alignment directions, is widely applied to generating, manipulating, and detecting novel optical fields with high quality, high efficiency, and unprecedented flexibility. It drastically enhances the capability of optical beam shaping and steering. Also, it settles a fundamental requirement in optics and photonics, bringing new opportunities in optical manipulation and even some uncharted territory [116–134].

References

1 Heikenfeld, J., Drzaic, P., Yeo, J.S., and Koch, T. (2011). Review paper: a critical review of the present and future prospects for electronic paper. *J. Soc. Inf. Disp.* 19: 129.
2 Fernández, M.R., Casanova, E.Z., and Alonso, I.G. (2015). Review of display technologies focusing on power consumption. *Sustainability* 7: 10854.
3 Omodani, M. (2004). 10.1: Invited paper: what is electronic paper? The expectations. *SID Int. Symp. Dig. Tech. Pap.* 35: 128.
4 Ota, I., Ohnishi, J., and Yoshiyama, M. (1973). Electrophoretic image display (EPID) panel. *Proc. IEEE* 61: 832.
5 Mürau, P. and Singer, B. (1978). The understanding and elimination of some suspension instabilities in an electrophoretic display. *J. Appl. Phys.* 49: 4820.

6 Comiskey, B., Albert, J.D., Yoshizawa, H., and Jacobson, J. (1998). An electrophoretic ink for all-printed reflective electronic displays. *Nature* 394: 253.
7 Yang, B., Hu, W., Zeng, Z. et al. (2021). Understanding the mechanisms of electronic ink operation. *J. Soc. Inf. Disp.* 29: 38.
8 Doane, J.W. and Khan, A. (2005). Cholesteric liquid crystals for flexible displays. In: *Flexible Flat Panel Displays* (ed. G.P. Crawford), 331–354. Wiley.
9 Yang, D. (2006). Flexible bistable cholesteric reflective displays. *J. Disp. Technol.* 2: 32.
10 Wang, C. and Lin, T. (2011). Bistable reflective polarizer-free optical switch based on dye-doped cholesteric liquid crystal. *Opt. Mater. Express* 1: 1457.
11 Tokunaga, S., Itoh, Y., Yaguchi, Y. et al. (2016). Electrophoretic deposition for cholesteric liquid-crystalline devices with memory and modulation of reflection colors. *Adv. Mater.* 28: 4077.
12 Jones, J.C. (2008). The zenithal bistable display: from concept to consumer. *J. Soc. Inf. Disp.* 16: 143.
13 Amos, R.M. and Jones, C. (2009). P-120: Optimizing the zenithal bistable display. *SID Int. Symp. Dig. Tech. Pap.* 40: 1577.
14 Spencer, T., Care, C., Amos, R., and Jones, J. (2010). Zenithal bistable device: comparison of modeling and experiment. *Phys. Rev. E* 82: 021702.
15 Dozov, I., Nobili, M., and Durand, G. (1997). Fast bistable nematic display using monostable surface switching. *Appl. Phys. Lett.* 70: 1179.
16 Joubert, C., Angele, J., Boissier, A. et al. (2003). Reflective bistable nematic displays (BiNem®) fabricated by standard manufacturing equipment. *J. Soc. Inf. Disp.* 11: 217.
17 Stalder, M. and Schadt, M. (2003). Photoaligned bistable twisted nematic liquid crystal displays. *Liq. Cryst.* 30: 285.
18 Liu, P., Tseng, M.C., Yeung, F.S.Y., and Kwok, H.S. (2020). P-136: A low voltage, switchable, bistable-twist-nematic display. *SID Int. Symp. Dig. Tech. Pap.* 51: 1889.
19 Chigrinov, V.G. (2014). *Liquid Crystal Photonics*. New York: Nova Science Publishers.
20 Muravsky, A., Murauski, A., Li, X. et al. (2007). Optical rewritable liquid-crystal-alignment technology. *J. Soc. Inf. Disp.* 15: 267.
21 Muravsky, A., Murauski, A., Chigrinov, V., and Kwok, H.S. (2008). 60.1: Optical rewritable electronic paper with fluorescent dye doped liquid crystal. *SID Int. Symp. Dig. Tech. Pap.* 39: 915.
22 Murauski, A., Chigrinov, V., Li, X. and Kwok, H.S. (2005). Optically rewriteable LC display with a high contrast and long life time. *IDW/AD'05*: 131
23 Li, X., Au, P., Xu, P. et al. (2006). P-153: Flexible photoaligned optically rewritable LC display. *SID Int. Symp. Dig. Tech. Pap.* 37: 783.
24 Ichimura, K., Suzuki, Y., Seki, T. et al. (1988). Reversible change in alignment mode of nematic liquid-crystals regulated photochemically by command surfaces modified with an azobenzene monolayer. *Langmuir* 4: 1214.
25 Chigrinov, V., Kozenkov, V., and Kwok, H.S. (2008). *Photoalignment of Liquid Crystalline Materials: Physics and Applications*. New Jersey: Wiley.
26 Yaroshchuk, O. and Reznikov, Y. (2012). Photoalignment of liquid crystals: basics and current trends. *J. Mater. Chem.* 22: 286.
27 Chigrinov, V., Prudnikova, E., Kozenkov, V. et al. (2002). Synthesis and properties of azo dye aligning layers for liquid crystal cells. *Liq. Cryst.* 29: 1321.
28 Hegde, G., Srivastava, A., Adhikari, A., et al. (2012). Recent developments in photoalignment technology: alignment properties of a thiophene based bis-hydrazone derivative. *IDW/AD'12*: 19: 1591. https://repository.ust.hk/ir/Record/1783.1-67303
29 Chigrinov, V., Sun, J., and Wang, X. (2020). Photoaligning and photopatterning: new LC technology. *Crystals* 10: 323.
30 Hu, W., Chen, P., and Lu, Y. (2019). Photoinduced liquid crystal domain engineering for optical field control. In: *Photoactive Functional Soft Materials* (ed. Q. Li), 361–387. Wiley-VCH.

31 Budaszewski, D., Srivastava, A., Wolinski, T., and Chigrinov, V. (2015). Photo-aligned photonic ferroelectric liquid crystal fibers. *J. Soc. Inf. Disp.* 23: 196.

32 Ertman, S., Srivastava, A., Chigrinov, V. et al. (2013). Patterned alignment of liquid crystal molecules in silica micro-capillaries. *Liq. Cryst.* 40: 1.

33 Pedersen, T.G., Ramanujam, P.S., Johansen, P.M., and Hvilsted, S. (1998). Quantum theory and experimental studies of absorption spectra and photoisomerization of azobenzene polymers. *J. Opt. Soc. Am. B* 15: 2721.

34 Pedersen, T.G., Johansen, P.M., Holme, N.C.R. et al. (1998). Theoretical model of photoinduced anisotropy in liquid-crystalline azobenzene side-chain polyesters. *J Opt. Soc. Am. B* 15: 1120.

35 Gaididei, Y.B., Christiansen, P.L., and Ramanujam, P.S. (2002). Theory of photoinduced deformation of molecular films. *Appl. Phys. B Lasers Opt.* 74: 139.

36 Pedersen, T.G. and Johansen, P.M. (1997). Mean-field theory of photoinduced molecular reorientation in azobenzene liquid crystalline side-chain polymers. *Phys. Rev. Lett.* 79: 2470.

37 Yaroshchuk, O.V., Kiselev, A.D., Zakrevskyy, Y. et al. (2003). Photoinduced three-dimensional orientational order in side chain liquid crystalline azopolymers. *Phy. Rev. E* 68: 011803.

38 Blinov, L.M. (1996). Photoinduced molecular reorientation in polymers, Langmuir-Blodgett films and liquid crystals. *J. Nonlinear Opt. Phys. Mater.* 05: 165.

39 Gibbons, W., Shannon, P., Sun, S., and Swetlin, B. (1991). Surface-mediated alignment of nematic liquid crystals with polarized laser light. *Nature* 351: 49.

40 Chigrinov, V., Pikin, S., Verevochnikov, A. et al. (2004). Diffusion model of photoaligning in azo-dye layers. *Phys. Rev.* 69: 061713.

41 Kiselev, A., Chigrinov, V., Pasechnik, S., and Dubtsov, A. (2012). Photoinduced reordering in thin azo-dye films and light-induced reorientation dynamics of the nematic liquid-crystal easy axis. *Phys. Rev. E* 86: 011706.

42 Schonhoff, M., Mertesdorf, M., and Losche, M. (1996). Mechanism of photoreorientation of azobenzene dyes in molecular films. *J. Phys. Chem.* 100: 7558.

43 Schadt, M., Schmitt, K., Kozinkov, V., and Chigrinov, V. (1992). Surface-induced parallel alignment of liquid-crystals by linearly polymerized photopolymers. *Jpn. J. Appl. Phys.* 31: 2155.

44 Dyadyusha, A., Marusii, T.Y., Reshetnyak, V.Y. et al. (1992). Orientational effect due to a change in the anisotropy of the interaction between a liquid crystal and a bounding surface. *JETP Lett.* 56: 17.

45 Hasegawa, M. and Taira, Y. (1995). Nematic homogeneous photo alignment by polyimide exposure to linearly polarized uv. *J. Photopolym. Sci. Technol.* 8: 241.

46 Chen, J., Johnson, D., Bos, P. et al. (1996). Model of liquid crystal alignment by exposure to linearly polarized ultraviolet light. *Phy. Rev. E* 54: 1599.

47 Hasegawa, M. (1999). Modeling of photoinduced optical anisotropy and anchoring energy of polyimide exposed to linearly polarized deep UV light. *Jpn. J. Appl. Phys.* 38: L457.

48 Nishikawa, M., Taheri, B., and West, J. (1998). Mechanism of unidirectional liquid-crystal alignment on polyimides with linearly polarized ultraviolet light exposure. *Appl. Phys. Lett.* 72: 2403.

49 Shi, Y., Zhao, C., Ho, J.Y. et al. (2017). Exotic property of azobenzenesulfonic photoalignment material based on relative humidity. *Langmuir* 33: 3968.

50 Shi, Y., Zhao, C., Ho, J. et al. (2018). High photoinduced ordering and controllable photostability of hydrophilic azobenzene material based on relative humidity. *Langmuir* 34: 4465.

51 Muravsky, A., Murauski, A., Chigrinov, V., and Kwok, H.S. (2008). Optical rewritable electronic paper. *IEICE Trans. Electron.* E91.C: 1576.

52 Muravsky, A., Murauski, A., Chigrinov, V., and Kwok, H.S. (2008). Light printing of grayscale pixel images on optical rewritable electronic paper. *Jpn. J. Appl. Phys.* 47: 6347.

53 Chigrinov, V., Murauski, A., Yu, Q., and Kwok, H.S. (2010). 15.1: Optically rewritable liquidcrystal technology: a new green epaper approach. *SID Int. Symp. Dig. Tech. Pap.* 41: 195.

54 Nersisyan, S., Tabiryan, N., Steeves, D. et al. (2010). Study of azo dye surface command photoalignment material for photonics applications. *Appl. Opt.* 49: 1720.

55 Chigrinov, V. (1999). *Liquid Crystal Devices: Physics and Applications*. Massachusetts: Artech House.

56 Chigrinov, V., Kwok, H.S., Takada, H., and Takatsu, H. (2005). Photo-aligning by azo-dyes: physics and applications. *Liq. Cryst. Today* 14: 1.

57 Kiselev, A., Chigrinov, V., and Huang, D. (2005). Photoinduced ordering and anchoring properties of azo-dye films. *Phys. Rev. E* 72: 061703.

58 Srivastava, A., Wang, X., Gong, S.Q. et al. (2015). Micro-patterned photo-aligned ferroelectric liquid crystal Fresnel zone lens. *Opt. Lett.* 40: 1643.

59 Ma, Y., Sun, J., Srivastava, A. et al. (2013). Optically rewritable ferroelectric liquid-crystal grating. *Europhys. Lett.* 102: 24005.

60 Chigrinov, V., Pozhidaev, E., Srivastava, A., Qi, G., Fan, F., Ma, Y., Lastochkin, A. and Kwok, H.S. (2011). Novel photoaligned fast ferroelectric liquid crystal display. *IDW'11*: 557.

61 Ma, Y., Wei, B., Shi, L. et al. (2016). Fork gratings based on ferroelectric liquid crystals. *Opt. Express* 24: 5822.

62 Ma, Y., Wang, X., Srivastava, A. et al. (2016). Fast switchable ferroelectric liquid crystal gratings with two electro-optical modes. *AIP Adv.* 6: 035207.

63 Srivastava, A., Hu, W., Chigrinov, V. et al. (2012). Fast switchable grating based on orthogonal photo alignments of ferroelectric liquid crystals. *Appl. Phys. Lett.* 101: 031112.

64 Wang, X., Yang, W., Liu, Z. et al. (2017). Switchable Fresnel lens based on hybrid photo-aligned dual frequency nematic liquid crystal. *Opt. Mater. Express* 7: 8.

65 Lin, X., Hu, W., Hu, X. et al. (2012). Fast response dual-frequency liquid crystal switch with photo-patterned alignments. *Opt. Lett.* 37: 3627.

66 Shteyner, E., Srivastava, A., Chigrinov, V. et al. (2013). Submicron-scale liquid crystal photo-alignment. *Soft Matter* 9: 5160.

67 Fan, F., Du, T., Srivastava, A. et al. (2012). Axially symmetric polarization converter made of patterned liquid crystal quarter wave plate. *Opt. Express* 20: 23036.

68 Fan, F., Srivastava, A., Chigrinov, V., and Kwok, H.S. (2012). Switchable liquid crystal grating with sub millisecond response. *Appl. Phys. Lett.* 100: 111105.

69 Hu, W., Srivastava, A., Xu, F. et al. (2012). Liquid crystal gratings based on alternate TN and PA photoalignment. *Opt. Express* 20: 5384.

70 Hu, W., Srivastava, A., Lin, X. et al. (2012). Polarization independent liquid crystal gratings based on orthogonal photoalignments. *Appl. Phys. Lett.* 100: 111116.

71 Sun, J., Srivastava, A., Wang, L. et al. (2013). Optically tunable and rewritable diffraction grating with photoaligned liquid crystals. *Opt. Lett.* 38: 2342.

72 Chen, P., Wei, B., Ji, W. et al. (2015). Arbitrary and reconfigurable optical vortex generation: a high-efficiency technique using director-varying liquid crystal fork gratings. *Photonics Res.* 3: 133.

73 Duan, W., Chen, P., Ge, S. et al. (2017). Helicity-dependent forked vortex lens based on photo-patterned liquid crystals. *Opt. Express* 25: 14059.

74 Wei, B., Hu, W., Ming, Y. et al. (2014). Generating switchable and reconfigurable optical vortices via photopatterning of liquid crystals. *Adv. Mater.* 26: 1590.

75 Wang, X., Srivastava, A., Fan, F. et al. (2016). Electrically/optically tunable photo-aligned hybrid nematic liquid crystal Dammann grating. *Opt. Lett.* 41: 5668.

76 Fan, F., Yao, L., Wang, X. et al. (2017). Ferroelectric liquid crystal Dammann grating by patterned photoalignment. *Crystals* 7: 79.

77 Fan, F., Srivastava, A., Du, T. et al. (2013). Low voltage tunable liquid crystal lens. *Opt. Lett.* 38: 4116.

78 Wang, X., Srivastava, A., Chigrinov, V., and Kwok, H.S. (2013). Switchable Fresnel lens based on micropatterned alignment. *Opt. Lett.* 38: 1775.

79 Wang, X., Fan, F., Du, T. et al. (2014). Liquid crystal Fresnel zone lens based on single-side-patterned photoalignment layer. *Appl. Opt.* 53: 2026.

80 Wang, X., Kumar, S., Tam, M.W. et al. (2016). Liquid crystal Fresnel lens display. *Chin. Phys. B* 25: 094215.

81 Chigrinov, V., Sun, J., Kuznetsov, M. et al. (2020). The effect of operating temperature on the response time of optically driven liquid crystal displays. *Crystals* 10: 626.

82 Sun, J., Srivastava, A., Zhang, W. et al. (2014). Optically rewritable 3D liquid crystal displays. *Opt. Lett.* 39: 6209.

83 Sun, J., Liu, Y., Liu, H. et al. (2019). Increasing rewriting speed of optically driving liquid crystal display by process optimisation. *Liq. Cryst.* 46: 151.

84 Guo, Q., Srivastava, A., Chigrinov, V., and Kwok, H.S. (2014). Polymer and azo-dye composite: a photo-alignment layer for liquid crystals. *Liq. Cryst.* 41: 1465.

85 Tseng, M.C., Yaroshchuk, O., Bidna, T. et al. (2016). Strengthening of liquid crystal photoalignment on azo dye films: passivation by reactive mesogens. *RSC Adv.* 6: 48181.

86 Yaroshchuk, O., Kyrychenko, V., Tao, D. et al. (2009). Stabilization of liquid crystal photoaligning layers by reactive mesogens. *Appl. Phys. Lett.* 95: 021902.

87 Chigrinov, V. and Kudreyko, A. (2021). Tunable optical properties for ORW e-paper. *Liq. Cryst.* 48 (7): 1073–1077. https://doi.org/10.1080/02678292.2020.1842924.

88 Sun, J. and Chigrinov, V. (2012). Effect of azo dye layer on rewriting speed of optical rewritable e-paper. *Mol. Cryst. Liq. Cryst.* 561: 1.

89 Geng, Y., Sun, J., Murauski, A. et al. (2012). Increasing the rewriting speed of optical rewritable e-paper by selecting proper liquid crystals. *Chin. Phys. B* 21: 080701.

90 Geng, Y., Yao, L., Chigrinov, V., and Kwok, H.S. (2014). Analysis of the rewriting time of liquid crystal optical rewritable e-paper. *Liq. Cryst.* 41: 610.

91 Geng, Y. and Yao, L. (2021). Effect of azimuthal anchoring energy on rewriting speed of optical rewritable e-paper. *Liq. Cryst.* 48 (6): 915–921. https://doi.org/10.1080/02678292.2020.1827311.

92 Sun, J., Srivastava, A., Chigrinov, V., and Kwok, H.S. (2014). P-125: Optimisation of the mechanical stability and rewriting speed of optically rwritable e-paper. *SID Int. Symp. Dig. Tech. Pap.* 45: 1453.

93 Wang, L., Sun, J., Liu, H. et al. (2018). Increasing the rewriting speed of ORW e-paper by electric field. *Liq. Cryst.* 45: 553.

94 Sun, J., Deng, K., Sang, J. et al. (2020). The effect of chiral dopant on the rewriting speed of optically driving liquid crystal display. *Liq. Cryst.* 47: 516.

95 Zhang, W., Tseng, M.C., Lee, C.Y. et al. (2016). P-76: Polarization-controllable light-printer for optically rewritable (ORW) liquid crystal displays. *SID Int. Symp. Dig. Tech. Pap.* 47: 1421.

96 Sun, J., Ren, L., Deng, K. et al. (2019). Greyscale generation for optically driving liquid crystal display. *Liq. Cryst.* 46: 1340.

97 Zhang, W., Sun, J., Srivastava, A. et al. (2015). Paper No S9.3: 3-D grayscale images generation on optically rewritable electronic paper. *SID Int. Symp. Dig. Tech. Pap.* 46: 40.

98 Wang, L., Sun, J., Srivastava, A. and Chigrinov, V. (2011). Thermal stability of gray-scale levels on optical rewritable electronic paper. *IDW'11* 11:1627.

99 Culbreath, C., Glazar, N., and Yokoyama, H. (2011). Note: automated maskless micro-multidomain photoalignment. *Rev. Sci. Instrum.* 82: 126107.

100 Wu, H., Hu, W., Hu, H. et al. (2012). Arbitrary photo-patterning in liquid crystal alignments using DMD based lithography system. *Opt. Express* 20: 16684.

101 Zhang, W., Schneider, J., Chigrinov, V. et al. (2018). Optically addressable photoaligned semiconductor nanorods in thin liquid crystal films for display applications. *Adv. Opt. Mater.* 6: 1800250.

102 Wang, X., Wang, L., Sun, J. et al. (2013). Autostereoscopic 3D pictures on optically rewritable electronic paper. *J. Soc. Inf. Disp.* 21: 103.

103 Wang, L., Sun, J., Wang, X. et al. (2012). P-24: Generation of 3D image on optically rewritable liquid crystal display. *SID Int. Symp. Dig. Tech. Pap.* 43: 1139.

104 Woods, A. and Harris, C. (2010). Comparing levels of crosstalk with red/cyan, blue/yellow, and green/magenta anaglyph 3D glasses. In *Proceedings of SPIE 7524, Stereoscopic Displays and Applications* XXI, 75240Q.

105 Faris, S. M. (1994). Novel 3D stereoscopic imaging technology. In *Proceedings of SPIE 2177, Stereoscopic Displays and Virtual Reality Systems.*

106 Srivastava, A., Zhang, W., Schneider, J. et al. (2017). Photoaligned nanorod enhancement films with polarized emission for liquid-crystal-display applications. *Adv. Mater.* 29: 1701091.

107 Yang, D. and Wu, S.T. (2014). *Fundamentals of Liquid Crystal Devices*. New Jersey: Wiley.

108 Ma, Y., Xin, S.J., Liu, X. et al. (2020). Colour generation for optically driving liquid crystal display. *Liq. Cryst.* 47: 1729–1734.

109 Crawford, G.P. (2005). Flexible flat panel display technology. In: *Flexible Flat Panel Displays* (ed. G.P. Crawford), 1–9. New Jersey: Wiley.

110 Zhang, Y., Sun, J., Liu, Y. et al. (2018). A flexible optically re-writable color liquid crystal display. *Appl. Phys. Lett.* 112: 131902.

111 Tixier-Mita, A., Ihida, S., Ségard, B. et al. (2016). Review on thin-film transistor technology, its applications, and possible new applications to biological cells. *Jpn. J. Appl. Phys.* 55: 04EA08.

112 Meng, C., Tseng, M., Tang, S., and Kwok, H.S. (2018). Optical rewritable liquid crystal displays without a front polarizer. *Opt. Lett.* 43: 899.

113 Zhao, X., Bermak, A., Boussaid, F. et al. (2009). High-resolution photoaligned liquid-crystal micropolarizer array for polarization imaging in visible spectrum. *Opt. Lett.* 34: 3619.

114 Zhao, X., Bermak, A., Boussaid, F., and Chigrinov, V. (2010). Liquid-crystal micropolarimeter array for full stokes polarization imaging in visible spectrum. *Opt. Express* 18: 17776.

115 Zhao, X., Boussaid, F., Bermak, A., and Chigrinov, V. (2011). High-resolution thin "guest-host" micropolarizer arrays for visible imaging polarimetry. *Opt. Express* 19: 5565.

116 Chen, P., Ji, W., Wei, B. et al. (2015). Generation of arbitrary vector beams with liquid crystal polarization converters and vector-photoaligned q-plates. *Appl. Phys. Lett.* 107: 241102.

117 Du, T., Fan, F., Tam, A.M. et al. (2015). Complex nanoscale-ordered liquid crystal polymer film for high transmittance holographic polarizer. *Adv. Mater.* 27: 7191.

118 Du, T., Schneider, J., Srivastava, A. et al. (2015). Combination of photoinduced alignment and self-assembly to realize polarized emission from ordered semiconductor nanorods. *ACS Nano* 9: 11049.

119 Chen, P., Lu, Y., and Hu, W. (2016). Beam shaping via photopatterned liquid crystals. *Liq. Cryst.* 43: 2051.

120 Ji, W., Lee, C., Chen, P. et al. (2016). Meta-q-plate for complex beam shaping. *Sci. Rep.* 6: 25528.

121 Ma, Y., Shi, L., Srivastava, A. et al. (2016). Restricted polymer-stabilised electrically suppressed helix ferroelectric liquid crystals. *Liq. Cryst.* 43: 1092.

122 Wei, B., Chen, P., Ge, S. et al. (2016). Generation of self-healing and transverse accelerating optical vortices. *Appl. Phys. Lett.* 109: 121105.

123 Chen, P., Ge, S., Duan, W. et al. (2017). Digitalized geometric phases for parallel optical spin and orbital angular momentum encoding. *ACS Photonics* 4: 1333.

124 Schneider, J., Zhang, W., Srivastava, A. et al. (2017). Photoinduced micropattern alignment of semiconductor nanorods with polarized emission in a liquid crystal polymer matrix. *Nano Lett.* 17: 3133.

125 Tam, A.M.W., Fan, F., Du, T. et al. (2017). Bifocal optical-vortex lens with sorting of the generated nonseparable spin-orbital angular-momentum states. *Phys. Rev. Appl.* 7: 034010.

126 Wang, L., Ge, S., Hu, W. et al. (2017). Tunable reflective liquid crystal terahertz waveplates. *Opt. Mater. Express* 7: 2023.

127 Shen, Z., Zhou, S., Ge, S. et al. (2018). Liquid-crystal-integrated metadevice: towards active multifunctional terahertz wave manipulations. *Opt. Lett.* 43: 4695.

128 Wei, B., Liu, S., Chen, P. et al. (2018). Vortex Airy beams directly generated via liquid crystal q-Airy-plates. *Appl. Phys. Lett.* 112: 121101.

129 Hu, Y., Fu, S., Wang, S., Zhang, W. and Kwok, H.S. (2019). Flatness and diffractive wavefront measurement of liquid crystal computer-generated hologram based on photoalignment technology. *Proc. SPIE 10841, 9th International*

Symposium on Advanced Optical Manufacturing and Testing Technologies: Meta-Surface-Wave and Planar Optics: 108410I

130 Zhou, Y., Yin, Y., Yuan, Y. et al. (2019). Liquid crystal Pancharatnam-Berry phase lens with spatially separated focuses. *Liq. Cryst.* 46: 995.

131 Gupta, S., Sun, Z., Kwok, H.S., and Srivastava, A. (2020). Low voltage tunable liquid crystal Fibonacci grating. *Liq. Cryst.* 47: 1162.

132 Shi, Y., Salter, P., Li, M. et al. (2020). Two-photon laser-written photoalignment layers for patterning liquid crystalline conjugated polymer orientation. *Adv. Funct. Mater.* 31: 2007493.

133 Zheng, Z., Lu, Y., and Li, Q. (2020). Photoprogrammable mesogenic soft helical architectures: a promising avenue toward future chiro-optics. *Adv. Mater.* 32: e1905318.

134 Zhou, S., Shen, Z., Li, X. et al. (2020). Liquid crystal integrated metalens with dynamic focusing property. *Opt. Lett.* 45: 4324.

9

Electrowetting Displays

Doeke J. Oostra

VP Business Development, Etulipa, a Miortech company, High Tech Campus 10, 5656 AE Eindhoven, The Netherlands

9.1 Overviews

Several excellent overviews on electrowetting display development have been written. Heikenfeld et al. [1] gave a detailed overview of the potential and limitations of many reflective display techniques, including electrowetting, and that same team also created an overview of application guidelines to create electrowetting displays in a scalable fabrication process [2]. In 2014 a review of paper-like displays was written by Peng Fei Bai et al., including a description of the capability of electrowetting displays [3]. Shui et al. present an overview of reflecting technologies, including electrowetting display technology [4]. Feenstra published an overview of the history and capability of developed electrowetting displays in the Handbook of Visual Display Technology in 2016 [5]. Mugele and Heikenfeld have written a comprehensive document on the fundamental principles and practical applications of electrowetting, published in 2019 [6].

Furthermore, in 2020 the commercial status of electrowetting microfluidics status was recapped by J Li and C-J Kim [7]. Etulipa is the first company that has developed electrowetting displays commercially used at the moment of writing this chapter, in 2020. In comparison with the above reviews, this chapter focuses on the technical considerations that determined successful introduction in the market and the manufacturing approach to create volume electrowetting displays that operate in the field reliably.

9.2 Introduction

It has taken several years since Sony's introduction of the e-reader in 2005 before other products based upon reflective displays like electronics shopping labels and outdoor digital information displays at bus stops slowly appeared in our daily life. This leads first to the question: what are the requirements for reflective displays to become ubiquitous in everyday life? And secondly, why is electrowetting display technology expected to make that happen?

The number of electronic displays in our daily world is continuously increasing. And ever since its invention, standard transmissive or emissive displays have been used. CRTs enabled a mass market for television in the homes. LCD and, more recently, OLED technology brought it into a mass-market where over 200 million tvs are sold per year worldwide. LCD has been a key enabler of the mobile device revolution, leading to yearly sales of over 1.5 billion smartphones word wide. LED displays have revolutionized the world of out-of-home advertisement with shipment quantities over millions of square meters per year.

E-Paper Displays, First Edition. Edited by Bo-Ru Yang.

On the other hand, reflective displays have been successful predominantly in one specific niche market segment: the e-reader. The e-reader became very popular because it created very attractive additional value for its user. The user suddenly had access to thousands of books at the tip of the finger that could be enjoyed in the same way as before. To enable that, the e-reader had to be very low power, and the display had to be comfortable to the human eye for using it for many hours in the same way as a text created by ink on paper are used without limitations. To pick a few examples: that implies readability in the sun, at the beach, and in low light conditions like the daily commute in the metro. Recently, reflective displays appeared in other markets like electronic shopping labels and bus stop information displays. From these market success stories, requirements for the successful introduction of a product based upon a reflective display technology in a specific market segment can be deduced and grouped into six categories.

9.2.1 Reading or Viewing Experience

First of all, the display should create a pleasant viewing or reading experience. It has been described [3] that human perception of a reflective display relates to two key characteristics: brightness and contrast. The perceived brightness is directly related to luminance, which is the amount of light coming from the display surface. Figure 9.1 gives the relation between lightness, the perceived white level for a human, versus the reflection of light from a perfectly reflecting surface. Note that the graph shows that humans are quite insensitive to different white levels but very sensitive to differences in the amount of black. To experience a high brightness, the display needs to have high luminance. Luminance is expressed in candela/m^2 or nits.

Contrast relates to the difference in lightness. Many definitions are possible for contrast. Mostly contrast ratio is used, being the luminance ratio, or the amount of light reflected between white and black [1, 3].

In addition, people are used to and expect a certain display of sharpness for clarity and detail of images. This user expectation translates into a requirement for pixel pitch depending on the distance between the observer and the display. The human eye has a discerning resolution of 30 line pairs per degree, which translates to a distance of 3 m as a limit to discern individual pixels of 1 mm. This leads to three typical pixel pitches necessary depending on three application distances. First, indoor devices like handheld and TV devices with a reading distance down to approximately 50 cm; secondly, information on street signs to be read from a standing or walking distance up to a few meters and thirdly, information to be read when passing from a larger distance of several tens of meters, for

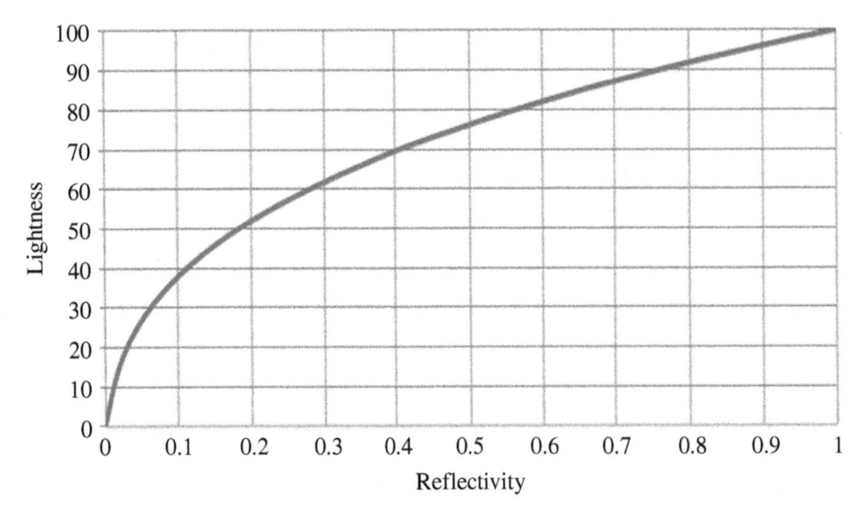

Figure 9.1 Relation between lightness, the perceived white level, and the percentage of light reflected from a perfectly reflecting surface [8].

example when driving in a car passing a shop or gas station. Based upon the human eye resolution, the pixel pitch should be in the order of a tenths of millimeters for the handheld; millimeters for the street furniture; and can be 10 or more millimeters for the road advertisement and information signs to perceive a smooth and high-quality image or text.

9.2.2 Full Color

Any newly introduced reflective display is expected to have the possibility of displaying high-quality colorful images. The perceived quality of color depends on the three prime parameters defining a color. These are: lightness, also referred to as "value,"; hue, also referred to as "color,"; saturation, also referred to as "chroma."

Together these define the color gamut, quantified, for example, by CIE 1976 with the parameters L*, a*, and b*. [9] Figure 9.2 gives Munsel's orb representation in which the north pole corresponds with pure white, L* = 100, the south pole with pure black, L* = 0, and the northsouth axis with gray scales from white to black. The equator represents all the different hues, the full saturated colors.

In reflective displays, colours can be created either by a diffractive approach, an RGB filter, or a CMY filter approach.

Heikenfeld et al. [1] have explained how in a reflective display, principally, only 1/3 of the maximum saturated color can be created with an RGB filter approach. The fundamental reason is that RGB filters cover a part of the surface and reflect only a specific part of the light spectrum. CMY is a much better approach for a reflective display because, in CMY, the color selection is made by subtraction. The three layers, cyan, magenta yellow, are positioned behind each other, and each layer filters out a specific part of the light spectrum. All incoming light contributes to reflection of the light with the defined spectrum.

This fundamental difference between RGB and CMY is illustrated in Figure 9.3 [8]. Whether reflecting a specific color or white, a CMY approach yields three times more light than an RGB approach.

This means that within the three parameters defining the colour space, as represented by Munsel"s orb, high lightness and saturated colours, can only be realized using a CMY approach. The third parameter, the hue will be defined by the quality of used materials like dyes or pigments.

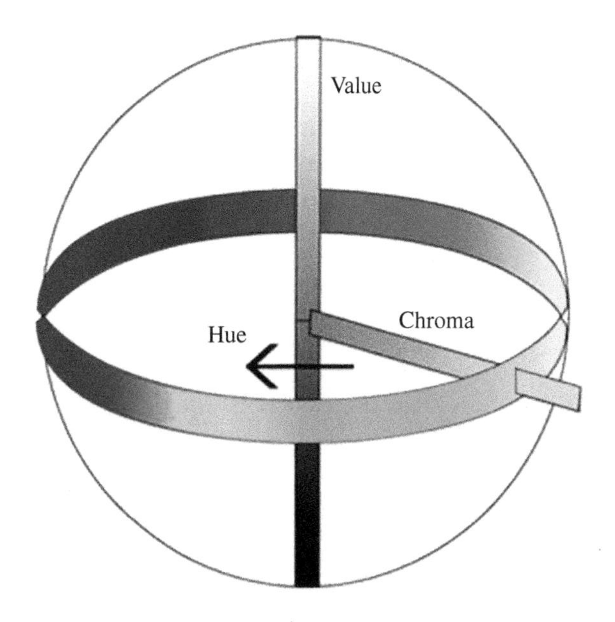

Figure 9.2 Munsel representation of the color space, with pure white at the north pole, and pure black at the south pole, and the hues along the equator. *Source:* By tsiaojian_lee – own work, public domain.

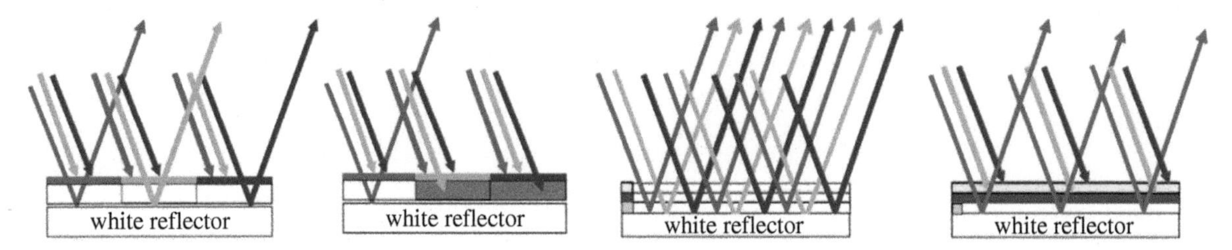

Figure 9.3 From left to right respectively: creation of white (most left figure) and red (second left figure) with Red, Green, and Blue color filters and creation of white (third figure) and red (most right figure) with Cyan, Magenta and Yellow color filters. RGB filter sequence is from left to right: Red, Green, Blue. CMY filter sequence is from top to bottom: Yellow, Magenta and Cyan. With RGB color filters 1/3 of light can be reflected, whereas with CMY filters 3/3 of light can be reflected [8].

9.2.3 Video Content

Video, the capability to display consecutive images is defined in frames per second, fps.

The eye can distinguish up to 12 individual images per second, and a frame rate of minimally 25 to typically 50 fps for HDTV quality content is required, which implies that the pixels in a reflective display have to change their status on a time scale of milliseconds.

9.2.4 Ultra-Low Power Consumption

Above, it was argued that battery lifetime and thus low power consumption are key parameters in the perception of the value of a reflective display device. Reflective displays can be a game-changer in many market segments because of their energy consumption. For example, common LCD displays will typically use 250 W/m^2, Out-of-Home LED displays can use up to 800 W/m^2 in the full sunlight to be readable for the targeted audience. Reflective displays will use a factor of 100 or less energy because they don't need to emit light. They need to reflect light.

9.2.5 Fitness for the Application

A reflective display has a key advantage in the outdoors compared to emissive displays due to its brightness, contrast for readability in daylight, and ultra-low power consumption that enables long operating time off-grid. These strengths have to be matched with the other outdoor application requirements to make the potential product successful. Excluding sparsely inhabited area's outdoor products still need to withstand rain, dust, wind and operate under large temperature variations, from temperatures below −30 °C up to above 50 °C even without taking into account additional induced heat from solar radiation. In addition, industrial applications will require compliance with industry standards.

9.2.6 Priced for the Application

Finally, the cost of manufacturing the product has to match with the value of the product as perceived by the customer. New display technology has a head start when it can make use of existing factory infrastructure and can make use of mature production processes that have been developed for high volume flat panel display manufacturing.

9.3 The Promise of Electrowetting Displays

Before going into a detailed discussion on how to successfully apply electrowetting, this section shortly summarizes the potential of the electrowetting display with respect to the parameters as discussed above.

The basics of the electrowetting switch are easily understood in the experiment, as indicated in Figure 9.4 : A photo is shown of a droplet of water that has been placed on a hydrophobic surface. A water droplet does not wet such a surface when hydrophobic. Only when a certain voltage difference is created between the water and an electrode below the hydrophobic surface, does the water droplet wet this surface, as shown on the right in Figure 9.4. This wetting is caused by the fact that a driving force exists in the presence of an electric field to flatten the droplet by storing energy in a capacitor created between the flattened droplet and the base electrode such that the sum of the surface energy and the electrical energy of the system is optimized.

Reflective displays can be designed based on this principle. Figure 9.5 indicates the cross-section of a stack build-up from sub-millimeter area's sandwiched between two glass plates. Each glass plate has a conducting ITO layer. The ITO of the bottom electrode is covered with a hydrophobic layer. The cells are filled with two liquids, predominantly with a colorless electrolyte, a polar solvent, and a small amount of a colored oil. Without electric fields, the surface tension of the electrolyte-hydrophobic surface interface is high. The oil will wet the hydrophobic layer with a thin layer and, create a colored layer in each cell, as illustrated on the left. In the presence of a voltage difference between the ITO layers, and thus an electric field between the polar solvent and the ITO layer below the hydrophobic layer, the polar solvent will wet the hydrophobic surface, indicated in Figure 9.5 on the right, and make the stack transparent. When a white reflector is mounted behind the stack, ambient light reflects from this reflector, and the observer will see a white surface. This is demonstrated in Figure 9.6, where a section of such electrowetting display is shown in a closed and open state. Either the colored layer absorbs the light, creating, in this case, a magenta color, or the colored oil layer is broken up into small droplets, and the system becomes transparent, thereby reflecting the white light from the reflector. Such electrowetting displays can combine high

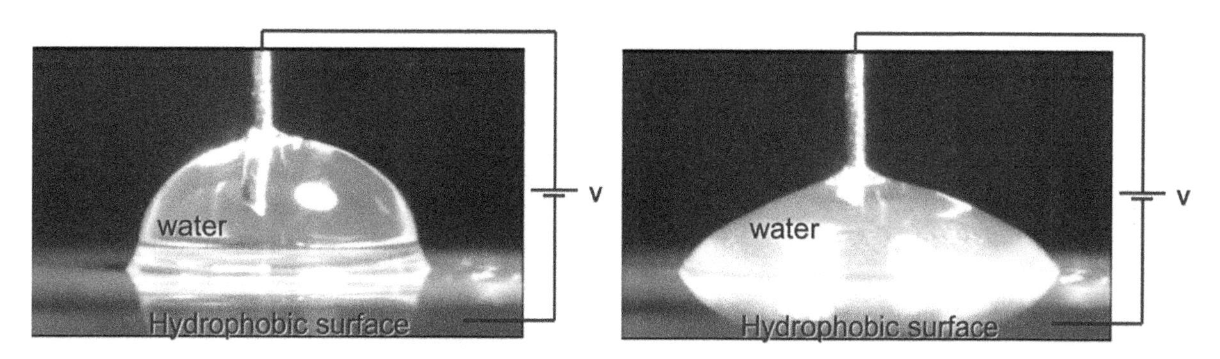

Figure 9.4 A droplet of water on a hydrophobic surface does not wet the surface (left) unless a voltage is applied between the water and the electrode below the hydrophobic surface (right) [8] D. Oostra, 2017 / Etulipa.

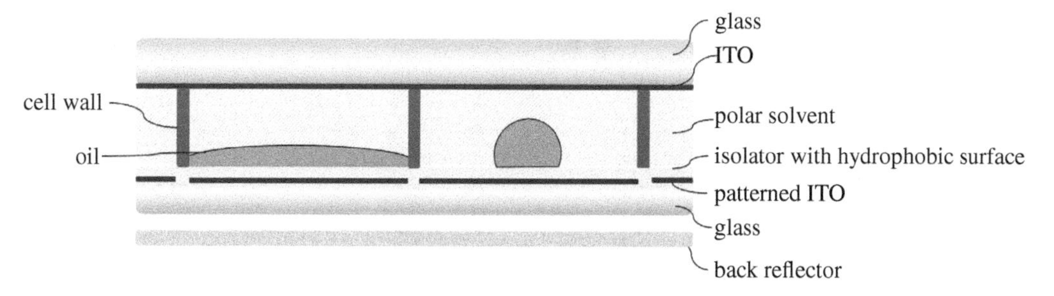

Figure 9.5 (left) A layer of oil on a hydrophobic surface wets the surface unless (right) a voltage is applied between the electrolyte and the patterned ITO electrode below the hydrophobic surface, which results in the oil contracting in a liquid droplet (right).

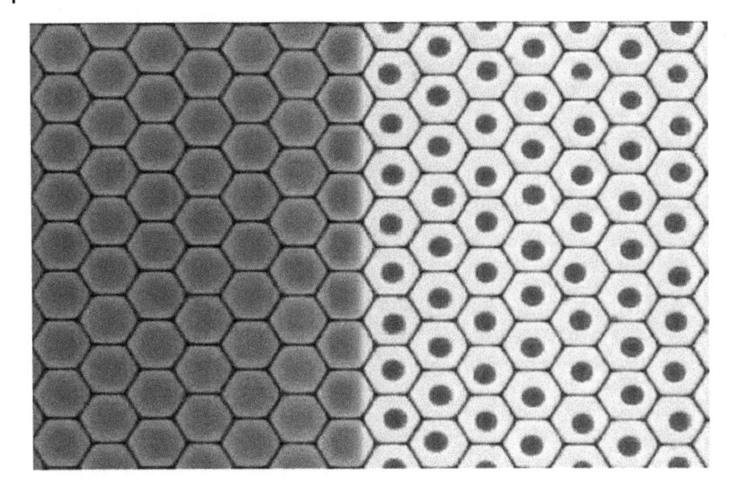

Figure 9.6 Electrowetting display, partly in closed state (left) and partly in open state (right). One pixel can be one cell, but one pixel can also be presented by many cells depending on the structuring of the underlying conductive transparent layer [8].

brightness with excellent contrast. Early demonstrators, made by Etulipa, illustrating the capabilities of the technology are shown in a video [10], of which a still is shown in Figure 9.7.

A full color electrowetting display is created by coupling three stacks with cyan, magenta, and yellow. Specific colors are created by switching the cyan, magenta, and yellow layers independently, as illustrated in Figure 9.8. For reference, Figure 9.8 also has the color wheel with cyan, magenta, and yellow as primary colors. Gray scales are created by switching layers to a partially open state. The in this way, made color demonstrators, as for example shown in Figure 9.7, demonstrate the color capability in a reflective display.

Switching between the transparent and closed light-absorbing state is caused by moving a colored fluid. Spreading, a droplet into a layer typically goes with a speed of many cm's/sec. With pixel sizes of a hundred micrometer, this implies switching speeds below 10 milliseconds, thereby enabling the possibility of a high-speed video with frame rates over 50 fps.

The electrowetting switch is caused by a change in voltage. Except for leakage currents, there is no driving current. This is the basis for extremely low power consumption in such displays. Figure 9.9 shows an Etulipa manufactured electronic changeably copy board (ECCB) with dimensions of $4 \times 8\,\mathrm{ft}^2$. Approximately 1/3 of this surface

Figure 9.7 A video still, Ref. [10], demonstrating high contrast in a reflective electrowetting display in the top right, and full saturated colors in the reflective electrowetting display in the centre under sunny weather conditions. *Source:* [10] Video Omroep Brabant T.V. Etulipa youtube channel.

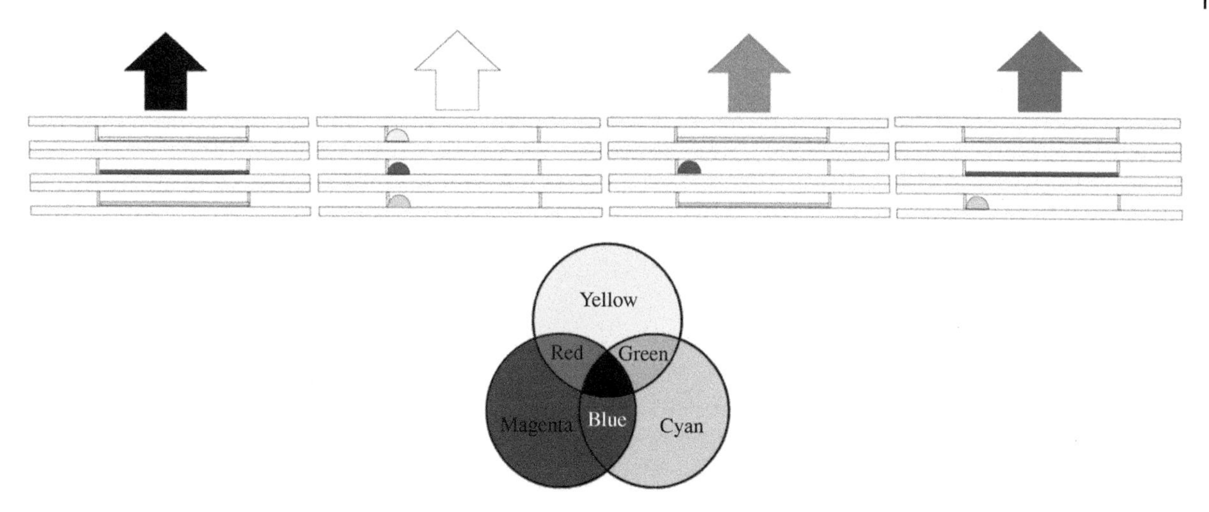

Figure 9.8 Three optically coupled electrowetting display bipanes can create any color [8]. Schematic indication on how to create any color with a CMY filter approach from white light. Filter sequence: Yellow (top), Magenta (middle) and Cyan (bottom) From left to right: 1) all closed, implies all light absorbed, leads to black; 2) all open, implies all light reflected, leading to white; 3) magenta open and cyan and yellow closed leading to green; and 4) cyan open and magenta and yellow closed leading to red. The color wheel shows how to create all color options using the cyan, magenta and yellow primary colors.

Figure 9.9 Etulipa's reflective electronic changeable copy board (ECCB) with electrowetting displays with dimensions of 4 ft × 8 ft running with a power consumption of less than 3 W. Energy for operation is provided by a solar panel with battery. *Source:* Photo made by Etulipa.

is an active area in the form of electrowetting displays. This ECCB in which the electronics have been optimized for lower power consumption consumes less than 3 W in operation, which translates to power consumption of less than 3 W/m^2 of the active area! The system can easily be run from an energy system consisting of a small solar panel and a storage battery.

In conclusion, the proposition of electrowetting displays is very attractive because they deliver bright-full-colors in reflective displays that can switch with video speed and are very low power. In addition, below, it will be shown that the electrowetting display architecture allows manufacturing in production lines very similar to common LCD display manufacturing lines, guaranteeing affordable product solutions for their applications.

9.4 History of Electrowetting Display Development

The electrowetting effect is the change in a solid-electrolyte contact angle due to an applied potential difference between the solid and the electrolyte. The concept to make use of this effect in a display was first published by Gerardo Beni and Susan Hackwood in 1981. In their paper, they demonstrate that the electrowetting effect can be used to develop reflective displays with very high response times and very low power consumption [11].

In the trailblazing paper of Hayes and Feenstra in 2003 [12] a simple, elegant display concept based on electrowetting was proposed. This work, performed at the Philips Research Laboratories, led worldwide to evaluate electrowetting technology for the application in reflective displays. A physical model describing the electro-optic behavior followed in 2004 [13]. Underlying physical wetting, and spreading mechanisms are explained [14] by Bonn et al. and by Mugele and Baret [15]. In 2006 the Philips group spun out of Philips with support of venture capital as the company Liquavista focusing on the development of reflective color e-paper video screens.

Parallel to the trajectories of developing electrowetting technology for application in displays, Berge et al. [16] developed electrowetting technology for application in liquid lenses since the nineties. The company founded by Berge in 2002, Varioptic, successfully developed and introduced into the mass market lenses that change focus based upon electrowetting effect thereby demonstrating the maturity of electrowetting technology in optical applications much earlier than in display applications.

In 2008, Blankenbach et al. reported on a bi-stable device based upon electrowetting and introduced the concept of a droplet-driven display [17]. The work was focused on large pixel displays for industrial applications like machine indicators and or outdoor displays. The technology was commercialized via Advanced Display Technologies Deutschland GmbH (ADT) [17].

At the same time, in 2008–2009, the Industrial Technology Research Institute (ITRI) in Taiwan presented their work on electrowetting technology. The work culminated in the presentation of a 6 in. and a 14 in. transparent electrowetting display [18, 19].

Also, in 2009 Heikenfeld et al. at the University of Cincinnati described the concept of a reflective device based upon the electro-fluidic effect, where fluids flow through orifices induced by the electrowetting effect [20]. Commercialization was started by creating the company Gamma Dynamics in 2011.

Electrowetting was evaluated following these pioneering research activities in many industrial laboratories, like AUO, Canon, Fuji, Motorola, Sony, and Xerox. Feenstra summarized a patent overview [5].

Independently from Liquavista in 2006, a small group of the original Philips team created Miortech, later rebranded to "Etulipa." This team focused on industrial applications for electrowetting display technology. In 2012 a dimming mirror with display capabilities was introduced for the automotive market as described in Feil's patent [21]. Subsequently, the Etulipa team has developed reflective out-of-home displays based on electrowetting display technology. The first product, an electronic changeable copy board, was installed in the field in 2019 [22]. This is a key milestone because it is the first time an electrowetting display has been validated for commercial usage in the field.

In the same time, the Institute of Electronic Paper Displays at the South China Normal University in Guangzhou (PRC), founded by Prof. Zhou Guofu, created a team to develop electrowetting display technology. Demonstrators were shown to the general public at the SID in 2017 [23].

In summary, after the publication of the work of the Philips team Hayes & Feenstra in Nature[12][5], many groups in the world evaluated the potential of the technology for commercial applications with many impressive results. The market of handheld e-readers grew fast in the 2007–2013 period, but the electrowetting display technology could not yet create inroads. Etulipa chose a different product market segment and developed electrowetting displays for the Digital Out of Home Display market, which resulted in the first product in the field in 2019. The key for this market success is fitness for the application.

9.5 Electrowetting Cells

A two-dimensional display consists of an array of individual pixels. The electrowetting display consists of a two-dimensional array of cells containing a colorless polar solvent and colored. The detailed build-up of an electrowetting cell as used in the Etulipa displays is explained in Figure 9.5. The cell has a top glass plate, further referred to as superstrate with a transparent ITO electrode. Below this transparent electrode, a transparent polar solvent is present to create an electric potential at the hydrophobic surface. Below the hydrophobic surface, an electric insulator is present above the substrate ITO electrode, which is deposited on the bottom glass, further referred to as the substrate. The polar solvent is transparent; the dye that absorbs a well-defined section of the light spectrum is absorbed in the oil that is present as a thin layer, creating an opaque state. When a voltage is applied between the superstrate ITO and the substrate ITO, the polar solvent wets the hydrophobic surface which causes the oil to contract into an oil droplet, thereby creating a transparent state. Note the difference when comparing Figure 9.4 and 9.5. In Figure 9.4 the droplet is a polar solvent spreading when voltage is applied. In Figure 9.5 the droplet is a non-polar oil that is created because the polar solvent forces the oil to contract.

Without any voltage present, the surface tension of the polar solvent with the hydrophobic surface is larger than the surface tension of polar solvent with oil plus the surface tension of oil with the hydrophobic surface:

$$\gamma_{oil,ps} + \gamma_{oil,s} < \gamma_{ps,s}$$

or using the spreading parameter:

$$S = \gamma_{ps,s} - (\gamma_{oil,ps} + \gamma_{oil,s}) > 0 \tag{9.1}$$

γ is the surface tension for respectively the oil-polar solvent (oil, ps), oil-surface (oil,s), and polar solvent-surface (ps,s).

Two items to be noted:

1) **Gravity**. Mugele and Heikenfeld [6] report on experiments with the possible effects of gravity and concluded that droplets have to be in the order of over a few millimeters before gravity plays a role. In displays, the individual cells are typically below 0.5 mm, therefore the effect of gravity can be neglected.
2) **The effect of the cell walls**. Note that when the cell walls have a surface tension equal or higher than the polar solvent's surface tension, the cell walls can be neglected. The polar solvent will always wet the walls. When the surface tension of cell walls becomes less than that of the polar solvent, that may play an undesired role. Depending on the external conditions, the oil may experience a driving force to creep up the cell wall surfaces, influencing the transmission. Etulipa has extensive knowledge in optimizing surface tensions, as analysed and evaluated by Marra [24]. Figure 9.6 illustrates a situation that does not experience this creeping up along the cell walls.

When an electric field is introduced over the insulator, a driving force exists that creates wetting of the hydrophobic surface by the polar solvent. This forces the oil to contract into a smaller volume of droplet shape. The contact angle of the droplet is determined by the Young-Lippmann equation, deduced for example in references [6, 25] which leads to a voltage (U) dependent wetting angle θ between the droplet and the surface given by

$$cos\theta(U) = \cos\theta_Y + (c_d/2\gamma)U^2 \tag{9.2}$$

$$c_d = \varepsilon\varepsilon_0/d \tag{9.3}$$

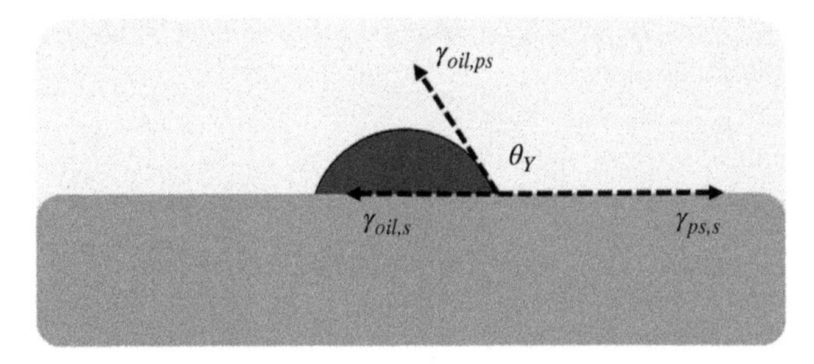

Figure 9.10 Illustration of equation of Young-Dupré: the contact angle, or wetting angle, is a resultant of the equilibrium created by the three surface tensions of the polar solvent-oil, polar solvent-surface and oil-surface interfaces.

c_d is the capacitance per unit area between the droplet and the substrate electrode, where d is the thickness of the insulator between the polar droplet and the substrate electrode. $\gamma = \gamma_{oil,\,ps}$

The angle θ_Y is determined by the Young-Dupré equation for droplets on a surface, as can be deduced from Figure 9.10:

$$\cos\theta_Y = \left(\gamma_{ps,s} - \gamma_{oil,s}\right) / \gamma_{oil,ps} \tag{9.4}$$

which value reaches −1 when the oil layer covers the hydrophobic surface.

9.6 Capabilities for Black and White

The quality of a reflective display cannot be defined in standards as created for emissive displays because the ambient light defines the luminance and contrast of a reflective display. In contrast, ambient light is considered absent during an emissive display measurement. Much better is to quantify the quality similarly to paper, using printing standards [1].

Three standards have evolved to describe the white level and black level on paper in printing: firstly, SNAP (Specifications for Newsprint Advertising Production) [26] and SWOP (Specifications for Web Offset Printing) that evolved into GRACoL [27]. Table 9.1 gives the values Lightness values according to the definition of CIE 1976 for acceptable white level and black level for these three standards, and related reflectivity level as deduced from Figure 9.1.

Table 9.1 Comparison of Lightness level and related luminance as given in SNAP, SWOP and GRACoL standards. N.B. For specific comparison measurement conditions defined in the mentioned standards have to be followed.

	L* black	L* white	Approximate R level black (%)	Approximate R level white (%)	Contrast ratio
SNAP 2009	30	82	6	60	10
SWOP 2007	19	90	3	76	25
GRACoL 2013	16	95	2	89	45

9.6.1 Black

The black level will be determined by the amount of light absorption in the oil layers. This absorption is determined by the thickness of the oil layer and its extinction coefficient α according $I = I_0 e^{-} \alpha^{l}$Where I_0 is the original light intensity, and I is the light intensity after transmission through the layer oil with thickness l. In the linear regime, the extinction coefficient is proportional to the concentration, c, of dye molecules or pigment particles in the oil solution, as expressed using Beer-Lamberts law as

$$I = I_0 e^{-\epsilon c l} \tag{9.5}$$

where ϵ, the molar absorptivity, is a constant indicating the effectiveness of the absorption per molecule or particle for dye and pigment, respectively. In the reflective display build up as in Figure 9.5, the light will transverse twice the oil layer; thus the reflectivity R will be equal to

$$\mathrm{R} = \left(\mathrm{I}/\mathrm{I}_0\right)^2 \tag{9.6}$$

For example, the SWOP requirement L* = 19 requires a Reflectivity 0.03. In a completely transparent system, this can be reached with an oil thickness, l, of approximately 4.5 μm using typical assumed dye molar absorptivities and concentrations. Note that a front glass surface without any measures to reduce Fresnel reflections reflects minimally 4%, thereby limiting possible black levels and possible contrast.

9.6.2 White

Two contributions limit the possible white level: the size of the droplets and the area covered by cell walls, although the effect of cell walls on the performance is relatively small, as will be argued below. In a first approximation in the black state, the volume of oil V equals the cell area multiplied with the oil layer thickness l. Note that in a more precise calculation of the oil volume, as done by Marra [24], the oil layer may have a varying thickness over the area depending on the influence of the surface tension of the cell walls on the oil surface shape and therefor is better be described by the shape of a spherical cap instead of a layer.

This oil volume will be in the droplet shape in the open state. To understand the performance, it is acceptable to consider the droplet as a half-sphere. This implies a contact angle $cos\theta(U) = 0$. The volume of a half-sphere with radius R is

$$A.l = V = \frac{2\pi}{3} R^3 \tag{9.7}$$

The spherical cap, approximated by the half sphere, covers an area $\pi.\ R^2$. The graph in Figure 9.11 indicates the relative reflectivity due to fraction of area not covered by the oil droplet for varying area size of the cell and three oil thicknesses. The transparency of the layer will be equal to 1- $\pi.\ R^2/$A. Obviously when the oil layer is thinner – by using a dye with higher molar absorptivity and with a higher concentration of dye – the area covered by oil droplets in the open, transparent state will be lower leading to higher white level. In addition, when the cell size increases, the possible white level will also be higher because relatively less area is covered by the droplets. To estimate the Reflectivity in the white state, two extremes can be considered:

i) The reflective layer is present immediately below the droplets. The system's reflectivity is proportional to the open area = 1- $\pi R^2/$A.
ii) The reflective layer is far below the droplets. In that case, the light will pass twice through a layer with droplets, and the system's reflectivity is proportional to (1- $\pi R^2/$A)^2.

Case i) is relevant for the laboratory experiments as reported by Hayes and Feenstra in which a white reflector is directly used to deposit ITO and define the pixel electrodes [12] and is relevant for related product designs with

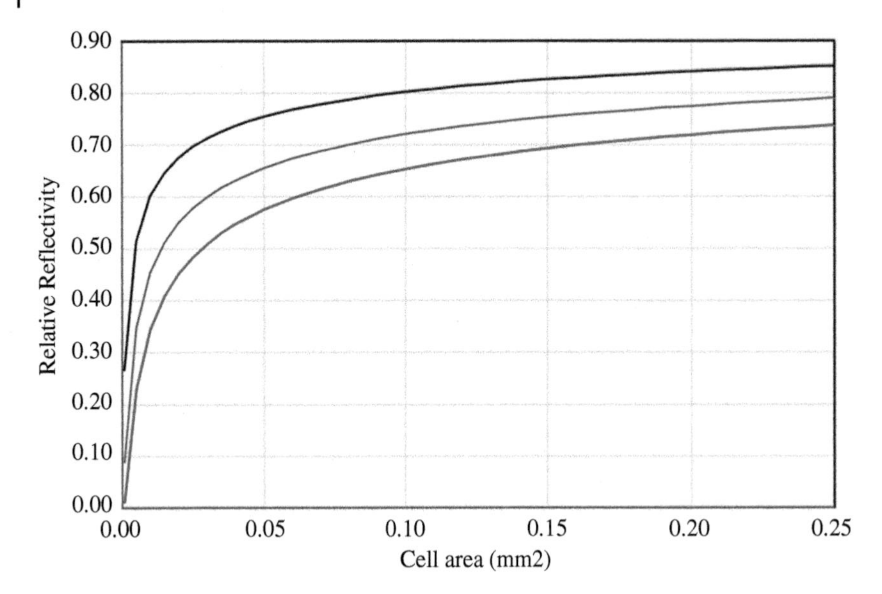

Figure 9.11 Calculated relative reflectivity as function of cell area under the assumption of a 4 μm (black, top curve), 7 μm (blue, middle curve), and 10 μm (red, bottom curve) oil thickness in closed state. Reflectivity refers to light reflected in the part of the spectrum absorbed by the oil. Relative reflectivity equals 1 when colored oil is absent. The relative reflectivity is calculated based upon assumption ii, where the reflective layer is far below the droplets.

the reflective layer in the stack [28]. Such an architecture does not allow any transparent display or a display with a backlight. In other words, the solution is reflective but cannot be transparent or transflective, as can be seen from Figure 9.12.

A commonly used architecture is as follows: A substrate glass with a thickness of typically 300–500 μm is directly covered by the electrode material, and the isolator with a hydrophobic surface layer. A white reflector is positioned below the substrate glass, as sketched in Figure 9.12(left). In the transparent, or white, state the hydrophobic surface is partially covered with the oil droplets. The pitch between the oil droplets is equal to the cell pitch, typically between 100 and 500 μm, whereas the substrate glass thickness is typically in the order of 300–500 μm. Taking into account that light comes in from any angle and the observer can look at any angle, the reflectivity is worst-case described by situation (ii).

Although this chapter focuses on reflective displays, the electrowetting display architecture also enables the creation of transparent displays and transflective displays. The latter creates the option for a backlight in

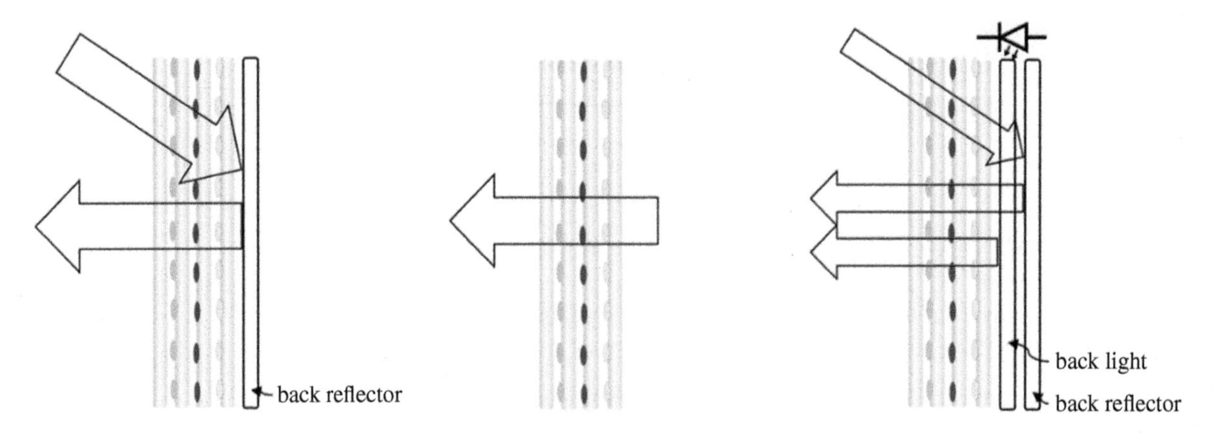

Figure 9.12 Illustrations of electrowetting display architectures to create reflective (left), transparent (middle) and transflective (right) displays.

conditions without any ambient light. Figure 9.12 gives schematic illustrations of these three options. In a reflective display, Figure 9.12(left), the light will traverse twice the oil layers of the display before reaching the observer. In a transparent display, Figure 9.12(middle), the light traverses the oil layers once. In the transflective display, Figure 9.12(right), the light from the backlight also traverses the oil layer only once. This transflective case requires that the back reflector separate from the electrowetting stack, not build inside the stack.

The cell design, as shown in Figure 9.6, as used in Etulipa's product designs, has a hexagonal shape and lateral pitch of 500 μm with an area, A, of approximately 0.2 mm^2. Under the assumption that in the open state, the oil contracts into a droplet with the shape of a half-sphere, the sphere radius will be approximately 80 μm, covering an area of 0.02 mm^2. For the cases (i) and (ii), this results respectively in reflectivity of 92%, or lightness L of 97%, and a reflectivity of 85%, or a lightness L of 94%, passing SNAP and SWOP and being very close to the GRACoL requirements.

9.6.3 Cell Pitch, Pixel Pitch and Effect on Maximum White Level

The above estimations are based upon an electrowetting cell with a pitch of 500 μm or 50 cells per inch. This is independent of the pixel pitch. The pixel is defined as the smallest electrically addressable unit and can, of course, be larger than an electrowetting cell, as illustrated in Figure 9.6.

Note that a finer cell pitch implies more oil droplets per area in the open state, which reduces the maximum white level. Also, the area covered by cell walls increases with increasing resolution, limiting the lightness either in the white or in the black level. The black level will be limited when the cell walls have highly transparent characteristics. The maximum white level will be limited when the cell walls have non-transparent characteristics.

The choice of dry film resist thickness will determine the possible cell wall thickness, which is a relevant parameter with respect to maximally possible transmission target value. The minimum cell wall width is determined by the possible aspect ratio between cell wall height and width in manufacturing. A high cell wall allows a larger oil droplet, which allows designs with larger cells, thus less droplets per area in the open state, and thus better transmission. Aspect ratio's up to 15 have been reported in dry film resist processing [29], implying that designs with dry film resist of 100 μm thickness can have a thickness less than 10 μm. Electrowetting cells with a relatively low oil volume in the cells, implying designs with a small cell pitch, can be designed with a dry film resist of, for example, 50 μm, which means cell walls with a thickness of less than 5 μm. The advantage or a relatively smaller cell is the faster switching speed, relevant for high speed video applications.

9.7 Capabilities for Video and Color

9.7.1 Video

The minimum time between closed and open state in an electrowetting cell is determined by the speed of the liquid contact line over the hydrophobic surface, which is easily over 1 cm/s [11] and can be up to 10 cm/s [20]. Figure 9.13 gives an overview of reported switching times to open state and closed state as a function of pixel area without considering any details like oil volume, applied voltage, and voltage waveform, surface type, and viscosity of liquids. Data are deduced from [2, 3, 5, 12, 30–34]. In general, the time to switch to transparent is expected to be shorter than switching from a droplet into a layer. The former process is electrically induced, whereas the time to switch back to a layer is completely determined by surface tensions. We can conclude that many groups report switching speeds that enable video speed with at least 50 Hz refresh rates.

9.7.2 Color

Above it was explained how an electrowetting display using a vertically stacked architecture in which cyan, magenta, and yellow layers subtract parts of the white light spectrum creates bright colors. This approach requires the absorption spectra of the dyes in the three layers to be independent and together cover the complete visible spectrum.

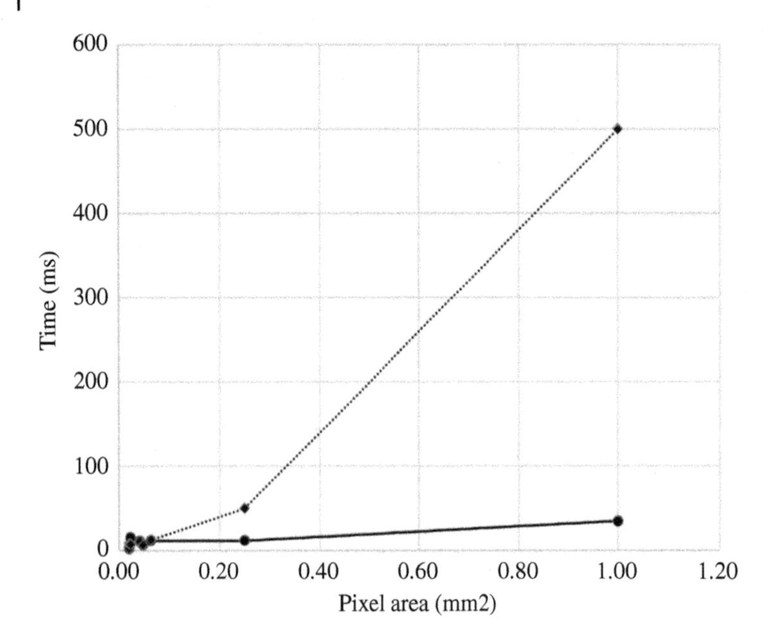

Figure 9.13 Overview of reported switching from references [2, 3, 5, 12, 30–33] and [34] interpreted by the author and summarized into a graph of switching time versus pixel area. Note that specific pixel forms will have an effect on the switching time. Dotted line connects values for time to switch from open to closed state. Solid line connects values for time to switch from closed to open state.

Furthermore, the molar absorptivity must be adequate to fulfill the required absorption in the thin layer of oil, and the dye or pigment should not lead to unwanted surfactant effects that influence the surface tensions between the materials in the cell.

Also, the lightfastness of the chosen dyes or pigments has to meet stringent criteria such that the display can withstand usage in the field for several years.

Several commercial institutions offer dyes that have been successfully used in electrowetting display development. For example, Zhou et al. [2] refer to Keystone as supplier of standard dyes that have been optimized for usage. Mitsubishi Chemical provided dyes to Liquavista that were used in their displays. Also, the South China Normal University (SCNU) group has reported on extensive dye development [35]. These dye developments led several teams independently to demonstrate bright color electrowetting displays. Etulipa has presented since 2017 demonstrators with excellent, bright colors and contrast [8]. Examples can be seen in Figures 9.7 and 9.14. In 2020 SCNU and Guohua released images and a video of high-speed full color 5.8″ electrowetting displays, shown in Figure 9.15 [36]. The ITRI team used pigments to present bright colorful high-resolution displays [19] as shown in Figure 9.16, and the Liquavista team presented high-resolution color displays, with a performance as for example shown in Figure 9.17 [30]. Figure 9.18 shows three color gamuts as reported by Liquavista [30], South China Normal University [37], and Etulipa [38], respectively.

9.7.3 Parallax

Note that the vertical stacking of the cyan, magenta, yellow layers can lead to the phenomenon of parallax. In a design indicated in Figure 9.8 with a glass thickness of 0.3 mm, the complete stack will be above 2 mm. When the pixels are much smaller than the stack thickness, parallax effects will deteriorate the possible image quality. The roadmap for optimization is clear: Feenstra & Hayes [30] suggest that the stack height can be reduced by having two dye layers in one cell. Furthermore, to optimize the thickness and lower the bill of materials, the superstrate

Figure 9.14 An example of Etulipa 10″ CMY electrowetting color tile (left) and an example of an Etulipa display build up from six 10″ CMY tiles (right). *Source:* Images made by Etulipa.

Figure 9.15 High-speed full color TFT electrowetting displays reported by South China Normal University and Shenzhen GuoHua Optoelectronics. *Source:* Photo courtesy of SCNU.

glass of the lower stack is proposed to be simultaneously substrate for the layer above. In addition, very thin glass or polymer thicknesses down to 0.1 mm can be applied. By fully using such optimizations, stack heights as low as 0.5 mm can be envisioned, allowing the design of high-resolution full color displays without parallax.

9.7.4 Application and Design

The application will drive the design. Optimization for high-speed video will require smaller electrowetting cells than architectures optimized for still images. Applications like an ECCB near a road will have high requirements

Figure 9.16 Examples of reported ITRI transparent 6″ (left) and 14″ (right) electrowetting color displays. *Source:* Photo's courtesy of Wiley Online.

Figure 9.17 Examples of Liquavista reflective electrowetting color displays. *Source:* Photo's courtesy of J. Feenstra.

on readability of still images that are switched only once per 5 or 10 seconds. In the latter case, the design can be optimized toward maximum contrast and toward maximum white level and black level instead of maximally possible switching speed.

9.7.5 Pixel Pitch and Driving Mechanism

As mentioned above, a pixel is defined as the smallest electrically addressable unit in the display. It is defined by the patterning of the ITO substrate. Such ITO area can encompass either one electrowetting cell or multiple electrowetting cells. Electrowetting cells may have spatial dimensions up to approximately 500 μm. This creates a freedom of design to optimize the substrate ITO pixel size compared to the electrowetting cell size, and

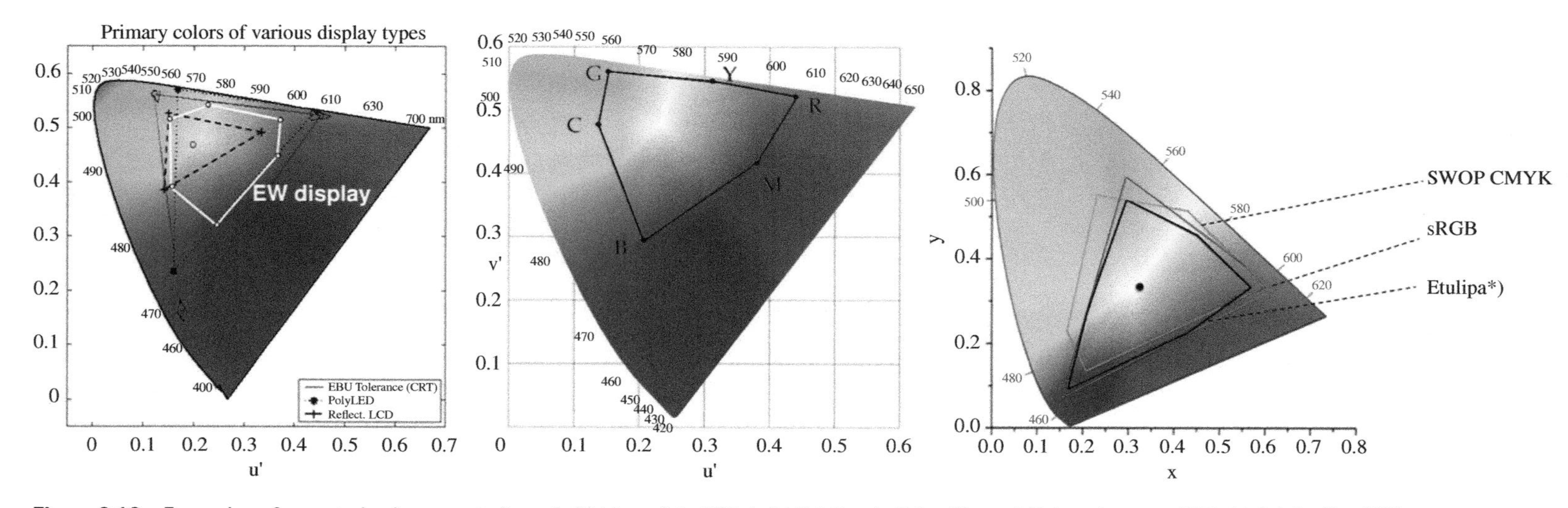

Figure 9.18 Examples of reported color gamuts from (left) Liquavista [30], (middle) South China Normal University team [37]; (right) Etulipa [38]. *Source:* Photo's of Liquavista and SCNU courtesy of J. Feenstra, and A. Henzen respectively.

subsequently, this freedom of design can be used to choose the optimum electronic driving approach. Generally, there are three ways to drive individual pixels in a display: direct drive, passive matrix, and active matrix.

Direct drive: in the direct-drive approach, every pixel on the substrate gets its voltage from a dedicated output of the driver chip. Figure 9.5 shows how the superstrate ITO and thus the polar solvent act as common. Since every pixel is connected via its own ITO address line to the driver chip, a space between the pixels has to be reserved in the active area on the substrate for these ITO lines. The approach is very cost-effective and is thus an excellent design choice for applications with relatively few or large pixels. In Figure 9.9, an example is shown of an application based upon this approach.

Passive matrix: in a passive matrix, the display uses a grid of thin conducting lines at the superstrate and thin conducting columns at the substrate. A, pixel state is defined at the interface of a specific line and column. In all reported electrowetting display designs, the polar solvent acts as an electrically conducting layer over the full display area, making a passive matrix approach impossible.

Active matrix using a thin film transistor (TFT) architecture: In a TFT design, the pixel state is determined by the TFT at the specific pixel, which is addressed via lines and columns. This solution is ideal for small pixel size and or large displays. TFT's are made by depositing semiconductor material on the display substrate, often glass. The materials are either amorphous Silicon (a-Si), poly-Silicon, or oxide materials like Indium Gallium Zinc Oxide (IGZO). A detailed overview of the benefits and disadvantages of each of these options with relevance for electrowetting is given in Ref. [39]. Key items: Poly-Si has higher mobility than a-Si but also has a higher leakage current. IGZO has much higher carrier mobility than poly-Si combined with a lower leakage current. Higher carrier mobility combined with low leakage currents enables high refresh rates combined with low power consumption and small dimensions.

The TFT covers a certain area of the pixel. This area reduces the transmittance and thus the reflectivity capability. By designing this area as small as possible, a large so-called aperture being the open transparent area in the absence of colored oil can be created. The transparency of the system can be optimized by aligning the TFT backplane with the electrowetting cell structure. A TFT made of transparent oxide material also improves the possible aperture in a TFT-electro wetting design.

S. Kwon gave in detail the process steps to realize a TFT-driven electrowetting device both with a-Si and with IGZO in the laboratory. TFT-driven electrowetting displays for the application of handheld mobile devices have been demonstrated by the Liquavista team [5, 30], the ITRI team [18, 19] and the SCNU team, as shown in Figures 9.17, 9.16 and 9.15, respectively. Fundamental studies on the driving of electrowetting for optimized response have been reported, for example, by [40–43].

9.7.6 Gray Scales

Three approaches for applying gray scales can be envisioned: (i) dithering; (ii) analog by voltage setting; and (iii) digital, by pulse width modulation (PWM) of the driving voltage.

In dithering, a grey scale is created on a certain area by setting some pixels in closed state and others to open state in that area. This approach allows only a minimal number of gray scales and it depends on the displayed content whether the resulting resolution is adequate for the application.

The voltage setting directly affects the area covered by a colored oil droplet as given in formula 2 and thus on the transparency and the white level. This is a very straightforward approach to set the gray scale level. Many studies [12, 13, 30, 31, 34, 40, 42] report graphs of transparency as function of the voltage setting. The measured curve can be used for calibration of gray scale setting versus the voltage setting.

Alternatively, a gray scale can be set by PWM. In Figure 9.19, the transparency of an active area as a function of the PWM duty cycle is shown. Duty cycle is defined as the fraction of time voltage within one refresh cycle. The curve shows a strong non-linear behavior which means that an equivalent set of gray scales can only be created by very fine steps in the duty cycle.

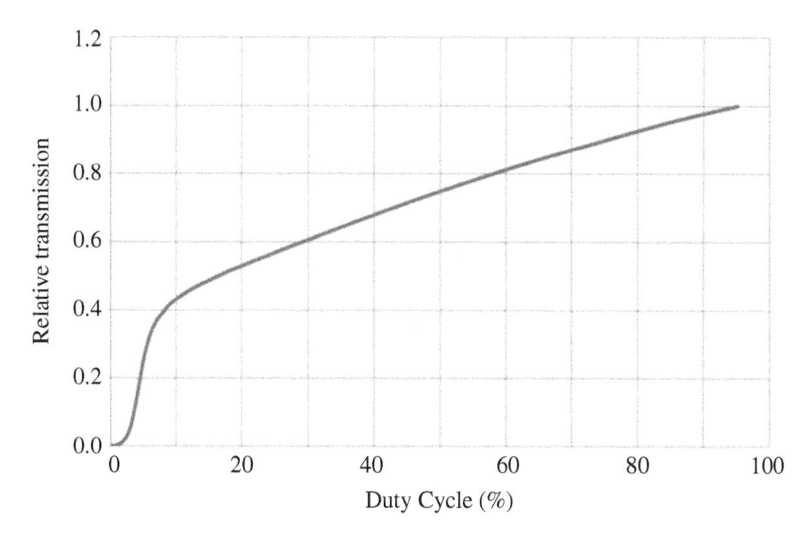

Figure 9.19 Example of light transmission as function of duty cycle of driving voltage using a pulse width modulation (PWM) approach.

The voltage and PWM have to consider the possible effects of hysteresis due to the different advancing or receding contact angles for an oil droplet at the same voltage setting [42]. Rui et al. summarized studies and a theoretical model to describe this phenomenon [44].

Electrowetting displays can be designed with a thickness of parameter d, the isolation layers in formula (3), such that the electro-effect can be created using voltages up to 30 V. That allows the usage of standard off the shelf drivers. Furthermore, above mentioned gray scale approaches can be used in parallel to optimize performance.

9.8 Driving

9.8.1 Power Consumption

A discriminating advantage of a reflective display compared to an emissive display is the extremely low power consumption which can be made to advantage in many applications. Beni and Hackwood [11] first noted that power consumption of an electrowetting display can be below a milli-Watt/cm^2. The origin of this very low power consumption is in the intrinsic nature of the electrowetting process. The oil droplets can switch by applying a voltage of typically below 25 V, without any need for driving currents. Power consumption can be further optimized by designing dedicated, very lower power driving electronics. Feenstra reported a power consumption of 4.3 mWatt/cm^2 for a display running at video rate [5].

Etulipa developed an electronic changeable copy board for off-grid applications optimized for running on a solar panel without any maintenance. The Etulipa ECCB's, as shown in Figure 9.9 have a power consumption of below 3 W/m^2 or 0.3 mW/cm^2, including driving electronics.

9.8.2 Driving Voltage

Power consumption is lowest when using a DC voltage or an AC voltage acting as a slowly alternating DC voltage, which means a frequency with a time constant much lower than the response time of the droplet. The limit for the maximum time without voltage change is caused by a secondary effect in the electrowetting cell referred to as backflow. Backflow is the effect that the contact angle between the oil droplet surface and the hydrophobic surface decreases with time reducing the effective electric field. This is observed as a sagging of the oil droplet, which implies a larger area of the hydrophobic surface is covered with oil, thereby reducing the display's transparency and contrast. The origin of backflow is in the fact that the effective electric field is reduced in a steady state because of polarization of the droplet due to free moving charge carriers in the oil. With AC driving, this backflow phenomenon can also appear as a flicker when the backflow time constants are comparable to the time constant of the AC pulse.

When using DC driving, a reset pulse with a time span of a few milliseconds, much shorter than the response time of the droplet, can eliminate the effective field reduction, and thus the backflow of the oil droplet. Implementing such a reset pulse is similar to a very light, imperceivable, gray scale setting in a PWM approach.

A disadvantage of using the approach of DC with reset-pulse is the possible occurrence of long-term polarization effects due to the mobility of charge carriers in the materials or at the interface between materials.

9.9 Architectures

9.9.1 First Generation

Hayes & Feenstra first paper [12] described a research approach in which a small pixel resolution is created by defining a cell wall structure in a polymer sheet using laser cutting that is subsequently glued onto a hydrophobic surface. After filling the boxes with oil and polar solvent, the ITO-covered superstrate is sealed on top, thereby creating boxes isolated from each other.

The Philips/Liquavista team subsequently created and developed a design in which hydrophilic cell walls are present on the hydrophobic substrate created using standard fabrication processes. A dry film resist is deposited on the hydrophobic surface. Subsequently, a pattern is defined with lithograpy, and etching is applied to create the desired cell wall structures. Variations on this approach can be envisioned; for example, processes have been described in which the hydrophobic layer is deposited after defining the cell structure [45]. A small volume of oil is created in each cell using a dipping process. Subsequently, the system is submerged in a polar solvent where the ITO-covered superstrate is mounted on top of the structure. The resulting oil layer or oil droplet in each cell is contained within its cell because of the hydrophilic nature of the cell walls. The architecture principle is shown in Figure 9.20 (left).

9.9.2 Second Generation

In the above-mentioned first generation manufacturing, a surface coated with a highly hydrophobic layer has to be patterned with a cell wall structure. However, the deposition of a film on a hydrophobic surface is a very tedious process. Effective manufacturing can be done by first activating this hydrophobic layer, for example, with an oxygen plasma process. After that plasma process step, standard film processing can, occur, and as the final step,

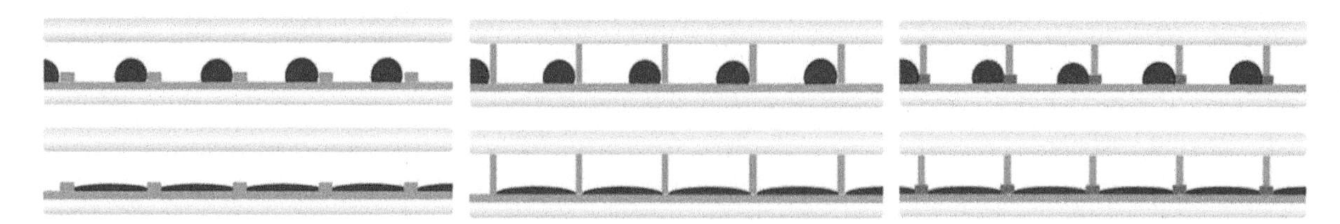

Figure 9.20 Illustration of electrowetting display architectures with above, transparent state and below, closed state of respectively Liquavista (left); Etulipa (middle); and ITRI (right).

an anneal step has to be applied to restore the original hydrophobic characteristics of the surface. In Etulipa, the quality of the restored surface was experienced to be inadequate for reproducible production. Randomly, ghosting effects were observed in the resulting display, which was explained to be due to residual damage at the surface layer after the annealing steps.

To eliminate any possibility of unwanted ghosting caused by residual damage after plasma and anneal process steps Etulipa created an architecture in which the cell walls are defined on the opposing superstrate. This second-generation architecture is shown in Figure 9.20 (middle). The superstrate consists of glass with a uniform ITO layer. It is straightforward to deposit a film on this ITO surface and subsequently pattern this film into the desired cell wall structure. The cell walls reach the substrate to keep the oil flow on the hydrophobic surface contained within the defined cell area, which automatically creates a very robust uniform spacer structure over the display.

9.9.3 Two Layers

Yun-Sheng Ku et al. [18] and Kuo-Lung Lo et al. [19] also concluded that it is challenging to consistently deposit uniformly a film on a hydrophobic layer. They noticed that the optimum materials for deposition were typically not very hydrophilic. This led to unwanted effects because the resulting cell walls did not block oil effectively from coalescing into neighboring cells, which resulted in some cells completely filled with oil, whereas other cells end up without oil, a phenomenon called "overflow." To eliminate the occurrence of this failure mechanism, the authors proposed a two-layer hydrophilic rib structure, as shown in Figure 9.20 (right). In this two-layer rib structure, the top structure has a highly hydrophilic nature. The bottom cell wall structure is used to define the oil volumes on the substrate and the top structure to eliminate overflow. Impressive 6″ and 14″ reflective electrowetting displays have been reported with this two-layer architecture [18, 19] as shown in Figure 9.16.

9.10 Manufacturing

The manufacturing process for creating electrowetting displays obviously depends on the chosen architecture. The Liquavista process to create the first generation electrowetting displays and an overview of the manufacturing steps to create a display with the double-layer structure, as used by ITRI, has been described earlier [5, 18, 19, 30]. This review outlines the process steps and the production flow to realize displays based on the second-generation architecture as used by Etulipa.

Two material flows exist in the Etulipa manufacturing line: one for the superstrate and one for the substrate, as indicated in Figure 9.21. The substrate and superstrate come together at the fill and seal station. After sealing and glue curing, test and measurement and final inspection take place before transfer to the mechanical assembly station to create assembled solutions that are shipped to the customer.

9.10.1 Preparation Superstrate

The superstrate is a sub-millimeter thick glass plate with a conducting ITO layer. Display quality glass with ITO with a defined square resistance and defined transparency can be bought from several suppliers.

Step 1) the ITO is patterned to create alignment marks such that the superstrate can be cut within microns to the required final display dimensions.

Step 2) Creation of cell structure. A dry film resist is deposited on the ITO layer. This dry film resist is patterned with the electrowetting cell wall structure, using the standard process steps of lithography, etching and annealing.

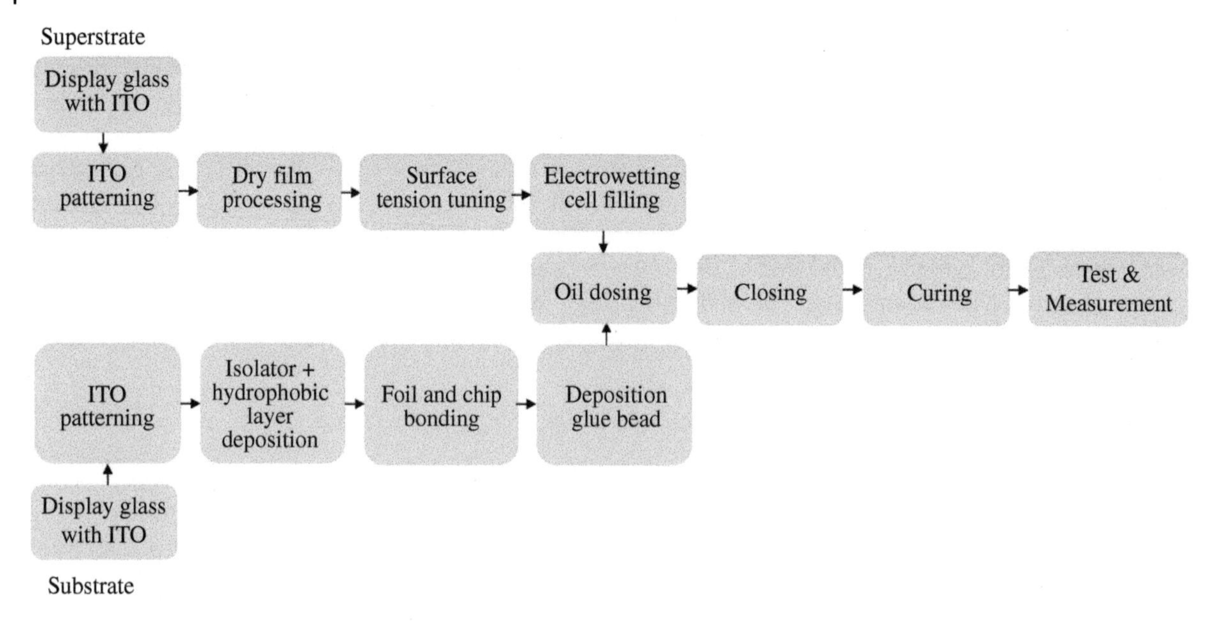

Figure 9.21 Basic steps of the process flow of Etulipa production line.

Note that either in ITO pattern or in the dry film resist pattern additional alignment marks can be introduced for aligning superstrate with respect to the substrate during the fill and seal process.
Step 3) Tuning. The surface tensions of the created cell walls are tuned, either by submerging in a chemical solution or by exposure to a plasma or UV-ozone. After surface tension tuning the superstrates can be stored as ready for fill and seal.

9.10.2 Preparation Substrate

The substrate is a sub-millimeter thick glass plate with a conducting ITO layer.

Step 1) The ITO layer is patterned to create the requested pixel structure and related address lines and switching elements.

Step 2) Deposition of the hydrophobic layer. A hydrophobic layer has to be created on the patterned ITO layer. Mugele and Heikenfeld [6] discuss several requirements and related options for this layer. The key requirements are that 1) the surface layer has to be hydrophobic and 2) the layer has to be inert to any de-composition or etching due to unwanted electrochemical processing during the operation of the display. As reported in the literature, the most common choice is a combination of a relatively thick inert barrier layer with on top a thin second layer with surface characteristics optimized for the targeted hydrophobicity level. The barrier layer can be an inorganic oxide, an organic layer made by deposition of a parylene or a combination made of stacked layers of such materials. After deposition of the barrier layer a fluoropolymer is deposited, either by spin coating or by dipping. This fluoropolymer is baked to create the hydrophobic surface layer. Two commonly used fluoropolymers are Cytop® and Teflon™ AF. Note that the total thickness of the barrier + hydrophobic surface layer has to be tuned carefully in thickness and uniformity to realize an electrowetting effect with voltage settings using off-the-shelf drivers, as can be seen from the formula's 2 and 3 above.

Step 3) Baking. The elegance in the generation 2 architecture is that after baking the hydrophobic layer, the substrate is ready for the fill and seal. No processing has to take place on the hydrophobic surface.

Step 4) Electrical connections. Locally the isolator is removed using a lift-off process step, and flat foil cable and driving chip are bond onto the designated ITO patterned area.

The substrate is ready for bipane manufacturing.

9.10.3 Bipane Manufacturing

Step 1) Filling. The electrowetting cells are filled with the transparent polar solvent. Options for the polar solvent are discussed in Mugele and Heikenfeld [6]. Examples that work very well are ethylene glycol, propylene glycol, or mixtures including these. The filling of the electrowetting cells on the superstrate can be done using a spinning process or a dipping process. The exact content in the cells is tuned using controlled time and temperature process steps to allow a certain amount of polar solvent to evaporate out of the cells. In a controlled environment, evaporation is an extremely uniform process, which leads to a very uniform fill level in each of the cells on the superstrate. The superstrate is now ready for the sealing.

Step 2) Deposition glue bead. The glue bead is placed preferably on the substrate to avoid interaction with the filling process. The glue must adhere to two fundamentally different surfaces: a hydrophilic superstrate and a hydrophobic substrate. To optimize the bonding quality after curing, the hydrophobic surface at the position of the glue bead can be made hydrophilic either by locally applying a plasma or UV-ozone process or by locally removing the hydrophobic surface material before deposition of the glue bead.

Step 3) Oil deposition. A few oil droplets containing the dye material are positioned on the surface of the substrate. The hydrophobic nature of the surface has to be such that the droplets do not wet the surface in the air but do wet the hydrophobic surface when in contact with the polar solvent. In other words, for successful filling using the Etulipa fill approach, the choice of materials has to be such that when in formula (1) material ps is air, then $S < 1$, and when ps is a polar solvent, $S > 1$.

Step 4) Closing. The superstrate is placed above the substrate and slowly brought in contact with the oil droplets. Upon contact between the polar solvent--filled cells, and the oil droplet, the oil spreads out over the surface, filling each cell's available space. This process is very similar to wet optical bonding as applied standardly in volume display manufacturing. When the glue bead has become fully in contact with the superstrate, the superstrate is slowly pressed until all excess oil has run out via the local stops at the corners of the formed bipane.

Step 5) Curing. The now compressed glue bead between substrate and superstrate is exposed to UV curing step. Subsequently, the bipane can be removed from the fill and seal station. The bipane is now ready for inspection.

9.10.4 Test and Measurement

For the observer or user of displays, there will be three types of requirements

1) The optical quality per pixel or area of pixels
2) The uniformity between pixels or pixel areas within the display
3) The uniformity between displays

The optical testing of reflective parameters in a high-volume production environment is rather challenging [46]. The measurement result of a reflective depends on three parameters: the incoming light, the device under test, and the measurement detector. Such measurement set-ups are available but mostly optimized for lab environments.

In a production environment, the measurement tools need to be fast, with easy-to-learn user interfaces and need to create data that can be transformed straightforwardly and unambiguously into information to control the product quality in the production line.

Because of these reasons, Etulipa chose for their production lines to use a transmissive measurement approach to record the performance of the bipanes. Transmissive measurements are quantitative, easily calibrated, and can be executed fast, independently of external lighting conditions. On every bipane coming off the production line for every pixel, the transmission value in the open and closed state is recorded, switch time to open and switch time from open to close. Results are stored per specified area of interest. The quantitative transmission measurements allow comparison of areas within the display, and comparison of displays produced over time.

9.11 Reliability

The core function of reliability analysis is to predict the product's failure rate in the field. Obviously, the conditions in the field are critical to any such prediction. Etulipa's reflective electrowetting displays are to be used in the outdoor environment, not shielded from any outdoor elements. Therefore design for reliability, highly accelerated life testing and reliability testing are fundamental parts of Etulipa's development programs.

A standardization for tests of reflective displays for outdoor application is not available. TÜV Institute standards [47] for outdoor electronics systems give useful hints for testing. Etulipa chooses to create reliability test programs following the guidelines defined in the International Electrotechnical Commission, for example, IEC 60068-2 [48], and the Military Standard Mil. Std. 810G [49]. Highly relevant for predictions of performance during several years in the field are the tests related to performance in cold, heat, damp heat, temperature change, both slow and shock, under solar radiation, under vibration, and in low air pressure. Depending on the specific customer application additional tests as defined in the mentioned standards can be applied. Etulipa electrowetting displays are standardly tested at varying humidity conditions with temperatures ranging from −40 °C up to above +70 °C. In addition they are tested in sunlight simulation chambers.

The measurements on products coming off the production line lead to a begin of life specification. The measurements of performance during and after reliability testing lead to an end of life specification. Together they describe the performance in such a way that a customer can understand what he is buying, and how it is expected to vary during its lifetime [50].

9.12 Failure Mechanisms

The most commonly observed and reported failure mechanisms in electrowetting devices relate to the below four phenomena. The probable root cause for each failure mechanism is given including the critical reliability test to determine sensitivity to the specific failure.

1) Dead pixels and bad area's
   Mugele and Heikenfeld [6] report in detail on the issue of electrochemical etching of the ITO in the substrate due to imperfect dielectric isolator or barrier between the hydrophobic surface and the ITO pattern. The etching of lines and columns may lead to dead pixels and cause gas bubbles formation in the active areas that show up as bad areas. The authors advise both on choices of barriers and on the choice of polar liquids to minimize the risks of failure.
   Critical test: high-temperature operation
   Critical parameter: non-functioning pixels and or gas bubble phenomena in the device.

2) Transparent non-switching spots in the active area
   The ITRI team refers to the failure of oil overflow [19]. Oil has moved from one cell into a neighboring cell. The original cell is void of oil and thus appears as a constant bright or transparent spot in the display. Note that oil can only flow out of a cell into a neighbor cell when parallel polar solvent flows back into that first cell. The root cause is related to non-optimally prepared cell walls. Cell walls have to be highly hydrophilic to avoid oil flowing over the walls. As a consequence, a safeguard was adopted by ITRI by the creation of a two-layer cell structure [18] to avoid this failure mechanism.
   Critical test: high-temperature operation
   Critical parameter: uniformity of display in the dark state
3) Discoloration and deterioration of switching speed due to solar radiation
   From LCD displays, it is widely known that extensive exposure to sunlight significantly deteriorates the quality of the display. A solution to avoid damage induced by solar light is using an UV-blocking layer in front of the active layer [51], which limits the transmission window to approximately 400 nm, thereby protecting dyes from exposure to harmful UV-A type radiation. Besides loss of color the deteriorating elements may lead to unwanted surface tensions in the electrowetting cell which can influence the transmission and switching behavior.
   Critical test: Solar radiation test
   Critical parameter: color of display and switch performance, time to open and close.
4) Sealant quality
   Seals need to have excellent adhesion to substrate and superstrate and are typically applied in a liquid environment. A deteriorating seal can lead to leaking liquids or evaporation of liquids, leading to transparent spots.
   Critical test conditions: high-temperature storage, temperature variation and humidity.
   For example, ITRI [19] reports high-quality seal performance under temperature shock conditions.
   Critical test: transparency in a closed and open state.

9.13 In Conclusion: Electrowetting Displays Have Reached Maturity

Electrowetting display development has been a very exciting journey. The technology progress over time can be understood using the concept of technology readiness. The European Union adopted the original NASA Technology Readiness Level scale and created nine levels as follows [52]:

Readiness level	Status
TRL 1	Basic principles observed
TRL 2	Technology concept formulated
TRL 3	Experimental proof of concept
TRL 4	Technology validated in lab
TRL 5	Technology validated in relevant environment (industrially relevant environment in the case of key enabling technologies)
TRL 6	Technology demonstrated in relevant environment (industrially relevant environment in the case of key enabling technologies)
TRL 7	System prototype demonstration in operational environment
TRL 8	System complete and qualified
TRL 9	Actual system proven in operational environment (competitive manufacturing in the case of key enabling technologies)

Figure 9.22 Bus stop in New York City with electrowetting displays indicating the time to arrival for next bus. The bus stop works off grid, powered by a solar panel. A reflective display designed by Etulipa was used to enable a low power solution. The electrowetting display has a 7 × 11 pixel architecture using a direct drive approach with pixels with 10 mm pitch. *Source:* Photo courtesy of Traffic Systems Inc. and Daktronics.

In the period 2003–2013, electrowetting display development has moved from experimental proof of concept, TRL3, to demonstration of system prototypes in the field, TRL7. Etulipa has developed and now produces electrowetting displays for application in the out-of-home digital display market. In 2019, customers have installed and tested systems in the field and qualified them for application, TLR8. Customers have chosen Etulipa's product for application in off-grid bus stop systems that deliver digital information for convenience of travelers and commuters, under all weather conditions, from fierce winterly conditions up to hot summer days. The system has been chosen in a commercial tender, and the first bus stops have been installed in the second half of 2020. The technology thereby reached TLR9. This application is shown in Figure 9.22.

Etulipa follows a product roadmap from monochrome displays for sharing information and messages readable on large distances toward high pixel resolution full color matrix displays. As an example of the latest developments, Figure 9.14 shows a full color electrowetting display matrix panel with dimensions of 44×42 cm^2 for application in full color matrix displays.

Acknowledgments

The creation of this chapter was only possible by working and learning together with the whole Etulipa team, and in particular by many discussions and white board sessions with Hans Feil, Rajan van Aar, Johan Marra, and Pieter van der Valk of the Etulipa team.

References

1 Heikenfeld, J., Drzaic, P., Yeo, J.-S., and Koch, T. (2011). A critical review of the present and future prospects for electronic paper. *J. Soc. Inf. Disp.* 19 (2): 129.

2 Zhou, K., Heikenfeld, J., Dean, K.A. et al. (2009). A full description of a simple and scalable fabrication process for electrowetting displays. *J. Micromech. Microeng.* 19 (6): 065029.

3 Bai, P.F., Hayes, R.A., Jin, M.L. et al. (2014). Review of paper-like display technologies. *Prog. Electromagn. Res.* 147: 95.
4 Shui, L., Hayes, R.A., Jin, M. et al. (2014). Microfluidics for electronic paper-like displays Lab on a Chip. *Lab Chip* 14: 2374.
5 Feenstra, J. (2016). Video-speed electrowetting display technology. In: *Handbook of Visual Display Technology* (ed. J. Chen, W. Cranton and M. Fihn), 2443–2458. Germany: Springer Berlin.
6 Mugele, F. and Heikenfeld, J. (2019). *Electrowetting Fundamental Principles and Practical Applications*. Germany: Wiley Weinheim.
7 Li, J. and Kim, C.-J. (2020). Current commercialization status of electrowetting-on-dielectric (EWOD) digital microfluidics. *Lab Chip* 10: 1705.
8 Oostra, D. (2017). Etulipa creates the opportunity for digital out of home displays to become as abundant as paper. Electronic Displays Conference, Nuremberg, March 16th 2017.
9 International Commission on Illumination (CIE). (1976).
10 Video Omroep Brabant T.V. Etulipa youtube channel. https://www.youtube.com/watch?v=D1mHKvGiPE4
11 Beni, G. and Hackwood, S. (1981). Electro-wetting displays. *Appl. Phys. Lett.* 38: 207.
12 Hayes, R.A. and Feenstra, B.J. (2003). Video-speed electronics paper based on electrowetting. Nature, Oct (2003).
13 Roques-Carmes, T., Hayes, R.A., and Schlangen, L.J.M. (2004). A physical model describing the electro-optic behavior of switchable optical elements based on electrowetting. *J. Appl. Phys.* 96 (11): 6267.
14 Bonn, D., Eggers, J., Indekeu, J. et al. (2009). Wetting and spreading. *Rev. Mod. Phys.* 81 (2): 739.
15 Mugele, F. and Baret, J.-C. (2005). Topical review Electrowetting: from basics to applications. *J. Phys. Condens. Matter* 17: R705.
16 Lenses go liquid. (2013). Interview with Bruno Berge European Patent Organisation. European Inventor Award, the finalists.
17 Blankenbach, K. and Rawert, J. (2011). Bistable Electrowetting Displays. *Proceedings of SPIE*.
18 Ku, Y.-S., Kuo, S.-W., Tsai, Y.-S. et al. (2012). The structure and manufacturing process of large area transparent electrowetting display. *SID Symp. Dig. Tech. Pap.* 62 (4): 850.
19 Lo, K.-L., Tsai, Y.-H., Cheng, W.-Y. et al. (2013). Recent development of transparent electrowetting display. *SID Symp. Dig. Tech. Pap.* 12 (4): 123.
20 J. Heikenfeld, N. Smith, M Dhindsa, et al. (2009). Recent Progress in Arrayed Electrowetting Optics. OPN January, p. 21.
21 Feil, H. (2010). Electrowetting optical element and mirror system comprising said element. Patent application WO2010/062163 A1.
22 Oostra, D.J. (2019). Reflective Electrowetting Displays for Out Of Home Display Applications. In *Proceedings of International Display Workshops*. Vol 26. EP2/DES4–2.
23 Institute of Electronic Paper Displays, South China Academy of Advanced Optoelectronics. (2017). SCNU. China Normal University Announces Latest Achievements in Color Video Electronic Paper. May 25, (2017).
24 Private communications with J. Marra. (2020). On the physics and chemistry of the electrowetting display.
25 Teng, P., Tian, D., Fu, H., and Wang, S. (2020). Recent progress of electrowetting for droplet manipulation: from wetting to superwetting systems. *Mater. Chem. Front.* 4: 140.
26 www.snapquality.com
27 www.idealliance.org
28 Yang, S.-L., Lin, Y.-H., Fan-Jiang, S.-C., and Shih, C.-H. (2011). Electro-wetting Display Panel, Patent no. US8,004,738 B2 Aug 23 (2011)
29 Wangler, N., Beck, S., Ahrens, G. et al. (2012). Ultra thick epoxy-based dry-film resist for high aspect ratios. *Microelectron. Eng.* 97: 92.
30 Electrowetting Displays. (2009). A paper by Johan Feenstra & Rob Hayes. Liquavista, May (2009).
31 Duan, M., Hayes, R.A., Zhang, X., and Zhou, G. (2014). A reflective display technology based on electrofluidics. *Appl. Mech. Mater.* 670-671: 976.

32 Wu, H., Hayes, R.A., Li, F. et al. (2018). Influence of fluoropolymer surface wettability on electrowetting display performance. *Displays* 53: 47–53.

33 Oostra, D. Etulipa, Private measurements.

34 Roques-Carmes, T., Hayes, R.A., Feenstra, B.J., and Schlangen, L.J.M. (2004). Liquid behavior inside a reflective display pixel based on electrowetting. *J. Appl. Phys.* 95 (8): 4389.

35 Li, S., Ye, D., Henzen, A. et al. (2020). Novel perylene-based organic dyes for electro-fluidic displays. *New J. Chem.* 44: 415–421.

36 News release provided by Electronic Paper Display Institute, South China Academy of Advanced Optoelectronics, South China Normal University. New breakthrough on full color video electronic paper display in China. Nov 25, (2020).

37 Yang, G., Tang, B., Yuan, D. et al. (2019). Scalable fabrication and testing processes for three-layer multi-color segmented electrowetting display. *Micromachines* 10: 341.

38 Oostra, D. (2018). Performance of reflective electrowetting displays in Out Of Home applications. ID TechEx, Berlin, April 11th 2018.

39 Kwon, S. (2010). Characterization and Fabrication of Active Matrix Thin Film Transistors for an Addressable Microfluidic Electrowetting Channel Device. University of Tennessee, Knoxville. Doctoral Dissertations 12–2010.

40 Zhou, M., Zhao, Q., Tang, B. et al. (2017). Simplified dynamical model for optical response of electrofluidic displays. *Displays* 49: 26.

41 Lin, S., Qian, M., Lin, Z., Guo, T. (2019). The Driving System of Electrowetting Display Based on Multi-Gray Dynamic Symmetry Driving Waveform. International Displays Workshop, Sapporo. EP2_DES4. p. 1407.

42 Yi, Z., Liu, L., Wang, L. et al. (2019). A driving system for fast and precise gray-scale response based on amplitude-frequency mixed modulation in TFT electrowetting displays. *Micromachines* 10: 732.

43 Luo, Z., Fan, J., Xu, J. et al. (2020). A novel driving scheme for oil-splitting suppression in electrowetting display. *Opt. Rev.* 27: 339–345.

44 Rui, Z., Qi-Chao, L., Ping, W., and Zhong-Cheng, L. (2015). Contact angle hysteresis in electrowetting on dielectric. *Chin. Phys. B* 24 (8): 086801.

45 Park, D.J., Noh, J.-H., Lee, B., et al. (2014). Electrowetting Display Device and Method of Manufacturing the Same. US Patent 8,780,435 B2 Jul 15th 2014.

46 Timmons, J., Failing, R.G., and Brewer, J.C. (2006). Meaningful Measurements and Metrics for Reflective Displays. ICIS '06 International Congress of Imaging Science.

47 Arndt, R. and Puto, R., TÜV SÜD America Inc. Basic Understanding of IEC Standard Testing For Photovoltaic Panels.

48 IEC 60068-2 Electronic Equipment & Product Standards. IEC

49 Mil. Std. 810G. (2008). Department of Defense Test Method Standard, Environmental Engineering Considerations and Laboratory Tests, 31 October 2008.

50 Wilkons, D.J. The Bathtub Curve and Product Failure Behavior. Reliasoft.

51 Neeff, R. (2020). Why Liquid Crystal Windows? Electronic Displays Conference, Nuremberg, Feb. 26th 2020.

52 European Commission Decision C (2017). European Commission EN Horizon 2020 Work Programme 2016–2017 Sect. 20. General Annexes. European Commission Decision C (2017) 2468 Apr 24th 2017.

10

Electrochromic Display

Norihisa Kobayashi

Group on Advanced Imaging Materials and Systems (G-AIMS), Graduate School of Engineering, Chiba University, JAPAN

10.1 Introduction

Electrochromic displays (ECDs) have been the subject of extensive research due to their excellent potential for application in novel paper-like display devices, so-called "e-paper". The EC device was first reported in the paper titled "A novel electrophotographic system" concerning to "display" in 1969, and anniversary 50 years have passed since then. Systematic and strategic research on EC materials and devices has increased since the 1980s. EC materials exhibiting clear color changes from colorless to the three primary colors (red, green, and blue, or cyan, magenta, and yellow) are required for full-color e-paper displaying devices. Recently, studies of electrochromism (EC) to realize full-color reflective display with flexibility, bendability, and wearability are motivated by this concept and requirements. Papers reported are extensively increasing worldwide. This chapter is a brief review of electrochromic materials and displays toward full-color reflective display and e-paper. This chapter also includes the mechanism, characteristics, and recent progress of various EC materials and future applications in addition to display to clarify the advantages of ECDs.

Phenomena in which reversible color tone changes are induced by a stimulus are referred to as "chromism," and have been long known. In recent years, "chromogenics," defined as various reversible optical changes, including color, scattering, and reflection, has been recognized as a distinct technical field. Chromism includes a variety of phenomena that are named based on the relevant stimulus: "photochromism" induced by photon stimulus, "electrochromism" by electric energy, "thermochromism" by heat, "solvatochromism" by solvent, and so on. Widely known examples of photochromism include the photoisomerization of azobenzene, spiropyran, and diarylethene derivatives, the photochemical reaction of rhodopsin in retinal tissue, and sunglasses that change optical density in response to light intensity; photochromism is also expected to find application in ultrahigh-density memory, photoresponsive devices and so on. Thermochromism is utilized in the well-known card thermometer with a, a black background that employs a cholesteric liquid crystal and the erasable gel pen (Pilot Frixion BallTM™), which employs a thermosensitive reaction of leuco dyes and developer. Solvatochromism is widely utilized as a pH indicator. These examples demonstrate that many chromic phenomena are utilized in our everyday activities.

Although electrochromism (EC) refers strictly to a reversible optical change induced by electric energy described above, it is more generally defined as a reversible color change induced by an electrochemical redox reaction. This color change is based on a change in the electronic state caused by electron transfer between the material and an electrode. The EC phenomenon is similar to the discharge and charge of a secondary battery, and thus, EC is

E-Paper Displays, First Edition. Edited by Bo-Ru Yang.

classified in a unique group different from electric field systems such as liquid crystal devices and electrophoresis. Tungsten oxide has been long recognized to exhibit color change, which has been known since the 1800s. Recently, many EC materials have emerged, such as inorganic, organic, and conducting polymer materials. These materials offer many advantages, including multi-color, low operation voltage, memory effects, etc. With the development of these materials, EC was expected to be applied to an imaging device called an ECD from as early as the 1980s.

The ECD is a light-receiving and reflective display. Unlike light-emitting displays such as the transmissive liquid crystal display (LCD), organic light-emitting diode (OLED), and the plasma display panel (PDP), the ECD offers high visibility under daylight conditions and a wide view angle and can reduce eye strain in the case of continuous use. Consequently, the ECD was expected to be applied to novel display systems. In fact, ECDs such as a 7-segment clock and an information board for displaying stock prices were test-marketed in the 1980s. However, the production of these items was discontinued because of the slow response time of the EC reaction, which occurs with the diffusion of substances (ions). Subsequently, LCD studied in the same period emerged for quick response displays such as TV and PC monitors.

Because of these factors, EC was applied to devices that do not necessitate a quick response time. In recent years, the use of natural energy has extensively collected attention from the perspective of environmental concerns, and the application of EC to a passive solar system for controlling the transmission of and for utilizing solar energy became an active area of focus. The passive solar EC system was expected to be applied to "curtain-less" smart windows, which control the amount of transmitted sunlight and increase air conditioning efficiency [1–3]. Smart windows have, in fact, already been put into practical use in Europe and the United States, as shown in Figure 10.1a. Another universal application of EC to automobile interior and exterior dimmable mirrors brought astonishing success in the EC field. The company also provided large-scale automobile privacy dimmable panels with EC for energy-conservation smart automobile systems, as shown in Figure 10.1b; these are already available on the market for motor vehicles. Figure 10.1c illustrates an aircraft light-modulating window called "dimmable window," installed in Boeing 787 Dreamliner as standard equipment, which is one of the most popular and eye-catching applications.

Nevertheless, it is still expected that EC will be applied to information displays. The slow response of EC previously precluded the development of EC-based information displays. However, in recent times, EC has re-emerged as a strong candidate for paper-like reflective displays, such as an e-paper, requiring memory property without quick response. The demand for novel paper-like information display media has re-initiated the study of EC for such display uses.

10.2 Structure of Electrochromic Display

Generally, an EC device is constructed by sandwiching an electrolyte solution between electrodes, similar to a battery, as shown in Figure 10.2. At least one of the electrodes is transparent, allowing observation of the color change of the cell. EC materials are dissolved in the electrolyte solution or attached as modifications to the electrode surface as a coloration layer. Although solutions, inorganic solids, polymers, gels, and ionic liquids have been used as the electrolyte layer, high ion conductivity materials are desired, as with a battery.

The EC device is a two-electrode cell composed of an anode and a cathode. The electrochemical reduction of a reducible material on one electrode should be accompanied by electrochemical oxidation of an oxidizable material on the other electrode with an equivalent amount of charge. Therefore, the EC device requires a counter material that compensates for the charge consumed on the counter electrode's coloration electrode. Since EC is a current-driven system like OLED, energy-saving ability is often debated compared to voltage-driven electrophoretic and liquid crystal systems. Even in ECD, consumed energy increases with increasing the number of writing and erasing processes. However, the EC with sufficient memory properties does not require energy to retain the image (just like to keep charged state in a secondary battery), leading a possibility to show very low energy

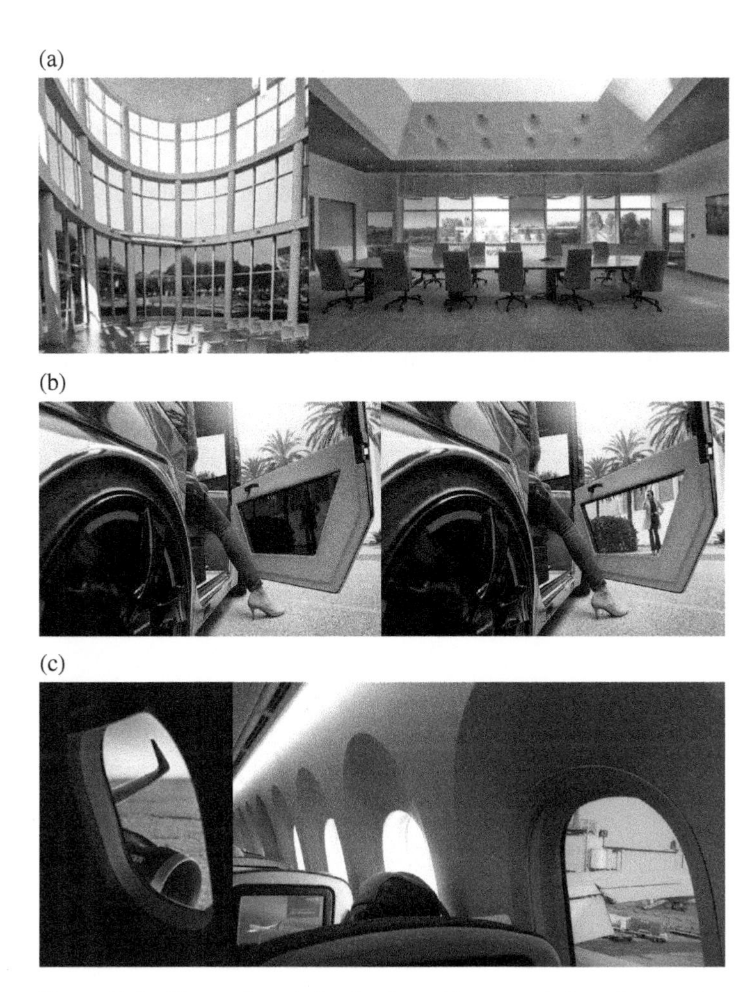

Figure 10.1 Electrochromic Applications: (a) smart windows, Source: Sageglass. (b) privacy dimmable panel Source: Rinspeed AG and (c) dimmable aircraft windows Jesse Echevarria/unsplash;Flickr/Payton Chung.

Figure 10.2 The typical structure and coloration mechanism of EC cell.

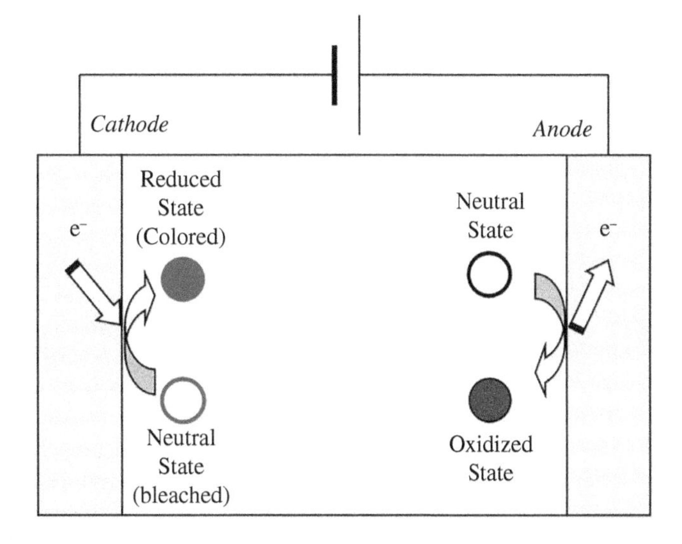

consumption rather than voltage-driven system if the number of writing erasing is not so frequent. Therefore, consumed energy depends on the number of writing and erasing processes. EC (Coloration) efficiency is commonly used to measure energy consumption when its value is compared with other EC values. EC efficiency is defined as the change in optical absorption per injected charge amount (so-called change in absorbance in coloration per injected charge density). Since charges required for writing in EC (mC/cm^2) are about 1000 times larger than that of LCD ($\mu C/cm^2$), the design of EC materials, counter materials, and device structure with and for high EC efficiency is considerably important to decrease energy consumption. The EC efficiency of EC devices is improved by employing suitable counter electrode systems that undergo electrochemical reactions that compensate for the charge consumed at the coloration electrode. If the counter material exhibits a color change associated with the electrochemical reaction, it may interfere with the color of the coloration electrode, and a white scattering/reflective layer must be introduced, especially in the case of a reflective display such as e-paper.

On the other hand, the counter electrode does not need to exhibit an electrochemical reaction for the charge compensation. For example, an electrochemical capacitor can accumulate the charge as a space charge by using an electric double layer. In fact, multiple colorations of anodic and cathodic EC was demonstrated in a single EC device with a hybrid capacitor architecture in which ITO nano-particle or porous carbon modified electrode was used as counter electrode. In this device, even large charge amount consumed by multiple redox pairs' EC materials was adequately compensated [4, 5]. In this way, many kinds of counter electrode systems become available, and an effective combination of the coloration electrode system and the counter electrode system is important for improving the characteristics of the EC cell.

10.3 EC Materials

10.3.1 Inorganic EC Materials

One typical inorganic EC material is tungsten oxide (WO_3) [6]. An EC cell employing WO_3 thin film was reported by S.K. Deb et al. in 1969 [7]. The redox process of WO_3 can be expressed in the following electrochemical reaction:

$$WO_3(\text{clear and colorless}) + xM^+ + xe^- \Leftrightarrow M_xWO_3(\text{blue})$$

When the WO_3 layer modified onto an electrode receives an electron from the electrode, a cation in the electrolyte is introduced into the WO_3 layer for charge compensation, as shown in Figure 10.3. Specifically, the WO_3 crystal has a perovskite structure comprised of W^{6+} and O^{2-}. A certain amount of W^{6+} is converted to W^{5+} by electrochemical reduction. WO_3 comprising the mixed-valence states of W^{6+} and W^{5+} is strongly colored by an intervalence transfer band. Cations having a small ionic radius (H^+, Li^+, etc.) can be introduced into the crystal lattice of WO_3 for charge compensation. Based on the film-forming method employed, quick response times of less than 100 ms can be achieved in the WO_3 film, along with long-term switching stability exceeding 10^7 cycles.

Many other inorganic materials [8] such as Prussian blue (PB), NiO, $Ir(OH)_x$, V_2O_5, and CeO_2 are also known to exhibit EC. These inorganic materials have already been applied to optical shutters, anti-glare mirrors, and smart windows based on their long-term switching stability and absorption in the infrared region. Recently, studies on PB are gradually increased from a viewpoint of color display because PB is known to expend its color variation by changing central metal ions. However, in the case of e-paper requiring low energy consumption and flexibility, inorganic materials may be disadvantageous. Depending on the film-forming method and cell structure, the coloration efficiencies of inorganic materials are within the range of 20–100 cm^2/C. EC efficiency is defined as the change in absorbance in coloration per injected charge density. These values are lower than those of organic EC materials and conducting polymers. Low energy consumption driving can be achieved by developing high EC efficiency inorganic EC materials and improving the cell structure.

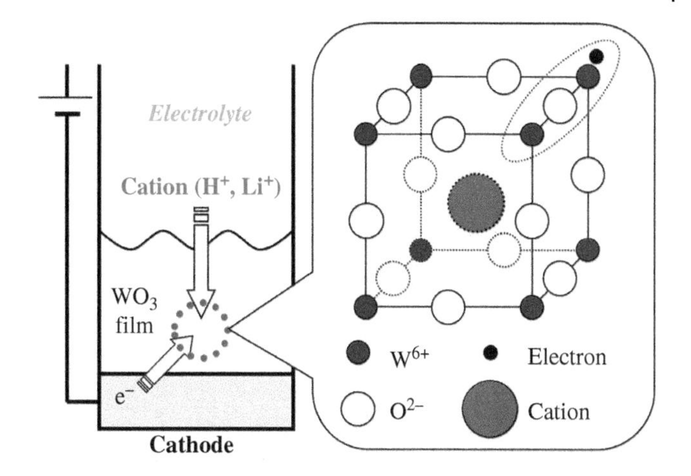

Figure 10.3 Schematic representation of coloration mechanism of WO^3 based EC material [7] / Optica Publishing Group / Public Domain.

One of the shortcomings in inorganic EC materials is color variation and multicolor presentation. A new strategy for color variation in inorganic EC comes true. Fabry-Perot nanocavity structure was fabricated by stacking the WO_3 EC layer and a partially reflective metal layer on the substrate as EC layer [9]. This structure enables strong interference resulting in various structural colors. The EC device with NiO counter electrode demonstrated vivid reflective color suitable for displaying. Multicolor representation, including transparent state, was reported using the intercalation of Zn^{2+} ion into the sodium ion stabilized vanadium oxide modified cathode [10]. Counter Zn electrode was located at the peripheral position of the device. Color switching between orange-light green-orange was demonstrated. Colorless transparent was not provided right now. These findings indicate that inorganic EC will be a potent candidate for full-color display, including e-paper.

10.3.2 Organic Electrochromic Materials

As previously described, inorganic EC materials, including metal oxide, are suitable for binary displays such as blue and transparent or white presentations. However, in the case of color displays, organic EC materials offer several advantages. Recently, increasing research efforts have been dedicated to the EC of organic compounds such as dyes, metal complexes, and conductive polymers from the viewpoint of multicolor and coloration efficiency. Examples of the application of organic EC materials to increase color palette leading to full-color e-paper, are emerging. Compared with inorganic oxide systems, organic EC materials offer color variety and EC efficiency advantages. Organic EC materials present various colors, including the three primary colors (such as red, green, and blue [RGB], or cyan, magenta, and yellow [CMY]) and colors intermediate between these. Organic EC systems, in which various colors may be derived from single materials and multicolor may be achieved by stacking the cells, are anticipated to be strong candidates for satisfying the rising demand for e-paper, such as in the case of black and white, multicolor and full-color representations.

Viologen derivatives, *N,N′*-dialkylated di-cations of 4,4′-bipyridine, exhibit various colorations upon variation of the quaternization agent. The viologen derivatives and analogous molecules are known as EC materials [11]. These organic materials show high EC efficiency compared to inorganic EC materials. The di-cationic derivatives of viologen can be dissolved in water and organic solvents depending on the molecular structure. The color change of viologen derivatives is induced by electrochemical reduction,. The reduced derivatives are deposited on the cathode due to the change in the solubility on turning from the di-cationic state to the mono-cationic state. However, like many other organic EC materials simply dissolved in an electrolyte solution, viologen derivatives do not exhibit quick response in common EC systems because the diffusion of the molecules and charge transfer between molecules is kinetically limited. In addition, viologen derivatives suffer from the disadvantage of bleaching because the colored states of the viologen derivatives sometimes crystallize and detach from the electrode as the colored precipitates.

To obviate the disadvantages of the viologen system, viologen derivatives were adsorbed onto the surface of titanium dioxide (TiO_2) nanoparticles modified on the electrode [12]. TiO_2 was prepared on the transparent electrode as a thin film of a few micrometers in thickness using the same method employed in the dye-sensitized solar cell. This TiO_2 thin film is porous, and its effective surface area is more than a thousand times the area of the geometrical electrode surface. Therefore, the viologen derivative adsorbed onto the TiO_2 surface has a high surface concentration, as shown in Figure 10.4. An experimental EC device constructed with a viologen derivative adsorbed TiO_2 nano-particle electrode exhibited long-term switching stability, more than several thousand times in -1.3 V driving. The charge transfer between TiO_2 and viologen is more efficient than that of the unmodified transparent electrode system; thus, the viologen-modified TiO_2 nano-particle system demonstrated quick response and high EC efficiency. In fact, the EC efficiency of the TiO_2 nanoparticle system was about 20 times higher than that of the systems without the TiO_2 nano-particles layer. The viologen-modified TiO_2 nano-particle system achieved a quick response time of several milliseconds due to the high mobility of charge compensation ions within the porous electrode the high electron transfer rate between viologen and TiO_2. This system has been widely studied and applied to the almost successful development of e-paper, such as an electronic book using the viologen system. For full-fledged applications to e-paper, prototype EC devices of the TiO_2 nano-particle system fabricated by the screen printing method were also reported. Thus, the TiO_2 nano-particle system has potential to further develop ECDs.

Recently, application to rollable and flexible subpixelated ECD was reported with viologens for wearable display. The TiO_2 nano-particle system shows pretty nice EC properties but is not good for stretchable and rollable application at present due to issues on stretchability and rollability of thermally connected inorganic TiO_2 nano-particles layer. As electrode, in addition to ITO on flexible PET substrate, single-walled carbon nanotube/Pd-coated Ag nanowire conducting bilayer on PDMS flexible substrate was also utilized for high chemical stability and excellent mechanical robustness against external strains. UV curable gels, including viologens, counter ferrocene derivative and the ionic liquid was pixelated through photomasks to fabricate subpixelated rollable ECD. Each subpixel was positioned side-by-side configuration just like color filters in an emissive display. The ECDs exhibit high durability even after 1000 cycles of mechanical bending tests at a bending radius of 10 mm [13, 14].

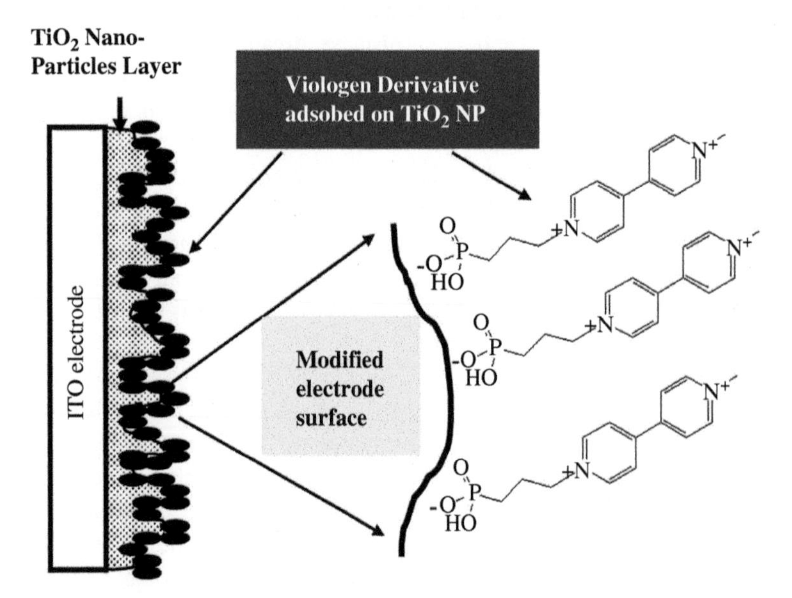

Figure 10.4 Schematic representation of viologen derivative-modified TiO_2 nano-particles modified electrode Adapted From [12].

Examples aiming for applying full-color e-paper have also been reported by the combination of three primary color EC. Since e-paper should be a reflective display, the three primary colors, cyan, magenta, and yellow (CMY), are suitable for full-color representation of e-paper as in the case of photography and printing. On this basis, the electrochemical properties of phthalate derivatives were evaluated. The three primary colors (cyan, magenta, and yellow) were electrochemically obtained in the ITO sandwich cell using diacetyl benzene, dimethyl terephthalate, and biphenyl dicarboxylic acid diethyl ester, respectively, as shown in Figure 10.5. Each color obtained with the various derivatives is regarded as one of the three primary colors based on the CIE 1931 colorimetric measurement [15]. It was revealed that the anion radical of each derivative generated at the cathode exhibited each three primary colors (in case of dimethyl terephthalate; monomeric structure of polyethylene terephthalate PET, shows magenta color), and that the coloration was affected by the supporting electrolyte and solvent. Red, green, blue, and black have also been achieved by stacking two or three primary color EC cells [16]. Figure 10.6 displays a photograph of the three-layered ECD. The color of this three-layered ECD was based on the subtractive color mixing process. Blue, red, and green pixels were obtained by Y + C, M + C, and Y + C combinations, respectively. The black color of this three-layered ECD was not adequate because the CMY subtractive color mixture process cannot fully represent a chromatic black. Therefore, a full-color reflective display based on the present system requires four layers (cyan, magenta, yellow, and black), as is the case with color printing. It was anticipated based on the facts mentioned above that the phthalate derivative-based EC cell should be a candidate for a multi- or full-color e-paper. Fabricating flexible EC cells has been demonstrated using an ITO coated flexible plastic electrode and a gel electrolyte. The flexible EC cell exhibited a clear color change, as is the case with the glass-electrode EC cell.

For the practical application, the cycle stability of the phthalate derivative-based sandwich EC cell is very important. In the case of a 2-electrode sandwich cell, a counter material is necessary to improve the cycle stability of the EC cell. Several counter materials such as ferrocene, carbon, nickel oxide, and so on in the phthalate-based EC cell were examined to improve cycle stability. Carbon is not electroactive but is effective in forming reversible

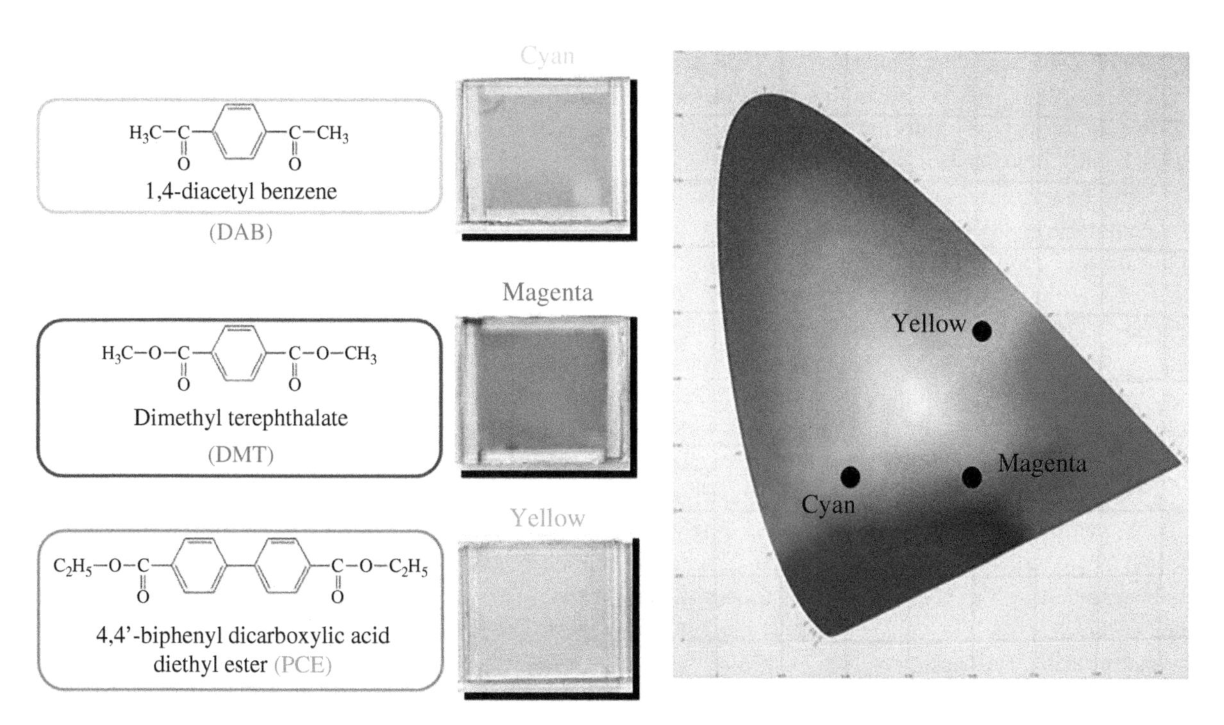

Figure 10.5 Digital camera images of colored state of terephthalate derivatives-based ECD. Colorimetry analysis of its colored states with CIE 1931 Yxy color diagram [15] / Royal Society of Chemistry.

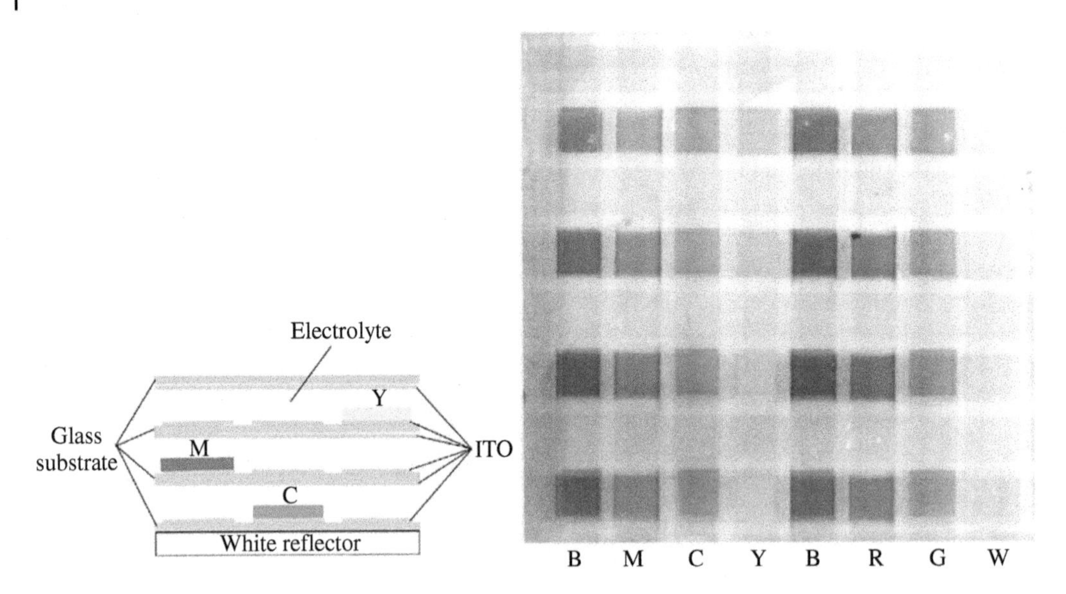

Figure 10.6 Digital camera image of the three-layered ECD with an 8 × 8 pixels resolution and passive matrix drive [16] / with permission of Elsevier.

electric double layer by ion adsorption and desorption on its surface. Therefore, it works as a capacitive layer to hold charges consumed at the working electrode. Particularly, nickel oxide was reported to work as an excellent counter material in phthalate-based EC cells. Cycle stability over 5000 coloring and bleaching cycles were reported [17]. This clearly indicates that even organic phthalate derivatives are not so weak in an EC cell by adopting suitable counter material and charge compensation system.

The memory property is also an important factor for e-paper applications to reduce energy consumption and realize continuous tone (grayscale) representation. If EC materials show excellent memory properties, it is easy to realize continuous tone because the EC material keeps a certain absorbance (coloration) without power supply, depending on the injected charges, as is the case with a variation of charging level in a secondary battery. However, the memory properties of phthalate derivatives are inefficient even in the open-circuit state of the EC cell. Application of continuous DC voltage with different magnitude to the EC cell can realize continuous tone, but this results in the consumption of a large amount of energy and a possibility of decomposition of the EC materials and cell. Therefore, a phthalate-based EC cell's continuous tone representation was realized by applying a suitable frequency rectangular wave voltage with various duty ratios instead of a DC voltage. According to a subtractive color mixture process, multiple colors, including intermediate colors between cyan, magenta, and yellow, were successfully demonstrated by stacking EC cells with a different tone (Figure 10.7) [18].

For full-color e-paper application of EC, innovative progress was made in 2011 [19]. Multi-layered EC consisting of one display unit was developed, and TiO_2 modified electrode system and three primary color organic dye-based EC were combined to realize full-color ECD. The white reflectivity was 70% at 550 nm, and multi-layered ECD showed 27% of color reproducibility compared with the standard color chart. Active matrix multi-layered ECD flexible panel using 3.5″ QVGA LTPS-TFT successfully expressed full-color image as shown in Figure 10.8. A possibility of flexible multi-layered ECD was also pointed out. This strongly supports that EC can be a technology for the real full-color e-paper.

Very recently, a new paradigm was reported in the organic EC system. A combination of leuco dyes, including fluoran and developer, was well known as a stimuli-induced colorant. This combination was commonly and widely utilized and commercialized in the thermochromic system, such as the erasable gel pen (Pilot Frixion Ball), thermal writable card, and so on, as previously mentioned. The developer works as a proton donor to stabilize the lactone

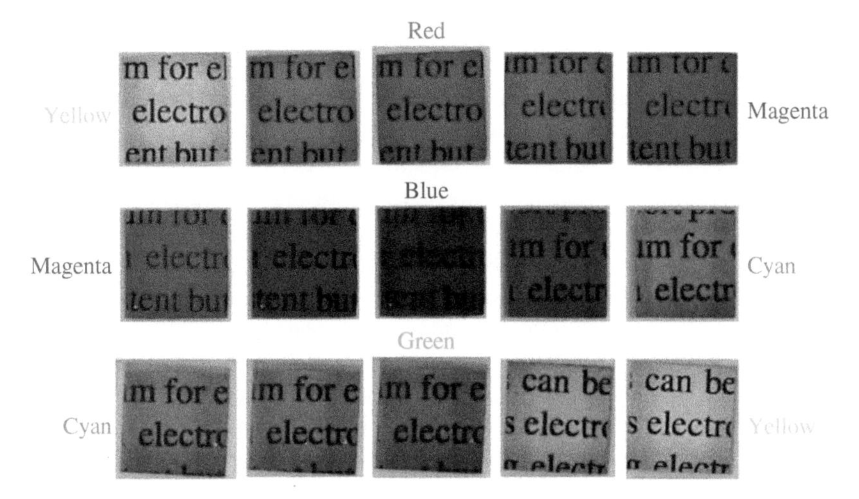

Figure 10.7 Color representation of stacked phthalate-based EC cell, two of which with different tone are stacked together. The coloration area of the EC cells was $1.0 \times 1.0\,\text{cm}^2$ [18].

Figure 10.8 Demonstrated image of active matrix 3.5" QVGA LTPS-TFT multi-layered flexible ECD panel Tohru Yashiro / Ricoh Company, Ltd.;Mikolaj Niemczewski / Adobe Stock.

ring-opening state. In a common organic low-molecule EC system, unstable radical is generated in either anodic or cathodic coloring molecules. This instability results in low cycling stability and less image retention, memory effect, properties. Certain leuco dyes themself have electrochemical activity and are easily oxidized. However, a structural change occurred after oxidation. Therefore, its reverse reaction to the neutral state required more negative potential. Cycling stability over several 10 times was not obtained. A new paradigm utilized the redox reaction of a developer to release proton, not to utilize the redox reaction of leuco dye. 2,3-Dimethylhydroquinon/benzoquinone (DMHQ) redox couple was reported to work as a proton donor and acceptor [20]. Gel electrolyte containing leuco dye showing suitable color and DMHQ was sandwiched between ITO electrodes to fabricated ECDs. ECD showing colorless transparent (T) and black (B) was incorporated with an optical see-through head-mounted display to fabricate an augmented reality display (ARD). The B state of ECD works well to enhance the visibility of displayed images under strong ambient light conditions. Display composed of conventional RGB color filter and ECD showing colorless transparent (T) and black (B) with white (W) reflector was demonstrated. Further, a side-by-side placed RGB-colorless EC color filter was stacked on the ECD showing T-B with W reflector, and color coordinates and color gamut were evaluated in the CIE 1931 color space [21]. A more attractive redox developer, urea derivative, was proposed, and impressive EC properties were reported. Combination of electroactive urea derivative and pH-sensitive rhodamine B derivative with closed ring structure gave excellent bistability (decay 7% in an hour), reversibility (over 10^4 cycles), coloration efficiency (430 cm^2/C) and concise color-switching time (2 ms), as shown in Figure 10.9 [22]. This combination showed nice light and thermal stability in-ring open and close states. The mechanism was based on intermolecular proton-coupled electron transfer (PCET) to increase proton concentration near dyes after oxidation leading to coloration. A prototype for a bistable electronic billboard and reader display device with high energy efficiency was demonstrated with this EC system. The same authors also designed and developed a superior bistable EC device with good overall performance, including bistability (>52 hours), reversibility (>12 000 cycles), coloration efficiency (≥1240 cm^2/C), and transmittance change (70%) with fast switching (≤1.5 seconds) [23]. This superior mechanism was based on concerted intramolecular PCET, which was achieved in the designed polymer with fluoran and redox-active proton donor unit. A prototype EC electronic shelf label was also successfully demonstrated. Taking variation of leco dyes, including fluoran structure and redox-active proton donor and resulting in excellent overall performance into account, this system can accelerate EC to various application fields. This will raise up the value of EC as a new wave and will be a potent candidate for energy saving display such as e-paper.

10.3.3 Conducting Polymers

Conducting polymers have been a focal prospect for film formation, multicolor property, and EC efficiency. The coloration efficiencies of the well-known conducting polymers, polybi-thiophene and poly(3-methyl thiophene), are 100 and 240 cm^2/C, respectively. The potential of poly(3,4-alkylenedioxythiophene) molecules, typified by poly(3,4-ethylene dioxythiophene) (PEDOT), and copolymers of the monomers of these molecules with other conjugate monomers have been recognized in conducting polymer systems in recent years. PEDOT possesses an ethylenedioxy group in positions 3 and 4 of the thiophene ring. Improved conductivity and redox switching stability have been achieved with PEDOT, derived from high electron density and reduced distortion of the thiophene ring due to the substitutional effect of the ethylenedioxy groups. Various colorations, including RGB, may be accessed using the copolymers of the alkylenedioxythiophene system by judicious variation of the combination of monomers [24]. Certain copolymers of this category exhibit coloration efficiencies higher than 1000 cm^2/C [25]. In addition, conducting polymers exhibit quick response and good memory property and thus, may be advantageous for electric power saving. Recent remarkable progress in EC of conducting polymers such as color variation, quick response, and so on are reported by academic scientists [26–29].

On the other hand, the conductivity and narrow bandgap of conducting polymers are debated to realize the fully-transparent state suitable for e-paper and display. Generally, transparent polymers are insulating, and most conducting polymers show some color in certain of their neutral and redox states. Therefore, conducting polymers may be disadvantageous to realize clear full-color representation.

For the application to stretchable, bendable, and wearable display, conducting polymer motivated EC researcher because of its processability and stability. Flexible and stretchable EC devices were demonstrated with PEDOT/PSS EC ink graphically printed on ITO/PET electrode using screen-printing or inkjet printing. For a 5 × 5 cm display, the full transition takes approximately 2.5 seconds with a maximum 3 V application. The active operation temperature ranging from −100 °C to +100 °C was achieved, and the displays are bendable up to a 7.5 mm radius [30]. A stretchable array of active-matrix (AM) ECD was fabricated by using poly(3-methyl thiophene) (P3MT) and Prussian blue (PB) electrodes with gel electrolyte on a deformable substrate. The ECD exhibited red, green and blue (RGB) depending on the applied voltage but not colorless state, and low power consumption of and 378 $\mu W/cm^2$ (10 seconds duration) and a high coloration efficiency of 201.6 cm^2/C at 1.0 V application driving.

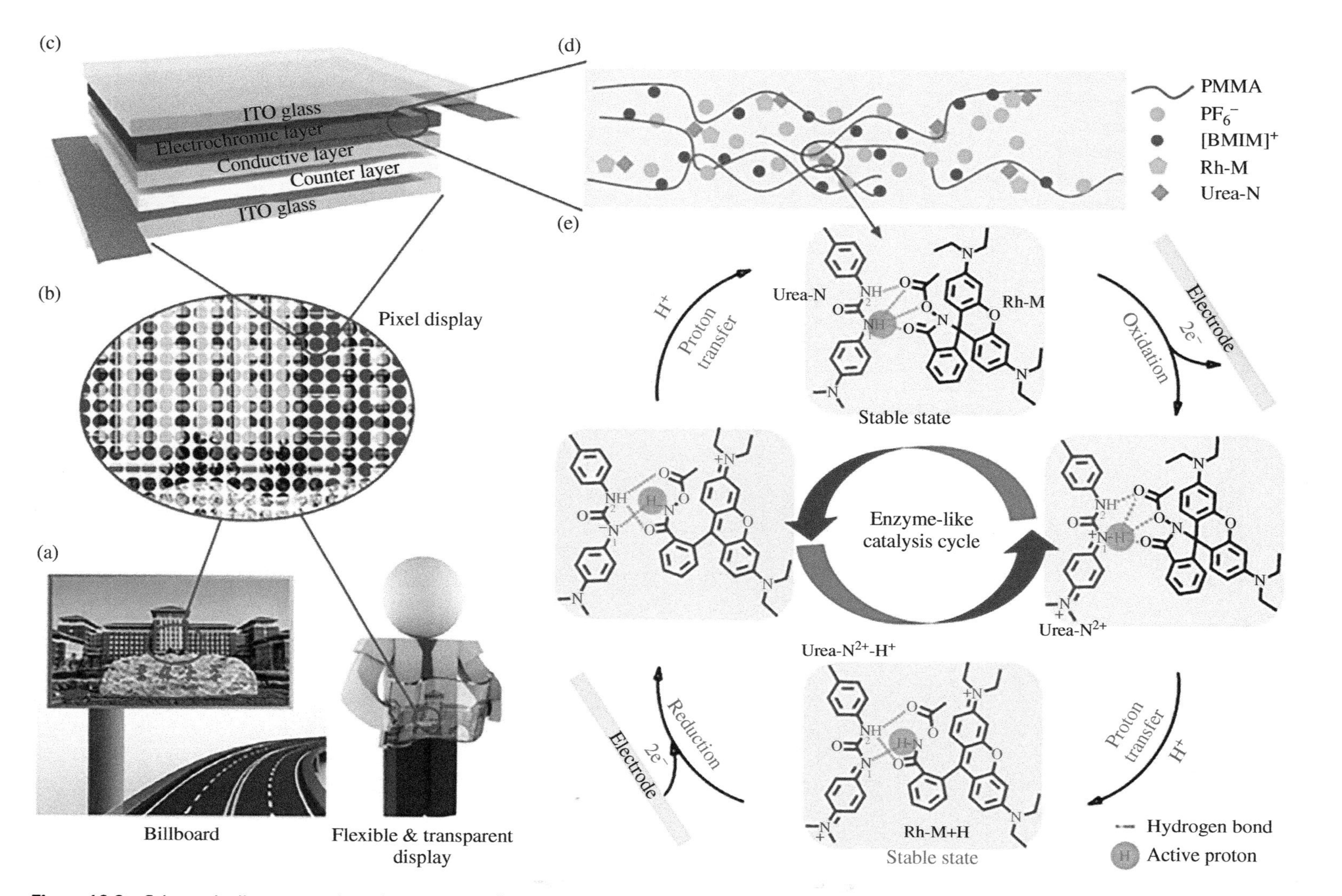

Figure 10.9 Schematic diagrams and mechanism of the "Urea-N+Rh-M" electrochromic system: Schematic of energy-saving billboard (left) and flexible transparent display (right), (b) Schematic of pixel display, (c) Structure of bistable electrochromic device, (d) The composition of the electrochromic layer and (e) The proposed mechanism (gray dashed line: hydrogen bonding between Urea-N and Rh-M; red circle: active proton that transferred between Urea-N and Rh-M) [22] Zhang, W., et.al, 2019 / Springer Nature / CC BY 4.0.

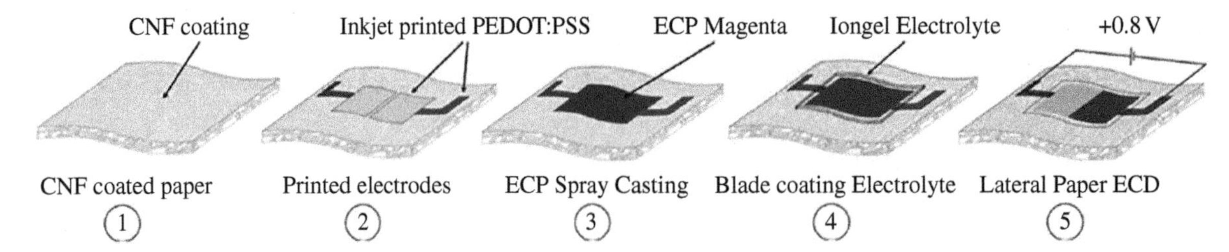

Figure 10.10 Fabrication process for lateral paper-ECDs showing inkjet-printed PEDOT: PSS electrodes, deposition of ECPs, and [EMI][TFSI]/PVDF-HFP ion gel electrolyte layer. Devices are operated by applying a 0.8 V bias across the two lateral pixels [31] / John Wiley & Sons.

A stretchable 6×6 a.m. ECD array for higher resolution display with patterned liquid metal GaInSn interconnection also shows mechanical stability under 30% biaxial stretching. These researches are targeting wearable display system. As thinking of environmental effects on both the processing and disposal of electronic devices become important rapidly, the ability to replace plastic and glass substrates with bioderived and biodegradable materials remains a major technological goal. According to this concept, a cellulose nanofiber-coated paper substate was used to fabricate ECD (Figure 10.10) [31]. PEDOT/PSS works well as the electrode on the paper substrate in this device. This paper ECD exhibited 9000 cycles ON and OFF EC stability even higher surface resistance of 460 ohm/sq as an electrode. This is probably attributed to the advantage of conducting polymer ECD, such as high conductivity and low driving voltage. The authors claimed that the paper ECD was combusted in the air leaving 3% of the initial mass at 600 °C, highlighting this approach as a promising route toward disposable displays. In some meanings, these results well match a concept of e-paper display.

10.3.4 Electrodeposition (Electroplating)

Electrodeposition can be described as an aspect of EC. A well-known electrodeposition process is the electroreduction of metal ions, commonly known as metal plating. The Zn electrodeposition system, in which electroreduction of Zn^{2+} is performed in solution, was reported in the 1920s. Subsequently, black and white presentation by electrodeposition of Bi was reported in 1995. The Bi electrodeposition system exhibits good contrast because colorless Bi^{3+} accentuates the white color of the white reflector. The EC efficiency for this Bi electrodeposition system is 75 cm^2/C at 700 nm, with a high white and black contrast ratio of 25 : 1 because colorless Bi^{3+} ions don't dull the color of the white coating layer. It was reported that the addition of Cu^{2+} to this system improved the reversibility of the electrodeposition and elution processes. A cell has been fabricated by sandwiching an electrolyte containing Bi^{3+} and Cu^{2+} between an ITO working electrode and a polyethylene/carbon counter electrode. Although degradation of the ITO electrode occurred with the use of the aqueous solution system due to the low pH of 1.5, degradation could be reduced by adjusting the driving voltage and driving mode. Interest in the Bi system is still active; 7 segment Bi-electrodeposition based ECD was reported in 1999 [32], and good long-term switching stability of Bi in a non-aqueous solvent system utilizing the EC of Bi was independently demonstrated [33]. Interestingly, a Bi system and cellulose combinationas were reported realizing paper-like feeling in the display [34].

Electrodeposition of silver ion (Ag^+) has attracted much interest and was reported in 1962. An Ag deposition-based EC display was also demonstrated. In 2002, a black and white presentation cell utilizing the electroreduction of silver ions was proposed to realize black and white color e-paper, as shown in Figure 10.11 [35]. This cell was fabricated by sandwiching a white-colored electrolyte layer consisting of TiO_2 and a gel electrolyte containing a silver salt between two electrodes. The reflectance of the white color based on TiO_2 was higher than 70% because the silver ion is colorless, and the contrast ratio was over 20 : 1. This cell exhibited a low driving voltage of less than 3.0 V and quick response of less than 100 ms. Improvement of the cell stability was attributed to using an ion-conducting polymer as the electrolyte.

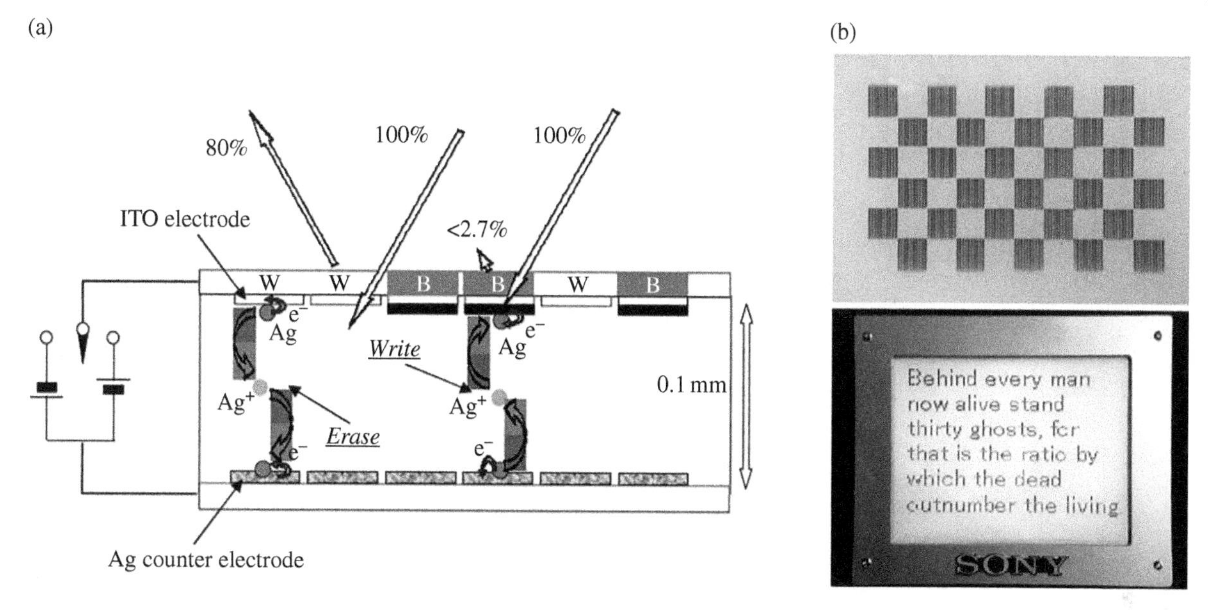

Figure 10.11 (a) Structure of Ag electrodeposition-based EC display by SONY, (b) Photographs of a passive matrix display (50 dpi, 160 × 120 pixels). Each square in the upper image consists of 10 × 10 pixels [35] K. Shinozaki, 2002 / JOHN WILEY & SONS, INC.

Exciting results were reported in Ag electrodeposition in 2012 [36]. Although the black color was generally generated by the electrodeposition of silver ions, an EC device based on silver electro-deposition that achieved three reversible optical changes (transparent, silver-mirror, and black) in a single cell were successfully demonstrated (Figure 10.12). The driving principle of this EC device is the exploitation of Ag nano-particle deposition on two different transparent electrodes: a flat indium tin oxide (ITO) electrode and a rough ITO particle-modified electrode. The EC material, consisting of a gel electrolyte in which Ag^+ is dissolved, is sandwiched between the two electrodes. The default state of this device is transparent, whereas by applying a negative voltage to either one of the electrodes, Ag is electrodeposited on the electrode surface. Following Ag deposition on the flat ITO electrode, the device becomes mirrored. Conversely, the device turns black when Ag deposition occurs on the ITO particle-modified electrode, which has a rough surface. This can apply to smart window systems as well as displays. The EC device shows pretty nice cycling stability over 20000 cycles (not examined more). Still, the EC device had poor color retention property under open-circuit state (memory effect) because of dissolution of deposited Ag metal by Cu^{2+} ions, which improves cycling stability as redox material at counter electrode and mediator to facilitate the oxidation of Ag deposit. This shortcoming was improved by introducing anion exchange membrane [37] or electroactive counter electrode to avoid Cu^{2+} ions [38]. The improved device achieved a longer retention time of the colored state (>4 hours). It is effective to maintain the coloring state without electric power for practical applications such as e-paper, smart window, etc.

Although progress toward multichromatic representation in full-color EC displays has been reported, control of the multichromatic state using inorganic EC devices has been a few reported except a combination of nanocavity structure as described above. The optical state related to the color of the metal electrodeposit on the electrode depends on the metal particle size, shape, and coalescence between particles. Suppose the size and shape of the Ag nano-particles electrodeposited can be controlled uniformly and homogeneously. In that case Localized surface plasmon resonance (LSPR) band of the Ag nanoparticles can be used as a tool to control multiple chromatic states because Ag nano-particles are known to show various colors based on their LSPR. On this basis, by changing voltage application method to Ag electrodeposition-based EC cell, a reversible multicolor changing phenomenon, from transparent to magenta, cyan, and yellow in addition to mirror and black states, based on controlling the size of Ag nano-particles during electrochemical deposition was successfully demonstrated (Figure 10.13) [39–41].

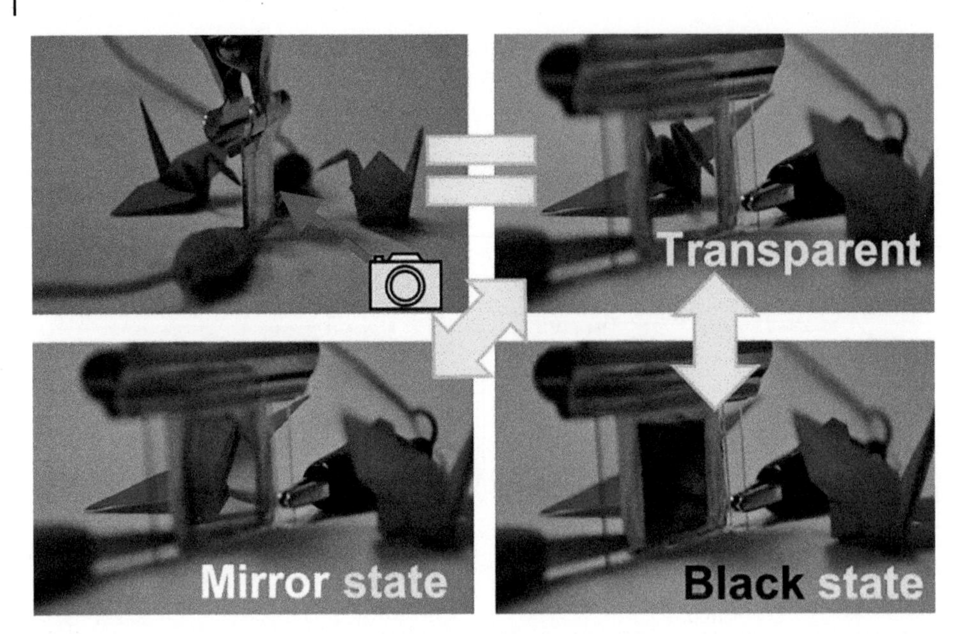

Figure 10.12 Digital camera images of Ag electrodeposition-based EC cell. Upper left; Side view, Mirror state (−2.5 V application), Transparent state (before voltage application), Black state (+2.5 V application) [36] S. Araki, et.al, 2012 / JOHN WILEY & SONS, INC.

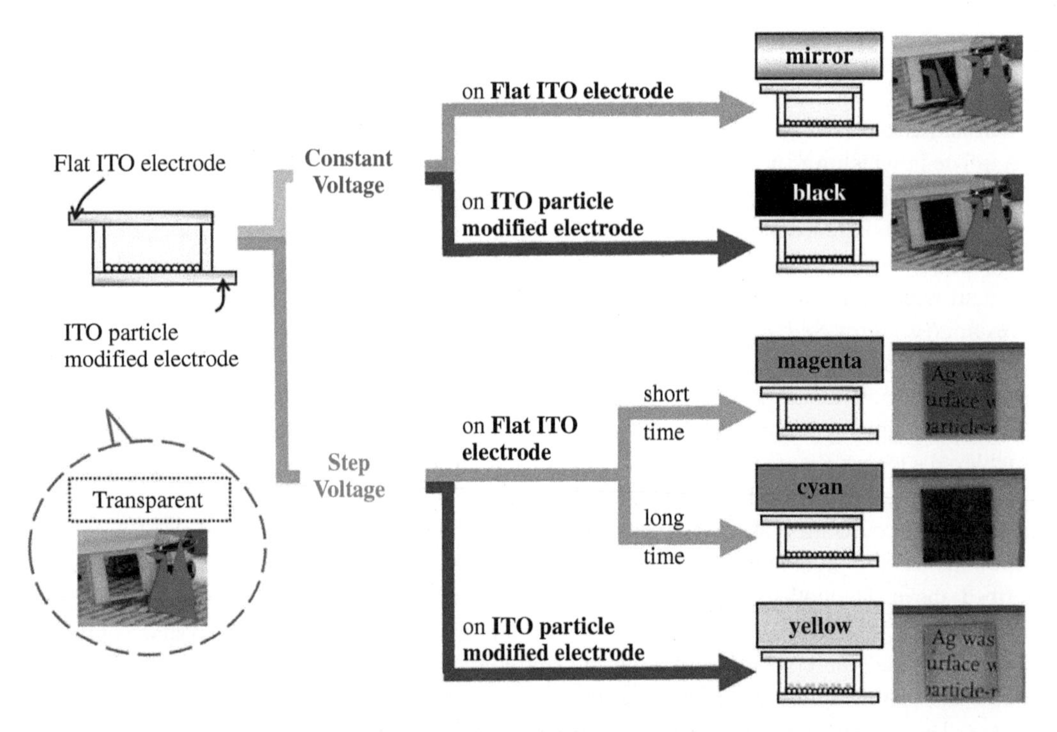

Figure 10.13 Digital camera images of Ag electrodeposition-based EC cell exhibiting BCMY and mirror states (6 optical states). The cell structure is flat ITO electrode/Ag based electrolyte/ITO particles modified electrode. Constant voltage application to flat ITO electrode (Mirror), to ITO particles modified electrode (Black), and voltage-step application to flat ITO electrode (Megenta to Cyan), to ITO particle modified electrode (Yellow) [39] A. Tsuboi, et.al, 2013 / JOHN WILEY & SONS, INC.

This clearly indicates that control of the morphology of Ag nano-particles can lead to dramatic changes in color, as their size and shape influence the LSPR band. To improve color purity and chroma value, studies from the viewpoint of diffusion rate of Ag^+ ions are conducted when Ag nano-particles are electrochemically deposited. Consequently, well-isolated Ag nano-particles were obtained due to the slow growth rate by using an electrolyte with a low concentration of Ag^+ ions, resulting in an improvement in the color quality of cyan and magenta [42]. Additionally, spherical Ag nanoparticles were deposited in the same device by optimizing their voltage application conditions, representing yellow and green colors. In particular, green coloration is a unique phenomenon because it can appear by combining two absorption peaks of LSPR. As a result of investigating the FDTD simulation, it was observed that the LSPR band in the long-wavelength region was originated from the effects of the connection between Ag particles. The multifunction of this LSPR-based EC display device could make it suitable for use in information displays and light-modulating devices such as e-paper, digital signage, and smart windows.

10.4 Summary

This chapter briefly reviews ECD technology, including cell structure, materials, and recent progress. The application of EC was mainly focused on the light-modulating system such as dimming windows and car mirrors, although display application was reported over 25 years ago. However, growing interest in full-color e-paper and energy-saving reflective display encourages research of EC and extends its possibility. Other several interesting materials and phenomena, rather than display applications, which are not covered in this chapter, are also emerging. These activate the research field and are believed to have the potential to generate unexpected attractive applications in the near future. We wish this chapter will help awaken interest for further investigations and applications of EC.

References

1 Rosseinsky, D.R. and Mortimer, R.J. (2001). Electrochromic systems and the prospects for devices. *Adv. Mater.* 13: 783–793.

2 Granqvist, C.G., Azens, A., Heszler, P. et al. (2007). Nanomaterials for benign indoor environments: Electrochromics for "smart windows", sensors for air quality, and photo-catalysts for air cleaning. *Sol. Energy Mater. Sol. Cells* 91 (4): 355–365.

3 Chua, M.H., Zhu, Q., Tang, T. et al. (2019). Diversity of electron acceptor groups in donor–acceptor type electrochromic conjugated polymers. *Sol. Energy Mater. Sol. Cells* 197: 32–75.

4 Liang, Z., Yukikawa, M., Nakamura, K. et al. (2018). A novel organic electrochromic device with hybrid capacitor architecture towards multicolour representation. *Phys. Chem. Chem. Phys.* 20: 19892–19899.

5 Liang, Z., Nakamura, K., and Kobayashi, N. (2019). A multicolor electrochromic device having hybrid capacitor architecture with a porous carbon electrode. *Sol. Energy Mater. Sol. Cells* 200: 109914-1-7.

6 Karaca, G.Y., Eren, E., Cogal, G.C. et al. (2019). Enhanced electrochromic characteristics induced by Au/PEDOT/Pt microtubes in WO_3 based electrochromic devices. *Opt. Mater.* 88: 472–478.

7 Deb, S.K. (1969). A novel electrophotographic system. *Appl. Opt.* 8: 192–195.

8 Monk, P.M.S., Mortimer, R.J., and Rosseinsky, D.R. (2007). *Electrochromism: Fundamentals and Applications*. Wiley.

9 Wang, Z., Wang, X., Cong, S. et al. (2020). Towards full-colour tunability of inorganic electrochromic devices using ultracompact fabry-perot nanocavities. *Nat. Commun.* 11 (1): 302. -1-9.

10 Zhang, W., Li, H., Yu, W.W. et al. (2020). Transparent inorganic multicolour displays enabled by zinc-based electrochromic devices. *Light Sci. Appl.* 9: 121-1-11.

11 Sen, S., Saraidaridis, J., Kim, S.Y. et al. (2013). Viologens as charge carriers in a polymer-based battery anode. *ACS Appl. Mater. Interfaces* 5 (16): 7825–7830.

12 Hagfeldt, A., Vlachopoulos, N., and Gratzel, M. (1994). Fast electrochromic switching with nanocrystalline oxide semiconductor films. *J. Electrochem. Soc.* 141 (7): L82–L84.

13 Kim, J.W. and Myoung, J.M. (2019). Flexible and transparent electrochromic displays with simultaneously implementable subpixelated ion gel-based viologens by multiple patterning. *Adv. Funct. Mater.* 29 (13): 1808911–1808920.

14 Kim, J.W., Kwon, D.K., and Myoung, J.M. (2020). Rollable and transparent subpixelated electrochromic displays using deformable nanowire electrodes with improved electrochemical and mechanical stability. *Chem. Eng. J.* 387: 124145–124160.

15 Urano, H., Sunohara, S., Ohtomo, H. et al. (2004). Electrochemical and spectroscopic characteristics of dimethyl terephthalate. *J. Mater. Chem.* 14 (15): 2366–2368.

16 Kobayashi, N., Miura, S., Nishimura, M. et al. (2008). Organic electrochromism for a new color electronic paper. *Sol. Energy Mater. Sol. Cells* 92 (2): 136–139.

17 Watanabe, Y., Imaizumi, K., Nakamura, K. et al. (2011). Effect of counter electrode reaction on coloration properties of phthalate-based electrochromic cell. *Sol. Energy Mater. Sol. Cells* 99: 88–94.

18 Watanabe, Y., Nagashima, T., Nakamura, K. et al. (2012). Continuous-tone images obtained using three primary-color electrochromic cells containing gel electrolyte. *Sol. Energy Mater. Sol. Cells* 104: 140–145.

19 Yashiro, T., Naijoh, Y. et al. (2011). SID2011 Digest, 42–45 and Proc. IDW2011, 2011: 375–378.

20 Kim, G.W., Kim, Y.C., Ko, I.J. et al. (2018). High-performance electrochromic optical shutter based on fluoran dye for visibility enhancement of augmented reality display. *Adv. Opt. Mater.* 6 (11): 1701382–1701392.

21 Ko, I.J., Park, J.H., Kim, G.W. et al. (2019). An optically efficient full-color reflective display with an electrochromic device and color production units. *J. Inf. Disp.* 20: 155–160.

22 Zhang, W., Wang, X., Wang, Y. et al. (2019). Bio-inspired ultra-high energy efficiency bistable electronic billboard and reader. *Nat. Commun.* 10: 1559-1-8.

23 Wang, Y., Wang, S., Wang, X. et al. (2019). A multicolour bistable electronic shelf label based on intramolecular proton-coupled electron transfer. *Nat. Mater.* 18: 1335–1342.

24 Thompson, B.C., Schottland, P., Zong, K. et al. (2002). In situ colorimetric analysis of electrochromic polymers and devices. *Chem. Mater.* 12: 1563–1571.

25 Rauh, R.D., Wang, F., Reynold, J.R. et al. (2001). High coloration efficiency electrochromics and their application to multi-color devices. *Electrochim. Acta* 46 (13): 2023–2029.

26 Bulloch, R.H., Kerszulis, J.A., Dye, A.L. et al. (2014). Mapping the broad CMY subtractive primary color gamut using a dual-active electrochromic device. *ACS Appl. Mater. Interfaces* 6 (9): 6623–6630.

27 Liu, D.Y., Chilton, A.D., Shi, P. et al. (2011). In situ spectroscopic analysis of sub-second switching polymer electrochromes. *Adv. Funct. Mater.* 21 (23): 4535–4542.

28 Beaujuge, P.M. and Reynolds, J.R. (2010). Color control in π-conjugated organic polymers for use in electrochromic devices. *Chem. Rev.* 110 (1): 268–320.

29 Amb, C.M., Dyer, A.L., and Reynolds, J.R. (2011). Navigating the color palette of solution-processable electrochromic polymers. *Chem. Mater.* 23 (3): 397–415.

30 Kim, D.S., Park, H., Hong, S.Y. et al. (2019). Low power stretchable active-matrix red, green, blue (RGB) electrochromic device array of poly(3-methylthiophene)/Prussian blue. *Appl. Surf. Sci.* 471: 300–308.

31 Lang, A.W., Österholm, A.M., and Reynolds, J.R. (2019). Paper-based electrochromic devices enabled by nanocellulose-coated substrates. *Adv. Funct. Mater.* 29 (39): 1903487–1903498.

32 Ziegler, J.P. (1999). Status of reversible electrodeposition electrochromic devices. *Sol. Energy Mater. Sol. Cells* 56: 477–493.

33 Imamura, A., Kimura, M., Kon, T. et al. (2009). Bi-based electrochromic cell with mediator for white/black imaging. *Sol. Energy Mater. Sol. Cells* 93 (12): 2079–2082.

34 Nakashima, M., Ebine, T., Shishikura, M. et al. (2010). Bismuth electrochromic device with high paper-like quality and high performances. *ACS Appl. Mater. Interfaces* 2 (5): 1471–1482.

35 Shinozaki, K. (2002). 5.5L: Late-news paper: electrodeposition device for paper-like displays. *SID Symp. Dig. Tech. Pap.* 33: 39–41.

36 Araki, S., Nakamura, K., Kobayashi, K. et al. (2012). Electrochromic materials: electrochemical optical-modulation device with reversible transformation between transparent, mirror, and black. *Adv. Mater.* 24 (23): OP122–OP126.

37 Kimura, S., Onodera, R., Nakamura, K. et al. (2018). Improvement of color retention properties of Ag deposition-based electrochromic device by introducing anion exchange membrane. *MRS Commun.* 8: 498–503.

38 Kimura, S., Nakamura, K., and Kobayashi, N. (2020). Bistable silver electrodeposition-based EC device with a Prussian blue counter electrode to maintain the mirror state without power supply. *Sol. Energy Mater. Sol. Cells* 205: 110247-1-6.

39 Tsuboi, A., Nakamura, K., and Kobayashi, N. (2013). A localized surface plasmon resonance-based multicolor electrochromic device with electrochemically size-controlled silver nano-particles. *Adv. Mater.* 25 (23): 3197–3201.

40 Tsuboi, A., Nakamura, K., and Kobayashi, N. (2013). Chromatic characterization of novel multicolor reflective display with electrochemically size-controlled silver nano-particles. *J. Soc. Inf. Disp.* 21 (7–9): 361–367.

41 Tsuboi, A., Nakamura, K., and Kobayashi, N. (2014). Multicolor electrochromism showing three primary color states (cyan–magenta–yellow) based on size- and shape-controlled silver nanoparticles. *Chem. Mater.* 26 (22): 6477–6485.

42 Kimura, S., Sugita, T., Nakamura, K. et al. (2020). An improvement in the coloration properties of Ag deposition-based plasmonic EC devices by precise control of shape and density of deposited Ag nano-particles. *Nanoscale* 12 (47): 23975–23983.

11

Phase Change Material Displays

Ben Broughton and Peiman Hosseini

11.1 Introduction

Phase change material (PCM) displays are an emerging display technology that can provide highly colored, high reflectivity, fast switching, fully bistable reflective displays, via a mechanism entirely new to display technology – reversible crystallization of optically absorbing, amorphous, ultra-thin films.

This mechanism, and the PCMs which enable it have been well developed for optical disk and computer memory applications, but until relatively recently had not been explored for their reflective display potential, despite the extremely fast switching and high refractive index change they exhibit. This was likely due to the microscopic areas switched in these devices and their limited optical contrast shown under broadband illumination. Interest in the use of PCMs in white-light reflective optical devices was initiated by the publication of Hosseini et al. [1], which demonstrated both macroscopic and nanoscopic thermal switching of PCMs incorporated into an ultra-thin film "strong interference" optical stack, for the first time translating the large change in complex refractive index into a striking change in reflected color. The remainder of this introduction will discuss in more detail the PCMs and switching mechanism, and the optical effect, which were combined in this development.

11.2 Phase Change Materials and Devices

Chalcogenide-based PCMs are a special class of functional materials that can switch between at least two phases: amorphous and crystalline, billions of times by means of simple optical or electrical energy pulses [2]. The two phases have very different optical and electronic properties and were first proposed to store information in 1968 by Ovshinsky et al. [3]. Interestingly, the first commercial application of PCM was in rewritable optical storage technologies such as CD-RW, DVD-RW, and BD-RE. More recently, Intel has begun the commercialization of the so-called Optane XPoint memory, a PCM-based memory technology with similar non-volatility to Flash Memories and speed performance close to that of DRAM [4].

PCM memory materials require pulses of different amplitude and speed to switch a device into the amorphous (RESET) or crystalline (SET) phase, as illustrated in Figure 11.1. Switching to the amorphous phase requires an energy pulse that is high enough to push the PCM above its melting point and short enough that it doesn't leave enough time for the material to recrystallize upon cooling. Several research groups have demonstrated pulses as short as picoseconds [5] in the electrical domain and femtoseconds in the optical domain [6]. Crystallization, on

E-Paper Displays, First Edition. Edited by Bo-Ru Yang.

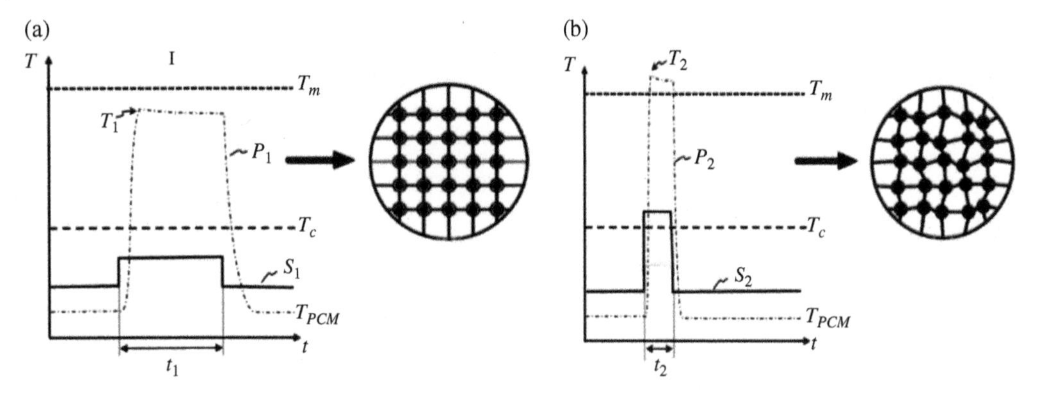

Figure 11.1 Diagram illustrating the switching signals (S_1 and S_2) input to, and consequent temperatures (T_1 and T_2) within, a PCM for (a) crystallization and (b) re-amorphisation, relative to the characteristic temperatures of the material (T_c and T_m for crystallization and melting respectively).

the other hand, is a slower process that requires the material to stay above the crystallization temperature for a minimum amount of time, often referred to as the crystallization time.

This crystallization time is an intrinsic property of the material and can be designed and optimized by varying its composition, doping, deposition conditions, and layer structure. Modern PCMs for data storage applications can crystallize in just a few nanoseconds without compromising the stability of the data, an essential requirement in-memory applications where data must be written, read, and erased in the shortest amount of time possible [7]. Designing a material that would take more than 10 years to crystallize at 80 °C but only a few nanoseconds at 150 °C was a remarkable feat of materials engineering. Although beyond the scope of this chapter, this peculiar growth behavior of PCM materials crystals has been previously linked to the extreme fragility of their supercooled liquid state [8] and their unique type of bonding mechanism more commonly seen in organic molecules [9]. The most successful commercial phase-change material is arguably the telluride $Ge_2Sb_2Te_5$, commonly known as GST, used as the starting composition for most optical disc and solid-state memories sold today [10].

11.3 Strong Interference in Ultra-Thin Absorbing Films

It can be seen from the description of PCM characteristics in the previous section that they exhibit several desirable properties for use as the light modulating medium in a display, but that as optically lossy, semi-metallic materials, the means to create a high contrast, and exceptionally high reflectivity device under broadband white-light illumination using them was not straightforward. The door to such improvements was opened by Kats and Capasso's demonstration of the strong interference effect in uniform, sub-wavelength thickness, similarly lossy metallic multilayers [11–14]. These developments showed that bright, highly colored, polarization insensitive, reflective multilayer stacks could be created by the deposition of a single ultra-thin (5–30 nm) layer of colorless absorbing material onto a metallic mirror layer, with the reflected color dependent only on the complex refractive indices of the mirror and absorber layer, and the thickness of the absorber layer. This effect results from the complex change in optical phase on reflection from, and transmission through the lossy layer, due to the large imaginary component of the complex refractive index ($\boldsymbol{n} = n + i\kappa$). This is illustrated in the images and phasor diagrams of Figure 11.2. Uniquely for interference-based reflection effects, the reflection spectra of these films have little angular dependence, and their visual appearance is almost indistinguishable from a dye-based selective absorption color film.

Such ultra-thin films can also be deposited on rough and flexible substrates without impacting the overall color or view-angle independence (in fact improving it if a less specular and more matte reflection appearance is desired) [15, 16].

The ability to tune the reflection spectrum of these strong interference films via the inclusion of a PCM with a controllable complex refractive index as the ultra-thin lossy layer was also demonstrated by Kats, Capasso et al.,

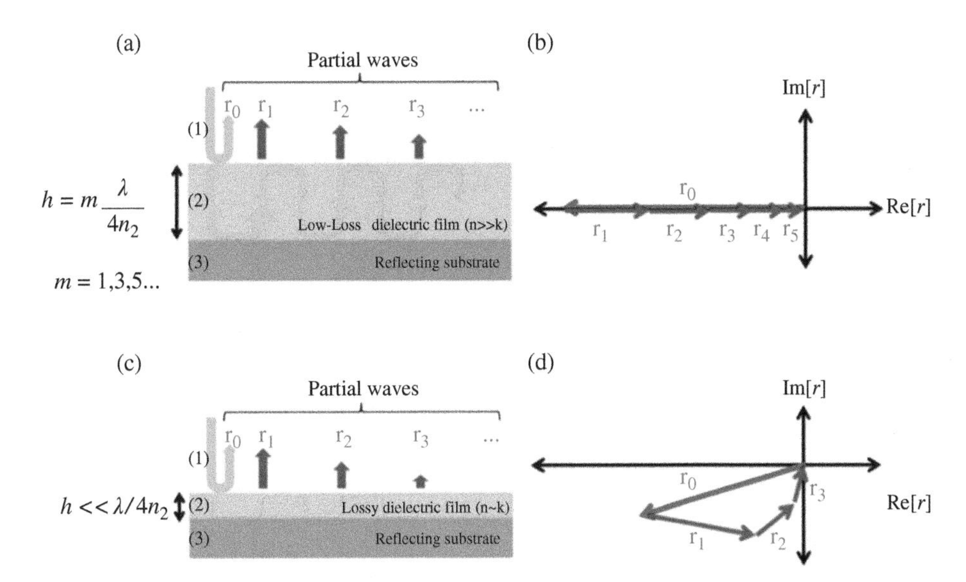

Figure 11.2 Illustrated reflection process and corresponding phasor diagram for conventional destructive interference in a quarter-wave thickness transparent layer (a–b) and strong interference in an ultra-thin absorbing layer (c–d). [15, 16].

using monostable, temperature transient vanadium dioxide as the active layer [11] and by Hosseini, Wright and Bhaskaran using the bistable, fast switching, commercially scalable PCM, GST [1]. Hosseini et al. also showed substantially improved reflected color control, saturation, and brightness resulting from the addition of transparent "spacer" and "capping" layers either side of the switchable absorber layer. As well as large-area one-time thermal switching of this improved color stack, they went on to demonstrate the ability to one-time write nanoscale images into the uniform films using conductive atomic-force microscopy (AFM) and reversibly switch nanoscale areas using conductive indium tin oxide (ITO) electrodes in a crossbar arrangement. The modified stack design, visual appearance, and AFM written image of [1] are shown in Figure 11.3.

It can be seen that with the attractive colors, electrical control, and reversible switching demonstrated in this paper, the prospect of using phase-change materials as the modulator in a new type of reflective display began to appear feasible. Further publications exploring the optics of PCM based thin-film stacks have demonstrated the possibility for intermediate switching control for greyscale [17] and multi-color switching via the inclusion of multiple, independently switchable PCM layers of different materials [18, 19], thickness, and/or position in the stack [20], via controlled structure in the PCM layer [21], or through finer temperature control of the switch to either partially crystallize or switch to an intermediate, metastable crystal structure [21–23]. These developments illustrate the range of possibilities in ultra-thin film PCM optical devices and open up the possibility for greater than binary image reproduction.

Aside from controllable color reflection for broadband visible light, there also exist further possibilities for PCM based optical devices, in very high contrast modulation of narrow-band light [24], and optical phase modulation for potential holographic applications [25, 26], as well as yet further optical control obtainable by patterning the PCM layer within a stack, or metallic features on the PCM layer, particularly with sub-wavelength features to create dynamically controllable metasurfaces [27–31].

11.4 Potential for High Brightness, Low Power Color Reflective Displays

Phase-change materials, therefore, having been previously developed for both all-optical data storage (in which reversible laser writing of microscopic areas and a reliably detectable optical contrast at a specific wavelength are utilized), and flash-type memory (in which they allow reversible all-electronic switching and read-out of nanoscale volumes of material with cycling lifetimes of billions or more), clearly have the potential for use in opto-electronic

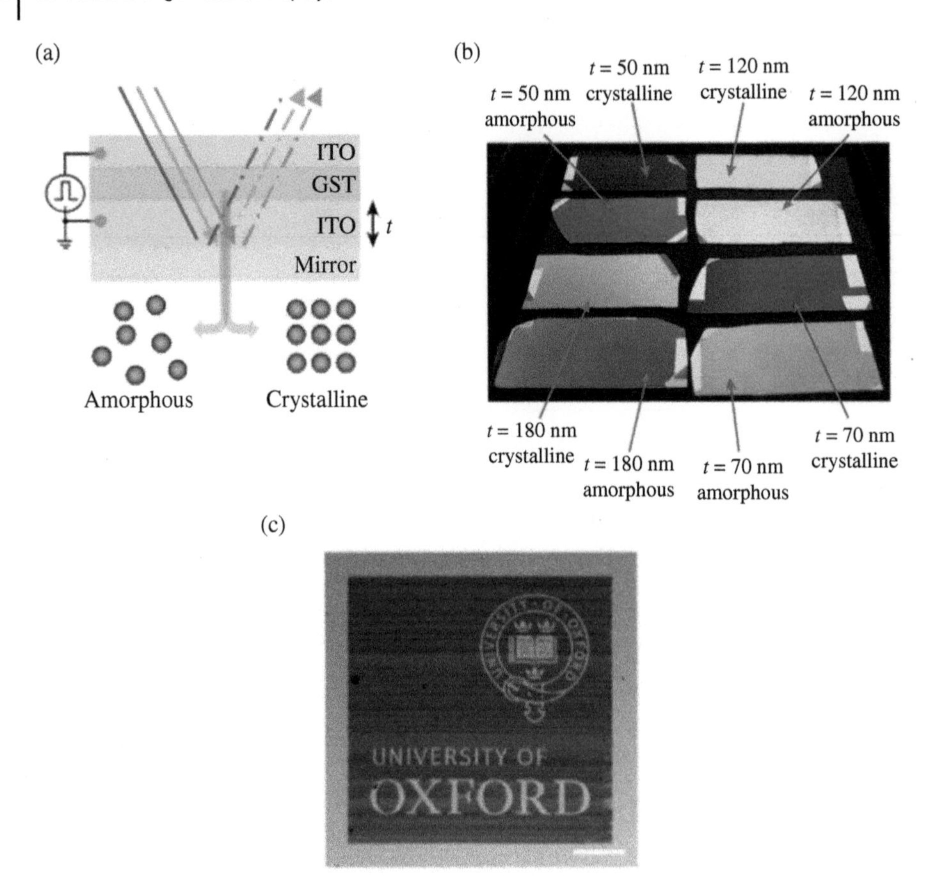

Figure 11.3 The Mirror-spacer-PCM-capping layer stack design (a), resulting spacer-layer thickness dependent color appearance of sample films thermally switched from the amorphous to crystalline state (b), and AFM written image with 10 μm scale bar (c) of [1] Hosseini. P, et.al, 2014 / Springer Nature.

devices in which electrical signals are used to control the state of the device, but observation of this state, or read-out, is via optical means. However, until recently no such devices had been experimentally demonstrated. This is likely due to the absence of any means to reversibly switch areas or volumes of PCM any larger than the microscopic active regions utilized in the memory devices, making a direct-view display of usable area impossible, and due to the limited optical contrast and poor color vibrancy observed in optical disc devices under white light illumination. However, with the ultra-thin film stack developments outlined in the previous section, the second of these demerits was removed, and the motivation to find solutions to the first limitation: larger area electrical control, consequently increased. Achieving this would enable a new class of color reflective display with a unique potential in several desirable characteristics:

- Very low power – long term bistability of the PCM crystallization state means zero power image maintenance, and the input powers required for switching PCMs in optical and electronic memory devices (10s of mJ/cm^2) [32], if scaled up to the volumes required for direct view reflective displays would still only require energies on the order of a few joules for a full image refresh. While this means overall display power consumption would be refresh-rate dependent, and high frame rate video would clearly require excessive power input to be sustainable, applications requiring fast but low-frequency image updates, such as smart-watches, fitness wearables, e-readers, and particularly electronic shelf-edge labels would be serviceable with long battery life capability.
- Fast update – the sub-microsecond switching crystallization and melt-quench re-amorphization mechanisms inherent to phase-change materials are extremely fast by display, and particularly reflective display standards.

The combination of very fast response and bistability is unusual in display technologies and useful particularly in reflective displays, commercially available options of which are either relatively slow to update (e.g., E-Ink) or require power to maintain even a static image (e.g., memory-in-pixel LCD). To date, only some bistable liquid crystal modes (e.g., ZBD, Ferroelectric LC) achieve both.

- Extremely thin and flexible – the PCM stacks of Figure 11.3a have a total thickness including the mirror layer of typically 500 nm. They can be deposited onto micron thickness flexible substrates or metal foils, illustrating the potential for extremely thin, tightly rollable, or multi-foldable displays.
- All-solid-state, inorganic material set – the semimetallic PCM alloys and metallic or dielectric intermediate layers are long-term stable and rugged to atmospheric exposure, so no glass sandwich or moisture barrier/encapsulation layers are needed, simplifying fabrication and further enhancing the thinness of a resulting display.
- Color changing pixel – the large change in the the complex refractive indexes of the PCM between states, combined with the solid interference-based effect, results in an optical switch that changes both the shape and amplitude of the reflection spectrum. This is a step toward the ultimate goal for a reflective display modulator of each pixel, providing control of both the peak wavelength and amplitude of reflection for full color control within each pixel.

However, despite this potential, significant outstanding challenges remain to solve the large area electrical switching limitation and realize a technically viable, commercially attractive PCM-based display. To address these challenges, development is needed in two key areas:

1) New PCM material development – despite the advanced level of development, and consequent scalability, proven high cycle-count lifetimes and commercial maturity of the overwhelmingly most utilized phase-change material, $G_2S_2T_5$, and the desirability of fast response materials for displays, the typically nano-second crystallization time of this material is the key limiting factor in its large area switchability. To be re-amorphized via the melt-quench mechanism, a PCM must have its temperature raised above its melting point to remove the crystal structure and then be cooled below its glass transition temperature in a sufficiently short period of time to prevent recrystallization during the cooling period. GST has melting and glass transition temperatures of approximately 600 °C and 180 °C respectively, depending on the exact composition [10, 33, 34]. In a passively cooled device, it is not possible to quench a volume of material larger than a few square microns in area, regardless of how thin the layer may be, in timescales shorter than microseconds. Realization of a display with a usable, visible area relies, therefore, on developing a PCM with much-increased crystallization time and ideally reduced melting temperature to reduce switching power requirements while retaining the optical contrast and high cycle count properties of GST.
2) Device development – given the thermally mediated nature of the PCM switch and the need to effectuate this uniformly over display pixels sized areas (tens of microns squared), an entirely new device architecture is needed for PCM based displays. A means to deliver short, high-intensity pulses of electrical energy to each pixel is required. It is also a means to convert this energy into heat within the pixel itself, efficiently and evenly over the pixel area, while still allowing passive cooling to occur quickly enough to allow amorphization. While thermo-optic and thermo-chromic effects in materials such as liquid crystals [35] and leuco dyes [36] are well known, and displays based on these have been previously developed, they have relied on either patterned wiring to selective display a fixed image [37–39] or laser write/erase to generate an arbitrary, re-writable image [40, 41]. To avoid these limitations, modification of the active-matrix backplane that forms the basis of the vast majority of high resolution, high image quality modern displays to allow fully electronic addressing of a thermo-optic medium is necessary. For compatibility with a phase-change material front-plane, this modification of the electronic backplane requires a shift from delivering signals of controlled voltage (in, e.g., liquid crystal and electrophoretic displays) or current (in, e.g., OLED displays) to an electrode that spatially defines each pixel, to delivering signals of controlled energy to a controlled volume of PCM defining the spatial extent of the pixel.

11.5 Solid-State Reflective Displays (SRD®)

To address the challenges and realize the display potential outlined above, the Solid-state Reflective Display (SRD) concept and device structure was proposed by Bodle Technologies. This display architecture is significantly different from both previous non-PCM display architectures and existing non-display PCM devices in the areas detailed in the four sub-sections below:

11.5.1 Microheater Switched PCM Pixels

To achieve reversible, large area switching of the PCM layer, a fundamental change was made away from the direct electronic switching of PCM memory devices [4, 42] and the ITO crossbar structure of [1] to adopt a buried microheater design. Directly switched PCM devices have a similar maximum simultaneously switchable area limitation as fast-crystallizing PCMs but in the opposite switching direction: the maximum volume which may be crystallized is restricted by the large (several orders of magnitude) electrical conductivity difference between the amorphous and crystalline states of the PCM material. While this difference is extremely useful in memory devices to allow low-power impedance-based read-out, when attempting to crystallize larger volumes of material, filaments of highly conducting crystalline material will form, lowering the electrical impedance of the device as a whole and funneling the ongoing current through the filaments [32]. Crystallization of the surrounding material relies on heat radiating from these filaments, inefficiently produced. This not only requires increased drive electronic complexity to account for the large and unpredictable mid-pulse electrical impedance change, but it also reduces the electro-thermal efficiency and increases the time of heating needed resulting in much increased overall power requirement.

Buried microheater PCM devices have been previously proposed for use in electrical switches and reconfigurable radio-frequency (RF) devices as a means to de-couple the electrical switching of a PCM from the highly state-dependent impedance of the PCM itself [43–45] and allow reversible switching of larger volumes of PCM material. Although optical contrast of the PCM was not the motivation in these devices, an optically visible change in the PCM over areas significantly larger than the data-dot size in re-writable optical disks was observed [45].

For such a buried microheater design to form the basis of a display pixel, switching of still larger areas is required (e.g., $50 \times 50\,\mu\text{m}$ for a 500 ppi direct view display). For a large optical contrast or strong color change, the strong interference PCM stack of Figure 11.3a, including the mirror layer, must be included [46, 47] (although the ITO layers on either side of the PCM layer may be replaced with electrically insulating dielectric layers chosen purely for their optical, thermal, and PCM encapsulating properties). This adds complexity and increases the volume and types of material subject to the thermal cycling, increasing the thermal mass of the device and, therefore, the total energy required for the switch. Increasing the thermal mass also increases the passive cooling time, which, as previously discussed, sets a lower boundary on the crystallization time a device-compatible PCM must exhibit to allow reversible, two-way switching. Nevertheless, reversible switching of $56 \times 56\,\mu\text{m}$ pixels comprising the mirror-based multi-layer stack with good optical contrast and sub-millisecond response time was demonstrated in 2017, the basic device design and resulting optical appearance for these first PCM display pixels being illustrated in Figure 11.4.

The key performance metrics for such a buried microheater PCM pixel are:

- The total required switching energy. This determines the energy to refresh the image, and therefore refresh rate dependent power consumption of any eventual display comprising these pixels. The principal determining factors of this are the pixel's effective thermal mass, and the PCM's melting temperature. These together determine how much material needs to be heated, and to what temperature to effect a switch, and it is desirable to minimize both.

(a)

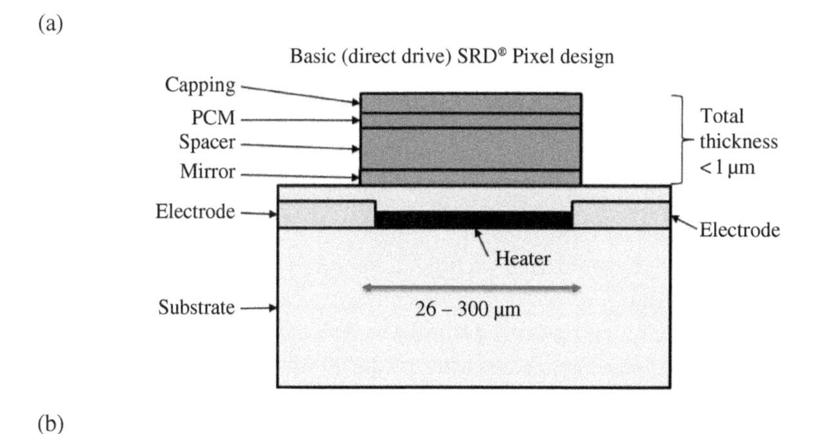

(b)

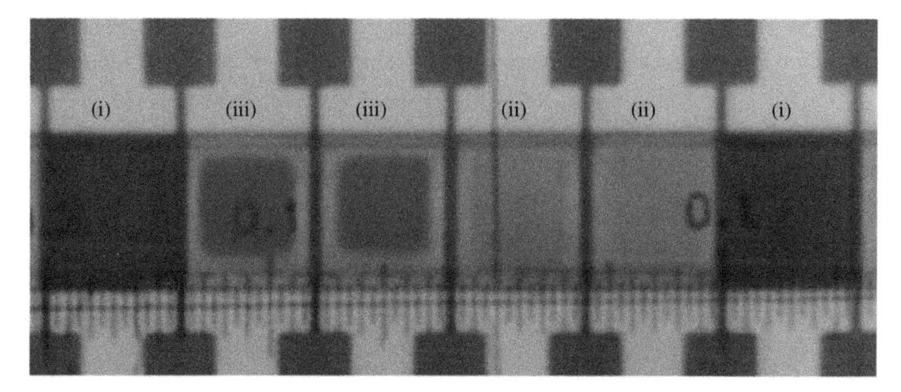

Figure 11.4 Illustrated device schematic (a) and resultant visual appearance in the as-deposited (i), crystallized (ii) and re-amorphised (iii) states [46] Broughton, B. et al. 2017 / John Wiley & Sons, Inc.

- The maximum instantaneous electrical power required to switch. This will determine the compatibility of a pixel design with any given drive electronics. To an extent, a trade-off exists with the total required switch energy, as a higher maximum instantaneous power allows the switching energy to be delivered in a shorter time period, reducing the degree to switch thermal conduction can heat material surrounding the pixel, reducing the effective thermal mass. However, reducing the maximum power reduces the demands placed on the backplane and drive electronics, simplifying the design and fabrication of the display.
- The heat uniformity over the area of the pixel. Switching requires the active PCM layer of the pixel to be raised to either the melting temperature or crystallization temperature of the material, and it is desirable to do this as uniformly as possible across the pixel area to maximize the optically active area while preventing overheating damage and wasted energy by excess heating within the pixel area, and unnecessary heating of material outside it.
- The time to passively cool from the PCM melting temperature to the crystallization temperature. As previously discussed, this is the ley limiting factor for compatibility for a PCM material composition with the display pixel. A faster cooling pixel increases the range of PCM alloys that may be selected from, thereby increasing the possibility of developing a compatible material with the desirable high optical contrast and cycle count.
- The electrical impedance of the heater. This determines the proportion of the input electrical switching pulse energy delivered to the pixel, which is lost to parasitic impedances in the complete electrical path from the source driver to the ground or return path. Given fixed parasitic impedances, it will always be desirable to maximize the resistance of the heater to maximize the proportion of the pulse energy which is dissipated in the heater itself. However, the practical limitation of a maximum voltage deliverable by the drive electronics means

that effectively impedance matching the heater to the parasitic elements in the current path will maximize the energy delivered to the heater, albeit at the expense of equal energy lost to the parasitics.

Further optimization of the pixel design to these targets, using electro-thermal finite element analysis (FEA), shows that, perhaps unintuitively, both the switching energy and the cooling time can be decreased by insulating the heater from the substrate [48], using vias through an insulating layer to make electrical connections and insulating the pixels laterally from their neighbors. Although this improved insulation reduces the rate at which the heat of switching dissipates, this effect is outweighed by the reduction in effective thermal mass, so less heat energy is present to dissipate after raising the pixel to the required switching temperature. Additionally, the FEA shows that using a serpentine heater design rather than the uniform plate of Figure 11.4 allows both the heat uniformity and electrical impedance of the heater itself to be optimized simultaneously. The material used for the heater and its thickness may be controlled to produce the desired electrical resistance, while its 2D geometry may be designed for optimal heat uniformity during the switching pulse, giving a degree of independent design control of these two important factors. The results of these FEA optimizations show that a $50 \times 50\,\mu m$ pixel may be designed to operate with switching energy of the order of a micro-Joule, cooling time of less than 10 μs, and heat uniformity at the PCM layer (defined as the proportion of pixel area at a temperature greater than the melting temperature, but less than 50 °C above this) of 99%. Output visualizations of such a design are shown in Figure 11.5.

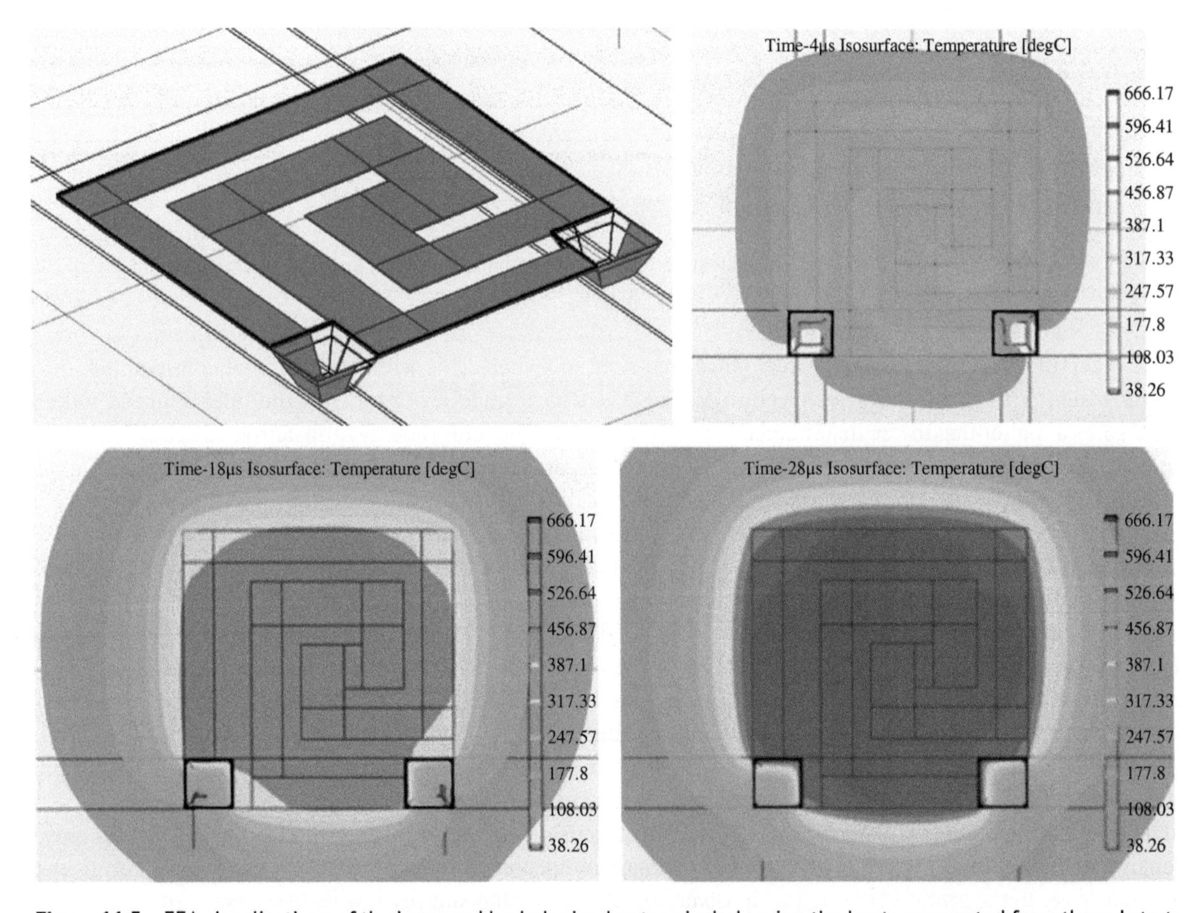

Figure 11.5 FEA visualizations of the improved buried microheater pixel, showing the heater separated from the substrate and electrical signal lines by an insulating layer, and the heat uniformity within the pixel area achievable with a serpentine heater design [49].

Finally, as well as the improved maximum switchable area, electrical control simplicity, and heat uniformity provided by the buried microheater pixel design, a further benefit is the complete separation of the electrical and optical portions of the device. The mirror, spacer, PCM, and capping layers may be selected solely for their thermal and optical properties, increasing the design freedom and allowing improved color control, and the drive electronics may be completely hidden behind the mirror layer, providing potential for very high pixel fill-factor, and an overall display appearance which may be blended seamlessly with passively reflecting surroundings [50].

11.5.2 RGB-to-Pale Pixel Color, Controlled by Layer Thickness

Although intermediate switching of PCM based optical devices has been demonstrated [17], they are significantly easier to control, and the nonvolatility of the switched-to state is more secure when operated as binary devices switching between fully crystalline and fully amorphous states. This poses a significant challenge to developing PCM-based pixels toward a full color reflective display. In addition to the lack of readily obtained greyscale, and despite the large and striking color differences exhibited between the two states of published PCM based optical devices, little has been demonstrated showing a switch to a black state (under broadband visible white illumination), as would be required for a high contrast display.

To attempt to overcome these limitations, and at least point the way to a route by which PCM displays may develop toward full color capability, it has been proposed to combine the color-changing PCM modulator with a switchable absorber acting effectively as a shutter layer over the reflective PCM device providing intensity modulation [51, 52]. Such a two-layer, dual modulator device could exhibit very good color selectivity, contrast, and brightness, as long as the color and reflectivity of the underlying PCM stack are sufficient to overcome any losses in the absorbing modulator when in the transparent state. To this end, it was established that, by controlling the thickness of the spacer, PCM, and capping layer of a PCM interference stack of the type in Figure 11.3a, it was possible to design three different stacks (all using the same materials), each exhibiting a switch from a pale, high overall reflectivity state, to a saturated, vivid color state of either red, green or blue [52]. The measured reflection spectra of these PCM interference stacks in their amorphous and crystalline states, and the resulting visual appearance of the stacks under natural sunlight, are shown in Figure 11.6.

This pale-to-vivid reflection capability opens the possibility to design a sub-pixelated display, with side-by-side RGB pixels, which overcomes the long-standing shortcoming of such an architecture in reflective displays – that the white state is a combination of reflections from the R, G, and B sub-pixels, each occupying less than one-third of the display area, so the maximum reflected white luminance is very limited and compromised directly with the color saturation of the individual sub-pixels. In any area-sharing sub-pixelated reflective display comprising sub-pixels of fixed reflection spectrum, to reflect a vivid, saturated color the sub-pixels must reflect a narrow band of R, G, or B wavelengths, and the narrower the band, the lower the reflectivity of the white state produced when they combine. This fundamental trade-off is circumvented in this design, as each of the R, G, B sub-pixels can be switched to a pale, high reflectivity state (ideally as close to white as possible) according to the input image data, so both a high reflectivity white state and vivid, saturated colors may be displayed simultaneously.

,To demonstrate this and ascertain the maximum reflected luminance and color gamut which could be expected from a sub-pixelated pale-to-R, G, or B, PCM display with additional top-shutter for greyscale control and black capability, the mock-up image of the parrot shown in Figure 11.7 was produced. Although only a static mock-up, the visual quality of this optically genuine demonstrator is a clear indication of the potential reflective color image quality attainable from a PCM based display specifically, but also any reflective display comprising sub-pixels capable of switching between black, white, and one of red, green or blue. The majority of attempts to overcome the color-gamut vs. white brightness compromise in color reflective displays have taken either the complex three-layer cyan-magenta-yellow to transparent approach or added a white sub-pixel at the expense of R, G, and B luminance. The ultimate solution has been held to be full color changing pixels that can switch from black to white to any color over their full area within a single layer [53], which to date has only been achieved using a

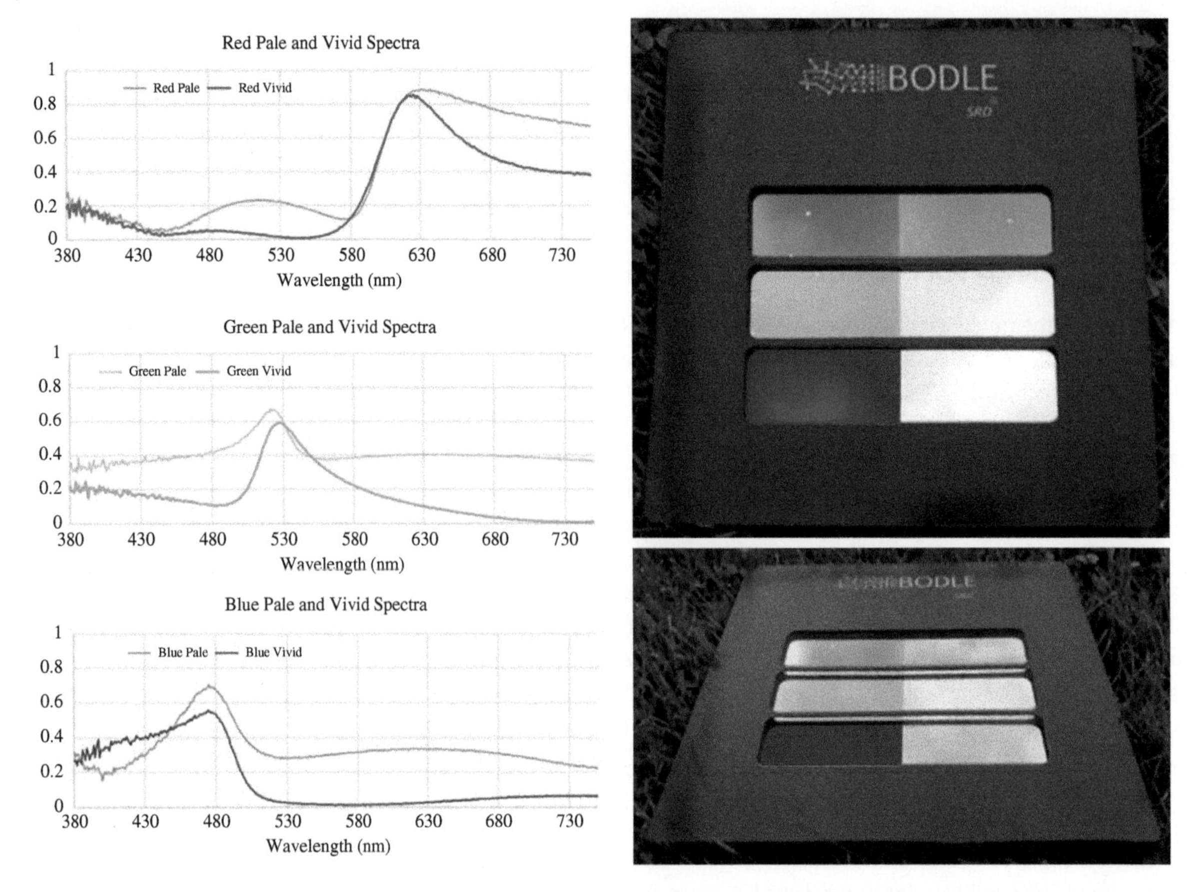

Figure 11.6 Reflection spectra of PCM interference stacks designed to switch between a saturated red, green or blue color and a high overall reflectivity pale color, and the visual appearance of these stacks under natural sunlight, at a viewing angle close to normal, and at high incidence, showing both the vibrancy of the reflected colors and the high degree of angular independence exhibited by the stacks [52] Talagrand, C. et al. 2018 / John Wiley & Sons, Inc.

four-pigment electrophoretic display with a complex drive scheme [54]. However, this result shows that a potentially much simpler method, with each sub-pixel capable of three-color states only (white, black, and one of R, G, and B) can provide color magazine standard reflected image quality.

The image quality of this demonstrator is quantified as having a white-state luminance reflectivity of $Y = 0.39$ (or $L^* = 69$) and color gamut of 61% of sRGB as shown in the reflection colorimetry measurements of Figure 11.8.

11.5.3 Phase Change Material Development

PCMs have a long history of development specifically targeting data storage applications and culminating in the commercialization of optical discs (CD-RW, DVD-RW, BR-RW) and, more recently, non-volatile solid-state memories [2]. However, it is important to highlight that the material requirements for display applications are considerably different from those of memory applications; a short comparison is presented in Table 11.1. The main differences are the speed of crystallization and the stability of the amorphous phase at room temperature.

Thermodynamically, the amorphous phase is a meta-stable state that will always crystallize given enough time. Data retention in memory applications is paramount, and years of R&D efforts were directed toward finding a

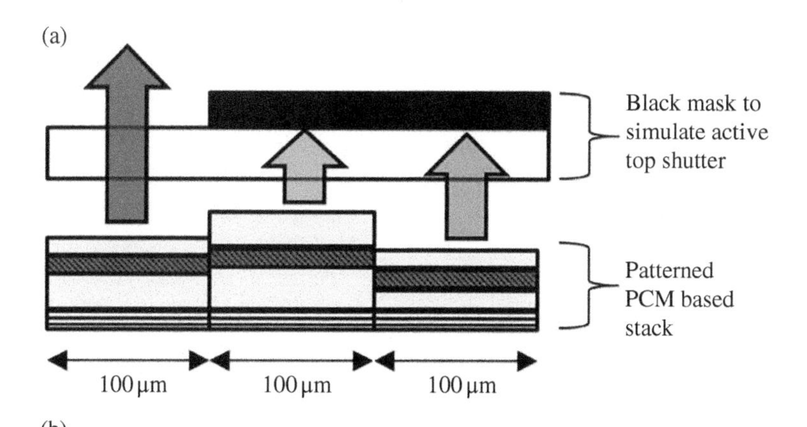

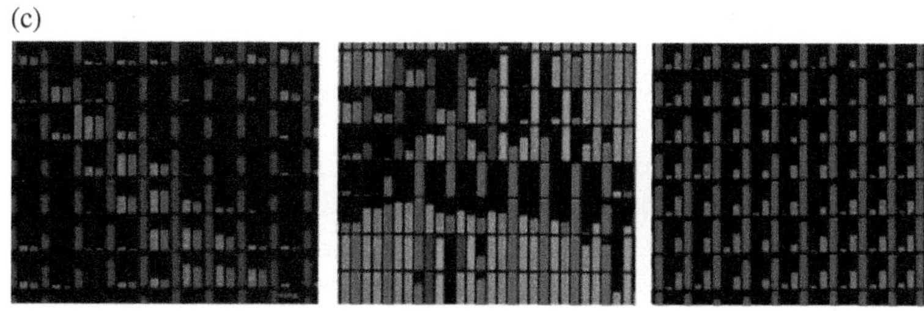

Figure 11.7 (a) Cross-section schematic of an RGB sub-pixelated PCM stack, with spatially patterned absorber top-shutter, (b) resulting visual appearance of a static demonstrator test image of a parrot using lithographically defined pixels of this type, and (c) microscope images of different regions of the parrot image, showing clearly the R, G, B sub-pixel stripe pattern, with each sub-pixel type switching as required by the image data between pale and vivid states, and the patterned absorber layer providing a spatial approximation to greyscale [52] Talagrand, C. et al. 2018 / John Wiley & Sons, Inc.

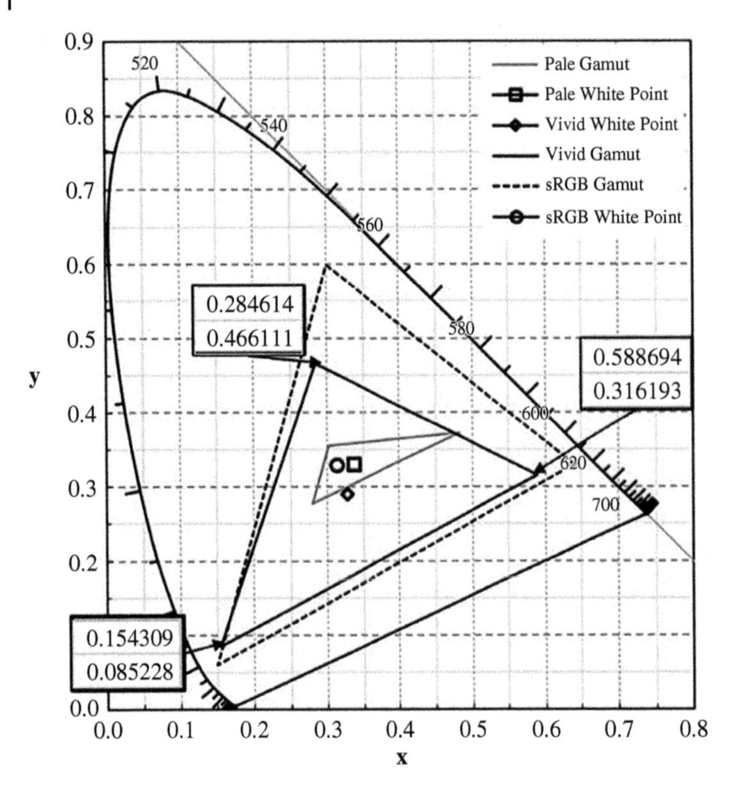

Figure 11.8 CIE 1931 chromaticity diagram showing the color gamut produced by the static demonstrator. A gamut area of 61% of sRGB is obtained with the R, G and B sub-pixels all switched to their vivid state, and a combined reflected luminance of Y = 0.39 (or L* = 69) is produced with the sub-pixels all switched to their pale state [52].

Table 11.1 Comparison between PCM requirements for data storage and display applications.

	Data storage applications	Display applications
Crystallization speed	<<100 ns	>40 us
Cycle-ability	> 10b cycles (ON/OFF)	>10 k cycles for Electronic Shelf Edge Labels; >1 M for e-Reader applications; >1b for video applications;
Stability of the amorphous phase at room temperature	10 years at 80 °C	>1 day at 80 °C
Melting point	>650 °C	<300 °C
Single active area size	<500 nm	(50 × 50) μm

PCM composition with high stability of the amorphous phase without compromising its crystallization speed. Using an ultra-stable PCM in a display product would be beneficial only for applications where a form of "extremely bistability" (>10 years at 80 °C) is needed. Bistability can be seen as an energy barrier that ensures display image retention at room temperature without an external energy source. If the energy barrier is too high, the power consumption of a display product at a video rate will inevitably become too high for practical applications. Designing a PCM with the right balance between stability and power consumption is one of the key challenges in adopting phase change technology in display applications.

Another fundamental requirement is the speed of crystallization and the ability to re-amorphise a PCM using a microheater device. Figure 11.4 shows a diagram of an SRD microheater electrically driven to high temperature in a short but finite amount of time. The pixel dimensions are roughly (50 × 50) μm which translates into a display with a very high resolution of 500 ppi. A microheater device of the corresponding size (50 × 50 μm) has a thermal time constant of roughly 10–20 μs and will passively cool down from slightly above the melting point of the PCM to room temperature in about 40–50 μs. If the crystallization speed of the PCM is shorter than the natural cooling time of the device, the PCM will always re-crystallize, making the amorphous phase impossible to access with a standard microheater device. It is important to highlight how moving to smaller pixels would naturally decrease the thermal constant of the microheater device and allow the use of faster crystallization materials [55]; however the display resolution would end up being in the thousands if not tens of thousands of ppi: too high for direct view applications.

Most widely studied PCM materials are often ternary and quaternary material systems with one or more chalcogen elements as their main components. Modifying the composition, even by a single-digit change in atomic percentage, might have dramatic effects on the material's functionality. Traditional material engineering approaches like serial synthesis -> test -> analyze loops are unbearably time-consuming and prone to errors. The simple audacity of journeying into a new PCM application is another source of uncertainty and unwanted complexity; very few research groups have, to date, published work on PCM-based reflective displays, information is scarce.

To design a PCM suitable for display applications in a reasonable timeframe, a high-throughput deposition approach with a combined automated test and analysis algorithm was developed. The project's main target was to explore one or more ternary systems of PCM materials and develop a model that would help identify materials with performance targeting display applications. The project was divided into three main areas:

1) Deposition. Ge, Sb, and Te were deposited on a single 2×2 in. sample using a compositional gradient approach. Depositing a film using a gradient creates tiny areas where the PCM has a unique composition. Controlling the deposition condition ensures obtaining an area almost as large as the entire ternary diagram of GST, deposited in one single run on a single sample.
   The samples is then probed by a static laser system designed to measure the functional properties such as:
   a) Crystallization speed (in μs).
   b) Optical power is required for crystallization and re-amorphization.
   c) Cycle-ability or how many times the material can be switched between the amorphous and crystalline phases without significant degradation.
   d) Optical contrast between the two phases.
2) The data is then moved into a structured database and saved for further analysis and visualization.

Such high-throughput compositional gradient approaches have been previously developed and utilized to explore PCM compositional space for, e.g., melting temperature, electrical resistivity and optical contrast of GST [34], but crystallization time measurements have only previously been published on a handful of individually fabricated compositions [56, 57]. An example result from this high-throughput experiment can be found in Figure 11.9. Here the crystallization speed of the material, as extracted by the static laser system, is displayed as a function of composition. A striking high crystallization speed "valley" can be clearly distinguished in the pseudo-tie-line between Sb_2Te_3 and GeTe. This is in agreement with previous literature on GST, and it is the tie line were the fastest compositions, including the aforementioned $Ge_2Sb_2Te_5$, are found.

The results clearly indicate how "slower" Sb-rich compositions, with crystallization speed in the order of tens of microseconds, are available to enable reversible microheater switching of direct-view display-sized pixels. These, and other newly developed, non-GST comprised, slow crystallizing PCMs were utilized in the display demonstrators detailed in the following sections, and further PCM composition development is continuing with the aim of increasing pixel cycle count lifetimes and optical contrast and reducing switching power requirements,

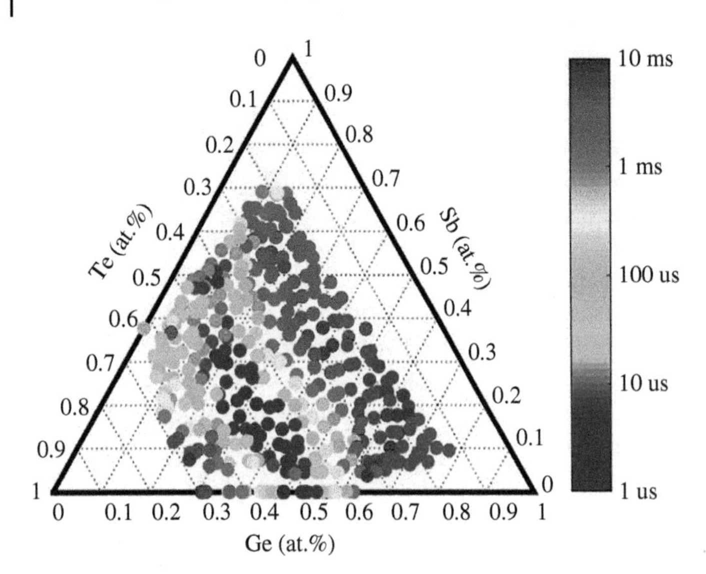

Figure 11.9 A high-throughput, combinatorial sample of the GST ternary diagram is deposited and then measured using a static laser system. The graph shows the minimum crystallization speed required by the material with regards to the composition.

with the aim of both improved PCM display performance and expanding the range of applications which PCM displays may be applicable. In fact, recently, a new class of PCM designed specifically for optoelectronic and photonic applications, based on Ge–Sb–Se–Te (GSST) and exhibiting high refractive index contrast but much-reduced absorption, has been reported [58–60], with reversible microheater switching of a 10 μm pixel demonstrated [19]. This additional example of application-specific material design further indicates the wide scope for new material development and the improved device performance and broader application space such discoveries may enable.

11.5.4 LTPS Backplane Development

To scale up from a single line of directly driven pixels to a large area array of matrix-addressed pixels, and thereby realize a high resolution, visible area, direct-view display, it is necessary to demonstrate the compatibility of the pixel design with a thin-film electronic backplane. To deliver a switching pulse to a given pixel in an array, via activation of row and column signals, each pixel must contain an electronic selector element to prevent dissipation of the drive signal to neighboring pixels in the array and ensure efficient delivery of the switching energy to the intended pixel. The binary, bistable nature of the PCM device means a passive-matrix drive of large arrays is possible, and PCM memory devices utilize an electrically non-linear ovonic-threshold switch (OTS) element as the selector in each bit region to make such passive arrays scalable and realize this [19, 20, 61], so this approach may well be practical for development of PCM displays also. However, active-matrix thin-film transistor (TFT) arrays are a highly mature technology underpinning the majority of displays of all types (liquid crystal, OLED, electrophoretic) in use today, making possible displays of remarkable pixel resolution and image size, and so demonstrable compatibility with a commercialized TFT backplane technology is a key advantage for any emerging display technology.

The results of the buried microheater pixel development outlined in Section 2.1 determine the requirements of any backplane technology for PCM pixel driving: an ideal in-pixel selector will be capable of high electrical current throughput with low impedance in the activated ON-state, have a high breakdown voltage in the OFF-state, and ability to withstand the thermal cycling inevitably experienced during repeated switching cycles. It was determined from the testing of single-pixel devices that a low-temperature polysilicon (LTPS) backplane would be capable of delivering the required switching power to microheater pixels, but that a customized design using the full 50 × 50 μm pixel area to maximize the transistor channel width over length ratio (W/L), and therefore ON-state current capability, of the LTPS selector, would be required. To date, Bodle Technologies has worked with

Xerox's Palo Alto Research Centre (PARC) and Tianma Microelectronics Co. Ltd. to develop LTPS TFT and LTPS diode backplanes. Then Taiwan''s Industrial Technology Research Institute (ITRI) developed a scalable TFT-based design [49].

With PARC, a design incorporating an unusual TFT geometry with laterally extended source line extended vias connecting to the polysilicon layer shared between neighboring pixels in a column, and an isolated drain contact island (also with extended vias), was adopted in order to minimize the path-length for current on the relatively high resistance doped polysilicon. This, combined with a highly meandering channel for maximal W/L, ensured a high ON current capability with low impedance. The design is shown in Figure 11.10a fabricated to the TFT level, and in Figure 11.10b with the microheater, connecting the drain island in each TFT to the drain mesh (common to all pixels) subsequently added, on top of a TFT planarization and thermal insulation layer which sits between the TFT and the microheater.

With Tianma, an alternative design using diodes rather than TFTs as the selector element was adopted. This design leverages the fact that the PCM switch's bistable, fast response nature removes any requirement for a sustained electrical current or voltage to be held on the pixel during the frame time, as with most other display technologies. Therefore an image may be written to an array with a sequence of very short pulses applied to the pixels in series, each pixel remaining in the state it was switched to while the other pixels in the array are addressed, and indeed after completion of the full array update. No TFT or storage capacitor is needed in the pixel; therefore, maintaining the addressing signal and a simple diode is sufficient to ensure each addressing pulse is delivered to the intended pixel via the corresponding row and column bus lines. This utilizes the fact that for current passed from any given row to column line to travel through any pixel other than the one at the primary intersection of these lines, it would have to travel through at least one other pixel in the reverse direction (i.e., through another diode but with opposing bias), and is therefore blocked.

Similarly to the PARC TFT design, LTPS was used as the semiconducting material for its high carrier mobility, and an interdigitated lateral p-n diode design, occupying the full 50 × 50 μm pixel area, was adopted to maximize the W/L of the junction (analogous to the channel in a TFT design) and thereby ON current. Microscope images of the Tianma diode backplane fabricated to the diode level (a), with microheater subsequently added (b), connecting the via formed at the end of each diode distal to its row line, to a via connecting to the source line, and with the mirror layer and PCM stack deposited (c) to complete the pixel fabrication, are shown in Figure 11.11.

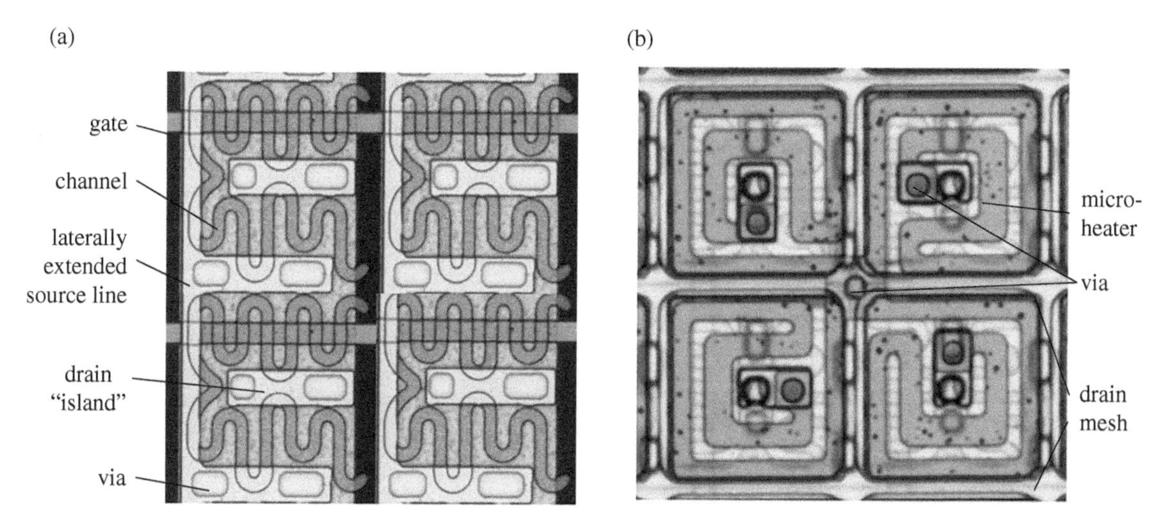

Figure 11.10 Microscope photographs of a 2 × 2 pixel region of the PARC LTPS TFT array before (a) and after (b) the serpentine microheater is deposited to connect the TFT drain via and ground mesh, showing the meandering channel design for maximized W/L within the 50 × 50 μm pixel area.

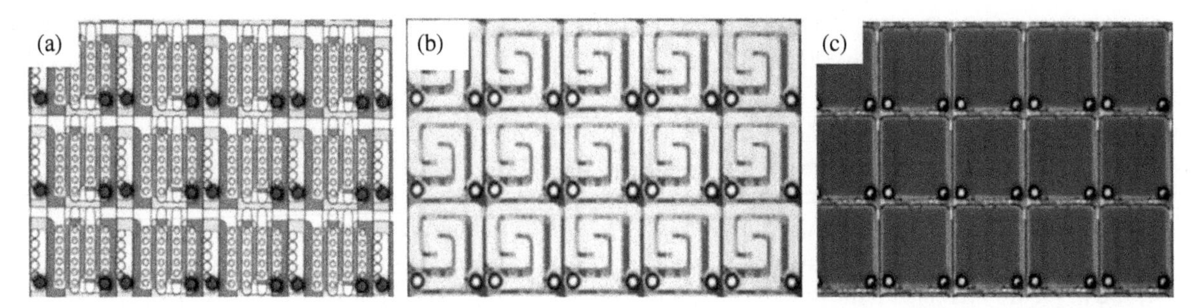

Figure 11.11 Microscope photographs of a 3 × 4 pixel region of the Tianma LTPS diode array before (a) and after (b) the serpentine microheater is deposited to connect the diode row and column vias, showing the interdigitated design for maximized W/L within the 50 × 50μm pixel area, and a further image (c) showing the same pixels after deposition of the PCM interference stack [49].

The primary electrical performance metric for these pixels is the maximal electrical power deliverable to the microheater. In simple terms, power is the product of voltage and current for a d.c. signal, this is determined by the maximum voltage which the pixels will withstand before a breakdown in the OFF condition (for the TFT design, being with the gate signal held at the voltage setting a hard OFF state, and for the diode, design being with a diode in reverse bias), and the current which this voltage will drive through the row and column bus lines, the TFT diode in its ON (or forward bias) state and the microheater in series. Having designed the row/column lines/mesh and the TFT or diode for minimum electrical impedance in the ON state, the current is increased by reducing the microheater resistance. Still, the share of pulse energy received by the heater, as opposed to being lost in the other resistive elements in the current path, is increased by increasing the microheater resistance. Optimal microheater impedance is therefore determined by impedance matching the heater to the other parasitic elements at the frequencies used for the signal pulses, and this may be achieved by choice of material, thickness, and 2D geometry, of the serpentine shape. Both the TFT and diode designs were found to be high power capable, delivering in excess of 100 mW to the microheater: more than enough to heat the PCM layer to its melting temperature for re-amorphization in microseconds.

Also, crucially for the operation of the display, the planarization and insulation layers between the TFT/diode and the microheater were designed to protect the semiconductor from thermal damage and minimize the effective thermal mass of the pixel by minimizing heat flow down from the microheater toward the LTPS backplane, thereby minimizing the time required for passive cooling and enabling a fast re-amorphization. The key functions of electrical power delivery, thermal ruggedness and fast passive cooling were realised by the backplanes in the first demonstration of reversible electro-optical switching of array driven PCM pixels in 2018. Figure 11.12 shows this array of 24 × 24 pixels, based on the Tianma diode backplane, with direct connections from each of the 48 signal lines to an FPGA programmed to deliver the required crystallization and amorphization pulses in sequence.

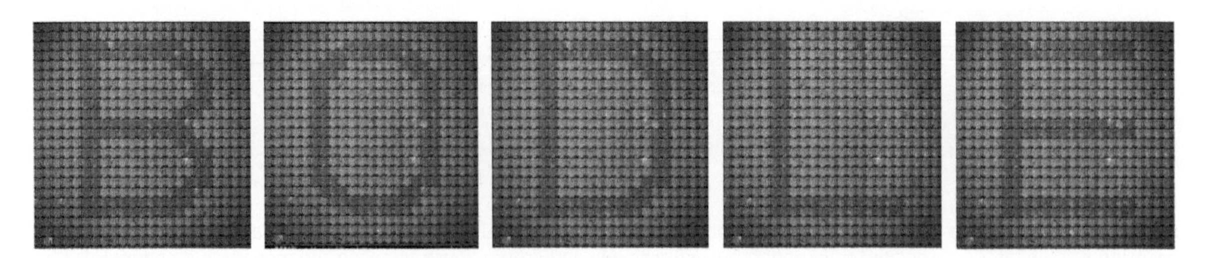

Figure 11.12 Microscope photographs of an array of 24 × 24 SRD pixels being sequentially updated with a series of letters [49] Castillo, S. G. et al. 2019 / John Wiley & Sons, Inc.

11.6 SRD Prototype – Progress and Performance

In a conventional active matrix display, source and gate drivers are used, typically in the form of integrated circuits (ICs) bonded to the array row and column lines as a chip-on-glass (CoG), to avoid having to connect all the signal lines to the display controller (as is required in the prototype of Figure 11.12) and enable large-high resolution arrays with only power, timing and a few serial image data connections needing to be delivered to the display from off the substrate. Due to the non-standard switching pulses of relatively high intensity and short duration required to activate the microheater pixels in SRD arrays, however, and to retain full freedom to experiment with the waveform shape in optimizing these signals, SRD prototypes were designed with large, very low ON resistance TFTs at the end of each row and column signal line, to enable these lines to be selectively activated by a standard gate driver IC CoG (Ultrachip UC8124C), with signals from this driver IC to the gates of the line-end TFTs. An activated row and column line pixel at the intersection may then be driven by switching pulses of arbitrary voltage amplitude and waveform, delivered directly from an off-glass FPGA. This array addressing scheme is outlined in Figure 11.13.

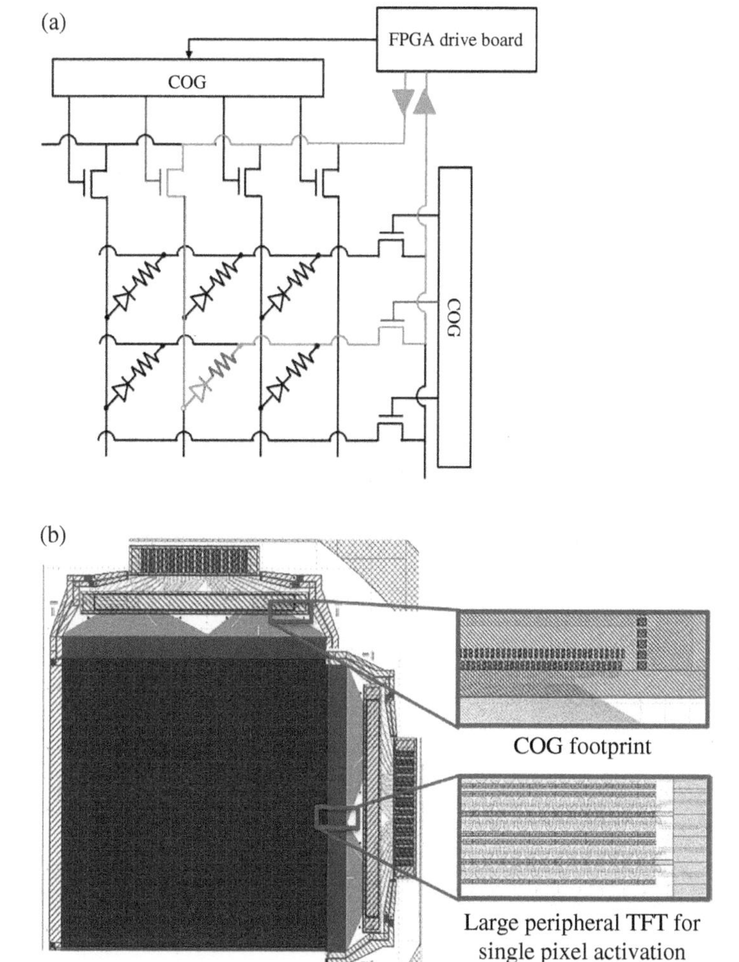

Figure 11.13 (a) Schematic illustration of the prototype array showing the low current (in blue) row/column selection signals produced by the CoG activating the gate of selected line-end TFTs, and high current, arbitrary pixel switching signals (in green) to and from the off-glass FPGA via the intended microheater, and (b) the complete 480 × 480 pixel CAD layout with detail showing the CoG footprint and large (1 mm × 50 μm) line-end TFTs.

Although this scheme limits the update of the display to a single-pixel raster and therefore a full image refresh is slow, it demonstrates clearly the capability of both diode and TFT LTPS backplanes to address large microheater-based PCM pixel arrays. It provides a valuable development platform for testing new PCM materials and optical stack configurations. It is envisaged once an optimum front plane material set is developed and the required switching pulses for the operation of this settled design are determined, SRD specific row and column drivers will be selected or designed, enabling simultaneous full-row addressing for a fast image update.

Fabrication of a complete SRD display is simplified by the electro-thermo-optical mechanism employed and the complete electrical separation this allows between the back and mid-planes (roughly defined as the LTPS matrix and the microheater and insulation layers, respectively) and the optical front plane. This allows the backplane to be fabricated in a dedicated LTPS display facility, using the same processes and mask-steps as a conventional LCD or OLED backplane fabrication before subsequent physical vapor deposition (PVD) of the frontplane. For Bodle's SRD prototypes, PARC and ITRI fabricated the LTPS backplanes up to the completed microheater layer on 4″ glass wafers and Gen 2 substrates, respectively, as shown in Figure 11.14. Tianma fabricated backplanes to the LTPS passivation layer on a commercial Gen 5.5 LCD production line, with vias left open for subsequent deposition and lithography of the microheater. These backplanes were then finished into complete prototypes by sputter deposition of the remaining dielectric, mirror, and PCM layers, with lithography steps to pattern the microheater (if necessary) and mirror layers, separating the pixels, followed by bonding of the CoG drivers and ribbon connectors. No mechanical protection or environmental encapsulation layers are necessary, as the PCM and inorganic capping layers are rugged to atmospheric exposure. Still, an optional diffuser layer may be applied to change the reflective appearance of the display from a metallic, specular reflecting surface to a diffuse, matte appearance if desired. As all additional layers required above the LTPS backplane are sputter deposited, solid thin-films, the entire display (not including substrate) can be fabricated with a thickness of less than two microns in total, making future development of an ultra-thin reflective display, with extremely small radius folding or rolling, a possibility.

11.6.1 Fully Functioning Display Demonstrators

Display prototypes fabricated according to the methods outlined above were demonstrated and evaluated, showing full write, erase and re-write capability of high resolution (480×480 pixels, 500 ppi), two color images.

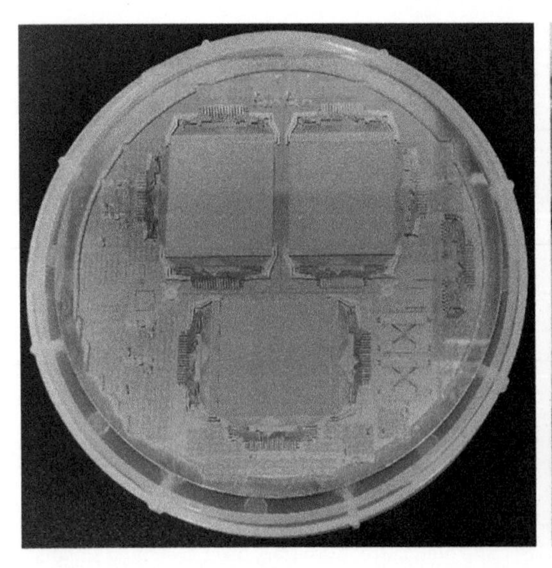

Figure 11.14 Examples of LTPS SRD backplanes for 1.3″, 480 x 480 pixel arrays, fabricated to the microheater level on 4″ glass wafers at PARC (left) and on Gen 2.5 pilot line at ITRI (right).

Figure 11.15 Photographs of the 480 × 480 pixel SRD display prototype, showing an image mid-update, partially written to the display (a and c), and a complete image retained after disconnection from the drive electronics, under natural indoor lighting (b) and outdoor sunlight (d).

Figure 11.15 shows photographs of the prototype connected by ribbon cables to the custom control FPGA, mid update, and after completion of the pixel raster, with the cables disconnected to demonstrate the vivid color appearance and zero-power image retention.

These fully functioning prototypes answer several key questions about the viability of any emerging display technology by virtue of showing verified compatibility with large-area backplane technology, fabricable at scale on a commercial display production line. They also show that several of the potential obstacles particular to making a PCM based display are surmountable, namely:

- Thermal efficiency and full-pixel area heat uniformity sufficient to allow single-pixel switching through a TFT or diode without thermal "crosstalk" or leakage affecting neighboring pixels.

- Compatibility of standard planarization/passivation layers with thermal switching, both in the intermediate layers' robustness and their ability to sufficiently insulate the TFT/diode to prevent thermal damage to those elements.
- Feasibility of PVD deposition of the optical frontplane subsequent to the backplane fabrication, with the required adhesion, layer planarization, and thickness control to produce the intended color appearance, providing a clear route to efficient and economical production.

However, they require further development to fully realize the potential of a PCM-based color reflective display and achieve the required optical and lifetime performance for commercial viability. Development to these ends will continue, in particular, targeting:

- Improved optical quality. The display prototypes exhibit color saturation and luminance below the levels observed in uniform thin films. It is believed this is due to the surface structure, exposed inter-pixel gaps, and imperfect planarization within the pixel area, causing uncontrolled reflections, diffraction, and scattering, which degrades the reflected color purity. It is expected that improved layer control and a black mask matrix between pixels can solve these issues.
- Reduced switching power and increased image refresh speed. The electrical power needed to switch a pixel is the ultimate determinant of refresh rate in a PCM display. Depending on the application, and display size, there will be a maximum power the source driver can deliver to the display as a whole, and this will be divided to switch as many pixels in a row simultaneously as possible. Due to the high instantaneous power needed to re-amorphise PCM pixels, relative to established display technologies, a typical power budget for a mobile display during the image update cycle of 1–3 W is unlikely to switch all pixels in a row of hundreds simultaneously. As PCM displays only need to deliver power to those pixels which are required to change state in an image update, and the power required to crystallise a pixel is typically ½ to ¼ that required to amorphise, driving a full row will not be necessary. Yet improved thermal engineering of the pixel material set and structure, in particular, toward reduced thermal mass and more complete energy delivery to the microheater itself, as well as potential improvements in the PCM materials themselves to lower the switching temperatures, is likely to make the difference between refreshing a watch or shelf-edge label sized display in under a second, and enabling video-speed refresh rates.
- Cycling lifetime. As discussed in Section 2.3, modifying the PCM alloy used in these SRD prototypes away from the highly developed, well-proven $Ge_2Sb_2Te_5$ archetype to increase the crystallization speed has allowed PCM microheater pixels to function, but this has been achieved at the expense of a reduction in the cycle count lifetimes observed in both optical and electrical switching of the material. There is an effectively infinite parameter space for PCM alloy compositional variation, small-proportion elemental doping and deposition conditions to explore, as well as scope for the addition of interface layers, known to stabilise PCM alloy compositions to provide much increased cycle counts in other applications, but not yet explored in PCM displays. So, although a material development effort to discover new alloys with pixel area compatible crystallizing speeds, high optical contrast, low melting temperature, and cycling lifetime in the range of 10^9–$10^{12,}$ as shown by GST, is necessary to yield a commercial display and expand the range of display applications PCM devices can address. This effort will need to be large and wide-ranging; there is good reason to believe a new composition ideal for display applications will be found.
- Full color capability. To realize the color image reproduction quality demonstrated by the static PCM reflector of Figure 11.7, which uses a fixed, spatially patterned black mask layer to provide high contrast and greyscale, in a dynamically updateable display, either a dual-layer display incorporating a switchable top-shutter on the array of two-state PCM pixels of the existing SRDs is required, or further development of the PCM device itself to enable a black state for high contrast, ideally as an additional third state to add to the vivid red, green or blue state and high reflectivity pale state produced by the stacks of Figure 11.6. Routes to such three or four state PCM devices have been shown [19, 20, 61], and several candidate technologies for a top-shutter with high transmission exist, including polariser free liquid crystal [62, 63], electrochromic [64], electrophoretic [65] and electrowetting [66] devices. As the base SRD display is fabricated on a single substrate and is extremely thin, scope exists to fabricate a dual-layer display using the SRD display as the active matrix substrate in a conventional LCD

or eWetting sandwich configuration, with minimal parallax issue with viewing angle and independent driving of both modulators from the same single TFT or diode matrix [67].

11.6.2 Potential Applications

If the development targets outlined above can be achieved, it is clear the SRD will have several unique attributes making it an attractive solution for multiple applications. Ambient light readable, direct view reflective displays requiring fast update and long periods of zero power image retention are the most suited application and of these electronic shelf-edge labels (eSEL) with their very low image refresh frequency requirements (1–2 updates per day), could be a readily addressable segment in which the SRD's vivid color capability and wide color design scope for matching branding could be advantageous. Similarly, smart watches, in which fast update is desirable but video display largely unnecessary, are a potential application in which the SRD's capability to produce striking, metallic-looking colors could be leveraged to allow new design directions for digital and hybrid digital/analog watch designs.

The SRD's long term zero-power image retention capability and extremely thin, conformable, plastic substrate compatibility could make it an ideal candidate for "embedded" displays for applications in which a display function is only occasionally needed, and therefore it is not desirable to compromise the design or form factor of the device to include one. In such applications, perhaps in connected "internet of things" objects, smart devices, status indicators in or on vehicles and signposts, smart mirrors, etc., the option to replace the flat, black rectangle of an inactive display with a shiny, optionally metallic looking, surface, color-matched to the body of the host device, from which high-resolution information could appear only when required, could be very attractive. In smaller, portable examples of such embedded displays, the battery and some control electronics could be removed and replaced with an antenna to allow a "passive display" updateable via power and data supplied by a separate RF/NFC device. Examples of applications for such a device are smart banking/ID cards or passports in which the displayed information could reflect current status or hold dynamic security information, and parcel tracking tags in which the display could be combined with sensors to provide an updateable but non-volatile, persistent visual record of temperature or mechanical shock experienced by the device.

Finally, the ultimate goal of any prospective reflective display technology is to replace printed media with a full color, magazine quality, video capable, flexible, rugged, thin, and light display. As outlined previously, to achieve this, the SRD requires a combination with an additional top shutter device which is a significant addition in terms of complexity and required development. Still, results to date show that SRD can deliver the color control and brightness capability to make this feasible and inherently possesses the other mechanical and electro-optical properties to make this a possibility.

11.7 Other Approaches

Besides the TFT-driven, direct-view, reflective e-paper type display, which the development outlined above has realized, other types of PCM-based display have been proposed, and initial progress toward these demonstrated.

The ability of phase-change material based thin-film stacks to modulate the phase and/or amplitude of reflected light of specific wavelengths, rather than the color of the broadband incident light, opens the possibility of using PCM devices for diffractive and holographic applications, and the most developed non-SRD PCM display developments to date aim to leverage this capability to improve the pixel density and therefore diffraction efficiency achievable in holographic displays.

Most notably, the capability to use layer-thickness determined R, G, and B modulating regions in a structure similar to that of Figure 11.7a to allow independently controllable diffraction of three (R, G, and B) incident wavelengths has been demonstrated [26]. As shown in Figure 11.16, by imposing a fine pitch (2 μm) lithographically defined and laser-written binary computer-generated hologram (CGH) pattern of amorphous and crystalline

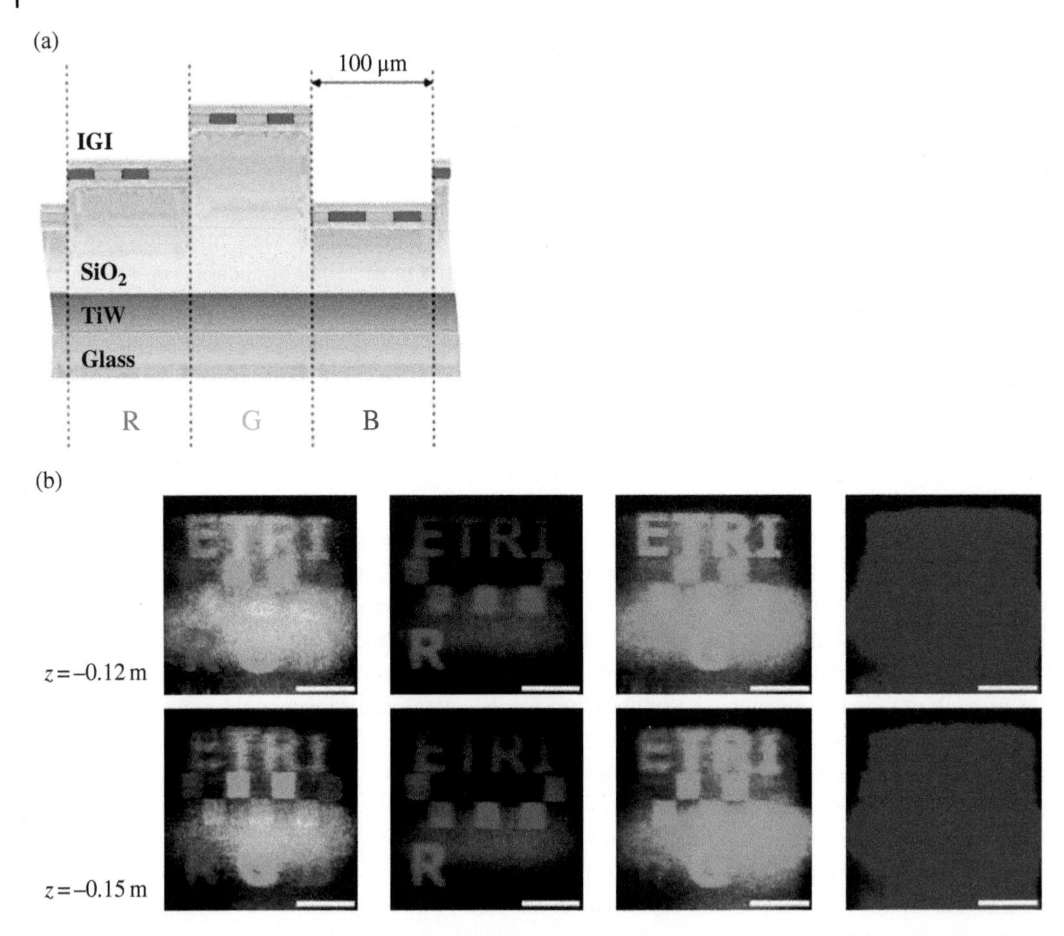

Figure 11.16 Structural schematic (a), and resultant projected image appearance (b) of the individual R, G, B and combined multi-color color holographic phase-change material based device of [26] Hwang, C. Y. et al. 2018 / Royal Society of Chemistry, showing the different focal planes of the alphanumeric and chequer-pattern portions of the image.

regions in the larger (100 μm) color specific sub-pixel regions of such a device, full color images can be reproduced with different image features in focus at different projection plane distances.

Taking such a holographic image modulator from a device that is re-writable using external laser input (in this case spatially controlled using a separate lithography step) to a dynamic, electronically updateable display requires extremely fine pitch pixel regions (ideally sub-micron) to achieve good diffraction efficiency and wide diffraction angles, the use of solid-silicon CMOS backplanes as utilized in high-resolution liquid-crystal-on-silicon displays has been investigated [58, 68–71]. As shown in Figure 11.17, switching of PCM pixels with an optically detectable result by CMOS transistors has been demonstrated, but doing this on a sufficiently large array if pixels with a sufficiently fine pitch to produce a high-quality holographic image still requires laser patterning of the PCM crystallization state within the pixels.

Nevertheless, the prospect of silicon wafer fabricated display backplanes combined with wafer post-processing to deposit the PCM optical modulator stack, resulting in high resolution, single substrate, all-solid-state inorganic displays with foundry-based manufacturing processes very similar to that already used for electronic memory chips, is highly attractive. Further development to fully electronic, reversible switching of high resolution, reconfigurable, computer-generated holograms can be expected.

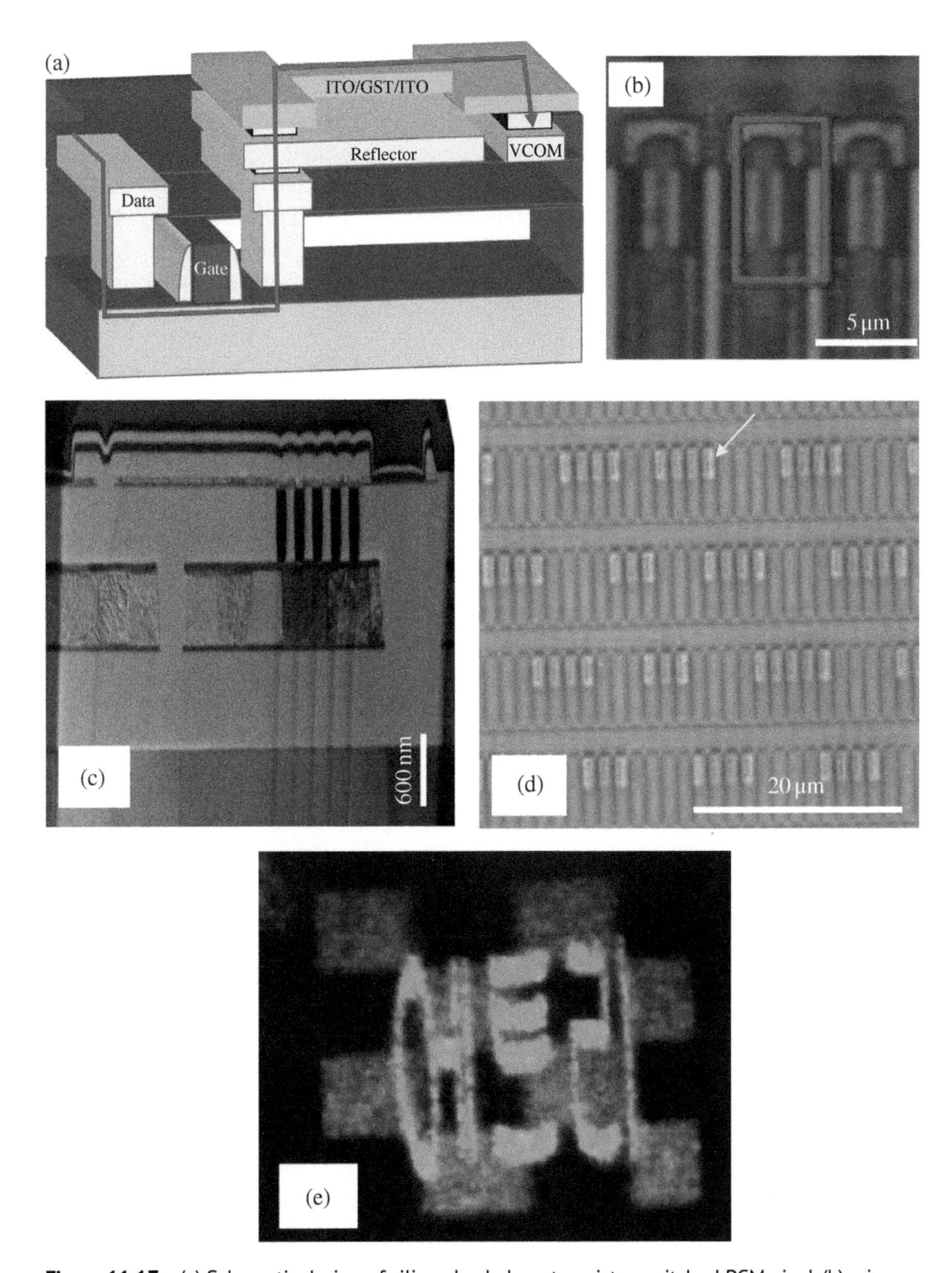

Figure 11.17 (a) Schematic design of silicon backplane transistor switched PCM pixel, (b) microscope image of such a transistor switched pixel, (c) TEM imaged cross-section of the device, (d) microscope image of a 1.5 μm pitch array of such pixels with laser-written crystallization pattern imposed into the active region of each, and (e) resulting replay image resulting from green laser illumination of the full 8kx2k array of laser-patterned pixels [70] Kim, Y.-H. et al. 2019 / Optical Society of America.

11.8 Conclusions

While it is clearly still the period of early development for PCM displays, hopefully, this chapter shows that basic feasibility of the technology has been established, compatibility with conventional TFT backplanes, and clear potential for high quality, bright, color reflectivity has been demonstrated.

The “SRD” implementation of a microheater switched, ambient reflecting, the direct-view device is currently the most developed example of a display using reversible crystallization of PCMs as the means of light modulation, but other display applications, particularly holographic projection types are possible, and in development.

Commercial success for PCM displays now rests on basic performance improvements for initial two-color devices, in particular cycle count, switching energy, and optical contrast. It is expected that further PCM material developments, taken from the hugely wide available range of possible compositions, will provide these necessary improvements and realize a new class of unobtrusive yet attractive, low-power, use-anywhere displays. Subsequently, further device developments may expand the range of addressable applications and enable dynamic, full-color reflected image reproduction.

Acknowledgments

The authors would like to thank the team at Bodle Technologies who contributed so much to the work outlined in this chapter: Lokeshwar Bandhu, Clement Talagrand, Graham Triggs, Tobias Bachmann, Sergio Garcia Castillo, Roberto Ricci, Katrina Didewar, Andrew Pauza, Richard Holliday, Mike Clary, Jill Liberman, Laura Leonard, Luci Bywater, TS Jen, Albert Tsai, Luke Lai and co-founders David Fyfe and Prof. Harish Bhaskaran. Additional thanks also go to Bodle’s consultants, Prof. C. David Wright and Noburu Yamada, who contributed greatly to the initial PCM alloy exploration, along with partners at the University of Southampton, Ioannis Zeimpekis and Prof. Dan Hewak, and Ilika Technologies Ltd., as well as our partners at both PARC: Bob Street and Jeng Ping Lu, and Tianma: Yao QiJun and Lu Feng, who made the LTPS backplane development possible.

References

1 Hosseini, P., Wright, C.D., and Bhaskaran, H. (2014). An optoelectronic framework enabled by low-dimensional phase-change films. *Nature* 511: 206–211.
2 Raoux, S. and Wuttig, M. (2009). *Phase Change Materials*. US: Springer https://doi.org/10.1007/978-0-387-84874-7.
3 Ovshinsky, S.R. (1968). Reversible electrical switching phenomena in disordered structures. *Phys. Rev. Lett.* 21: 1450–1453.
4 Choe, J. (2017). Intel 3D XPoint Memory Die Removed from Intel OptaneTM PCM (Phase Change Memory). TechInsights http://www.techinsights.com/about-techinsights/overview/blog/intel-3D-xpoint-memory-die-removed-from-intel-optane-pcm
5 Loke, D., Lee, T. H., Wang, W. J. et al. (2012). Breaking the speed limits of phase-change memory. *Science* 336: 1566–1569.
6 Schlich, F.F., Zalden, P., Lindenberg, A.M., and Spolenak, R. (2015). Color switching with enhanced optical contrast in ultrathin phase-change materials and semiconductors induced by femtosecond laser pulses. *ACS Photonics* 2: 178–182.
7 le Gallo, M. and Sebastian, A. (2020). https://doi.org/10.1016/b978-0-08-102782-0.00003-4). Phase-change memory. In: *Memristive Devices for Brain-Inspired Computing* (ed. S. Spiga, A. Sebastian, D. Querlioz and B. Rajendran), 63–96. Elsevier.
8 Salinga, M., Carria, E., Kaldenbach, A. et al. (2013). Measurement of crystal growth velocity in a melt-quenched phase-change material. *Nat. Commun.* 4: 1–8.
9 Shportko, K., Kremers, S., Woda, M. et al. (2008). Resonant bonding in crystalline phase-change materials. *Nat. Mater.* 7: 653–658.
10 Yamada, N. (2012). Origin, secret, and application of the ideal phase-change material GeSbTe. *Phys. Status Solidi (B) Basic Res.* 249: 1837–1842.

11 Kats, M.A., Sharma, D., Lin, J. et al. (2012). Ultra-thin perfect absorber employing a tunable phase change material. *Appl. Phys. Lett.* 101.

12 Kats, M.A., Blanchard, R., Genevet, P., and Capasso, F. (2013). Nanometre optical coatings based on strong interference effects in highly absorbing media. *Nat. Mater.* 12: 20–24.

13 Kats, M.A., Blanchard, R., Ramanathan, S., and Capasso, F. (2014). Thin-film interference in lossy, ultra-thin layers. *Opt. Photonics News* 25: 40–47.

14 Kats, M.A. and Capasso, F. (2016). Optical absorbers based on strong interference in ultra-thin films. *Laser Photonics Rev.* 10: 735–749.

15 Schlich, F.F. and Spolenak, R. (2013). Strong interference in ultrathin semiconducting layers on a wide variety of substrate materials. *Appl. Phys. Lett.* 103.

16 Kats, M.A. and Capasso, F. (2014). Ultra-thin optical interference coatings on rough and flexible substrates. *Appl. Phys. Lett.* 105.

17 Ríos, C., Hosseini, P., Taylor, R.A., and Bhaskaran, H. (2016). Color depth modulation and resolution in phase-change material nanodisplays. *Adv. Mater.* 28: 4720–4726.

18 Vermeulen, P.A., Yimam, D.T., Loi, M.A., and Kooi, B.J. (2019). Multilevel reflectance switching of ultrathin phase-change films. *J. Appl. Phys.* 125.

19 Ji, H.K., Tong, H., Qian, H. et al. (2016). Non-binary colour modulation for display device based on phase change materials. *Sci. Rep.* 6.

20 Yoo, S., Gwon, T., Eom, T. et al. (2016). Multicolor changeable optical coating by adopting multiple layers of ultrathin phase change material film. *ACS Photonics* 3: 1265–1270.

21 Wei, T., Wei, J., and Liu, B. (2019). Multi-color modulation based on bump structures of phase-change material for color printing. *Opt. Mater.* 98.

22 Jiang, Y.-L., Tang, T-A., Fan, Y. et al. (2018). ICSICT-2018: 2018 *14th IEEE International Conference on Solid-State and Integrated Circuit Technology (ICSICT): proceedings*, Qingdao, China.

23 Liu, H., Dong, W., Wang, L. et al. (2020). Rewritable color nanoprints in antimony trisulfide films. *Sci. Adv* 6 http://advances.sciencemag.org.

24 Cueff, S., Taute, A., Bourgade, A. et al. (2020). Reconfigurable Flat Optics with Programmable Reflection Amplitude Using Lithography-Free Phase-Change Materials Ultra Thin Films. arXiv:2009.02532.

25 Lee, S.Y., Kim, Y.H., Cho, S.M. et al. (2017). Holographic image generation with a thin-film resonance caused by chalcogenide phase-change material. *Sci. Rep.* 7: 41152.

26 Hwang, C.Y., Kim, G.-H, Yang J.-H. et al. (2018). Rewritable full-color computer-generated holograms based on color-selective diffractive optical components including phase-change materials. *Nanoscale* 10: 21648–21655.

27 Zheludev, N.I. and Kivshar, Y.S. (2012). From metamaterials to metadevices. *Nat. Mater.* 11: 917–924.

28 Ding, F., Yang, Y., and Bozhevolnyi, S.I. (2019). Dynamic metasurfaces using phase-change chalcogenides. *Adv. Opt. Mater.* 7: 1801709.

29 Karvounis, A., Gholipour, B., MacDonald, K.F., and Zheludev, N.I. (2016). All-dielectric phase-change reconfigurable metasurface. *Appl. Phys. Lett.* 109: 051103.

30 Carrillo, S.G.C., Trimby, L., Au, Y.-Y. et al. (2019). A nonvolatile phase-change metamaterial color display. *Adv. Opt. Mater.* 7: 1801782.

31 Cao, T., Wang, R., Simpson, R.E., and Li, G. (2020). Photonic Ge-Sb-Te phase change metamaterials and their applications. *Prog. Quantum. Electron.* 74: 100299.

32 Liang, G., Zhang, K., Zhai, F. et al. (2011). Comparison of optical and electrical transient response during nanosecond laser pulse-induced phase transition of Ge2Sb2Te 5 thin films. *Chem. Phys. Lett.* 507: 203–207.

33 Yamada, N. (2009). Development of materials for third generation optical storage media. In: *Phase Change Materials* (ed. R. Simone and M. Wuttig), 199–226. US: Springer https://doi.org/10.1007/978-0-387-84874-7_10.

34 Guerin, S., Hayden, B., Hewak, D.W., and Vian, C. (2017). Synthesis and screening of phase change chalcogenide thin film materials for data storage. *ACS Comb. Sci.* 19: 478–491.

35 Raynes, E. P. (1983). Electro-Optic and Thermo-Optic Effects in Liquid Crystals. Philosophical Transactions of the Royal Society of London. Series A, Mathematical and Physical Sciences. 309 https://about.jstor.org/terms

36 Horiguchi, T., Koshiba, Y., Ueda, Y. et al. (2008). Reversible coloring/decoloring reaction of leuco dye controlled by long-chain molecule. *Thin Solid Films* 516: 2591–2594.

37 Siegel, A.C., Phillips, S.T., Wiley, B.J., and Whitesides, G.M. (2009). Thin, lightweight, foldable thermochromic displays on paper. *Lab Chip* 9: 2775–2781.

38 Liu, L., Peng, S., Wen, W., and Sheng, P. (2007). Paperlike thermochromic display. *Appl. Phys. Lett.* 90: 213508.

39 Park, J., Ha, B.H., Destgeer, G. et al. (2016). Spatiotemporally controllable acoustothermal heating and its application to disposable thermochromic displays. *RSC Adv.* 6: 33937–33944.

40 Kahn, F.J. (1973). Ir-laser-addressed thermo-optic smectic liquid-crystal storage displays. *Appl. Phys. Lett.* 22: 111–113.

41 Kaino, Y., Kurihara, K., Shuto, A. et al. (2019) Laser-Addressed Full-Color Photo-Quality Rewritable Sheets Based on Thermochromic Systems with Leuco Dyes. https://doi.org/10.1002/(ISSN)1938-3657.display-week-2019.

42 le Gallo, M. and Sebastian, A. (2020). An overview of phase-change memory device physics. *J. Phys. D. Appl. Phys.* 53: 213002.

43 Lo, H., Chua, E., Huang, J. et al. (2010). Three-terminal probe reconfigurable phase-change material switches. *IEEE Trans. Electron Dev.* 57: 312–320.

44 Wang, M. and Rais-Zadeh, M. (2014). Directly heated four-terminal phase change switches. In: *IEEE MTT-S International Microwave Symposium Digest*, 1–4. Institute of Electrical and Electronics Engineers Inc https://doi.org/10.1109/MWSYM.2014.6848367.

45 El-Hinnawy, N., Borodulin, P., Wagner, B. et al. (2013). A four-terminal, inline, chalcogenide phase-change RF switch using an independent resistive heater for thermal actuation. *IEEE Electron Device Lett.* 34: 1313–1315.

46 Broughton, B., Bandhu, L., Talagrand, C. et al. (2017). Solid-state reflective displays (SRD ®) utilizing ultrathin phase-change materials. *Dig. Tech. Pap. – SID Int. Symp.* 48: 546–549.

47 Hosseini, P., Bhaskaran, H. and Broughton, B. (2019). Optical device with thermally switching phase change material US20190064555A1.

48 Hosseini, P., Broughton, B. and Bandhu, L. (2019). WO2019081914A1 Display apparatus.

49 Castillo, S.G., Feng, L., Bachmann, T. et al. (2019). Solid state reflective display (SRD®) with LTPS diode backplane. *Dig. Tech. Pap. – SID Int. Symp.* 50: 807–810.

50 Bhaskaran, H., Hosseini, P., Broughton, B. and Holliday, R. (2020). US20200033644A1 Display Apparatus.

51 Broughton, B., Gallardo, D., Heywood-Lonsdale, E. and Hopkin, H. (2018). US20180017840A1 Solid state reflective display.

52 Talagrand, C., Triggs, G., Bandhu, L. et al. (2018). Solid-state reflective displays (SRD®) for video-rate, full color, outdoor readable displays. *J. Soc. Inf. Disp.* 26: 619–624.

53 Heikenfeld, J., Drzaic, P., Yeo, J.-S., and Koch, T. (2011). Review paper: a critical review of the present and future prospects for electronic paper. *J. Soc. Inf. Disp.* 19: 129.

54 Telfer, S.J. and Mccreary, M.D. (2016) A Full-Color Electrophoretic Display †. https://doi.org/10.1002/j.21.

55 Xu, M., Slovin, G., Paramesh, J. et al. (2016). Thermometry of a high temperature high speed micro heater. *Rev. Sci. Instrum.* 87: 024904.

56 Cheng, H.Y., Raoux, S., and Jordan-Sweet, J.L. (2014). Crystallization properties of materials along the pseudo-binary line between GeTe and Sb. *J. Appl. Phys.* 115: 093101.

57 Raoux, S., Cheng, H.Y., Caldwell, M.A., and Wong, H.S.P. (2009). Crystallization times of Ge-Te phase change materials as a function of composition. *Appl. Phys. Lett.* 95: 071910.

58 Zhang, Y., Chou, J.B., Li, J. et al. (2017). Broadband Transparent Optical Phase Change Materials.

59 Zhang, Y., Chou, J.B., Li, J. et al. (2019). Broadband transparent optical phase change materials for high-performance nonvolatile photonics. *Nat. Commun.* 10: 4279.

60 Wang, J., Li, Q., Tao, S. et al. (2020). Improving the reflectance and color contrasts of phase-change materials by vacancy reduction for optical-storage and display applications. *Opt. Lett.* 45: 244.

61 Lyu, Y., Mou, S., Bai, Y. et al. (2017). Multi-color modulation in solid-state display based on phase changing materials. In: *2016 13th IEEE International Conference on Solid-State and Integrated Circuit Technology*, ICSICT, 165–167. https://doi.org/10.1109/ICSICT.2016.7998868.

62 Yang, D.-K. (1071). Review of operating principle and performance of polarizer-free reflective liquid-crystal displays.

63 Masutani, A., Roberts, T., Schüller, B. et al. (1071). A novel polarizer-free dye-doped polymer-dispersed liquid crystal for reflective TFT displays.

64 Yu, X., Chang, M., Chen, W. et al. (2020). Colorless-to-black electrochromism from binary electrochromes toward multifunctional displays. *ACS Appl. Mater. Interfaces* 12: 39505–39514.

65 Mccreary, M. D., Paolini, R. J., Smith, J. A., et al. (2019). Variable Transmission Electrophoretic Films †.

66 Heikenfeld, J. and Steckl, A.J. (2005). High-transmission electrowetting light valves. *Appl. Phys. Lett.* 86: 1–3.

67 Castillo Garcia, S., Broughton, B., Hosseini, P. and Talagrand, C. (2020). US20200202804A1 Display.

68 Hendrickson, J., Liang, H., Soref, R., and Mu, J. (2015). Electrically actuated phase-change pixels for transmissive and reflective spatial light modulators in the near and mid infrared. *Appl. Opt.* 54: 10698.

69 Kim, Y. H., Hwang, C.Y., Kim, G.H. et al. (2019) Towards 1.5 micrometer pixel pitch holographic display using Ge2Sb2Te5 phase change material. in Proceedings Digital Holography and Three-Dimensional Imaging 2019 M4A.2. Optical Society of America (OSA), https://doi.org/10.1364/DH.2019.M4A.2.

70 Kim, Y.-H., Cho, S.-M., Choi, K. et al. (2019). Crafting a 1.5 μm pixel pitch spatial light modulator using Ge 2 Sb 2 Te 5 phase change material. *J. Opt. Soc. Am. A* 36: D23.

71 Hwang, C.-S., Yoo, C., Cho, J. et al. Ultra High Resolution Display for Digital Holography.

12

Optical Measurements for E-Paper Displays

Karlheinz Blankenbach

Pforzheim University, Display Lab, Tiefenbronner Str. 65, 75175 Pforzheim, Germany

12.1 Introduction

There are two types of direct view display technologies (beside projection): Displays that modulate their reflection of ambient light and emissive or permanently backlit displays which convert electrical power to light. The fundamental characteristics of emissive displays like luminance and color are measured under darkroom conditions. That is relatively easy to perform compared to measurements on reflective displays, which must be illuminated. The main challenge here is that the results depend on the actual set-up and geometry of the illumination, the display (e.g. angular reflection characteristics), and the measurement device. This applies as well for emissive displays when measuring their ambient light performance. A very widely used and comprehensive display measurement standard is the SID IDMS [1] (especially Chapter 11 "Reflection Measurements"). For example, additional background information on display measurements with explanations can be found in, e.g. [2]. These two references contain all the information necessary to understand and perform detailed reflective display measurements.

The structure of this chapter is as follows:

- The fundamentals of the physics of reflection are the basis for understanding the procedures for reflective measurements.
- Examples of measurement set-ups such as those using a sampling sphere point out challenges and solutions for performance data acquisition.
- The section "Display Quality Parameters" introduces the most essential criteri for judging display quality: Luminance, contrast ratio, graygrey scale, and color reproduction.
- Application relevant parameters such as switching speed, lifetime, viewing angle, and front-light are briefly listed with some references.

There is a wide range of reflective display technologies with different principles and optical characteristics, typical sample data and references are provided here. An example is switching speed, which is relatively slow for electrochromic and electrophoretic technologies compared to, e.g. electro-wetting displays that can achieve video speed.

E-Paper Displays, First Edition. Edited by Bo-Ru Yang.

12.2 Fundamentals of Reflection

When incident light hits a (reflecting) object, a portion of this light is reflected. The parameter "reflectance" refers to the ratio of reflected incident optical power. The simplest case illustrating the reflection characteristics is a narrow beam that hits an ideal mirror: The incident angle is equal to the angle of the outgoing (specular reflected) ray (see Figure 12.1 left). However, typical surfaces of reflective displays have diffuse to semi-glossy characteristics. Therefore, the incident light is reflected in various directions with different intensities. For an ideal diffuse surface, the intensity of the reflected light is independent of the incident and outgoing angles (visualized as semicircle). Please note that reflections happen not only in two-dimensions as illustrated but also all directions (3D) as illustrated in Figure 12.1 right: A tiny ray of a laser is reflected in all directions (hemisphere). The reflected light is represented by a polar plot, in which the distance from the origin to the plot line at any angle θ The intensity measured is measured in all (3D) directions and coded by grayscale (no reflection: black; maximum: white). Semi-glossy surfaces reflect more light in the specular direction (center of the plot) than at other angles.

Figure 12.2 shows the typical reflectance characteristics of diffuse e-paper for white and black (lower reflectance as for white). The semi-glossy curve is typically due to the front surface (glass, plastic). However, the actual behavior of displays is more complex as they consist of many different layers with different optical characteristics that contribute to the reflectance characteristics [3]. That is why it is usual to present the results of measurements on a complete display. Other geometric optical effects occur when the display is curved (e.g. flexible displays).

A typical set-up for measuring the angular reflection distribution is illustrated in Figure 12.3 left: The light from a small source strikes the surface at an angle θ_s measured relative to the normal line (dashed). The reflected light

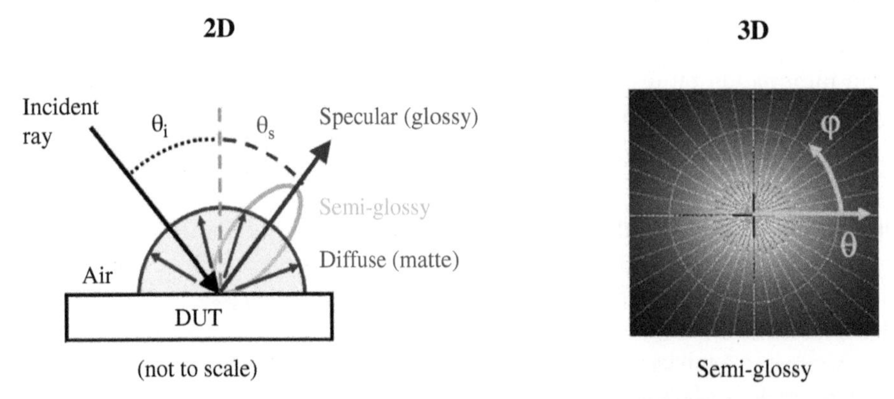

Figure 12.1 Fundamentals of reflection for incident and reflected light in two-dimensional view (left) and a measurements of a semi-glossy surface in 3D (right) as polar plot with intensity represented by gray levels.

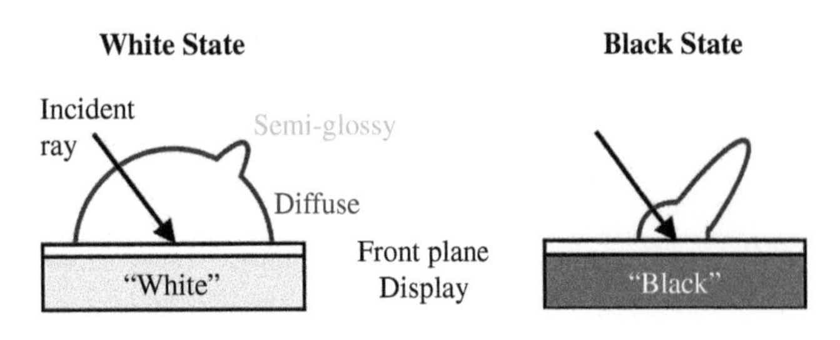

Figure 12.2 Examples for bright (left) and dark (right) state of diffuse flat e-paper display material with semi-glossy surface. The reflections are illustrated as 2D but occur in 3D (hemisphere) for diffuse case.

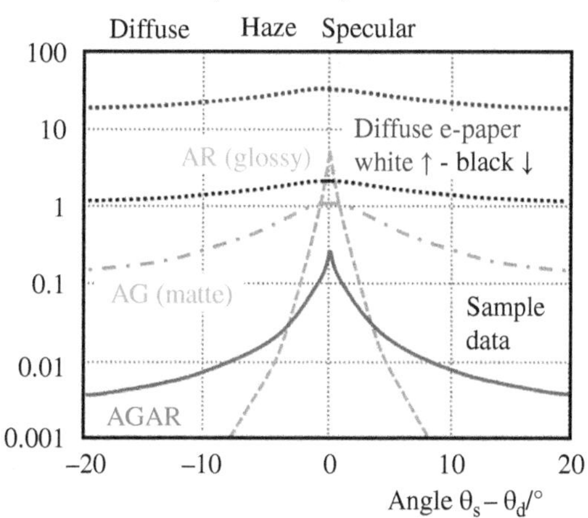

Figure 12.3 Measurement set-up for angle resolved reflectance measurement (left) and typical characteristics (right) for various surfaces and a diffuse display.

is captured by an optical detector (typically a luminance meter for displays), which can be moved around the point of incidence; its (actual) angle is θ_d. The result is plotted as reflectance (percentage of luminance referring to a standard, e.g. diffuse white standard or black glass) over the angular difference θ_s-θ_d. Such measurements are often termed a Bi-directional Reflectance Distribution Function (BRDF, [4, 5]). The region around 0° is called the specular range. Next to the specular peak is the "haze" range, and toward larger angles from specular, we are in the diffuse region.

This results in the sample curves in Figure 12.3 right (top to bottom): A diffuse e-paper in its white state (blue/dotted curve) with a reflectance r in the range of 30% and a small glossy peak, for e-paper in the black state (black/dotted curve) the reflectance is smaller, but the character is typically the same. A glossy display (cyan/dashed curve – for example, most smartphone screens) has a narrow peak in the specular range whose height can be reduced by anti-reflection (AR) coatings like those used for glasses. A larger half-width is the characteristic of a matte surface, which acts as an anti-glare layer (green/dotdash curve, AG). When AR and AG are combined, the reflection reduction is highest (red/solid curve, AGAR). A perfect mirror has a narrow peak at 0°.

Therefore, it is evident that the total reflection properties of reflective displays are given by the characteristics of the electro-optic material (e.g. electrophoretic) and all the other surfaces of the display (such as reflection reduction coating, glass, transparent semiconductors). The reflectance value is a major parameter by which to judge the legibility of reflective displays (see below). When a display is in use, dirt, dust, and fingerprints will change its reflection characteristics. The front surface coating can also degrade through damage (abrasion testing). To summarize, the reflectance characteristics must be measured for a given display, as there is hardly any simulation or extrapolation from other displays possible.

12.3 Reflection Measurements Set-Ups

Measuring the reflectance of displays over a range of angles as described above is not practical for characterizing, e.g. the legibility on a reflective display. The performance of displays under ambient light can be measured using set-ups such as those shown in Figures 12.4, 12.6, and 12.8. It is inevitable that the results of these measurements

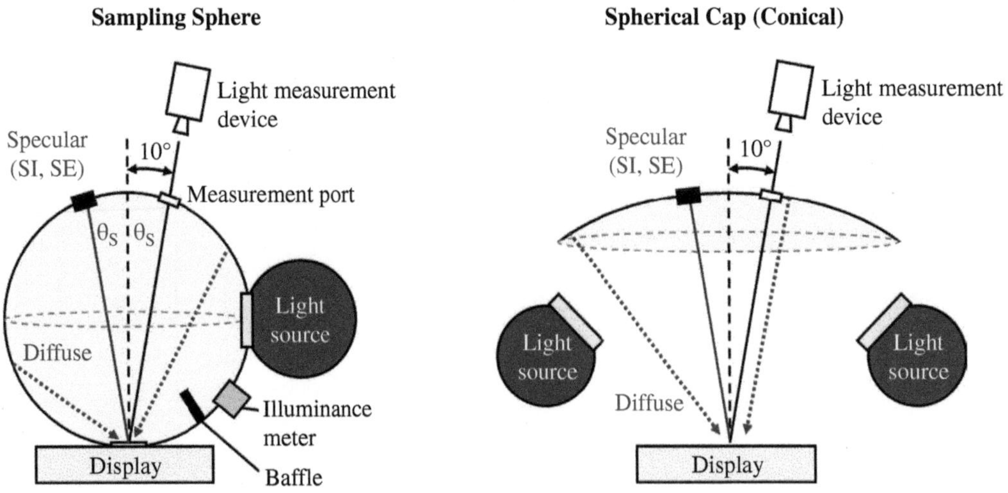

Sampling Sphere

Spherical Cap (Conical)

Light measurement device

Specular (SI, SE)

10°

Measurement port

θ_S θ_S

Light source

Diffuse

Illuminance meter

Display

Baffle

Figure 12.4 Reflective displays measurement set-ups: Sampling sphere (left) and spherical (conical) cap (right). The light measurement device "points" to the specular port, whose surface has a huge impact in the result (see Figure 12.5).

strongly depend on the set-up geometry, of which many other variations exist as well [1, 6, 7]. Therefore, only values acquired using the same setup can be compared. The examples described here refer to flat displays: the measurement of curved ones is more complex, see, e.g. [8].

The most widespread set-up for measurements with simulated ambient light is a sampling (or integrating) sphere (Figure 12.4 left) with a diameter in the range of 25 cm. Its interior consists of a diffuse white coating, which homogenizes the light from the source to provide near-ideal diffuse illumination. The baffle blocks light from the source from hitting the display directly. The illuminance is measured either by a fixed meter (as shown) or indirectly by using a white diffuse reflectance standard in place of the display under evaluation, and the same a luminance meter at the measurement port as used for the display evaluation.

A small port (e.g. 3 cm diameter) at the bottom of the sphere allows the light to illuminate the display under test. Light reflected from the display is measured by luminance or a color meter (imaging detectors are also possible) at an angle of 5–10° off normal to avoid specular reflections of the detector lens on display. There is an exchangeable port on the opposite side with either white (specular included, SCI) or black matte (specular excluded, SCE) finish; the impact is discussed below. The influence of the diameter of the specular port is significant, as can be understood from Figure 12.3 right.

The "spherical cap" is similar but less costly approach is the "spherical cap" (Figure 12.4 right). A major difference compared to the integrating sphere is that only light within a cone of about 30° falls onto the display. An advantage is that there is more flexibility on the display arrangement than the sampling sphere. For example, as there is no physical contact between the measuring equipment and the display, curved displays and displays integrated in housings are more easily handled using the spherical cap.

In contrast to reflective displays that do not generate light, emissive displays or those with an amount of light emitted by the display raises the illuminance inside the sampling sphere. Therefore, the light source has to be adjusted when defined illuminance values are required, as may be the case for full-screen black and white contrast ratio measurements.

The effect of placing the SCI and SCE caps onto the specular port is pictured in Figure 12.5 for a glossy, a semi-glossy, and matte display. The outer (black) circle denotes the border of the measurement port from the viewpoint of an imaging photometer from outside the sphere. The gray area is the display behind its corresponding port. Let us start with the "specular included" cap in place, with a glossy (mirror-like) display surface: The coating of the

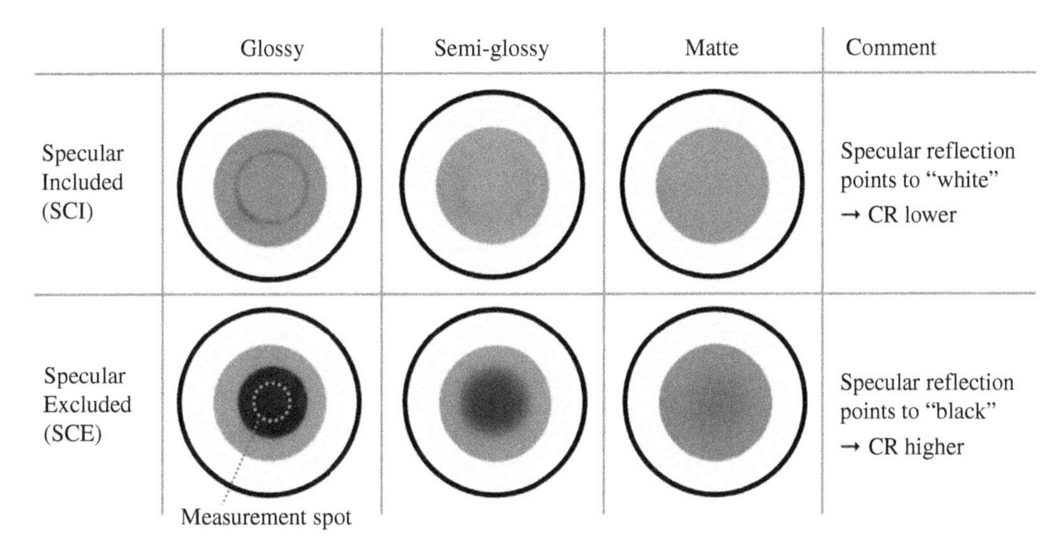

Figure 12.5 Effect on the measured reflections of typical surface characteristics depending on the specular port paint (c.f. Figure 12.4): White diffuse (top, same as, e.g. sampling sphere) and black diffuse (bottom) material.

port is the same as the interior of the sampling sphere. The faint gray ring arises from the cap's imperfect join and port. Therefore there is no noticeable difference between the reflections inside and outside of the specular direction of the cap (top). Now we change the cap to one with a black coating or a light trap (SCE): The specular reflection in the glossy (or "mirror-like") display surface is now dark (bottom) as a result of pointing to the black area of the sphere.

The outcome is that the contrast ratio (see below) is lower for SCI as the glossy reflection of the front surface adds to the (diffuse) reflection of the display. This is not the case for SCE. The more matte (or "diffuse") the surface is, the less noticeable is the effect of the black area of the sphere. Reflections of objects with a small angular size can also be used to visually judge the reflection characteristics: A glossy (mirror-like) display shows sharp edges which will be blurred for semi-glossy and are hardly noticeable for matte (diffuse) surfaces. Summarizing: It is essential to report on the nature of the specular port for sampling sphere and corresponding measurements.

Another configuration for measuring reflections from displays is provided in the standard MIL-L-86762 and illustrated in Figure 12.6 left: The values for the diffuse illuminance E and the specular luminance Ls can be individually adjusted to the typical requirements of the application such as E = 50,000 lx and L_s = 25 000 cd/m^2. This makes this set-up suitable for "simulating" light conditions encountered in various applications. Please note that for this set-up and most other ambient light measurements, the reflected light is measured as luminance (unit: cd/m^2, a photometric unit) used below for image quality evaluation.

A luminance (or color) imager, as shown in Figure 12.6 left, is recommended for all measurements with ambient light, as the image-like acquisition (see inset) documents the whole measurement, and the analysis is performed using this "image" and placing measurement areas or line profile. This reduces the risk of unwanted effects such as non-uniform illuminance over the display area or specular stray light. Those devices consists of an imaging chip (similar to a digital camera), dedicated color filters, and calibration to achieve absolute measurements [9].

The photo in Figure 12.5 right shows a display with reflections of various types of light sources. Information on the display modulates a diffuse reflection and can be read in areas where it is not overwhelmed by bright light sources' mostly glossy (specular) reflections. Those specular reflections typically originate from the front surface (front-plane), and the contrast can be optimized by reflection reduction methods like anti-reflective or anti-glare coatings.

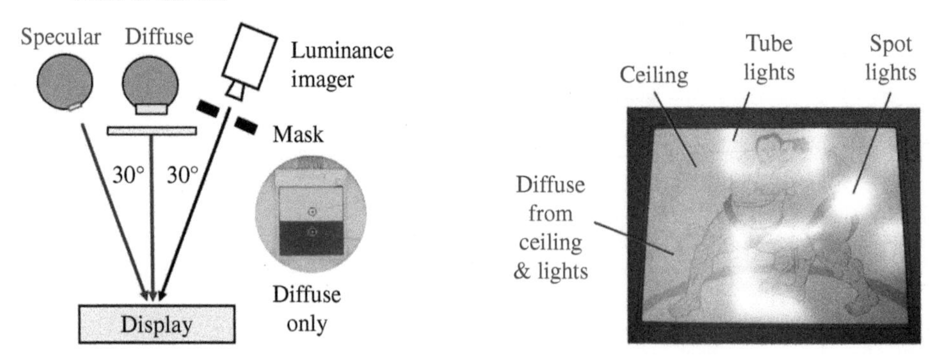

Figure 12.6 Measurement set-up (left) with luminance (or color) imager and sample photo for diffuse and specular conditions; reflection on s semi-glossy e-paper display (center).

Figure 12.7 Visual comparison of an electrophoretic display and a reflective LCD for diffuse (top) and mainly oblique (bottom, from the right) lighting conditions.

Figure 12.7 illustrates the difference in white state reflectance of an electrophoretic e-paper display (left) and a reflective LCD (right). The white reflectance of the latter is significantly smaller than that of the electrophoretic display. The white paper at the background (above or beneath the e-paper display) appears brighter than either of the two displays. The photos show that a higher reflectance value is preferable for many applications, e.g. appearance.

12.4 Display Image Quality Parameters

To ease the description of image quality, the following paragraphs refer to luminance (as for emissive and front- or backlighted reflective displays) instead of reflectance. The optical appearance of displays is mostly measured by luminance (and derived parameters) and color (if applicable). Common references for all-optical measurements are, e.g. [1, 2]. It is essential for image quality measurements for reflective displays to use the same setup, such as a sampling sphere with the specular included condition. Comparisons can only be made for exactly the same setup.

12.4.1 Luminance

Illuminance E (unit lx), reflected luminance LReflected (unit cd/m^2) and the diffuse reflectance factor r_D (dimensionless, see Figure 12.7), also referred to as the luminance factor) are related for diffuse reflective displays by the following equation:

$$L_{Diffuse} = \frac{r_D E}{\pi} \tag{12.1}$$

Luminance measures the amount of light into a certain direction (here, measurement device or eye) emitted or reflected by a display or surface. The luminance of a display has to be adjusted to the eye adaptation to avoid dazzling or too dark images. The advantage of reflective displays is that this condition is fulfilled automatically (except for dark environments) when their reflectance is above 20% (the higher, the better) of a diffuse white standard or white paper. If the reflectance is low, e.g. 10%, the displayed image appears dark, and colors are hardly perceived or distinguished. The challenge for reflective color displays is to achieve a high reflectance in combination with a large color gamut (see below, [10]).

12.4.2 Contrast Ratio

A widely used measure for legibility is the contrast ratio [11]. It is defined by the luminance ratio of white (bright) to black (dark) content under darkroom conditions. If we illuminate the display under test, luminance is reflected from the display (mainly front surface but also transparent electrodes, etc.). This adds to both the bright and dark luminance:

$$CR_{Dark} = \frac{L_{White}}{L_{Black}} \xrightarrow{+reflections\ of\ ambient\ light} CR_{Bright} = \frac{L_{White} + L_{Reflected}}{L_{Black} + L_{Reflected}} \tag{12.2}$$

If we assume that the black luminance is significantly smaller than the value for white and the reflected luminance (here: of the surface, see Figure 12.2), Eq. (12.2) simplifies to

$$CR_{Bright} \approx \frac{L_{White} + L_{Reflected}}{L_{Reflected}} = \frac{L_{White}}{L_{Reflected}} + 1 \tag{12.3}$$

For displays with diffuse reflection characteristics as is typical for most e-paper technologies, we can now substitute Eq. (12.1) into (12.2):

$$CR_{Bright} \approx \frac{r_D E}{\pi L_{Reflected}} + 1 \tag{12.4}$$

The goal is to design the contrast ratio as high as possible or reasonably. Therefore, the reflectance of the display (here: diffuse reflectance factor r_D) should be maximized (e.g. 0.5 or 50%, this determines as well the visual appearance, see Figure 12.7) and the reflected luminance LReflected (of the front surface) minimized. The latter strongly depends mainly on the measurement geometry (see above), the front surface's reflectance characteristics, and

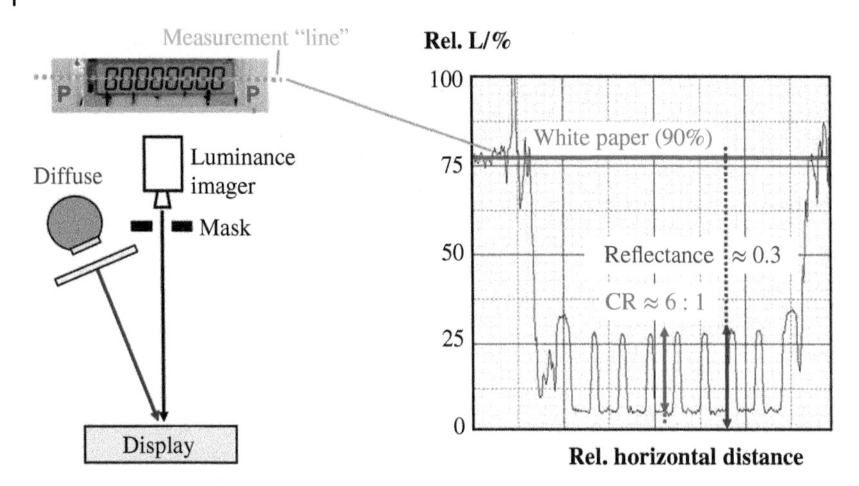

Figure 12.8 Example of a character contrast ratio measurements using an imager (left) and the result plotted as relative luminance (right) compared to white paper.

reflection reducing layers. The contrast ratio can be measured from sequential measurements made on the same display area (e.g. first white than black) or at neighboring locations, which is more similar to human perception (see inset Figure 12.7 left). An imaging luminance (or color) meter is required for the latter. The typical contrast ratio CR for reflective displays is in the range of 10 : 1. ISO standard 15008 specifies a minimum CR of 2 : 1 for readability in the presence of bright ambient light.

Figure 12.8 illustrates the reflectance – luminance relationship for a numeric electrophoretic display and a measurement set-up (left) similar to the daylight case of ISO 15008. Only diffuse (non-specular) light is captured by the luminance imager. The graph on the right shows the measurement result as relative luminance as a function of position along the line from P to P (red, left). The display is surrounded a by white paper at each side, which acts as a reference with a measured reflectance of 90% of a white diffuse standard. The peaks at the left and right sides are caused by glossy reflections from the, display's edge seal. At bottom, the "modulated" waveform results from successive white and black regions with differing reflectance. The white state reflectance is about 30% (0.3), and the black is ~5%. This results in a measured intra-character contrast ratio of about 6 : 1. The graph also demonstrates the advantages of using an imaging device: Inhomogeneities incl. Those from the illumination are easier to detect than when a spotmeter is used. If the measurement area of the black segments is small, viewing glare can significantly reduce the measured contrast ratio.

For a pure diffuse illumination as used in Figure 12.8, Eq. (12.2) has to be modified as the specularly reflected luminance is zero. Therefore, only the different luminance values for black and white have to be considered. They depend according to Eq. (12.1) on the diffuse reflectance factor r_D, which has then to be indexed for black $r_{D,K}$ (K for black to avoid confusion with blue), and white $r_{D,W}$. As the illuminance E is the same for both cases, the fraction reduces to

$$CR_{Bright} \approx \frac{r_{D,W}}{r_{D,K}} \tag{12.5}$$

This proves that the contrast ratio for diffuse illumination of reflective displays is independent of the illuminance value.

The next step is to measure the contrast ratio for different illuminance levels, as shown in Figure 12.9. According to the equations above, CR should be independent of the illuminance E. This is proven by the measurements of

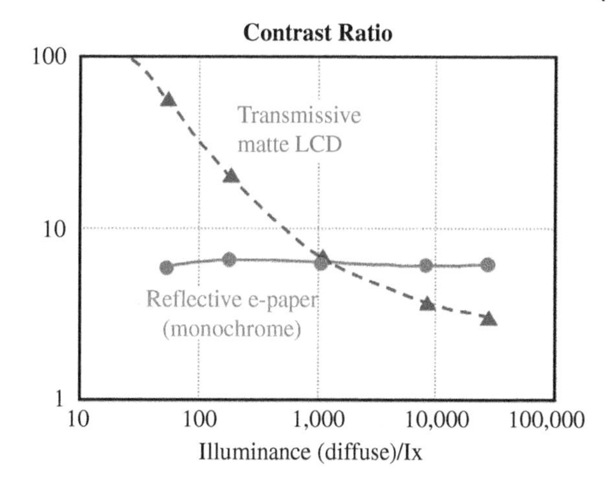

Figure 12.9 Comparison of the contrast ratio for diffuse illumination of a color LCD (red/dashed curve, "emissive") and a reflective electrophoretic display (green/solid curve) at various illuminance levels.

the reflective e-paper display. The slightly lower value for 50 lx is caused by reduced measurement accuracy for the black-state luminance under dark conditions.

So, reflective displays are even readable in bright sunlight. Emissive displays, exemplified here by a color LCD, show high CR values under dark conditions, but the reflected luminance increases with the illuminance. Therefore, the contrast ratio drops as illuminance rises, and in this case, is lower than for the reflective display when the illuminance is above 1,000 lx.

12.4.3 Gray Scale

Many reflective displays can render (color) gray scales. To match human perception, the luminance L – gray level GL relation has to be non-linear. A typical function is L ~ (rel. GL)γ with γ termed the gamma coefficient. Its typical value is 2.2 and has to be measured for evaluation. This is done by measuring the luminance of 10 or more equidistant gray levels; it is recommended to make successive measurements at the same point on display to avoid the effects of illumination or inhomogeneity of the display. Luminance and gray levels are normalized to 1 and, e.g. plotted on a double logarithmic chart. The slope is the gamma value. This procedure is described in detail in [1, 2]. Deviations of the gamma curve result in the incorrect rendering of gray levels on display, which is most significant for graphic images and GUI elements

The number of available gray levels is a key issue for monochrome and color images. The gray scale resolution of reflective displays is typically lower than for emissive shows. Many reflective display technologies have 4 to 5-bit resolution instead of 8 bit for emissive displays. The impact on color is discussed below.

12.4.4 Color

Most modern content to be reproduced by displays is colored. Monochrome displays are only accepted for a few applications such as meters or e-books. The number of colors is typically between 4096 (3 × 4 bit, see above) and 32768 (3 × 5 bit), either of which is significantly fewer than for emissive displays for which a 24-bit color depth is standard, resulting in 16.7 million colors. The other challenge for reflective displays is to render vivid colors, expressed by a large gamut in color measurements. The gamut and color coordinates are described in CIE (Commission International de l'Eclairage) color spaces such as CIE 1931, 1976 UCS, and CIE LAB. The principles, definitions, equations, charts, and measurement methods can be found in, e.g. [1, 2, 12].

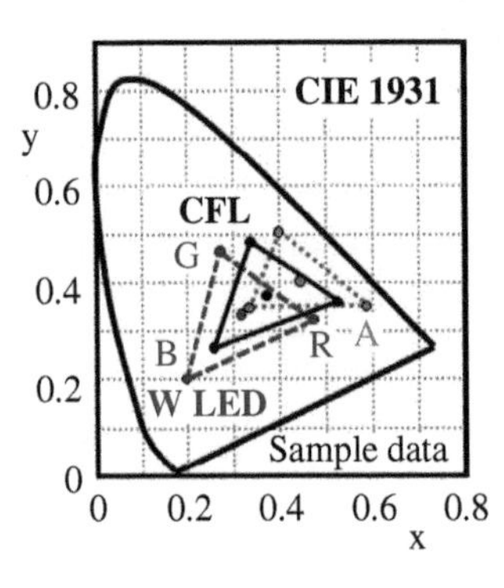

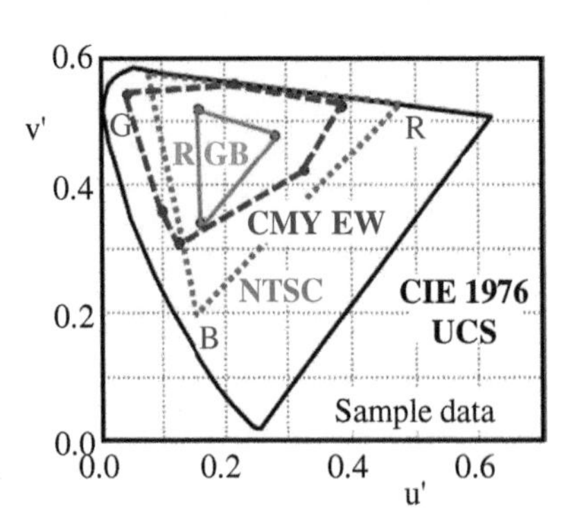

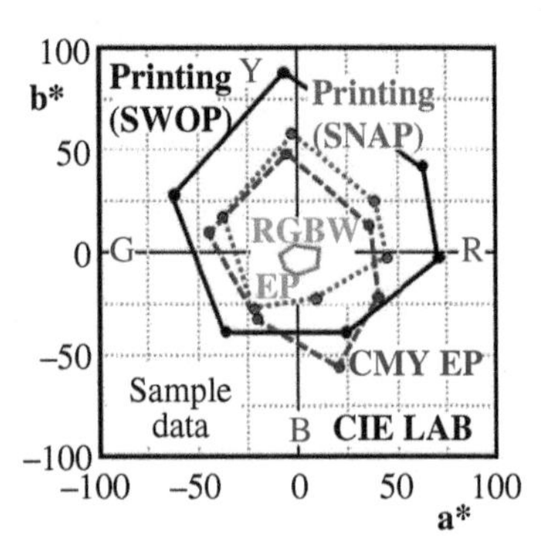

Figure 12.10 Color performance of reflective displays in various color spaces (here two-dimensional): Left: Dependency of display RGB color co-ordinates on the light source, plotted in the CIE 1931 color space; see main text for details of the sources. Center: Color gamut of RGB and CMY e-paper in CIE 1976 UCS. Right: Color gamut of RGBW and CMY e-paper in comparison with printing standards plotted in the CIE LAB color space.

The method to render color is very different for e-paper displays compared to emissive types. An extended overview and comparison were published by Heikenfeld et al. [10]. Basically, the use of an RGB color filter (like for color LCDs) in an e-paper panel results in low reflectance or low gamut (see Figure 12.6 right) when the density of the filters is lowered. Vertically stacked CMY (printer-like, [13, 14]) displays achieve a larger color gamut at the expense of increased cost since three switchable layers are needed instead of one. Electrophoretic and similar displays can use either a color filter array (RGB, W) or colored particles; the latter currently results in a long image update duration (~10 seconds).

Figure 12.10 shows three examples demonstrating various aspects of color reproduction of reflective displays. The left chart uses the CIE 1931 color space, and the color coordinates x, y are plotted for the red (R), green (G), blue (B), and white (filled circles) points of the display, here RGB electrokinetic samples were used. All color measurement devices provide such x, y co-ordinates as an output option. The co-ordinates of RGB are connected by lines for each light source. The white point is represented by a circle inside the corresponding triangle of the same color/filled circle. The chart clearly shows that the color reproduction of reflective displays depends strongly on the light source. Three examples are provided here: A white LED (W LED), a cold cathode fluorescent tube (CFL, neon lamp), and an incandescent light source (A, CIE illuminant A). The third co-ordinate is the luminance, which is usually not plotted.

The CIE 1976 uniform color space (UCS) (white point 0.20/0.47, luminance not shown) is recommended for modern display evaluation and is shown as a 2D plot in the centre panel of Figure 12.10. The red/dotted triangle illustrates the NTSC color TV standard. The blue/dashed hexagon show the color gamut of an electro-wetting display where cyan, magenta, and yellow layers are stacked (subtractive primary colors parallel those used in printing). The green/solid triangle refer to a reflective display that incorporates an RGB mosaic filter. The measured points span a triangle or hexagon, whose size corresponds to the gamut and is often expressed as a percentage of the NTSC area. The larger the gamut (area) is, the more vivid are the reproduced colors.

CIELAB is better suitable for judging the color performance of reflective displays and conventionally printed products than CIE 1931 or CIE 1976 UCS: It is based on lightness L* (equals the perceived brightness) instead of luminance L. The color coordinates are expressed by a* and b*. Figure 12.10 right plots CIE LAB as two-dimensional (2D) graph for a* and b*. Also, three-dimensional (3D) charts, as provided in, e.g. [13], are used to compare color gamut. However, 3D plots are difficult to compare by vision except for their size.

The printing standards SNAP (Specification for Newsprint Advertising Production, red/dotted) and SWOP (Specification for Web Offset Publications, black/solid) are used in Figure 12.10 right to compare printing requirements to the color gamut of reflective displays. The sample data show that RGBW color filter (CFA) displays (green/solid line) have a smaller gamut compared to stacked CMY (blue/dashed) approaches and to printing standards.

All color measurements must take into account the spectral characteristics of reflective displays and light sources. However, the spectral dependencies cannot by derived "backward" from CIE coordinates. For example, a white point of 0.33/0.33 can be achieved with various light sources like a white LED, RGB LED, fluorescent tubes, and RGB lasers. However, the three lasers (RGB) do not emit any light in, e.g. the yellow region, so any yellow particle of a reflective display appears black. Another example is white (W) LEDs: They consist of a blue LED as emitter and phosphors which (activated by blue) emit either yellow or red and green light. The CIE color coordinates of different W LEDs can be the same, but the spectral characteristics are different. Therefore, spectral corrections must be made if reflective displays are evaluated using reference light sources different from those practical lights used for illumination of those displays in ordinary use.

Summarizing: The color image quality of reflective displays is typically lower than for emissive displays. However, the color performance of reflective displays should preferably be expressed by 3D color spaces such as CIE LAB (used in the printing industry)nd lightness as a parameter. Many improvements are underway to reach applications currently occupied by LCDs and LEDs.

12.5 Temporal Parameters

Another important characteristic of displays is their temporal behavior. This extends from the short durations required for image update to a timespan of years for the lifetime of a display in an application. The references [1, 2] provide details on definitions and measurements.

12.5.1 Switching Time

An update of content on display should not be noticeable. The switching time (often called response time) is usually measured by a photodiode and an oscilloscope. The duration is defined for an It is defined as the time required to switch the content from black (K) to white (W) or vice versa (W ↔ K) for the optical transition between 10% and 90% of the white luminance. A response time in the range of 10 ms or less is labeled as "video speed" and can be achieved by several reflective display technologies such as electro-wetting [15] as well as by emissive technologies. Reflective LCDs (mostly Twisted Nematic type) are mostly monochrome and show switching speeds in the range of 50 ms at 25 °C and 1,000 ms at −20 °C.

Display technologies using moveable particles show mostly longer update durations, partly due to complex driving waveforms used in electrophoretic devices to avoid any retention of "ghost" images carried over from the previous frame. Figure 12.11 shows the two image update methods for an industrial electrophoretic display: Full (left) and partial (right) refresh. A complete update without accumulating artifacts over time requires multiple different images with "opposite color" (here black or white) and lasts several seconds. This "flashing" or "flickering" is rated as annoying by observers. E-book readers perform this task in the range of 0.5 seconds, some of them without switching to the inverse image. This can be achieved for parts of the image for industrial displays and is called "partial image update" (right). Depending on the technology and the quality of the waveforms, previous content is not fully "deleted," so a complete refresh is needed after several partial image updates. Color electrophoretic displays with many different particles (e.g. CMYWK) require about 10 seconds to complete an image update.

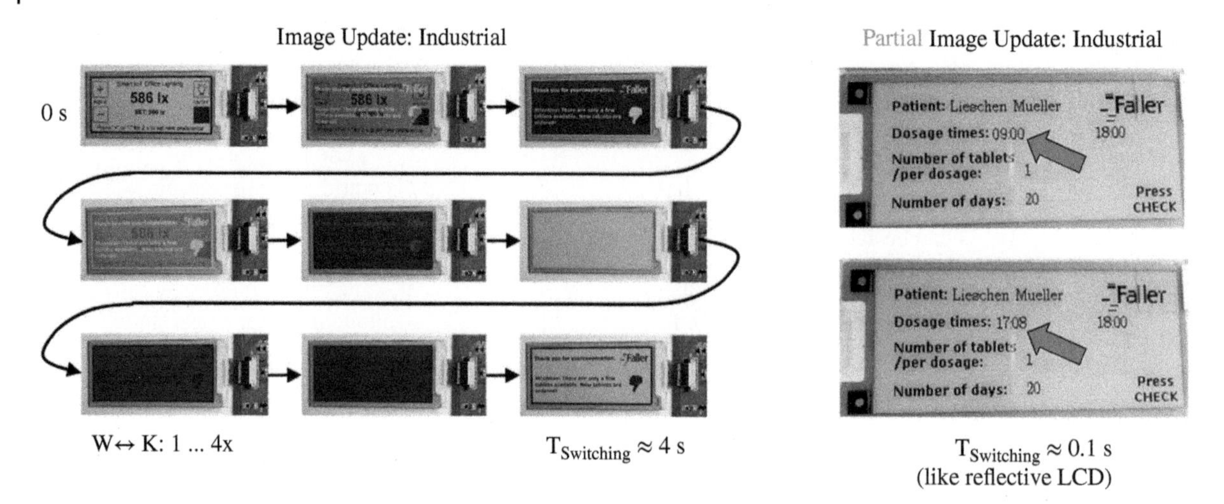

Figure 12.11 Standard updating of a professional electrophoretic display (left, whole content) and partial image update (right, orange arrow: time).

12.5.2 Bi-stability

Some e-paper technologies such as electrophoretic are bi-stable: The content is written to the display once and persists over hours to years without any power or refresh (see above); emissive displays are usually updated at 60 Hz even when the content does not change. This makes the e-paper displays "green" as no power is required, either for light generation or image data update by, e.g. the graphics adapter or panel electronics.

Some applications need to measure the temporal characteristic of the bi-stability, e.g. the time taken for the contrast ratio of a character (e.g. white background vs. black character, see above and insets "1" in Figure 12.12) to fade to a defined threshold, e.g. 2 : 1. Furthermore, the color of bi-stable displays can change over time without refresh. A typical effect of tri-color (black, white, and, e.g. red) electrophoretic displays is that the white background becomes reddish over 100's of hours without refresh.

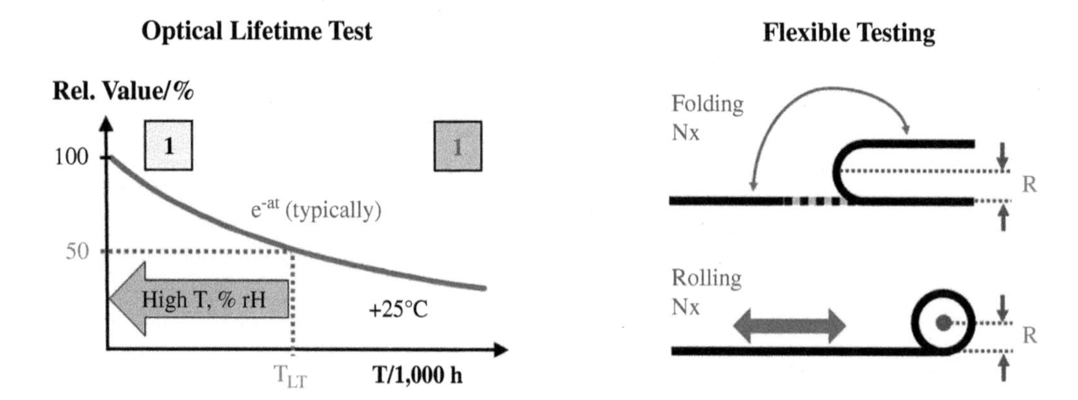

Figure 12.12 Lifetime testing: Left: Typical reduction of reflectance (e.g. measured as luminance) over time Right: Repeated folding and rolling of flexible display.

12.5.3 Lifetime

Like all other technical products, displays should reach a defined lifetime or operational time required for the intended application within the defined temperature and humidity range. As there is a huge variety of e-paper display technologies, the fundamental principle is illustrated in Figure 12.12 left: The lifetime is defined as the duration taken for a chosen parameter to reach half of its initial value. This could refer to, e.g. reflectance, luminance, or contrast ratio. The insets with a black "1" in a square show a potential degradation mechanism: The light gray (white) background darkens over time, e.g. due to chemical processes. So the essential image quality parameters degrade. Typical lifetime data refer to operation at 25 °C and under moderate humidity (25 °C, ~50% rH, also named "room conditions"). Many degradation effects follow an exponential decay. The lifetime is significantly reduced compared to room conditions by high temperatures (e.g. 60 °C) and/or high humidity (e.g. 90%) [16].

12.5.4 Flexible Displays

Many e-paper displays can be made flexible, in which category we include foldable or rollable devices (Figure 12.12 right). When the display is repeatedly mechanically flexed, degradations are likely to occur. Examples are cracks in the sealing and layers, leading to the breakage of data lines to pixels. The smaller is the bend radius R, the more challenging is the development and manufacturing of flexible displays. The quality criterion is the number N of folding or rolling operations until a failure occurs or the minimum number without failure.

12.6 Further Topics

Besides the above-described methods and criteria for display and image quality, many other measurements and tests (see [1, 2]) exist, such as

- Uniformity of the displays in terms of reflectance, luminance, color, etc.
- Ghosting of content into neighboring areas
- Luminance and color of potential front- and backlights [17]
- Viewing angle degradation when the display is observed from oblique angles incl. stacked displays [8, 18]
- Defective pixels of a display
- Power consumption
- Operation in a harsh environment incl. vibration and shock

12.7 Summary

Optical measurements of reflective displays are more challenging than for emissive technologies. That is because reflective displays, must be illuminated, and the results depend on the measurement set-up. Therefore, comparisons of datasheets and measurement results require special attention. The most important parameter of the reflective display is their reflectance, which affects the perceived image quality. The reproduction of color images is, for many technologies, not so good compared to emissive displays. It is often not required to perform all possible evaluation methods and measurements but concentrates on the most essential application.

Glossary incl. Abbreviations

BRD	FBi-directional Reflectance Distribution Function
CIE	Commission International de l'Eclairage
CR	Contrast ratio
E	Illuminance
Gamma	Exponent of luminance gray level dependency
Gamut	Area within a 2D color space which can be rendered by a display
GL	Gray level (digital value)
L	Luminance
R	Reflectance (factor)
SE	Specular excluded
SI	Specular included

References

1 Society for Information Display (SID). (2012). Information Display Measurements Standard. https://www.sid.org/Publications/ICDM.aspx.
2 Blankenbach, K. (2016). Section XI display metrology. In: *Handbook of Visual Display Technologies*, (ed. J. Chen, W. Cranton and M. Fihn), 3071–3226, 3263–3282; 3313–3334. Heidelberg: Springer.
3 Becker, M.E. (2006). Display reflectance: basics, measurement, and rating. *J. Soc. Inf. Disp.* 14: 1003–1017. https://doi.org/10.1889/1.2393025.
4 Kelley, E.F., Jones, G.R., and Germer, T.A. (1998). Display reflectance model based on the BRDF. *Displays* 19 (1): 27–34.
5 Hertel, D. and Penczek, J. (2020). Evaluating display reflections in reflective displays and beyond. *SID Inf. Disp.* 2: 14–24.
6 Hertel, D. (2018). Application of the optical measurement methodologies of IEC and ISO standards to reflective E-paper displays. *SID Symp. Dig. Tech. Pap.* 49: 161–164. https://doi.org/10.1002/sdtp.12509.
7 IEC 62679-3-1. (2014). Electronic Paper Displays: Optical measuring methods, part 3-1.
8 Hertel, D. (2012). Viewing angle measurements on reflective E-paper displays. *SID Symp. Dig. Tech. Pap.* 43: 46–49. https://doi.org/10.1002/j.2168-0159.2012.tb05705.x.
9 Becker, M.E. and Neumeier, J. (2016). Imaging light measurement devices. In: *Handbook of Visual Display Technologies* (ed. J. Chen, W. Cranton and M. Fihn), 3285–3312. Heidelberg: Springer.
10 Heikenfeld, J., Drzaic, P., Yeo, J.-S., and Koch, T. (2011). Review paper: a critical review of the present and future prospects for electronic paper. *J. Soc. Inf. Disp.* 19: 129–156. https://doi.org/10.1889/JSID19.2.129.
11 Kelley, E.F., Lindfors, M., and Penczek, J. (2006). Display daylight ambient contrast measurement methods and daylight readability. *J. Soc. Inf. Disp.* 14: 1019–1030. https://doi.org/10.1889/1.2393026.
12 Nose, M. and Yoshihara, T. (2012). Quantification of image quality for color electronic paper. *J. Soc. Inf. Disp.* 20: 624–631. https://doi.org/10.1002/jsid.128.
13 Henzen, A., Zhou, G., and Zhong, B. (2020). Color reproduction in reflective displays: stacked CMY. *SID Symp. Dig. Tech. Pap.* 51: 719–721. https://doi.org/10.1002/sdtp.13969.
14 Blankenbach, K., Rawert, J. (2011). Bistable electrowetting displays. In Proceedings of SPIE 7956, Advances in Display Technologies; and E-papers and Flexible Displays, 795609 (3 February 2011). https://doi.org/10.1117/12.871039.

15 Deng, Y., Tang, B., Henzen, A., and Zhou, G.F. (2017). Recent progress in video electronic paper displays based on electro-fluidic technology. *SID Symp. Dig. Tech. Pap.* 48: 535–538. https://doi.org/10.1002/sdtp.11695.

16 Wu, M., Luo, Y., Wu, F. et al. (2020). Lifetime prediction model of electrophoretic display based on high temperature and high humidity test. *J. Soc. Inf. Disp.* 28: 731–743. https://doi.org/10.1002/jsid.869.

17 Wang, X., Qin, G., Tan, J. et al. (2015). A study on the front light guide used in color reflective LCDs. *SID Symp. Dig. Tech. Pap.* 46: 466–468. https://doi.org/10.1002/sdtp.10424.

18 Hertel, D. (2013). Viewing direction measurements on flat and curved flexible E-paper displays. *J. Soc. Inf. Disp.* 21: 239–248. https://doi.org/10.1002/jsid.163.

Index

E-Paper Displays, First Edition. Edited by Bo-Ru Yang.

d

q

r

S

t

u

v